BEYOND THE SKY
ROCKET SCIENCE

THE DAWN OF THE
SPACE AGE

WALTER SIERRA

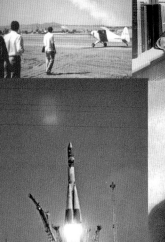

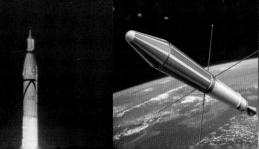

COVER ART:

Top row: (1) Otto Lilienthal preparing for flight with his second "glider," August 1984; (2) Otto Lilienthal, circa 1896

Second row: (1) Orville Wright makes world's first powered heavier–than–air flight on December 17, 1903; (2, 3) Wilbur and Orville Wright, respectively, 1903.

Third row: (1) Historic photo of Robert Goddard beside his 1926 liquid– fueled rocket; (2) Robert Goddard while professor at Clark University ; (3) World's first successful JATO launch, August 16, 1941; (4) Theodore von Kármán in Paris, 1961.

Fourth row: (1) Sergei Korolyov's R–7 rocket launches world's first satellite, Sputnik 1 into the darkness from Baikonur on October 5, 1957; (2) artist's impression of Sputnik 1 in orbit; (3) launch of Korolyov's Vostok rocket from Baikonur; (5) Sergei Korolyov in his prime.

Fifth row: (1) Test launch of V–2 rocket from Peenemünde, Germany, October 1943; (2) Wernher von Braun's Juno I launches America's first satellite, Explorer 1 on January 31, 1958; (3) artist's impression of Explorer 1 satellite in orbit; (4) liftoff of von Braun's Saturn I rocket (the eighth Saturn I flight) on February 16, 1965; (5) Wernher von Braun official portrait, 1960.

To order additional copies of this book, contact:
Xlibris
1-888-795-4274
www.Xlibris.com
Orders@Xlibris.com

Life is a leaf of white paper
Whereon each one of us may write
His word or two, and then comes night.

Greatly begin! though thou have time
But for a line, be that sublime –
Not failure, but low aim, is crime.

James Russell Lowell (1819-1891)

DEDICATION

To be revealed in *The Never–Ending Frontier*

PREFACE

OK, I'll admit it. I am a rocket scientist, have five college degrees, and have already written two ponderous tomes for my fellows. All told, it took eight years of research to write them. But in 2004 I decided that was not enough. I needed to get the word out, and in the process abolish the typical stereotype about "rocket scientists". So I have spent the last 12 years writing a series of books that anyone with a basic education can read. I have tried to make them as entertaining and readable as possible, while communicating to a lay audience what rocketry is all about. It's not just about rocketry. Rockets (an overused term) were conceived, designed, built, and operated by people like you and me. People who had emotions, aspirations, dreams, and euphoric moments of success. I can assure you that when Neil Armstrong first set foot on the moon on July 20, 1969, the whole world was captivated and watching every move. If you find the subject of rocketry "boring" or too hard to understand (I honestly hope this isn't the case) you can skip over those parts. But don't skip too much, because throughout the entire series I have tried gradually and painlessly (I promise) to build up your grasp of the drama behind humankind's technological progress over the last 100 years. Go back and reread the parts you skipped over with better understanding. I implore you to give it a try, and please read the books in sequential order—they are one unified whole.

Both of us are on a journey toward an unfathomable future. For the past half–century no human being has ventured outside low Earth orbit, let alone go back to the moon. But that is going to change. Drastically and within your lifetime. Humans are not just going to explore the planets in our solar system, but venture out to the stars. Not on world ships that take a century, or several centuries, to get there; but on faster–than–light spaceships that take a few years, or less. History teaches us that technological breakthroughs don't come gradually—they come in a flash—typically after years of hard work, trial and error, and numerous failed experiments. Then within a few decades engineers such as myself make the impossible happen for everyone on the planet. The inexorable pace of technological advancements is accelerating. You need only look at the nearest laptop, the internet which puts the entire world at your fingertips, those hundreds of microprocessors and embedded systems in your car, or your cell phone.

Today, quantum teleportation—a process by which the exact state of an atom or photon can be transmitted instantly over arbitrarily large distances—is real and has been proven by experiments. In 2013 researchers succeeded in demonstrating complete quantum teleportation of photonic *quantum qubits*.

Quantum teleportation requires the objects in question to be "quantum entangled", meaning that they share the same so-called "quantum state". Think of it this way (I humbly ask for forgiveness from knowledgeable physicist readers): two identical twins share the same features like eye and hair color, height, etc. One of the twins is on Mars and the other is enjoying a day at the beach in Florida. The one on Mars changes his hair color from blonde to dark brown. What's cool is that the hair on his Earth-bound twin *instantaneously* changes to dark brown also!

Scientists have been experimenting with quantum teleportation since 1997, and have progressively extended the teleportation distance. In 2012 scientists transferred quantum information over 143 km in space, and in 2015 researchers at the U.S. National Institute of Standards and Technology teleported information encoded into particles of light over 100 km of optical fiber.

This advance could lead to devices called "quantum repeaters". They would be like the repeaters we currently use in our networks, which receive a signal and then retransmit it at a higher level, to make our information travel around the whole world. A quantum repeater would do the same thing, but with quantum information, and they could potentially extend the reach far enough to build an entire "quantum Internet", which would be faster, more efficient, and more secure than the networks we rely on today. While we're still a long way off from building a whole network made up of entangled light particles, the potential is pretty great.

However, there is a catch. There's no known way in today's physics for the twins to actually talk to each other, or send any kind of messages back and forth instantaneously. Ordinary communications are restricted to light-speed. So until someone comes up with a breakthrough, instantaneous communications are out; moreover, physical teleportation of ordinary objects is impossible (quantum teleportation involves quantum-entangled ensembles of atoms too small to see under an ordinary microscope).

Maybe that's why we haven't been successful in the search for extraterrestrial intelligence. We simply don't have the tools to listen in.

Faster–than–light (FTL) travel can't be far behind. The universe is already doing FTL, you probably just haven't heard about it. Right after the Big Bang ("right after" means 0.000000000000000000000000000000001 second) the universe expanded at FTL speeds according to generally accepted "inflation" cosmology. Averaging things out on more meaningful timescales, the universe expanded at around 112 times light speed over the first 379,000 years, give or take. It's

still moving at FTL speeds today, possibly at many times light speed. But it's not slowing down, astronomical observations have proven it's speeding up! So FTL is real. How does the universe do it, and how can we use FTL to get to the stars?

The universe itself isn't really "moving" at FTL speeds at all. It is *space* that can move at FTL speeds, or practically at any speed it wants to without violating any "known" physical laws. Let's be careful here though. We haven't the foggiest notion about what 68% of the universe (the "dark energy" part) is made of, and are very uncertain about another 27% (the "dark matter" part), leaving less than 5% (the material part) that we can vaguely understand. The part you can see, touch, and feel (stars, planets, gas, your body) is even less, a paltry few percent. Humans are in the dark about physics (pun intended) and what makes the universe tick, so there is a heck of a lot of "unknown, new" physics yet to be discovered!

People around the world, including me, are really, really trying to find out how we can move space instead of things so that the first baby steps towards FTL can happen. So far, astronomical observations have proven that there are thousands of planets in the tiny corner of the Milky Way galaxy we inhabit. And that is only the ones we've been able to discover, there are many more we've missed with the relatively crude telescopes at our disposal. Some of the planets look a little like Earth, and it's only a matter of time before astronomers discover an Earth twin that has life on it.

I hope you join me and many others in a quest to get out of the cradle we call Earth, get started on human colonization of the solar system, and uncover the secrets that nature is hiding from us to enable routine FTL to the stars and visit the intelligent lifeforms that are surely waiting for us out there. Reading the *Saga of Rocket Science* series will be a good start in making you part of the world's educated citizenry that can help make it happen. Find out more at www.waltersierra.com.

It would be impossible to personally thank all those without whose help this book could not have been written. The acknowledgments and credits section towards the back is only a start. Here I would like to humbly acknowledge just three of the organizations that have made it possible:

Wikipedia, at http://en.wikipedia.org/wiki/Main_Page

Google at www.google.com

Xlibris at http://www.xlibris.com

Walter Sierra
Orlando, Florida
May 2017

CONTENTS

SERIES INTRODUCTION

Beyond the Saga of Rocket Science series of four closely related books provides an amply–illustrated, overarching perspective to a broad, non–technical audience of the entire panorama surrounding the development of rockets, missiles, and space vehicles as we know them today and what the exciting future holds. The books are sequential and form an integrated whole.

Beyond the Saga of Rocket Science
- The Dawn of the Space Age
- Avoiding Armageddon
- In Space To Stay
- The Never–Ending Frontier

The panorama includes fascinating personalities, personal feelings and vignettes, political events, the ambiance and perspective of countries possessing rocket power. The series covers the entire timeframe from 228 A.D. through the present day and into the far future in describing what really happened when the dreams of rocket pioneers were fulfilled, or when they perished in disaster and flames. It blends fact–based dialogue among key, real characters with enthralling vignettes and vivid on–scene descriptions of the successes, the traumas, the failures and disasters, and triumphs encountered during the roller coaster ride of rocket development. The books describe complex aerospace technologies— from aerodynamics to rocket engine turbines–- in terms anyone can understand, using pictures and clear explanations instead of equations and dry dissertations of technical facts and engineering drawings. The reader will understand in lurid detail exactly what went wrong during catastrophic failures, and what went right when rocket engineers and scientists experienced the euphoria of success.

In bringing to life the drama and complex interplay among scientists, engineers, and politicians that gave birth to the launch vehicles and spacecraft we take for granted today, *Beyond the Saga of Rocket Science* harnesses surrounding events as never before: political maneuvering, the drama of spies and counter–spies, the feelings that key characters—all of whom are real— experienced, the society and political structure, life– changing crises, and more. You'll be taken behind the scenes and come away with an understanding underlying what happened, what caused it to happen, and what

nearly happened. The third and fourth books also offer a rare glimpse, based on the best knowledge of physicists and scientists the world over, of what lies ahead for humankind as it ventures into space, the 'final frontier'.

This book begins with exciting tales of the earliest developers of rudimentary rockets and the deadly battles they fought in China between 200 and 1600 A.D.

You'll experience what early stalwart pioneers had to do to make the controlled explosions we call rockets work. Men like Robert Goddard, Wernher von Braun, Sergei Korolyov, and Valentin Glushko succeeded against all odds with miraculous feats of engineering.

You'll be introduced to the basics of rocketry and aerodynamics, in easy to understand words and pictures instead of equations. The Wright Brothers—Wilbur and Orville—pioneered powered flight, and presaged the early innovators who would found the giant aerospace conglomerates of today. Every rocket and aircraft that flies today depends upon the crucial developments in the science of aerodynamics, led by Ludwig Prandtl in Germany and Theodore von Kármán in the United States.

World War II gave the biggest impetus ever to advancing rocket science and related technologies. In the ensuing decades many unsung space heroes on both sides of the Cold War's Iron Curtain matured the field to where it is today. You'll see how the Soviet Union, led by Korolyov and Glushko, captured an early lead over the United States and established many space firsts: world's first satellites, first man and first woman in orbit, first interplanetary spacecraft, first space station. The U.S. had its own rocket experts—Wernher von Braun and his team of 114 captured WW II German rocket scientists. America mounted a counterattack. Von Braun's team spearheaded the Mercury, Gemini, and Apollo space programs. The United States won the Moon Race against its rival when Neil Armstrong and Buzz Aldrin landed on the Moon in July 1969.

Picking up where the first book leaves off, *Avoiding Armageddon* provides a detailed (and fascinating!) behind the scenes look at the Cold War conflict where it really counted: the *military* space race. Despite the public's perceptions at the time, in this sphere it was the United States that held a technological lead in several respects: more accurate delivery systems, more nuclear warheads (at least until 1978), unmatched overhead surveillance and photographic technology on spy planes and satellites, a better bombing fleet, quieter and better equipped submarines (not to mention more of them).

It is a dichotomy that U.S. intelligence authorities in America's more open, democratic society knew more about their secretive rival's true capabilities than military leaders on both sides would have liked to admit.

It's a common misconception that a nuclear World War III would have obliterated the human race. Hardly. The facts don't support this assertion. Nuclear targets were (and still are) mostly confined to the Northern Hemisphere; furthermore, military facilities and critical infrastructure would be the primary focus. Many smaller cities and towns would be spared, and large swaths of open farmland in America and Siberia would escape relatively unscathed. Admittedly, both superpowers and their allies possessed enough nuclear firepower to devastate—but not obliterate—each other, inflicting hundreds of millions in mass casualties.

In consonance with the overarching theme of *Beyond the Saga of Rocket Science*, you'll gradually *absorb* (a better term than learn) knowledge about the different kinds of rockets and the corresponding rocket families– on both sides of the Iron Curtain—including the Atlas, Delta, Titan, Soyuz, and Proton. *Avoiding Armageddon* (assuming you've read *The Dawn of the Space Age*, which is highly recommended) delves a little deeper into rocket combustion physics, storable propellant chemistry, what makes up solid propellants and how they work, and orbits. Yes, orbits. They're important, because after all that is the whole purpose of launch vehicles: to get payloads into orbit. The kind of orbit is very important, so the differences are clearly explained and illustrated in terms anyone should understand. Don't be intimidated by sophisticated–sounding terms like *ballistic coefficients*, hopefully you'll absorb the essentials without having to worry about any math.

The most important payload rockets carried during the Cold War was nuclear bombs. You'll see the differences (clearly explained) between different kinds of nukes, the distinction between A–bombs and H–bombs, and their delivery systems—principally land–based missiles, submarines, and aircraft; although howitzers and battleships can be used as well.

Avoiding Armageddon disentangles and clearly explains the labyrinthine Soviet organizational and design bureau structure. To make things even more complicated, sometimes they would purposely obfuscate names and organizations in an attempt to deceive prying eyes from Western intelligence agencies. In fact, it's a wonder the Soviets *themselves* could understand their system! Soviet media would also routinely hide their space failures behind a veil of secrecy, use different names, or misrepresent a space mission's true intent until its success was assured.

Despite these handicaps Soviet rocket engineers and scientists were somehow able to perform miraculous feats of rocket science and engineering with only a fraction of the resources available to their American counterparts. A long list of hardworking leaders came to the forefront in this era; some of the principal ones were design bureau chiefs Sergei Korolyov, Vladimir Chelomei, Mikhail Yangel, Vladimir Utkin, and Viktor Makeyev; and propulsion bureau chiefs Valentin Glushko, Aleksei Isayev, and Nikolai Kuznetsov. Their excellent work maintained the Soviet Union's status as a 'peer competitor' to the United States throughout the decades of the Cold War, and in fact prevented it from becoming a Hot War through mutual strategic deterrence.

In the civilian sphere, America woke up after Korolyov's well–publicized Sputnik success in 1957, and relentlessly pursued space supremacy. The richer nation, buoyed up by a more efficient capitalist economy, soon caught up with and arguably surpassed the Soviet Union, sometime in the mid–1960s, never to relinquish the lead again. While pursuing military applications of rockets in a serious way, both nations scrambled to be the first to put humans on the moon—the United States openly, the Soviets secretly. Although the U.S. ostensibly won the Moon Race with with Armstrong and Aldrin on July 20, 1969, the U.S.S.R. was *not* that far behind.

Unbeknownst to the general public (and undoubtedly a great surprise to the reader), the Soviets built their own Moon rocket, the N1, as a counterpart to the American Saturn V. They launched it four times and ...came so close...to matching America's accomplishments. In fact, design bureau leaders had laid plans for building lunar colonies and even manned missions to Mars to outpace the United States had the Space Race continued.

In Space To Stay, the third book in the spell–binding *Beyond the Saga of Rocket Science* series, gives a thorough exposé of the U.S. Apollo and Space Shuttle programs, representing America's side of the Cold War Space Race. You will *be there* as Gus Grissom, Ed White, and Roger Chaffee perished and got incinerated inside their locked Apollo 1 capsule; when Jim Lovell, Jack Swigert, and Fred Haise didn't know if they would make it back alive aboard a freezing lunar module on Apollo 13; and as Neil and Buzz experienced the euphoria of being the first humans to land on the Moon, while

Mike Collins in lunar orbit and an anxious world looked on. You'll see the same panoramic vistas of the lunar landscape and the beautiful blue marble we call Earth as the astronauts saw. You are taken inside the *Challenger* Space Shuttle as it caught fire and disintegrated in flight; and what seven brave astronauts felt like as they plunged to their deaths in a basically intact crew cockpit. You'll understand exactly what failed and how it failed on both the *Challenger* and the ill–fated *Columbia* space shuttles, and why another seven astronauts aboard the *Columbia* felt no pain despite their grisly annihilation during reentry.

Space experts thought of the Apollo missions as just a precursor to more ambitious plans to come. Knowledgeable experts inside and outside NASA knew that the manned moon landing space spectaculars and ostentatious plantings of the U.S. flag were just the first step in human exploration—and many would argue eventual *exploitation*—of the cosmos. American space advocates, including the incomparable Dr. Wernher von Braun, championed a logical, step-by-step long–term space program in line with the Soviet approach. Follow–on extended-stay lunar missions would enable permanently manned lunar habitats in the 1970s–1980s. There was a lot of moon left to explore, after all its surface area is about the same as North and South America combined. One or more permanent Earth–orbiting space stations, and perhaps in lunar orbit as well, would enable the first of a series of manned Mars missions as early as the mid–1980s, a manned Venus flyby mission, mining of valuable ores on the Moon and near–Earth asteroids, and eventual construction of city–size rotating habitats in free space. *All of these space dreams were technically feasible.* But what happened? Why hasn't the first human been born on Mars yet?

In Space To Stay explains why. You'll even learn how President Nixon's Watergate fiasco, the Vietnam War, the demands of social programs, and a plethora of other developments drew America's attention away from space in the post–Apollo era. A disenchanted von Braun and his cohorts were forced to abandon any thoughts of an aggressive long–term space program. No human would even get out of low Earth orbit—much less set foot on the Moon again—for over 50 years!

In the heydays of the Moon Race America had many fantastic military space plans, over a wide, well–financed, and totally secret front. Besides orbiting hundreds of reconnaissance spy satellites, the country

planned manned military space stations, maneuverable manned spacecraft, the establishment of scientific and military bases on the Moon, and anti–satellites for knocking Soviet surveillance satellites out of the sky. Lucrative secret contracts were let for the development of a well-equipped Manned Orbital Laboratory (MOL) to enable astronaut spies to take better and more detailed photographs of the Soviet Union and its allies than competing spy planes and surveillance satellites. The MOL orbital spy station was equipped with a sophisticated camera the size of a car. The magnificent camera system and optics—an advanced set of folded mirrors tucked into the station—were so far ahead of their time that a nearly identical configuration is still in use today! Tied to NASA's Gemini Program the United States had a secret "shadow" Gemini, complete with spacecraft, 17 highly trained MOL astronauts, and support personnel to boot.

The Air Force built a boilerplate MOL, attached a refurbished NASA Gemini capsule atop it, and quietly conducted an unmanned test launch in 1966. However, proponents of competing approaches argued that equally capable reconnaissance systems and optics could be mounted on high–flying spy planes and satellites, and that they'd end up being cheaper than having to keep military men in space. Technology had advanced to the point that real–time operators on the ground—like their space–borne counterparts—could make snap judgments on which targets were of interest and zoom in on them as long as the automated systems in orbit and on the ground functioned properly. Which they usually did. Reconnaissance aircraft like the U-2 "Dragon Lady" and SR-71 Blackbird were even more mission–flexible. So the MOL program was cancelled in 1969.

The Soviet Union countered America's MOL and then some with their Almaz military space station. They surpassed the MOL, which never housed astrospies in orbit, by launching *five* Almaz space stations, including three extended–stay missions with two military cosmonauts. Like the U.S., Russia has continued launching spy satellites to this day. They followed Almaz with the Yantar, Orlets, Resurs, and Persona unmanned reconnaissance satellites.

Today's enormously expensive International Space Station (ISS)—after many fits and starts—finally established a permanent human presence in space in the 21st century. But by no means has it been the only inhabited space station. It was the Soviets—not the Americans—who orbited the world's first space station:

the Salyut. This accomplishment gave sagging Russian morale a much-needed boost after losing the moon race to the United States. In fact, for the next 30 years the Russians dominated space station development, with six additional Salyut stations and the larger Mir space station. The latter remained operational for 15 years (1986-2001) until the long–delayed ISS finally became a reality.

In Space To Stay describes the spacecraft required to operate, maintain, and resupply the ISS in orbit. When the Space Shuttle was retired in July 2011 after 135 flights, Russian Soyuz spacecraft took over the ISS servicing and astronaut replacement role. Soyuz vehicles are used to launch manned Soyuz spacecraft as part of the Soyuz program, as well as to launch unmanned Progress supply spacecraft to the ISS; and for commercial launches marketed and operated by *Starsem*, a European/Russian conglomerate. Spacecraft don't just arrive at the ISS and magically find a docking port, like one would spot a parking space in a parking garage on Earth. *In Space to Stay* uses plain language, easily understandable terms, and lucid explanations accompanied by clear illustrations to explain what spacecraft have to do and the systems they use to do it: orbital phasing and maneuvering, precise attitude control and stabilization, orbital rendezvous, approach, docking maneuvers, docking, and undocking.

During the Cold War and the years leading up to President Reagan's Strategic Defense Initiative (SDI), the Soviets were concerned that the U.S. Space Shuttle could place massive experimental laser weapons into orbit that could destroy enemy missiles from a distance of several thousands of kilometers. Their reasoning was that such weapons could only be effectively tested in actual space conditions and that in order to cut their development time and save costs it would be necessary to regularly bring them back to Earth for modifications and fine-tuning. Heck, for all they knew the Shuttle, endowed with a 1,500 nautical mile (1726 mi or 2778 km) cross–range maneuver capability during reentry, could make a sudden dive into the atmosphere to drop nukes on Moscow! So in the early 1970s they began developing the equally capable Energia–Buran space shuttle system. In November 1988 a specially–designed Energia booster rocket lifted the Buran on an inaugural unmanned test flight (unlike the Shuttle it could fly with or without humans onboard). Buran orbited the Earth twice and made a flawless automatic landing at the Baikonur Cosmodrome in Kazakhstan. But the Soviet Union was crumbling, and the Buran's first flight would be its last. To further seal Buran's doom, the Union of Soviet Socialist Republics (USSR) formally ceased to exist on December 26, 1991.

In Space To Stay brings together as never before all the difficulties plaguing rocket development—the politics, federal budgeting and funding, the lack of vision and consensus, and public apathy. The book describes, in everyday terms anyone can understand, why today's advances in high–speed aircraft technologies go hand in hand with building the reusable launch vehicles of the future. You'll understand esoteric terms like Richard Whitcomb's "area rule", "gravity losses", the difference between ramjet and scramjet propulsion systems, "aerospike" nozzles; and how the failure of NASA's X–33 showcased why the Holy Grail of reusable rockets able to launch payloads to orbit in a single stage has been so difficult to achieve even with today's best technologies. *In Space to Stay* will equip you with the wisdom needed to help build a public basis of support for the right kind of space programs going forward. The world is in space to stay. But where are we going in the future? Don't just be a passive bystander. After thoroughly absorbing this book, you should be motivated to make a difference and make your voice heard. *You the public have a job to do!*

The Never Ending Frontier is the fourth book and culminates *Beyond the Saga of Rocket Science* by taking you on a journey from the present to the future. It begins by explaining how the former Soviet Union steadily matured its Intercontinental Ballistic Missiles (ICBMs), Submarine–Launched Ballistic Missiles (SLBMs), civilian launch vehicles, and tactical missiles across a broad front to reach approximate parity with the United States, only to see it all crumble with the Soviet empire's collapse in late 1991. The book explains how Russia is making an admirable comeback today after throwing off the yoke of communist rule.

There are only three countries in the world today that are capable of manned spaceflight: the United States, Russia, and China. While Russia is recovering from the USSR's collapse and the U.S. hasn't sent anybody beyond low Earth orbit since 1972, China's military and civilian space programs are surging ahead. Unencumbered by the nuclear arms reduction treaties that hobble the United States and Russia, China has secretively embarked upon an aggressive ballistic missile development program. With a practically inexhaustible source of personnel,

China is forming additional missile units, qualitatively upgrading missile systems, and developing methods to counter ballistic missile defenses. China has the world's most diverse fleet of Short Range Ballistic Missiles (SRBMs), and is expanding its strategic, tactical, and land attack cruise missiles as well. The Chang'e and Shenzhou spacecraft are steadily increasing China's reach in unmanned and manned space exploration, respectively.

Countries around the world have also embarked on their own space programs, and many new foreign space launch systems have come into existence in the last few decades. The European Space Agency, a consortium of 21 member countries, has diverse programs; including participation in the ISS, launch and operations of unmanned exploration missions to other planets and the Moon, Earth observation, science, telecommunication satellites, and design of launch vehicles, principally the Ariane heavy lifter which launches out of ESA's major spaceport in Kourou, French Guiana. India developed its Satellite Launch Vehicle (SLV), Augmented Satellite Launch Vehicle (ASLV), Polar Satellite Launch Vehicle (PSLV), and Geosynchronous Satellite Launch Vehicle (GSLV). Japan introduced the H-IIA and H-IIB. Brazil developed the Veículo Lançador de Satélites--VLS-1, -2. South Korea has the Naro launch vehicle. Britain has its Arrow launch system for smaller payloads. Smaller countries (in terms of spacecraft) have also joined the fray.

Since World War II both the United States and Russia have worked on ballistic missile defense systems (BMDS) to prevent enemy ICBMs, SLBMs, and their smaller cousins Medium and Intermediate Range Ballistic Missiles (MRBMs/IRBMs) from reaching their targets. During the Cold War both countries built elaborate systems of spy satellites, radars, and computer networks to scan the skies for ballistic missiles and their deadly nuclear warheads. In *The Never Ending Frontier* you'll learn all about how these BMDS were developed, how they work, the different kinds of BMDS, and what the future holds.

In February 1983 President Ronald Reagan announced the Strategic Defense Initiative, a large–scale research and development effort with the grand goal of building an impregnable BMDS against a massive Soviet nuclear attack. Reagan's "Star Wars" program, so called because it would require huge investments in space–based BMDS elements, funded R&D in five broad areas:

1. Systems Analysis and Battle Management
2. Surveillance, Acquisition, Tracking, and Kill Assessment
3. Kinetic Energy Weapons
4. Directed Energy Weapons
5. Survivability, Lethality, and Key Technologies

The Star Wars BMDS would deploy a gauntlet of multiple tiers of defense. Space–based killer laser beams would slice into enemy ICBMs and SLBMs before they could dispense their lethal payloads, hypersonic interceptor missiles would engage any missiles in space that made it through, and shorter range attack interceptors would kill surviving reentry vehicles before they could reach their targets. A massive network of ground and space–based radars, sensors, and observation platforms would maintain a watchful eye over the entire battlespace both on the ground and in space. By the late 1980s this grand scheme was deemed too expensive to deploy; nevertheless it had two beneficial effects: it helped bankrupt a Soviet Union struggling to bolster its strategic arsenal to keep up, leading to its demise; and a foundation was laid for the more reasonable BMDS we have in place today. The U.S. Missile Defense Agency operates a network of ground and sea–based sensors, radars, and interceptor missiles that can adequately defend against the more realistic threat of *limited* ICBM, SLBM, IRBM, or MRBM attacks by rogue nations.

The world–wide space industry has traditionally been dominated by government–funded communications and observation satellites. *The Never–Ending Frontier* describes how in recent decades a few courageous space entrepreneurs have given birth to a multifaceted commercial age in space, including launch vehicles and spacecraft. Fledgling private sector space companies like SpaceX, Orbital Sciences, Virgin Galactic, and Bigelow Aerospace are facing daunting technological and business challenges as they democratize space with privately–owned, financed, and operated launch vehicles and spacecraft. NASA and the DoD continue to play a key role by giving promising companies development money and serving as a guaranteed customer while the industry improves its products and develops economies of scale. The progress seems slow until the industry takes off, but then watch out!

History shows us how it works. The U.S. government helped create the commercial airline industry, in part by purchasing airmail services which jump-started the market for air transportation. Commercial airlines, in turn, invested private capital in designing and building the airplanes. This system rapidly drove down the cost of air travel for all segments of society in the U.S., opened up new billion-dollar markets, and created hundreds of thousands of high-paying jobs. The DoD was the primary funder and customer of many microchip manufacturers in the early 1970s. Soon "Moore's Law"—that the number of transistors on integrated circuits doubles every 18 months— kicked in. Computer processing speed and memory capacity zoomed out of sight. Increased competition yielded astounding improvements in chip capabilities and pricing, leading to a proliferation of superfast and comparatively cheap computers, laptops, and hand–helds. The DoD also funded and kick started the ARPANET, the Advanced Research Project Agency's operational packet switching network that became the progenitor of today's ubiquitous global Internet. So there you have it.

Today over 88 companies around the world have or are participating in private space ventures. The competition is fierce and relentless. Some have gone out of business, others are non-profit, a few have merged or been acquired, and new ones are constantly being added. It is only a matter of time until a formidable commercial space presence takes shape.

In *The Never Ending Frontier* you'll discover why "Kondratiev Cycles" presage an end to rocket science as we know it today; most likely within the lifetime of the average reader. In fact, if history repeats itself, rocket science will not enter a slow death spiral. It will be a rather quick death, lasting a decade or two at most. The book describes what kind of "propulsion" systems will supersede rocket science in the 21st century and what they will look like, and it will blow your mind!

In just a few years NASA's Kepler space telescope discovered there are thousands of planets orbiting other nearby stars. But given Kepler's limited technology there are probably hundreds of planets missed for everyone found. Today's James Webb Space Telescope, the CHEOPS (CHaracterizing ExOPlanets Satellite) and TESS (Transiting Exoplanet Survey Satellite) are greatly expanding the search for exoplanets and the origins of life. The Milky Way is teeming with billions of planets that are about the size of Earth, orbit stars just like our sun, and exist in the so-called "Goldilocks zone" — not too hot and not too cold for life. Astronomers using NASA data have calculated for the first time that in our galaxy alone there are almost nine billion stars with Earth-size planets in the habitable temperature zone.

That's more Earth–like planets than there are people on Earth!

So where does this leave us? According to most astronomers, the chances are 100% that intelligent extraterrestrials are out there someplace in the vast universe. But where are they, and why hasn't ET paid us a visit? Or have they, and we just don't know about it because a government–sponsored supersecret cabal is hiding a deep, dark secret? Perhaps they're waiting for the next Kondratiev cycle to take hold so that humanity is mature enough? An equally daunting possibility is that we indeed *are* alone.

Humanity is still evolving today, although at a relatively slow pace according to most anthropologists. So where are we headed? Good questions, and there are as many answers as there are science fiction books, movies, and TV shows discussing this topic. *The Never–Ending Frontier* raises some tantalizing possibilities. A scary one is that humanity could someday face extinction as a species, as have over 99% of the species that have ever lived on earth. The book uses straightforward statistical methods to reveal that we will probably become extinct as a recognizable race by 2750 A.D. The rapidly increasing pace of technology development all across the board could reduce this statistically–based prediction by hundreds of years! Think about how computers have proliferated all over the planet in just a few decades. Computers have already demonstrated they can beat the world's best chess players, and the top champions on the Jeopardy! TV game show. The well–known Predator movie series dramatizes the most dreaded prediction, whereas the Star Wars movies showcase artificially intelligent robots as mankind's indispensable servants.

The Never Ending Frontier reveals another thought–provoking possibility. The Organians were a race of beings featured in the 26th episode of the original Star Trek TV series. In a surprising end to the story, these extraterrestrials had evolved to pure conceptual beings, an essence of thought with no corporeal bodies! Could that answer the perplexing "where are they?" question. That Darwin's natural evolution paired with exponentially accelerated technology development inexorably drives all intelligent species in the universe through a relatively short technological stage—perhaps 1,000 years—and into such a mind–boggling spiritual state? Intelligent aliens are out there, all right. We just don't see them watching our every move because they reside in Star Trek's "Q–Continuum," an extra–

dimensional plane of existence inhabited by the "Q", a race of immortal, omniscient, and omnipotent beings possessing the ability of instantaneous matter–energy transformation and teleportation. It is harder to believe that they possess as well the ability of time travel because unless there is some inviolable protocol we're not aware of, this violates the causality principle that you can't go back and kill your parents, so that you were never born.

One thing is certain. The earth is doomed. Eventually we'll *have* to leave our planet or face extinction. As the sun brightens and heats up during the coming millennia eventually Earth will become uninhabitable. How soon? Over the next 1.5 billion years increasing solar luminosity will increase Earth's average temperature to about 104ºF (40ºC), double today's average of 58ºF (20ºC). In about 2.8 billion years the last life on Earth, single-celled, heat-loving organisms in isolated pools of hot, salty water will die. Today's global warming will only hasten these dreadful timetables, but what awaits us *Beyond the Saga of Rocket Science* will rescue us in the end.

The Epilogue provides thought-provoking answers to four perplexing questions that have dogged humanity since antiquity: (1) Does God really exist? (2) Why are we here? (3) Why does evil seem more powerful than good if God's all-powerful? What happens after we die?

INTRODUCTION

The story of rocketry from the clumsy Chinese fireworks through the birth of controlled chemical combustion, the growth and development of rocket science into full–fledged guided missiles through today's state of the art maintenance–free systems was one of trial and error; a pattern of devoted human endeavor studded with many failures and fewer heartening successes, acknowledging each failure and profiting from it, and providing an efficient deterrent to aggression. The story of the world's rocket pioneers is one of improvisations, of making do with what was available in materials and components, and of feeling the way as explorers into the unknown, uncharted realm of rocketry.

Audacious engineers and scientists, principally in Germany (Wernher von Braun), the Soviet Union (Sergei Korolyov and Valentin Glushko), and the United States (Robert Goddard and Theodore von Kármán) had to blaze the trail through a wilderness of dynamics, aerodynamics, and electronics as applied to guided missiles. The German V2, Soviet Soyuz, and American Corporal rockets paved the way for manufacturers and military personnel to follow with the designing, fabrication, and operation of more refined, sophisticated second and third–generations of such missile systems. As Boris Chertok, a renowned Soviet guidance and control engineer said, "The road to space was paved with combat missiles."

Today, we take it for granted that missiles and rockets are going to reliably launch and perform their mission, and not explode on the launch pad. But in the 1940s and 1950s rocketry was still a very new science. Just as with the world's first fearsome atomic weapons, engineers and scientists had to work by trial and error, with explosions in the air, failures of control surface actuators, gyroscopes that didn't work right, fuel and oxidizer valves that leaked, onboard electrical circuits that caught fire, and so on. Early rocket launches were often plagued by one failure after another. Some rockets failed to take off; for example immediately after ignition the electrical circuit would inexplicably reset itself. Some started to take off, only to immediately and disastrously fall on their tails, destroying the launch pad. Others successfully took off, but then exploded at an altitude of several kilometers due to fires in the tail section, or they crashed due to control system failures, or they broke up in the air due to aerodynamic heating of the oxidizer tanks, for example. At the time there were no textbooks or manuals on how to ensure the successful launch of rockets. Early–on, guidance systems were primitive and there were no multi–channel telemetry systems.

A point of confusion arises tracing the history of rocketry back before 1045 A.D. Chinese documents record the use of "fire arrows ," a term which can mean either rockets or an arrow carrying a flammable substance.

The earliest solid rocket fuel was a form of gunpowder, and the earliest recorded mention of gunpowder comes from China late in the third century before Christ. Bamboo tubes filled with saltpeter, sulphur and charcoal were tossed into ceremonial fires during religious festivals in hopes that the noise of the explosion would frighten away evil spirits. In all probability more than a few of these bamboo tubes were imperfectly sealed and, -instead of bursting with an explosion- simply went skittering out of the fire, propelled by the rapidly burning gunpowder. Some clever observer whose name is lost to history may have then begun fire arrow experiments by deliberately producing the same effect as the bamboo tubes which leaked fire. Certainly by the year 1045 — 21 years before William the Conqueror would land on the shores of England — the use of gunpowder and rockets formed an integral aspect of Chinese military tactics.

By the beginning of the 13th century, the Chinese Sung Dynasty, under pressure from growing Mongolian hordes, found itself forced to rely more and more on technology to counter the threat. Chinese ordnance experts introduced and perfected many types of projectiles, including explosive grenades and cannon. Rocket fire–arrows were certainly used to repel Mongol invaders at the battle of Kaifeng in 1232 A.D. The rockets were huge and apparently quite powerful. According to a report: "When the rocket was lit, it made a noise that resembled thunder that could be heard for five leagues (about 15 miles). When it fell to Earth, the point of impact was devastated for 2,000 feet in all directions." Apparently these large military rockets carried incendiary material and iron shrapnel. The rockets may have included the world's first combustion chambers, for sources describe the design as incorporating an "iron pot" to contain and direct the thrust of the gunpowder propellant.

Rocket technology seems to have arrived in Europe around 1241 A.D. Contemporary accounts describe rocket–like weapons being used by the Mongols against Magyar forces at the battle of Sejo which preceded their capture of Buda (now known as Budapest) on Dec. 25, 1241. Accounts also describe Mongol's use of a noxious smoke screen — possibly the first instance of chemical warfare.

Rockets appear in Arab literature in 1258 A.D., describing Mongol invaders' use of them on February 15 to capture the city of Baghdad. Quick to learn, the Arabs adopted the rocket into their own arms inventory and, during the Seventh Crusade, used them against the French Army of King Louis IX in 1268.

By 1300, rockets had found their way into European arsenals, reaching Italy by the year 1500, Germany shortly afterwards, and later, England. A 1647 study of the "Art of Gunnery" published in London contains a 43–page segment on rockets. The Italians are credited, by the way, with adopting military rockets for use as fireworks — completing the circle, so to speak, of the bursting bamboo used at the Chinese festivals 1,900 years earlier.

The French Army has traditionally been among the largest, if not THE largest, army in Europe and was quick to adopt rockets for military operations. Records from 1429 show rockets in use at the siege of Orleans during the Hundred Years War against the English. Dutch military rockets appear by 1650 and the Germans' first military rocket experiments began in 1668. By 1730, a German field artillery colonel, Christoph Fredrich von Geissler, was manufacturing rockets weighing 55 to 120 pounds.

As the 18th Century dawned, European military experts began to take a serious interest in rockets — if only because they, like the Magyars 500 years earlier, found themselves on the receiving end of rocket warfare. During the 18th century both the French and British began wrestling for control of the riches of India. In addition to fighting one another, they also found themselves frequently engaged against the Mogol forces of Tipu Sultan of Mysore. During the two battles of Seringapatam in 1792 and 1799, rockets were used against the British. One of Tipu Sultan's rockets is now displayed in the Royal Ordnance Museum at Woolwich Arsenal, near London. Tipu Sultan's father, Hyder Ali, had incorporated a 1,200 man contingent of rocketeers into his army in the year 1788. Tipu Sultan increased this force to about 5,000 men, about a seventh of his total Army's strength.

Profiting from their Indian experience, the British, led by Sir William Congrieve, began development of a series of barrage rockets ranging in weight from 18 to 300 pounds. Congrieve–designed rockets were used against Napoleon. Surprisingly, Napoleon seems to have made no use of rockets in the French Army, but it must be remembered Napoleon was an artillery officer and may have simply been too hide–bound a traditionalist to favor new–fangled rockets over more familiar cannons. The scope of the British use of the Congrieve rocket can be ascertained from the 1807 attack on Copenhagen. The Danes were subjected to a barrage of 25,000 rockets which burnt many houses and warehouses. An official rocket brigade was created in the British Army in 1818.

Rockets came to the New World during the War of 1812. During the Battle of Bladensburg on August 24, 1814, the British 85th Light Infantry used rockets against an American rifle battalion commanded by U.S. Attorney General William Pickney. British Lieutenant George R. Gleig witnessed the Americans' response to the new threat — "Never did men with arms in their hands make better use of their legs," he wrote. A few weeks later (September 13-14) British warships sent a relentless downpour of shells and rockets onto Fort McHenry in Baltimore Harbor. Francis Scott Key, a 35-year-old American lawyer, witnessed the attack and would later pen the famous words "...and the rockets' red glare..." which became part of the U.S. national anthem. On December 4, 1846, a brigade of rocketeers was authorized to accompany Maj. Gen. Winfield Scott's expedition against Mexico. The Army's first battalion of rocketeers — consisting of about 150 men and armed with about 50 rockets — was placed under the command of First Lieutenant George H. Talcott.

A rocket battery was used on March 24, 1847 against Mexican forces at the siege of Veracruz. On April 8 the rocketeers moved inland, being placed in their firing position by Captain Robert E. Lee (later to command the Confederate Army of Northern Virginia in the Civil War). About 30 rockets were fired during the battle for Telegraph Hill. Later, the rockets were used in the capture of the fortress of Chapultepec, which forced the surrender of Mexico City. With typical lack of foresight, as soon as the fighting in Mexico was over, the rocketeer battalion was disbanded and the remaining rockets were placed in storage. They remained in mothballs for about 13 years — until 1861 when they were hauled out for use in the Civil War. The rockets were found to have deteriorated; however, so new ones were made.

The first recorded use of rockets in the Civil War came on July 3, 1862, when Maj. Gen. J.E.B. Stuart's Confederate cavalry fired rockets at Maj. Gen. George B. McClellan's Union troops at Harrison's Landing, Virginia. No record exists of the Northerners' opinion of this premature "Fourth of July" fireworks demonstration. Later in 1862 an attempt was made by the Union Army's New York Rocket Battalion — 160 men under the command of British–born Major Thomas W. Lion — to use rockets against Confederates defending Richmond and Yorktown, Virginia. It wasn't an overwhelming success. When ignited, the rockets skittered wildly across the ground, passing between the legs of a number of mules. One detonated harmlessly under a mule, lifting the animal several feet off the ground and precipitating its immediate desertion to the Confederate Army! The only other documented use of rockets is at Charleston, South Carolina, in 1864. Union troops under Maj. Gen. Alexander Schimmelfennig found rockets "especially practical in driving off Confederate picket boats, especially at night."

As an interesting sidelight, the author Burke Davis, in his book "Our Incredible Civil War," tells a tale of a Confederate attempt to fire a ballistic missile at Washington, D.C. from a point outside Richmond, Virginia. According to the author, Jefferson Davis witnessed the event in which a 12–foot–long, solid–fueled rocket, carrying a 10–pound gunpowder warhead in a brass case engraved with the letters C.S.A., was ignited and seen to roar rapidly up and out of sight. No one ever saw the rocket land. It's interesting to speculate whether, almost 100 years before Sputnik, a satellite marked with the initials of the Confederate States of America might have been launched into orbit!

The military appears to have remained underwhelmed with the potential of rockets. They were employed in fits and starts in many of the brushfire wars which punctuated the otherwise calm closing days of the late Victorian Era. If the military was lukewarm to rockets, another profession welcomed them with open arms. The international whaling industry developed rocket–powered, explosive–tipped harpoons which were most effective against the ocean–going leviathans. During the First World War, rockets were first fired from aircraft attempting to shoot down enemy hydrogen gas–filled observation balloons. Successes were rare and pilots resisted being asked to fire rockets from the highly flammable cloth and varnish-covered wings of their biplanes. The French were the principal users of aerial rockets, and used a model developed by a Naval lieutenant, Y.P.G. LePrieur.

On March 16, 1926, Robert Goddard launched the first successful rocket using liquid propellants in Auburn, Massachusetts. Before World War II it was the Germans who led the world in rocket science by harnessing Goddard's pioneering work while he was for the most part ridiculed in the United States. During the war two German missiles alone—the infamous supersonic V–2 and its subsonic winged counterpart, the V–1—were responsible for about 18,000 deaths and twice as many injuries in Western Europe. Some 12,000 forced laborers and prisoners were killed in producing over 36,000 missiles.

After the war the victorious Soviet and American superpowers fought over captured German rocket technology, scientists, and engineers. The ensuing Cold War served as the principal impetus for the maturation of rocket science in both countries. By far the largest share of the space budgets in both nations has been taken up by *military applications* of rocket science across the board: Intercontinental Ballistic Missiles (ICBMs), Submarine–Launched Ballistic Missiles (SLBMs), and their smaller cousins Medium and Intermediate Range Ballistic Missiles (MRBMs/IRBMs). Rockets and missiles were developed as weapons of war first, before peaceful spinoffs were explored. As the Cold War unfolded two men rose to positions of prominence in their respective countries: Dr. Werner von Braun in the United States and the lesser–known but equally talented Sergei Pavlovich Korolyov in the Soviet Union. Korolyov partnered with Valentin Petrovich Glushko to lead Soviet rocket engineering and design, culminating with the launch of the world's first satellite—Sputnik 1— on October 5, 1957 (not the widely advertised October 4). On November 3 the half–ton Sputnik 2 put a female dog (Laika) into orbit. Virtual (but unwarranted) panic arose in the United States at perceived Soviet supremacy in missiles and space.

In contrast to the Soviets, the U.S. used German hardware *and* a captured team of 114 German expatriates led by von Braun to jumpstart an equally impressive space program. On January 31 the United States countered the Soviets by launching the first of the Explorer series of satellites. Von Braun's team was transferred from the Army to NASA after its founding in July 1958. However the USSR preserved its ostensible lead in the Cold War space race by orbiting the world's first human, Yuri Gagarin, on April 12, 1961. The U.S. public and news media perceived a widening "missile gap" between the two Cold War adversaries. But military leaders on both sides knew that if there was such a thing, it was decidedly in favor of the United States with better, more accurate missiles as weapons of war and many more nuclear warheads to boot.

The so–called "Mercury Seven" astronauts of the Mercury Program put the von Braun–led U.S. civilian space program back on track. In two years (May 1961– May 1963) six astronauts successfully completed six solo missions in space. NASA quickly followed Mercury with Project Gemini. Gemini was a series of Earth-orbital missions with two-person crews, where astronauts practiced orbital maneuvers, rendezvous, docking, undocking, and reentry while close to Mother Earth. Such maneuvers would be necessary to land men on the moon. The Gemini Project consisted of two successful unmanned test flights, followed by 10 manned flights.

The U.S. was pulling ahead. But the Soviets would not go down easy, as we will see in lucid detail in the next book in the series, *Avoiding Armageddon*.

开始火箭

"What I hear, I forget
What I see, I remember
What I do, I understand"

Confucius, circa 520 B.C.

"All warfare is based on deception. Hence,
when able to attack, we must seem unable;
when using our forces, we must seem inactive;
when we are near, we must make the enemy
believe that we are far away; when far away,
we must make him believe we are near. Hold
out baits to entice the enemy. Feign disorder,
and crush him."

Sun Tzu, circa 502 B.C.

BATTLE OF CHENCANG

Late March, 228 A.D. Zhuge Liang was leading his vanguard of 5,000 troops over the crest of the last hill. A feeling of requited revenge surged through his veins. Through the misty predawn darkness he could see the flaming torches like so many fireflies at the southwest and southeast watchtowers. An unobstructed view was afforded by the sparse vegetation. The 80,000 inhabitants of Chencang (in today's Baoji city of North China's Shannxi Province) would not feel so secure, he thought, if they knew he had amassed a well–equipped force of 20,000 seasoned troops to attack and invade their fortress city.

Five pennant bearers and their *sheng* players took their stations, spread out across 2 kilometers alongside the vanguard. Ru Heng, the Shu State's principal military advisor, strode his horse alongside General Liang's decorated mount. "We have followed the precepts of the immortal Sun Tzu," he whispered. "When the enemy advances, we retreat. When the enemy halts, we harass. When the enemy seeks to avoid battle, we attack. When the enemy retreats, we pursue." As each division assembled with its mobile tower and scaling siege ladders, its commander rode his decorated mount to the center. "Sound the attack!" barked the General. Each commander dispersed to lead his division and ordered his *sheng*, a mouth organ made of seven bamboo pipes, and the drums to sound the advance of thousands of foot soldiers and the scaling equipment. First the center, then the flank divisions. As prearranged, the mounted cavalry and war chariots were held back in reserve. General Liang and his advisors watched.

The Chinese had long–since perfected sword–making and shields to a fine art. Bronze blades honed to a sharp edge on both sides. Warriors trained to skewer an unprotected enemy with one blow. Highly accurate long distance crossbows. Armored shields able to ward off the sharpest arrows. Cavalry equipped with long spears or bows and arrows. War chariots would be pulled by a team of four horses, with three men: a driver, an expert swordsman for close–in protection, and a bowman able to accurately kill from many meters.

Minutes later, Bohai Ch'üan braced himself for a counter–attack. He saw that three other courageous warriors had already mounted their long mobile siege ladders—known as 'cloud ladders'—onto the thick, impenetrable walls of Chencang city. His was the fourth. As he started climbing, he raised his shield as an example to the others, who began pushing him up from behind. Up one step. Two steps. Bohai expected a barrage of arrows from the defending Wei warriors, commanded by General Hao Zhao. Anytime now. Then, halfway up, he saw the disaster out of the corner of his eye. His younger brother's hastily erected ladder was too short to reach the top of the fortified wall. Chongkun Yao–ch'en looked on aghast as a Wei swordsman from on top gave a tremendous whack to the left side of his ladder,

1

splintering it so that it began to sway uncontrollably. Chongkun briefly locked eyes with his older sibling Bohai, whose look turned to horror as the young man, three warriors, and the broken ladder fell aback on top of the vanguard of attacking troops.

"Chongkun! …No!" Bohai shouted.

Li Fa regarded his aged, but venerable mother Qiaohui and his other exhausted and famished family members with a look of utter dismay. "Mother, please don't exchange your two youngest granddaughters to their deaths" he pleaded. Once before, Li had witnessed a desperate exchange of children among besieged city dwellers. The revolting stench of death permeating the city. Then, the horror of eating one another's children, splitting and cooking their bones. One could never, of course, eat one's <u>own</u> progeny. But kill a life, in order to save a life? By tradition, the youngest females went first.

By 220 A.D, the Chinese Han Empire had practically ceased to exist. During the preceding 50 years of Tung Han Rule, North China became subject to invasion from different sides; and as was observed by several philosopher–statesmen, administration had become corrupt and ineffective. Powerful regional officials were able to establish themselves almost independently of the central government. Rivalry between consorts' families and eunuchs led to a massacre of the latter in 189, and the rebel bands that arose included the Yellow Turbans, who were fired by beliefs in supernatural influences and led by inspired demagogues. Soldiers of fortune and contestants for power were putting troops in the field in their attempts to establish themselves as emperors of a single united China. By 207 Ts'ao Ts'ao had gained

control over the north and became the generalissimo and protector of the Han dynasty. Had he not been defeated by Sun Ch'üan at the battle of the Red Cliff, which later became famous in Chinese literature, he might well have succeeded in establishing a single dynastic rule. Other participants in the fighting included Tung Cho, Liu Pei, and Chu–ko Liang. The situation was resolved in 220 when Ts'ao P'ei, son of Ts'ao Ts'ao, accepted an instrument of abdication from Hsien–ti, last of the Han emperors (he had acceded in 189). Ts'ao P'ei thus became the legitimate heir and first ruler of a dynasty

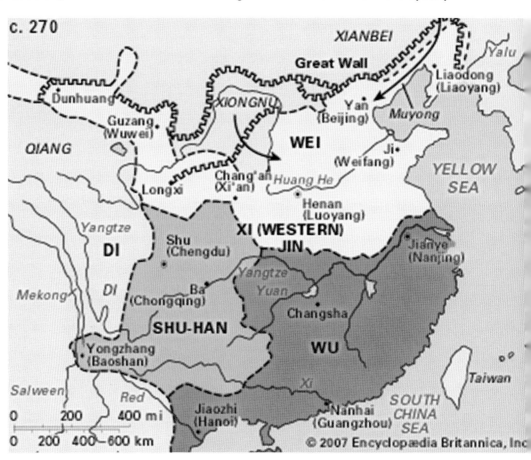

China in the Three Kingdoms period (circa 270 A.D.) Xi Jin province was associated with the Wei Kingdom

that would become known as Wei, whose territories would stretch over the northern part of China and whose capital was at Lo–yang. Soon afterward, two competing military leaders proclaimed themselves emperor. In the far interior, Liu Pei was declared emperor of the Shu–Han dynasty in 221, thereby maintaining the fiction that as a member of the Liu family he was continuing its rule of the Han dynasty, albeit in the restricted regions of Shu in the southwest (capital at Ch'eng–tu). In the southeast, protected by the formidable barrier of the Yangtze River, the kingdom of Wu was formed, with its capital at Chien–yeh, under the initial dispensation of Sun Ch'üan. Thus ensued the short and turbulent period of the San–kuo ("Three Kingdoms")—Wu, Shu–Han, and Wei—as the period from 220 to 280 has come to be described. Wei was the strongest of the three, but even it hardly represented any real political power. It was the middle Kingdom, Shu–Han, led by prime minister Liang, that was attacking the Wei state by besieging Chencang.

General Hao Zhao had mobilized the entire population of Chencang. He created a three–tiered line of defense. The first echelon were able–bodied men who with abundant provisions and sharp weapons awaited the enemy, many stationed atop the city wall. Close behind, the second echelon were able–bodied and armed women, who had worked with the men to dig pits and moats around the city, backed up by earthworks. Last, and a ways back, stood children and old and feeble men and women who had fed, watered, and guarded the livestock and dwindling supplies of food near the city center. In those days the populace living in the core of Chinese cities would get used to the disgusting smell of pigs, excrement, and rotting food. The human sense of smell has an amazing power to accustom itself to the most putrid odors over time. But no one had counted on the cunning Shu State to mount their attack so soon, before the defenders could finish the pits. And an ongoing drought had any made moats impossible to fill.

Now the interminable siege seemed to be nearing a grisly end. Li Fa could already hear the ear–piercing war trumpets and ominous drumbeats getting louder and louder as the attacking Shu warriors advanced. No doubt, their brightly colored attack banners were waving in the early dawn sky. "They will be bringing scaling ladders," cried his father, Minsheng. "But our expert swordsmen can knock them off one by one before they can surmount the walls," countered the wizened Qiaohui. "Mother, our spies have reported that they possess movable, multi–platformed towers which can be placed against our fortified city walls. Much harder if not impossible to knock down, and we can expect their warriors to have shields," said Li.

"But the thing I fear the most," said Minsheng, "are the mobile protective 'tortoises' designed to shield the Shu tunnelers. And with our earthen works and pits only half finished!"

Lei Yen–Hsi stood with two guards outside the guarded compound in the middle of the besieged city. "O.K., are you about finished with the last one?" he asked the men feverishly working inside. "As our leader Hao Zhao has directed, my skilled technicians are fixing the last batch of arrows," replied Cheng Qi. "Hurry then, have everyone in your troop carry these to the southwest wall where we expect the Shu to mount their attack," directed Lei. Each man began carrying the biggest load of fire arrows he could manage in his rucksack.

Weeks before, Cheng Qi had approached General Hao–Zhao with his proposal. "Great leader, I have developed an ingenious solution to break the Shu siege," Cheng had said. "Outside the north wall I have discovered a deposit of a black, gooey substance that ignites and burns. I have been able to roll up balls of this sticky goo and have affixed one to an arrow. I have brought one here to demonstrate." "Proceed."

Cheng's assistant handed him a long arrow, loaded with a black ball of tar at the front, just behind the sharp arrowhead. Cheng plunged the ball into the General's campfire, taking care not to burn the arrow shaft itself. After some seconds, a flame enveloped the ball. Cheng pulled it out, then waved the flaming arrow point back and forth above his head. To General Zhao's amazement, the flame did not extinguish itself, but instead kept burning brighter and brighter. Finally, Cheng put the arrow down and extinguished the flame by piling earth on it.

Chinese Fire-Arrows

It took centuries for the Chinese to evolve fire arrows into the fire rockets shown here

"Bring the arrow here," Zhao ordered. He eyed it carefully, turning it over and over, then thought a minute. "Bring me a handful of dry, matted straw," he told his aide–de–camp. He then shook the dirt off the oily tar, and mashed the straw all over and into the soft tar ball. "Now try it" he said. Cheng took the arrow, and again carefully plunged the now bigger ball into the flames. A significantly larger flame soon enveloped the whole forward end of the arrow. Again, it did not extinguish as he waved it to and fro above his head, but this time it burned even more vigorously than before.

"Hand me that bow," Cheng directed his assistant. Taking the bow, he carefully strung the arrow, and took aim at a pile of wood by General Zhao's campfire. The arrow hit the wood pile dead on, but upon striking the first limb, the arrowhead and flaming ball broke off the shaft, skittering to the side instead of sticking into the wood as it should have. Cheng stepped forward and examined both ends of the broken arrow. Sure enough, the flaming tar had burnt almost completely through the arrow shaft in the time it had taken him to perform the demonstration and shoot the arrow.

After a minute, Hao Zhao said, "you must soak the arrow shaft in water, let it absorb all the water it can, then affix the oily tar and straw to it so that it shall withstand the heat without breaking. Also, mix the tar and straw beforehand." "Yes, my Lord," Cheng replied. "I also think it wise to have the arrow smiths make the arrow shafts somewhat thicker on the forward end. This will prevent the shaft from breaking so soon." "Henceforth, we will call these fire arrows," proclaimed the General.

The next day, Cheng Qi returned to General Zhao's camp, this time with three prepared fire arrows. His assistant held a flaming torch. Cheng had found that tar also served to keep torches burning much longer than before. Zhao's aide had prepared a pile of wood as the practice target. "Now, draw your bow and let us see if your improvements will work," said Zhao. Cheng marked off 50 paces from the target. He lit the first arrow from the torch. He took the bow, drew it as tightly as possible, took aim, and…missed the target completely. At least the tar ball stayed lit as the arrow skittered along the ground beside the wood pile. Cheng drew the second arrow, took aim, and this time only managed to hit the top of the wood pile a glancing blow. General Zhao looked on with disdain.

Hands trembling, Cheng grabbed the third and last arrow. He silently breathed a short prayer to the overarching heaven (*tian*), considered then to be the sky, and to the personal deity *Shangdi* ("Supreme Deity") who lived there. Drawing the bow more carefully and a little less tightly this time, he stood absolutely still for several seconds. With a firm belief in the Mandate of Heaven given to the worthy Wei dynastic ruler (Ts'ao P'ei), he breathed a quick prayer. *Oh Great One, may this fire arrow be guided by Thee to save our people*. The arrow struck right in the middle of the wood pile. At first, just a small flame appeared. But the flame stayed lit, and grew. And grew. And grew until the entire woodpile was engulfed in flames. General Zhao smiled.

The rucksack bearers rushed the badly needed fire arrows to the southwest city walls. Others had already lit hundreds of flaming torches. Vicious fighting was already taking place atop the walls. One, then two, then more Wei defenders fell on the ground from on top. One had his head almost completely severed, a large pool of bright red blood forming on the ground. Nevertheless, the torches and fire arrows made their way up ramps and ladders to the platforms on top. None too soon. The accomplished Wei archers started sending salvos of fire arrows upon the unsuspecting Shu troops. Their cloud ladders started catching fire.

A flaming arrow struck young Chongkun Yao–ch'en square on the back as he lay on the ground, his broken ladder beside him. Soon he was writhing in pain as the flames engulfed his clothes, the sticky black ooze almost impossible to get off. Bohai Ch'üan continued his perilous ascent. "Good God in heaven!" he exclaimed to himself. The early morning sky around him seemed aflame with fire arrows. Men were catching fire and falling to the ground. But worst of all, many of the wooden siege ladders were catching fire as rocket after flaming rocket hit home. Several bowmen had concentrated their fire on the first three–platform seize tower before it could be moved against the wall. Men atop the tower struggled to put out the flames, but to no avail as fire arrow after fire arrow hit home.

Finally, Bohai clambered up to the top, sword drawn, shield up. He gutted the first Wei archer before he could draw his bow again. Then two defending swordsmen were on him, gaining purchase from favorable higher positions on the platform. As he desperately fended off blow after blow, he heard General Liang's war trumpets sound retreat behind him. Too late. The two swords struck him almost simultaneously on the neck, neatly severing his head. Blood spewed from his neck as he dropped. His lifeless head fell and unceremoniously rolled on the ground behind the city wall. All was lost for the Shu-Han State.

EMPEROR XIAOWU

In mid–November of 383 A.D., King Fu Jian of Qianqin State in North China had assembled an invading army of 870,000 men to attack the Jin State. Under the command of major General Fu Rong, they had reached the north side of the Fei River (today's Wei, a western tributary of the Huang Ho, or "Yellow" River) in Eastern Jin. The Wei River is sharply defined to the south by the abrupt clifflike northern face of the Tsinling Mountains, so mounting an attack from the south was difficult, if not impossible. Chinese riverbanks are often heavily festooned with vegetation right down to the water level. In many places the Wei and Yangtze riverbanks also slope down at an impossible angle, so that it seems they form a V–shape, with the bottom of the V resting on the river bottom.

In other places, limestone pinnacles line the riverbanks, abruptly rising up out of the ground like so many dragon teeth. To the newcomer, it seems impossible that plants, bushes, or trees can even grow along such steeply sloped ground. But they do. The usually heavily silt–laden Wei was still high after the summer monsoon rains, and now more navigable. Fu Jian began to set up defenses at the riverside, hiding his forces in heavily wooded areas. Across the river, the eastern Jin emperor was poised to resist the invading northern nomads with 80,000 warriors.

Emperor Xiaowu stood resplendent in imperial costume, and cut an imposing figure. His yellow skin was a pleasant suntanned brown. Slightly slanted eyes with low eyelids. The fore half of his head was shaved. In compensation, the remainder was uncut and gathered into a long queue flowing down his back like a waterfall. His black hair was coarse, vigorous, and never curled. He wore a broad–sleeved, bright blue and yellow jacket with pleated apron around the waist. His royal jacket displayed the 12 imperial symbols, designated as imperial insignia and emblematic of the universe in microcosm and thus his universal sovereignty. "Generally, operations of war require 1000 fast four–horse chariots, 1000 four–horse wagons covered in leather, and 100,000 mailed troops," decreed General Xie Shi, quoting the brilliant Chinese military strategist Sun Tzu, who wrote *The Art of War* in 475–480 B.C. "True, our troops do not equal his," replied General Xu Yan. "Fu Rong is overconfident and can be defeated. We must strike the enemy as swiftly as a falcon strikes its prey. We await the right moment to strike. Our movements must be well regulated," said the third General, Xie Xuan.

The Emperor maintained his royal bearing and motioned silence. After some minutes, he said, "every engagement is a bloody and destructive test of physical and moral strength. Whoever has the greater sum of both left at the end is the victor. Our forces are outnumbered, so deception and maneuver are key. Fu Rong will expect us to mount an attack across the river, but he will not meet us at the water's edge. It will be advantageous for him to allow half our force to cross and then strike. Bring me HoYen–Hsi."

"How well prepared is our stock of fire arrows?" asked the Emperor. "Your Majesty, I have prepared at least 1,000," replied Yen–Hsi. By this time, the Chinese had improved upon the fire arrow. Chief technician Yen–Hsi had perfected the tapering of the arrow shaft to act as

a simple guidance system that kept the arrow headed in one general direction as it flew through the air. A three–feathered tail also helped stabilize the arrow in flight. Yen–Hsi had also unsuccessfully tried to combine the Chinese invention of gunpowder with fire arrows. For centuries the Chinese had tossed bamboo tubes filled with saltpeter, sulphur and charcoal into ceremonial fires during religious festivals in hopes the noise of the explosion would frighten evil spirits. Yen–Hsi had tried fastening a gunpowder–filled bamboo tube to a fire arrow for explosive effect, but the added weight greatly decreased the arrow's range and accuracy. The tubes also failed to explode unless they were immersed in fire.

Emperor Xiaowu told his three generals, Xie Shi, Xu Yan and Xie Xuan, "As there is a great disparity of strength, we have hardly any hope of victory if we start a face–to–face battle. First, to deceive him I will send a royal messenger to Fu Rong. He shall tell him, 'You are setting up defenses along the river, so it is quite obvious that you are planning for a long war. But as you are far from your country and supplies cannot be timely guaranteed, you are no doubt putting yourselves in a very disadvantageous situation. Why don't you let your troops retreat a few hundred yards so that we can cross the river to fight a decisive battle with you?'"

The emperor continued, "King Fu Jian and General Fu Rong will think that they can feign retreat, then attack us as the major part of our force is in the river and vulnerable. But beforehand, you shall send a contingent of your best arrow men across the Fei, under cover of darkness, around and behind the enemy's flanks to set up a 'pincer' maneuver. They will carry the 1,000 fire arrows. Others shall bring our double kettledrums, war trumpets, and megaphones. Then in unison we strike with the fire arrows, simultaneously beat the kettledrums, and sound the war trumpets behind the enemy's back while he prepares to attack us in the river. In the midst of all this confusion and the fog of war, all of your warriors and your heralds with megaphones shall shout, "Retreat! The King is defeated!"

"A great plan, your highness! And the best part is that your messenger can report back King Fu Jian's reaction so that we will know if and how we can mount this subterfuge," commented General Xie Xuan. He added, "we will first engage them with the *ch'i* operation (unexpected, strange, or unorthodox), then we'll win with the *cheng* (more obvious, frontal attack). In other words, distractive efforts are necessary to ensure that decisive blows may be struck where the enemy is least prepared and where he does not anticipate them."

The message was soon taken to the enemy king. He laughed derisively, showing his hand. Later to his councillors he said, "How silly those generals are! How dare they wade across the river to fight against a troop of 870,000 men! They surely overrate themselves. Let's retreat so that they can come across. But we will return and wipe them out when they are in the middle of the river."

Presently, the feigned retreat started at dawn, and the Eastern Chin generals began deploying forces into the Wei River as if to attack, with signal men, banners, and drums at the front. As the retreat continued, Emperor Xiaowu's clandestine plan was put into effect. The fire arrows engulfed many of the enemy Qianqin tents and the men themselves in flames. Panic ensued as the kettledrums, trumpets, shouts of defeat, and fire arrows took effect. As the first wave of Qianqin warriors ran back faster and faster, the entire force got out of control. General Rong attempted to mount a counterattack, but it was too late. His troops were already in a thorough confusion and no one would hear his order. This invading Titanic was at last sunk by a much smaller group of fighters, thanks in part to advances in fire arrows.

King Fu Jian's mistake lay in the fact that he only knew that an army in water is easy to defeat. Yet, he should have also known that when two armies confront each other, the one who first retreats tends to lose.

BATTLE OF KAIFENG

Early October, 1232. Ogodei Khan was resplendent in luminous golden robe, trimmed in bright royal red, mounted on a pure white steed. His large saber swung slowly back and forth on his left, and an equally deadly dagger dangled on the right side. For this risky adventure, he had doffed a burnished bronze helmet, complete with pinnacle on top. His ornaments befitted his appointment as chosen successor and chief administrator over his father's vast empire.

The great consolidator Genghis Khan had died in 1227 in Xi Xia, near its capital city Ningxia, but not before ordering his soldiers to take no prisoners. Genghis's burial site remained a closely guarded secret. Ogodei was Genghis's chosen successor. He brought an administrator's touch to his father's vast empire. Though his younger brother, Tolui, surpassed him in bravery, Ogodei was the most able and even–tempered of Genghis's four principal sons. He knew how to marshal authority and how to handle his willful brothers. Under Ogodei the Mongols penetrated as far west as they ever would, clamping the notorious Mongol yoke over much of Russia. He spent his first years quelling new assertions of sovereignty in the Muslim Middle East, invading Korea, and —most importantly—now faced the task of completing the Mongol conquest of northern China.

Ogodei had prepared his fearsome Mongol cavalry to advance on the Jin capital of Kaifeng on a moment's notice. Kaifeng stood on the south side of the Huang (Yellow) river, in today's Henan Province. Each horseman carried a full complement of weapons of war: lances and sabers as well as bows and arrows. As customary, Ogodei had directed a screen of outriders to ride far ahead and serve as an early warning system for his army. Horsemen without equal, Mongol scouts could gather intelligence at a speed unrivaled in their time, traveling at a rate of 160 kilometers a day, undeterred by vast expanses of uncharted terrain. Legend has it that a Mongol stallion of exceptional quality could cover over 250 kilometers without food or water.

"Kaifeng is our principal target to enable our conquest of the Jin Empire and northern China," Ogodei shouted to his lancers. The Khan's guard consisted of 12 trusted Mongol lancers, riding in high saddles with short stirrups, giving them leverage for powerful thrusts. "Chaghatai, see to it that our right flank falls back to cover the baggage train, and get them moving!" Two of Ogodei's three principal brothers had already died honorably in battle, the brave

Tolui in 1231 and Jochi in 1227. Only Chaghatai remained to accompany the attacking Mongol horde.

The Mongol army totaled 100,000 fighting men on the move, divided into tumens, or divisions of 10,000 warriors, followed well behind by 40,000 or more family members. For the attack on Kaifeng, the brothers Ogodei and Chaghatai had decided to converge two tumens coming from different directions, one swinging in from the northwest and another from the northeast. A third tumen was delayed because they had to detour to the west in order to ford a narrower passage through a tributary of the Wei River. Originally, they were to converge on Kaifeng from the southwest, crushing it between the three tumens. A fourth tumen, led by Ogodei's and Chaghatai's half–brother Kashi, was still engaged in attacking Koryo (today's Korea), and was scheduled to arrive in mid–1233 to perform mop up actions. The four tumens would then join forces and together complete the Mongol conquest of northern China.

Based upon past successes in war with the Chinese, and advance scouts sent previously to estimate the population of Kaifeng, the brothers and their ministers of state had decided to press the attack with just two tumens. Besides, the men were getting impatient, and would be hard to control without fighting.

The Chinese defenders of Kaifeng, led by General Changming, had had time to deploy earthen works and dig pits surrounding the most likely Mongol approaches to the city, northwest to northeast. At first, Changming had been thrown into a panic upon learning that his spies had spotted the Mongol scouts a week earlier. Then he learned that the first tumen was still 90 kilometers away and advancing at only 9 kilometers per day. And he had breathed the prayer of Chang Tsai.

"Heaven is my father and Earth is my mother, and even such a small being as I finds a central abode in their midst. Therefore that which fills the universe I regard as my body and that which directs the universe I consider as my nature. All people are my brothers and sisters, and all things are my companions."

Changming meditated on the omnipresence of ch'i ("vital energy"), the oneness of li ("principle"; comparable to the idea of Natural Law) and the multiplicity of its manifestations, which is created as the principle expresses itself through the "vital energy."

The superior man understands that "life entails no gain nor death any loss," he thought. "The universe

is a unity, with myriad aspects, and all existence is an eternal integration and disintegration. Ch'i ("vital energy, matter") is identified with the Great Ultimate, the ultimate reality. My ch'i is influenced by yang (male) elements, it floats and rises, dispersing its substance. Man was once ch'i, like all other aspects of the universe, and he has an original nature that is one with all the things of the world. I am a man. My physical nature derives from the physical form into which my particular ch'i has been dispersed. My moral self–cultivation grows as I conscientiously do my duty as a member of society and as a member of the universe. The moral man does not try to prolong or extend his life."

So was Changming comforted. *Life entails no gain nor death any loss. As a moral man I will not try to prolong or extend my life. Ogodei and his Mongol invaders can decapitate, hang, skewer, quarter, or burn me alive and it does not matter. It <u>does not matter</u>. My ch'i will return to the universal Godhead from whence it came.*

Days later, the advancing Mongol tumen had been able to accelerate its pace to 10 kilometers per day. But General Changming's chief of armaments, Ch'en Hao, had a surprise waiting for them. He introduced a key innovation into Chinese warfare. Barrage after barrage of not just fire arrows, but *fire rockets*, the first use of rockets in warfare in the history of mankind. Rockets did not just suddenly appear based upon somebody's eureka moment. Rather, in a scenario to be repeated many times over, one advance built on another over many years to slowly evolve, adapt and perfect rocket technology from very crude beginnings. Very crude beginnings indeed.

Mongolian scouts could gather intelligence as a speed unrivaled in their time, traveling up to 100 miles a day. Fully deployed, however, the Mongols moved slowly. Their armies, which might total 100,000 men with perhaps four times that number of family members in tow, each would have equaled a midsize city in population. When a great khan like Odowei went to war, as depicted here, his tumen traveled well behind those of the strike force.

Large as they were, Mongol armies prevailed over even larger forces in China, the Middle East, and Europe. History had rarely seen their like in battle; they combined cunning with advanced training, logistics, and weaponry. With technology borrowed from Persians, Arabs, and Chinese, they became terrifyingly efficient at siege warfare, and city after city fell before their onslaught.

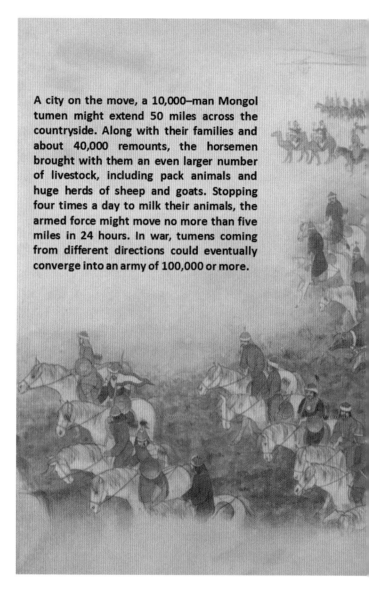

A city on the move, a 10,000–man Mongol tumen might extend 50 miles across the countryside. Along with their families and about 40,000 remounts, the horsemen brought with them an even larger number of livestock, including pack animals and huge herds of sheep and goats. Stopping four times a day to milk their animals, the armed force might move no more than five miles in 24 hours. In war, tumens coming from different directions could eventually converge into an army of 100,000 or more.

Reportedly as early as the first century A.D., the Chinese had a simple form of gunpowder made from saltpeter, sulfur, and charcoal dust. To create explosions during religious festivals, they filled bamboo tubes with a mixture and tossed them into fires during religious festivals. Just when true thrusting rockets first appeared in China is unclear. The concepts of propulsive thrust and action–reaction were unknown, much less understood. Perhaps the first true rockets were accidents. It is probable that more than a few of these bamboo tubes were imperfectly sealed and, instead of bursting with an explosion, simply went skittering out of control, propelled by the gases and sparks produced by the rapidly burning gunpowder. Some clever observer whose name is lost to history may have then begun experiments to deliberately produce the same effect as the bamboo tubes which leaked fire. The first primitive solid propellant 'rockets' appeared when the Chinese discovered that these gunpowder tubes could launch themselves just by the power

The Khan's Ger

Principal wife
Khan
Ministers of state
Visitors
Brazier
Offering table
Khan's sons
Guards
Lesser wives and children

Portable court, the khan's golden ger was a strictly governed administrative center when it was set up for the night after a day of trundling across the steppes. The ger was always positioned so that its door faced south, and visitors were made to stand on the west wall.

produced from the escaping gas. These were the first and most rudimentary 'rockets' in the history of humankind.

By the late 10th century in the Sung Dynasty (960–1279), the Chinese had adapted gunpowder to their fire arrows. The bamboo tubes, filled with gunpowder, had one end closed and the other end open. A small hole was left for the blasting fuse. When ignited, the rapid burning of the gunpowder produced fire, smoke, and gas that escaped out the open end and produced a

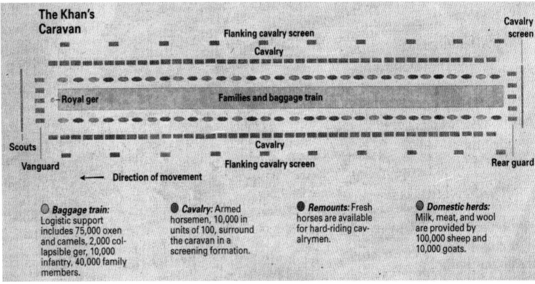

The Khan's Caravan

Cavalry screen
Flanking cavalry screen
Cavalry
Royal ger
Families and baggage train
Scouts
Vanguard
Cavalry
Flanking cavalry screen
Rear guard
← Direction of movement

⬤ **Baggage train:** Logistic support includes 75,000 oxen and camels, 2,000 collapsible ger, 10,000 infantry, 40,000 family members.

⬤ **Cavalry:** Armed horsemen, 10,000 in units of 100, surround the caravan in a screening formation.

⬤ **Remounts:** Fresh horses are available for hard-riding cavalrymen.

⬤ **Domestic herds:** Milk, meat, and wool are provided by 100,000 sheep and 10,000 goats.

9

thrust. They attached the tubes to long sticks or fire arrows so that bowmen could attack enemy troops. Centuries before the Chinese had learned that the arrow shaft and tail feathers acted as a simple guidance system that kept the rocket headed in the right direction as it flew through the air. A long stick could accomplish the same purpose. This was believed to be the embryonic form of modern rockets. By 1232 the Chinese could make 'explosive fire arrows' by carrying the gunpowder directly inside an arrow, and the thunderous explosion of the gunpowder could frighten away the enemy. The battle of Kaifeng was to prove the formidable psychological effect on the Mongols of these arrows of flying fire as weapons of destruction.

Ch'en Hao had all the new rockets prepared, as General Changming ordered. He and his assistants rushed to parcel out 10,000 (the Chinese word for many) rockets to the frontline defenders of Kaifeng. Most went to the northern approaches to the city where they expected the Mongols to mount their attack. "Guowei, hurry with your helpers and carry these 350 rockets to the west ramparts. We cannot allow the men there to fight only with lances, spears, and conventional arrows. Jianguo, you and your men carry the remaining 300 rockets to the east rampart," directed Ch'en Hao. "Yes, sir," said Guowei and Jianguo as they rushed off.

"Great leader, the vanguard of mounted Mongol warriors is but 30 li (a Chinese li is 500m) distant, and the mass of cavalry is not far behind," breathlessly reported the recently recruited Changming spy. "Thank you, we are within a few hours of having our defenses and fortifications ready," replied Changming, smiling. "And what was your name?" "Tsai Shen Fu, sir."

"Can you describe to me the Mongol Khan's caravan?" asked Changming. "…Um…the cavalry is flanking a large baggage train. They are grouped in units of a hundred," said Tsai Shen, hesitantly. "They surround the caravan in a screening formation. I saw huge masses of remount horses and domestic herds, mostly sheep and goats. The baggage train includes many oxen and camels, collapsible gers, several thousand infantry, and countless family members, and…" "How long is the entire formation?" interrupted Changming. "I could not tell, sir. I could not see the end of it in the dusty distance, but I know it is many li. It looks like a giant centipede snaking its way across the desert." "Thank you. Anything else?" questioned Changming. Again, Tsai Shen hesitated. "Uh…uh. Um, well, sir, there is one more thing. Right behind

the vanguard of riders I saw the royal ger. It is very large, about 30 meters across, and has two rows of 10 oxen each pulling it on four large wooden wheels. It also has a standard of horsetails on display." "What color are the horsetails, are they white or black?" asked Changming. "Uh, sir they are black, they look like beautiful black tresses." "Thank you for this information, Tsai Shen, you are very observant," said Changming. "It was nothing, sir" replied Tsai Shen. "I have one more duty for you. I need you to find out, any way you can, how many guards the Khan has at the entrance to, and around, his royal tent. For this information, you will receive a grand reward," intoned Changming. "Uh, sir, may I ask how much reward?" Tsai Shen asked. "Well, not paper money. I will give you twenty of these gold pieces," replied Changming, holding up a gold coin imprinted with the image of Li-tsung, then emperor of the Nan (southern) Sung. "You can take this invoice for 20 gold pieces to my treasurer upon successful completion of your task," he added.

Less than 10 kilometers away, Odowei's tumen continued its steady advance, black horsetails on standards indicating attack. That night, under cover of darkness, Changming's spy Tsai Shen had penetrated deep into the Mongol Khan's caravan. Assuming a prone position, he steadied his improvised spyglass made from rolled up paperboard. He could see that there were two guards stationed on each side of the large royal ger's entrance, and three additional guards spaced around each side, for a total of 10. The great Khan Odowei and his brother Chaghatai, their principal wives, and lesser wives with children would be sleeping inside, he thought. Getting up, Tsai Shen lost his balance and tripped on a rock.

"What was that?" whispered Ulaankhuu, of the Khan's lead guard. "Go check it out," he directed Batzorig, his assistant. Slinking in the darkness towards the sound, Batzorig spotted a man crouching in the shadows of the camel herd, on the side away from the guards' campfire. "Halt! Who goes there?" he called out. All of a sudden, Tsai Shen took off running towards safety, the younger and more nimble Batzorig catching up with him from behind. As Batzorig caught him by the ankle, both he and Tsai Shen tumbled to the ground. After subduing him and tying his hands behind his back with camel rope, Batzorig marched Tsai Shen back to the main guard camp.

"What the?..Isn't that our spy Batbold Aldar?" incredulously asked Ulaankhuu, the other guards standing beside him. "Why, it *is* Batbold! Why were you trying to get away? What are you doing here, Batbold? Did not Chaghatai send you off two days ago to spy and report back on the Kaifeng defenses? You are a day overdue!" he exclaimed. "Please, please, I was just …" pleaded the double-crossing Batbold (alias Tsai Shen Fu). Ulaankhuu cut him off, "Silence! Guards! Search him!" They soon found Changming's incriminating invoice stuffed in the left sleeve of his tunic. Ulaankhuu was furious. "You bastard! You know that treason is punishable by death! You three, bind him well, and take this traitorous fool to the royal felt tent," he directed three of the guards.

At daybreak, the main assault started. Two human skulls mounted on pikes. Two sets of booming double kettledrums on camelback heralded the Mongol charge, as was customary. First, the mounted cavalry would have to unleash a deadly hail of arrows to overcome the first line of defense, dug into the pits on Kaifeng's northern approach. They and long lancers would follow, maneuvering over the earthworks and around embedded, sharp–pointed shafts to press the attack. The mounted bowmen and long lancers would impale any remaining defenders and push past them into the city proper. Here, sabers and daggers would complete the deadly work of dismembering any of the population that resisted.

Ch'en Hao demonstrating how to light a fire rocket

Changming and his commanders braced their troops for the Mongol assault. They feigned a normal, two–tiered defensive perimeter. A front line of sharp–pointed long shafts dug into earthworks, with expert bowmen crouched in trenches and pits right behind. A second line with armored foot troops and spearmen. What the Mongols did not know was that the majority of the bowmen were camouflaged behind the second tier of armed foot troops, ready to unleash a deadly barrage of explosive fire arrows on their unsuspecting foes.

"Get ready!" shouted a very nervous Ch'en Hao to his fellow soldiers. "I remind you of the instructions to fire and keep firing with all you've got as soon as the flags go down." The first Mongol cavalry were already attacking and penetrating the unfortunate defenders in the forward earthworks and trenches. Through test firings, Ch'en Hao had determined that the optimum range for his rocket–propelled fire arrows was from 280 to 320 meters on charging enemy forces. He had also experimented with angling the tail feathers to counter the forward turning moment produced by the rocket tubes, and keep the trajectory of explosive fire arrows more or less on course. "FIRE!" he screamed as the flagmen signaled.

Ten minutes later, not one, but three successive charges of Mongol cavalry had been repulsed. Horses lay bleeding, their sides split and the guts coming out. Manes were burnt off. The worst part was the men. Many Mongol horsemen lay on the ground and were barely recognizable as human beings. Stomachs and intestines ripped out. Skin and flesh completely burned down to the bone. The stench of death everywhere. Inevitably, the kettledrums stopped signaling attack and gongs sounded a retreat. Even the Mongols had never experienced such mayhem.

For the second time in Chinese history, fire arrows and rocket technology had proven their worth The gunpowder tube compared well with the modern propelling system. The sharp arrowhead, with its piercing power of destruction, compared favorably to the warhead of a modern rocket. Tail feathers helped to stabilize the arrow, just like the modern stabilizing system, and the barrel was similar to the body part of modern rockets.

"I said I want four–horse war chariots tied to his legs on each side," ordered Ogodei Khan. He had consulted with Chaghatai and they decided on a quick, more merciful death for the treacherous counter-spy Batbold. Two–horse chariots on each side would only prolong his death agony. After all, he was still a Mongol, albeit an unworthy one. The women tied his leg bindings so tight that his skin turned red between the ropes. Other volunteers securely fastened the upper and lower parts of his legs to the chariots on each side of him. "At my command!" shouted Ogodei. Batbold Aldar trembled with fear.

"Go!" The horses pulled, and Batbold's bulging eyeballs almost popped out of their sockets. "Pull, you maggots!" shouted Ogodei as each chariot driver whipped at the horses. At last, Batbold screamed, that terrible death scream that no one wants to hear. Both of his muscular legs were at an impossible right angle to his torso. All of a sudden, the left leg came completely out of his hip socket, tendons streaming, flesh and blood spewing everywhere. Batbold kept screaming out, writhing in agony as the chariot dragged his limb away. The war chariot still holding his right leg and torso took off, dragging behind what was left of him on the rocky steppe. "Two kilometers on this rocky ground should do it" said the lead driver. "Naw, I think just one will do" answered his fellow standing to his right on the chariot. When they returned, there was nothing left of Batbold but his skull and skeleton, barely held together by scraps of skin and tendon. Ogodei would make his skull into a drinking cup, as his brother had done.

Hungary in April 1241. Even the Teutonic knights were no match for the Mongol cavalry. Mongols used rocket–like weapons at the battle of Seio which preceded their capture of Buda (present–day Budapest) in December, 1241. Austria was spared when the invaders retreated to the steppes upon news of Ogodei's death in Mongolia the same month. Then on February 15, 1258 Mongol invaders used rockets to capture the city of Baghdad. Quick to learn, the Arabs adopted the rocket into their own arms inventory and used them during the Seventh Crusade against the French Army of King Louis IX in 1268. Back in China, Kublai Khan, of Marco Polo fame, would rule from 1260–1294 as founder of the Yuan dynasty. Kublai Khan was destined to be the greatest of the khans after his grandfather, Genghis Khan.

In 1368 Zhu Yuanzhang headed a peasant revolt that overthrew the Mongol Yuan dynasty and, as the Hongwu emperor, established the Ming dynasty (1368–1644). He moved the capital to Jinling in Jiangsu province and

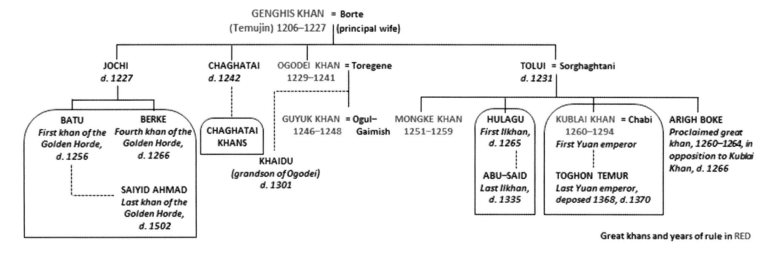

Genghis Khan bequeathed portions of his empire to his sons, and armies to support them. By Kublai's time these lands were four nearly autonomous khanates ruled by Genghis's grandsons: Berke's khans of the "Golden Horde," the khanate of the house of Chaghatai, Hulagu's Ilkhanate, and Kublai's own eastern kingdom. Kublai Khan ruled under the auspices of his Yuan dynasty, which lasted until peasant uprisings spawned the native Ming dynasty.

Following the battle of Kaifeng, the ingenious Mongols produced rockets of their own. Later they would go on to capture Kaifeng and complete their rampage of China: the Jin Empire was captured by the Mongols in 1234. The capital of the Mongol empire after 1235 became Karakorum, east of a tributary of the Orhon river by today's town of Har Horin in central Mongolia. After defeating Russia in 1238, the Mongols stunned Europe with a two–pronged attack into Poland and

called it Nanjing. Dadu was renamed Beiping ("Northern Peace") and was placed under his son's rule. On Zhu's death (1398) the throne passed to his grandson in Nanjing, but his son, Zhu Di (also called the *Yongle* emperor), who ruled Beiping, usurped the throne. In consequence, in 1403 the city was renamed Peking ("Northern Capital"), and in 1421 it was officially made the capital city of the Ming dynasty. City walls were erected and moats, palaces, and temples were built. The old city of Dadu, including its palaces, was largely demolished. The new city was situated farther southwest, which left the northern part of the Mongol city derelict while at the same time slicing off one gate from the east and west walls, respectively.

Before the year 1300, Chinese rockets found their way into European arsenals. Many records describe rocket experiments throughout the 13th to the 15th centuries. Rocket technology reached Italy by the year 1500, Germany shortly afterwards, and later England where they were used as gunnery weapons. However, the crude rockets of this time could not keep up with developments in other types of weaponry in Europe. Large cannons were first used in warfare about 1350. An early giant cannon, the Dulle Griete, fired 320–kilogram balls in the Battle of Ghent in 1411. Heavy handguns called muskets were first used in Italy in the early 1500's and aided Hernán Cortés in his conquest of Mexico in 1519–1521. Cortés's harquebusiers also used an early type of portable gun supported on a tripod by a hook, or on a forked post. Early muskets were up to two meters long and weighed 18 kilograms or more. They fired either single round balls or balls with smaller lead balls called buckshot, and were loaded from the muzzle. A cord match set off the powder charge. Even though the first muskets were so inaccurate that it was difficult to hit a target more than 90 meters away, they steadily improved and became more accurate at short ranges than rockets, which were hard to control in flight.

Consequently, rockets fell into a time of disuse as weapons of war in 16th century Europe, though they were still used for fireworks displays. A German fireworks maker, Johann Schmidlap, invented the "step rocket," a multi–staged vehicle for lifting fireworks to higher altitudes. A large sky rocket (first stage) carried a smaller sky rocket (second stage). When the large rocket burned out, the smaller one continued to a higher altitude before showering the sky with glowing cinders. Schmidlap's staging idea is basic to all rockets today that go into outer space. By adopting military rockets for use as fireworks, the Europeans completed the circle, so to speak, of the bursting bamboo used in Chinese festivals 1,700 years earlier.

QI JIGUANG

September, 1553. "Captain Jiguang reporting for duty, Commissioner," Qi said with a gracious bow. Returning his bow, Regional Military Commissioner Chao Weischeng said, "please sit down," offering him a plush chair. Lighting his pipe, Commissioner Weischeng began with a brief review of the captain's past accomplishments.

"I have been reviewing your records carefully. Your family has a long military history, Qi". Both men knew that Chinese given names carried particular significance. In this case, *Qi* represented the qualities of fine jade, magnificence and eminence. "Your father Qi Jingtong ably served his Highness, our Emperor Zhu Yuanzhang, as a key military leader; and he died gloriously in battle. As the founding Emperor of the Ming ("Brilliant") dynasty, he bestowed upon your family the hereditary post of Commander–in– Chief of the Dengzhou Garrison (a district of present–day Penglai)."

His father had died when Qi Jiguang was only seventeen. His siblings were still young and someone had to take over the household. He soon married the well-educated Wang Jia Li and entrusted domestic affairs to her. They shared much kindness and affection in their monogamous marriage, much mutual helpfulness, and constant cooperation in the busy functioning of their home. The Chinese word for "wife" meant "equal", and Wang preserved her own name after marriage. Her prestige rose with the number and predominantly male sex of her children. At that time, the Chinese proceeded on the assumption that the purpose of a moral code was to transform the chaos of sexual relations into an orderly institution for the rearing of children.

Puffing contentedly on his pipe, Weischeng continued. "I see from your military records that you continued your family tradition by building up our defenses at the garrison. While under the Emperor we have managed to drive the Mongols north beyond our Great Wall, we have been deficient in emasculating the Mongolian power. They continue to pester our northern border with frequent raids. As you know, East Mongolian troops led by the abominable Dada [Altan Khan, a Tatar] broke through our defense and laid siege to Peking [present–day Beijing]. Even though at 22 you were one of our youngest candidates for the martial arts section of the imperial examination, you were mobilized as your final trial to defend our capital. You displayed extraordinary valor and military ingenuity during the battle, which saw the defeat of the invaders. Then, more recently, you courageously led your troops against East Mongolian raiders to help in the defense of Jizhou [southwest of present–day Beijing]. Three years, I believe." Captain Jiguang graciously bowed his head in acknowledgement. "No sir," he clarified. "Four years, from 1548 to last year. The Mongolians always seem to attack in the spring time."

"Tell me, what military strategy did you use?" The Captain replied, "Commissioner, my military tacticians and I recently succeeded in perfecting a new kind of jet rocket that has proven very effective in deterring these attacks, they can even pierce the armor of enemy troops." Commissioner Weischeng's eyes widened in amazement. "Jet rocket? I have never heard of such a thing. What kind of innovation is this?" Captain Jiguang went on, "Well, sir, we have invented three different kinds of rocket. A rocket is a device for accelerating arrowheads with the use of gunpowder" "But how does *that* work?" quizzically asked the Commissioner. "We pack gunpowder inside a hollow cylindrical power flask, carved out from hard wood. We open one end, close the other end, then stick a fuse into the open end, into the gunpowder and then…" The Commissioner interrupted, "how tightly do your pack the gunpowder?"

"Not very tightly. We found that if you pack it too tightly it is harder to set off. If you pack it too loosely the rocket does not have enough power." "So it is a compromise, eh?" intoned the Commissioner. "Yes. At some intermediate packing density, it seems to have the maximum thrust to accelerate the arrows." "How interesting," observed the Commissioner. "And the arrows are attached how?" he asked. "We attach the power flask rigidly, using strong twine, to the rear of each arrow but behind the tail feathers." "And then?" "And then we position multiple rockets on wood racks. We have arrowmen hold the rear of the rockets to point them at the enemy, then soldiers light the fuses all at the same time and get out of the way. By making all the fuses exactly the same length and lighting them all on signal at exactly the same time, we can launch a salvo of many arrows almost simultaneously." "The arrowheads are the usual sharp spear shapes, I suppose?" asked the Commissioner. "No. Only some are. We can launch metal arrowheads composed of knives and swords as well. Up to three different kinds of jet rocket, depending upon the military situation." Finally, the Commissioner asked, "how many rocket arrows can you fire at once?" "We can position and fire up to 20 rocket arrows on the racks that I have built so far. I can make the racks bigger and bigger, but then it becomes a matter of logistics and being able to repeat the firings without everybody tripping over each other. It is better to build multiple wood racks instead. So if you want us to fire 40 rockets at once, we would use two wood racks." "What if I wanted to shoot 200 rockets arrows at once against a loosely deployed enemy force?" queried the Commissioner. Captain Jiguang replied, "in that case, we would have to fire rockets from 10 wood racks simultaneously, and time synchronization becomes a problem. This would be difficult, if not impossible for my soldiers to do. It would be better to use less racks, but repeat the firings. Using two wood racks at once, we could launch five salvos of 40 rockets each, in about two minutes from the same location. Synchronizing more than two racks complicates the logistics. Of course, you could fire a barrage of 1000 rockets at the enemy, but you would have to space out 25 pairs of wooden holding racks and keep them all going with a continuing resupply of rockets. You could never get them to fire all at once. And pretty soon, you would run out of rockets."

"I am impressed," said Commissioner Weischeng brightly. "Let us call your flying weapons 'flock–of–bee fire arrows' after the sting of bees from a disturbed bee hive." He paused, then added "I have consulted with

Wokuo pirate raids, mid–1500s

LIAONING

Beijing

HEBEI

KOREA

Dengzhou
Laizhou

Hanseong

SHANXI

Ji'nan
SHANDONG

JAPAN

Hakata
Hirado

KYUSHU

JIANGSU

Bōnotsu

HENAN

ANHUI

Yangzhou

*Main routes
of wokou raids*

CHINA

Nanjing

Suzhou

Hangzhou

Ningbo

HUBEI

ZHEJIANG

Taizhou

Wenzhou

Ryūkyū
Is.

Nanchang

HUNAN

JIANGXI

FUJIAN

Fuzhou

Xinghua

Guilin

Xiamen
Zhangzhou

Taiwan

GUANGXI

GUANGDONG

Guangzhou

■ *Areas affected by
early wokou raids*

■ *Areas affected by
later wokou raids
(1540-1565)*

*Drawn by Yu Ninjie
yuninjie@hotmail.com*

my military advisors. Based upon your past record of achievement, we would like to promote you to the position of Assistant Regional Military Commander of Shandong's defense force against Japanese pirates. You will report directly to me." "I am pleased to consider your kind offer," replied Captain Jiguang. "May I have three days to consider this?" he asked. "No," said the Commissioner. "The pirates have been continuously harassing and attacking our ships before they can even reach safe harbor in the coastal cities of Dengzhou and Laizhou. Worse yet, Japanese pirates are colluding with their Chinese counterparts and expanding their forces. We call them Wokuo pirates. I can only give you one day to decide. You must be ready to assume your duties tomorrow." "Yes, sir" replied Qi. "I will let you know this afternoon." "That is all" said Commissioner Weischeng. "Dismissed."

When now Commander Qi Jiguang took over the commandership of Shandong's coastal defense, he had less than 10,000 troops at hand, though the recorded strength was 30,000. Furthermore he was plagued with deserters, most of whom were young and strong men who could find a living elsewhere, leaving behind the old and the weak. The troops also lacked training and discipline, while the defense works were dilapidated due to years of negligence. But Jiguang disciplined his troops and reinforced the defense works so well that the pirates, seeing strong resistance in Shandong, had to move southwards to seek more vulnerable targets.

Japan at this time was in a state of great unrest. The Sengoku Period saw all of Japan plunged into small–scale regional civil wars. Many defeated samurais as well as impoverished workers and farmers turned to piracy. They often occupied offshore islands near the coast of China and raided Chinese coastal cities. The raids seriously impeded China's economy and trade, not to mention the mass killings and lootings. Unlike today's Hollywood vainglorious portrayal, pirates were often desperate, drunken men who dressed in tatters and wasted food and money as soon as they got either. Most pirates rarely lived long. They often turned to piracy in protest against oppressive conditions at home or on merchant vessels, yet they could seldom control themselves when they had chosen the alluring life of a free, unrestrained outlaw. The Wokuo pirate problem in China intensified during the mid–16th century.

In the fall of 1555, Qi Jiguang was sent south to the province of Zhejiang where the Japanese and Chinese pirates were a persistent threat. By 1558 they had penetrated inland past the coastal cities of Ningbo and Hangzhou, and frequently attacked the city of Taizhou in central Jiagsu province. Jiguang had been able to repulse each attack. However, to the north, other Wokuo raiders had managed to sail their junks from the Yellow Sea, past the mouth of the Huang Ho (Yellow) river, into China's interior, then up the silt–laden Wei river to disembark near Cengang (in today's Baoji city in east central Shaanxi Province). There, a consolidated force of pirates and ne'er–do–wells prepared to attack the city because it controls a pass along the Tsinling Mountains between the Wei river valley and the upper Han river. It was becoming more common to have more native Chinese than Japanese pirates participate in these attacks.

"How many carts of multi–shot rockets do you have," General Yú Dàyóu asked recently promoted General Jiguang. "My ordnance experts have assembled 40 wooden carts. We plan to deploy 20 carts widely spaced in the front line, with another 20 carts between them for firing later, as needed. We have a few thousand rocket arrows. My force of 3,500 men are equipped with Japanese Katana swords, which we have named Wodao ('sword/knife of the Wo people')". "Let me see your Wodao sword," said General Tan Lun. He held the sword, which he saw bore a strong resemblance to a Japanese Tachi or Odachi, with a handle about 25 cm long and a gently curved blade 80 cm long. He carefully ran his fingers over the razor–sharp edge, almost cutting himself in the process. "I have seen Wokuo raiders easily break our pole weapons with these," Jiguang said. "General Dàyóu and I copied the design, then we had our sword craftsmen make replicas for our foot troops. I was able to equip half of my force of 4,000 men with Wodao swords." "I command a force of 2,000 foot troops, 150 four–horse chariots, and 1,000 long lancers," added General Tan Lun.

The three Chinese generals were in conference in Jiguang's tent. Jiguang was to lead the combined forces attack on the Wokuo pirates from the valley to the east of Cengang. Their objective was to entrap them since Cengang

Left to right: Generals Yú Dàyóu, Qi Jiguang, and the boyish–looking Tan Lun preparing to lead their troops into battle, October 19, 1558

is surrounded by mountains on the other three sides, and then annihilate them. "If we are able to hold the critical points on strategic roads, the enemy cannot get out," General Jiguang had said. He paraphrased, "when a cat is at the rat hole, ten thousand rats dare not come out. When a tiger guards the ford, ten thousand deer cannot cross."

Before dawn the next day, the three Chinese generals were preparing their combined forces attack. First, General Jiguang's rocket–bearing shock troops were to launch many salvos of rockets at the enemy pirates. Then General Lun was to lead a cavalry force of four–horse chariots and long lancers. Generals Dàyóu and Jiguang were then to conduct mop–up operations with well–equipped armored foot troops.

"Make sure that each cart is fully loaded with rockets," Jiguang ordered his four ordinance technicians. Each technician with his assistants was in charge of the care and feeding of five multi–shot rocket carts. His chief ordnance expert Kun Chong came running up, out of breath. "My General, I tested a new innovation, and it seems to work!" he exclaimed. "I have been pondering over the last few months how to improve the guidance system for our rockets. Two nights ago, I had a dream. I saw a large arrow flying through the air with two wings attached to it, in the shape of a crow. So, yesterday I attached wings to one of our long arrow rockets, one in the shape of a crow, and another in the shape of a ball. Attaching one of our power flasks, I lit a fuse to a standard gunpowder load. Like always, it took off. After repeated trials, my technicians and I found that the hardest part was in keeping the wings attached to the arrow in flight. For some reason, the crow shape got a lot more range than the ball shape. I don't know why. They both had the same gunpowder load in the cylinder."

"What were the ranges you got?" General Jiguang asked. "Well, approximately 420 meters with the ball, and about 500 meters with the crow wing shape," Chong replied. Jiguang took a meditative pose for a minute. Then two minutes. After three minutes, Chong queried, "General?" "Why can't you load gunpowder inside the crow, or the ball itself?" Jiguang asked. "That way, when the rocket arrows hit their target, they will explode and kill even armored troops." "I hadn't thought of that," replied the ordnance expert. "Also, have you thought about enlarging the arrow shafts and hollowing the rear part out, so that your technicians can load more gunpowder inside and set the whole assembly off? Then you could use a fuse to link the powder that pushes the rocket forward with that in the body of the ball or crow." "No, I hadn't thought of that,"

responded Kun Chong. "But if we did that I bet we could get a several li's range out of the rockets."

Once again, the rockets demonstrated their worth. As the battle wore on, the devastating results of multiple fusillades of rocket arrows on the Wokuo pirates proved even better than expected. The rocket technicians and assistants became better and better at synchronizing their firings with practice, so that barrages from three and even four of the rocket–carrying carts were fired in synchrony. The resulting salvos of 60 to 80 rocket arrows at once demoralized the pirates. The five morale–busting "losses" described by the famous fifth century B.C. Chinese military strategist Sun Tzu were played out on the Wokuo pirates as they suffered an ignominious defeat: loss of order, loss of courage, loss of confidence, loss of cohesion, and loss of warfighting plan.

The foot troops of generals Dàyóu and Jiguang only had to round up the ragtag remnants of the Wokuo raiders. The majority of the pirates lay dead or dying on the ground.

Late that afternoon, 200 of the surviving ragtag ruffians were herded together under guard. "Please spare my men, you may take my life," pleaded Takeshi Yuudai on behalf of the captured pirate force. Five pirate captains had been separated out, their hands bound, and tied to trees in full view of the others. The three Chinese generals Qi Jiguang, Yú Dàyóu, and Tan Lun glared at them in silence. "You would not have spared the population of Kaifeng, you hypocrite," countered General Dàyóu. General Lun's left arm lay heavily bandaged and motionless at his side, having taken a glancing sword strike during the battle. A direct blow from a Wodao sword would have severed it. He had bled profusely. The three Chinese leaders turned their backs on the prisoners and conferred for some minutes. Then General Jiguang marched in front of the whole assembly. With his characteristically loud, booming command voice he said, "it has been decided. You five," he wagged his Wodao sword at the bound captains. "Select four prisoners each from your forces, and bring them forward for punishment. If you refuse, then I will have the lot of you sentenced to death by hanging."

Takeshi and the other four pirate commanders were unbound and forced to single out the 20 condemned men from the others. They chose mostly the oldest and most wounded of the prisoners. "Guards, take the 180 prisoners to the holding pen. They shall be shipped to our northern frontier and the Great Wall, to serve his Majesty the Emperor as slave labor for the rest of their wretched lives," said commanding General Jiguang. To the guards he commented, "in its construction we have gone beyond the simple tamped earth method, reinforced by willow reeds. Now we are using bricks, with a very effective sticky rice and burnt lime mixture to hold them together. Guard towers and parapets are spaced from two li down to one li in strategic places. The soldiers can even drive chariots atop its paved road…You," he said, motioning to ten other trusted guards, "rebind the captains securely, and bring them forward."

The General did not mention that invaders could sometimes either break through a breach in the wall, or simply bribe their way through. By 204 B.C. China had constructed 10,000 li (5000 kilometers) of the Great Wall to keep out invaders from the north. The wall also served to keep rebellious peasants in. For centuries the wall was in a continual state of expansion, repair, and restoration. By the time of the Ming Dynasty (1368–1644) the wall, counting its branches and other secondary sections, would stretch some 7,300 kilometers from the sea in the east to the Gobi desert in the west. Even today, the Great Wall is the only human–built structure easily visible from Earth orbit. At times the wall would engage as many as ten million laborers, one–fifth of China's workforce, who broke their backs constructing, improving, or repairing it. Thousands died in its construction from starvation, overwork, fatigue, or illness. Those who complained or ran away were usually killed. The Great Wall was called the "long graveyard" because some 400 people were buried alive in it.

A half hour later, the 20 pirates were lined up in five rows of four each. Instead of death, five mutilating punishments were to be exacted upon them. As arranged, General Lun had a fire going, four branding irons heating up fiercely in the blaze. "Bring the first row forward and restrain them," he commanded. The guards brought four pirates forward. "Bind them at the feet, knees, and arms in preparation for tattooing." Other guards brought the four red–hot branding irons forward and positioned them. "Go!"

The smell of sizzling flesh was drowned out by the prisoners' piercing screams. Red blood oozed from the men's foreheads as their flesh burned and smoked with a pungent odor. Finally, the screams turned to pitiful moans as they were led away. Each would bear the stamp of the Imperial seal upon his forehead forevermore.

"Bring the next four forward," commanded General Dàyóu. As an expert swordsman, it was decided to have him select his four finest warriors for the next mutilation. "Swords up!" he shouted. "…Down!"

Three of the blows precisely cut off the nose of three of the unfortunate prisoners. Since the nose of the fourth pirate was rather flat, the sword strike cut off not only his nose, but also skinned part of his forehead, lips, and chin. Again, the agonizing screams and blood.

"Next!" Dàyóu shouted. "Take off their leg bindings. You, you, and you, raise up their pants legs, and have them kneel so that their kneecaps are cleanly exposed." His guards complied as ordered. His four swordsmen wiped their Wodao swords clean of flesh and blood. "Swordsmen come forward. Ready, now! … Swords up! … Down!" he exclaimed.

This time, all four sword blows were clean. The crunching noise of broken kneecaps was heard simultaneously with horrendous screams, at the top of the men's voices. Assistants brushed away each of the four severed kneecaps, while avoiding the flailing upper thighs of the mutilates. More blood and gore. The four men had to be carried off this time, lower legs dangling uselessly below what was left of their profusely bleeding knees.

Again. "Next!" said General Dàyóu. This time, three of the four hapless pirates wiggled and struggled against their bindings, but to no avail. What could possibly be worse?, they all thought. "Remove their shoes and stockings. Keep their leg bindings on. Move their foot bindings midway up the calf, and make doubly sure they are bound as tight as can be. Roll up their pants legs about a hand's breadth above the ankles. Have them lie down, face up to watch their fate," commanded the General. "Swordsmen forward … Ready! … Now down!" he shouted, even louder than before.

Eight cleanly severed feet. One pirate wet his pants, while his fellow farted, then lost control of his bowels, befouling himself. To keep them alive, several assistants had to tie tourniquets just below the calf, ten or so centimeters above their stumps. Arterial bleeding and shock would bring a quick end to their lives, and General Jiguan wanted them kept alive. The four howling miscreants were led away, agonizing screams fading in the distance. If the bleeding could not be stopped, cauterizing was the only alternative. It was not unusual to have a prisoner faint from the pain of having his bleeding leg stump thrust into the fire. Medics would have to weigh the balance between loosening tourniquets too soon, or having gangrene set in.

This time, a more severe punishment awaited the last four. Although they had sustained injuries of varying severity, and most of them did not have many years left anyway, they moaned miserably when General Lun pronounced their fate. In medieval China, and Japan for that matter, castration (or for women, claustration) was just about the most severe punishment short of death. Soldiers carried the last four unwilling victims into the medical battle staff's tent. The men panicked at the sight of the razor–sharp cutting knives. Tight bandages were wound around their thighs and lower abdomen. While one soldier held each unlucky pirate by the waist, two others loosened his leg binds and held his legs firmly down to prevent movement. Both penis and testicles were then swiftly cut off with a small curved knife, as close to the body as possible. The pirates were not treated to the customary comfortably low bed, nor bowls of nerve–stunning tea, nor desensitization with baths of hot pepper water. Just bandages to keep the mutilates from bleeding to death—which would mercifully end their suffering. In five minutes it was over. The men knew their lives would be over too. Prisoners who had been involuntarily rendered sexually impotent by mutilation or removal of the external genitals were generally shunned in old Chinese culture, where both male and female fecundity were highly prized.

In medieval China eunuchs, males who had been rendered sexually impotent by voluntary mutilation or removal of the external genitals, served mainly in the Imperial court as palace menials, harem watch–dogs, and spies for rulers. As personal attendants to the reigning sovereign ruler, especially to guard harems, the eunuchs were in a better position than even the most powerful minister to curry favor, exercise influence, and accumulate wealth. Many eunuchs participated in political intrigue, and some amassed considerable power in state affairs, serving the Imperial Presence, the ladies of his royal family, and his thousands of concubines, all hidden together in the "Great Within" behind forbidding palace doors. Low–ranking eunuchs wore a long grey robe under a shorter dark blue coat, and had to wear their official hats and boots when on duty. In olden times, high–ranking palace eunuchs wore ornate robes of brilliantly embroidered colors.

Sometimes eunuchs played upon the fierce female rivalries, jealousies, and raw ambitions prevalent in the harem. There, several hundred ladies competed for the

attention of the Emperor – their only road to wealth and power for themselves, their clans, and their hoped–for princely sons. More than one eunuch joined forces with a scheming empress or concubine in dark plots to do away with the heir apparent and place her own son or favorite in line for succession. If the intrigue was successful, the conspiring eunuch was in a position to usurp enormous authority.

One particularly ambitious eunuch, Gan Zhang ("daring bowyer"), was gently brushing Chuang–Mu's vigorously flowing thick black mane across her swan–like neck, shoulders, and behind her back. She was very tired after a long night of sex with many sex–starved soldiers. She earned her reputation as the most wanton concubine in the Emperor's harem. An occasional tryst with the aging Jiajing Emperor Zhū Hòucōng would simply not do. To curry her favor, Zhang would often clandestinely let in her suitors one by one past the palace guards. He usually managed to bribe them, but one could never tell if a new guard was filling in, or if one of the more greedy guards thought he was not being paid enough. Two other concubines, Chang–Juan and Lin Yao looked on.

"Yan Song continues to stand in the way of my plan to influence the Mandarin in Hangzhou to appoint me as his deputy," said Zhang annoyingly. "I have cooked up a plan to dispose of him once and for all." He put down the brush and the comb. "Would you help me?" he plaintively asked Chuang–Mu. "It involves sexual intrigue and will only add to your pleasure. It will greatly help if you three can encourage him to over–drink at the upcoming Emperor's banquet." "Yes, the Emperor has decided to move it up, so now it is to take place in three days," Lin Yao chimed in. "What is in it for us?" she asked. "And for me?" Chang–Juan added. The trio of concubines looked at him expectantly. "I will ensure that all of

A portrait of the seductress Chuang–Mu (left), whispering the latest gossip with Chang Juan and making love (right)

you escape from this prison, as my well–remunerated assistants in Hangzhou. Chang–Mu, I hear that Hangzhou harbors a plethora of lascivious soldiers and bureaucrats." He paused to let it soak in. "I have already talked this over with Liwei and his assistant. They both agreed that our regular payments out of Emperor Zhū Hòucōng's treasury had earned me this position. Remember, I too want to escape from this prison." Chang–Mu contemplated for a few minutes, while the others assumed a lotus pose.

"Okay, I'm in. But only if a quick escape is arranged, and if we have a backup plan in place," she stated. The four conspirators quietly talked over Gan Zhang's well–conceived plan. He patiently answered all their questions to their satisfaction. Now they were all in on the dirty, dark secret.

"Damn him!" exclaimed Prime Minister Yan Song. He eyed his clan of court and district officials. "We cannot afford for him to take credit for his victories against the Japanese pirates at Taozhu, Haimen Garrison, and Taizhou. I want you to start a campaign to vilify him," he ordered his second–in–command, Twen–Ch'ang. "You will start a rumor in the eunuch hierarchy serving the Imperial court that General Jiguang is secretly conspiring with the Japanese pirates in Taizhou Prefecture. His successes are emboldening our enemies in the Imperial palace, and we have to act now." Minister Song was an extremely corrupt official who abused his power. Every year, forty percent of the wages meant for troops guarding the frontlines would end up in his pocket. As a result, damaged defense works were not promptly repaired and acts of desertion were rampant. Yan Song and his cronies ostracized those who opposed them, like Qi Jiguang. Besides covering up one another's acts of corruption, his followers also blamed their inability on others, especially those whose capabilities threatened their positions.

That evening, Yan Song had to force his bulging stomach from the banquet table. The rich man's dinner of 16 courses was too much, even for the gluttonous minister. He was past the point of satiety and any more food or drink would only add to his discomfort. He motioned his favorite concubine Chang–Juan to accompany him to the luxurious harem. "My Lord, I have taken a cold." She feigned a cough. "Besides, my menstrual cycle has just started. I am impure." She paused. "But Chuang–Mu stands here ready to take my place and satisfy your every need." With a nod of his head, he agreed.

Chuang–Mu ("Goddess of the bedroom and sexual delights") bowed to him in a demonstration of obeisance.

Her chi–fu, or "dragon robe," was a straight, kimono–sleeved affair with closely fitted neckband that continued across her voluminous breasts and down to her underarm closing on the right side, the long tubular sleeves terminating in horsehoof cuffs. Her large skirt cleared the ground for easy walking, and her entire costume was elaborately patterned with specified arrangements of dragons, clouds, mountains, and waves, to which were added an auspicious Buddhist motif. Yan Song's ch'ao–fu robe was also patterned like hers, but his was more commodious, and was slit front and back as well as at the sides to facilitate riding. His high rank was clearly differentiated by a capelike collar and flaring set–on epaulets, which gradually narrowed and were carried down under his arms. Over this whole costume he wore a bright yellow stole–like vest.

They left for the principal boudoir, with Chang–Juan and Lin Yao quietly and softly swishing along behind them. As they entered the boudoir, Chuang–Mu lit several fragrant, sweet smelling incense candles. As pre–arranged, Chang–Juan and Lin Yao took their stations nearby the closed door, engaging in small talk. Like many royal courtesans of the time, Chuang–Mu was well versed in philosophy and the arts, and began to make polite conversation. As she let her shiny long black hair fall to her neck and shoulders, he could see that she was ravishingly beautiful. Her lotus blossom perfume smelled divine. She had applied her fragrance where it would last longer on her skin—the inside of her wrist, behind the ear, chest, neck, behind the knee, and inside the elbow. These pulse points are the areas of skin where the blood flow is the strongest and the skin is the warmest. Fragrance rises, so the scent she applied on the back of her knee would last well into the night. They both reclined on the comfortable, padded low bed with silk sheets, making sweet talk. He caressed her soft breasts through her silky smooth gown, then gently massaged them. She expertly encouraged his passion, though his mind was foggy from too much wine. She disrobed and helped him untie his sash, then they both slid naked onto the bed and embraced.

He ran his fingertips over Chuang–Mu's satiny smooth, well–rounded buttocks. *Wow, I'm in heaven*, he thought. She hugged him more tightly, then locked her hands behind his back. She usually liked being on top in the reverse cowgirl position, but this time she used all her strength to roll them over, putting him into the missionary position (so–called because missionaries held it to be the least reprehensible), as Chang had ordered. As he passionately penetrated her she quietly reached under the pad where the weapon was hidden, as had been prearranged. Her heart pounding, she grabbed the sharply curved knife with her right hand. With expert timing between his thrusts, she slipped it vertically under him, so that his own weight would assist in the disembowelment. He loudly shouted in pain. The few within earshot outside the boudoir regarded the noise as typical of courtesan sex play, and went on about their business.

Suddenly Gan Zhang burst forth from his hiding place behind the curtains, grabbed a silk pillow, and covered Yang Song's panic-stricken face to smother his screams. Chang-Juan tried to grab his arms, while Lin Yao threw herself on top of him to keep him from getting up. With all her strength, Chang-Mu continued slicing open his abdomen with repeated twisting motions using both hands on the knife's handle. She sustained a deep cut on her right index finger as the handle became slippery with Yan Song's blood. The bed linens became soaked in blood, and then soiled by his foul–smelling intestines as they tumbled out. At last his quivering body lay still.

Minister Song's own treachery had done him in. It was easy to make enemies in the royal court, particularly with scheming concubines conspiring with eunuchs. It was not two difficult to sneak his lifeless body out of a hole that Zhang had busted out near the floor by a wall. Shorn of ears and eyes, his body ended up being thrown into a latrine in secret. In contrast to the wicked Yan Song and his like, Qi Jiguang's father, Qi Jingtong, had cultivated in his son a yearning for knowledge as well as a firm set of morals, patterned after the teachings of the venerable Confucius. But the damage inflicted by Song had been done. As Mark Anthony once said at Caesar's funeral oration: "The evil that men do lives after them, but the good is oft interred with their bones." Not only was Qi Jiguang not credited for his valor at Cengang, but he was almost demoted over the slander that he had liaised with Japanese pirates.

After his decisive victory at Cengang in 1558, Qi Jiguang, with the support of his old friend Commissioner Chao Weischeng, overcame Yan Song's treachery with fatal blows to the pirates at Taozhu, Haimen Garrison, and Taizhou. He also oversaw the construction of forty four vessels of various sizes to be used against pirates at sea. In 1559, he was waging a month–long battle against Japanese pirates in the Taizhou Prefecture. By this time, more than 2,000 rockets were equipped on ten Chinese warships, and 4,760 rockets came into use among Chinese infantry and cavalry troops. Such a great number was unprecedented both in Chinese and world military history. The other countries of the world would only learn of this kind of weapon some 240 years later.

"Turn about and launch a broadside cannonade at the closest one!" the ship's captain shouted to his crew. Two heavily armed Japanese pirate vessels were simultaneously attacking Qi Jiguang's flagship, the *Pan Gu*, one on each side. Sails a flying, she turned about. The cannoneers had readied the ship's twenty bronze cannons, ten on each side, and loaded them with gunpowder and heavy iron balls. As the Pan Gu came a broadside to the pirates, they lit the fuses on the cannons almost simultaneously. With a deafening BOOM the cannonballs went flying towards their target. Chinese ship cannon could accurately reach targets at 300 meters. Through the smoke, General Jiguang saw that about five cannonballs had found their mark. With at least two gaping holes, the pirate junk was starting to take on water. But worst of all, one of the balls had hit their powder storage bin amidships, and the subsequent explosion set their junk afire. Many of the pirates started to leap overboard to escape the flames.

"Shoot them!" Jiguang commanded his 30 fusiliers. "Our provisions are running low, and with the 200 buccaneers we already have in the holds, we simply don't have room for more prisoners. Regardless, you are doing them a favor. They could never swim back to land this far out to sea, and besides the sharks are out." Deadly fusillades of bullets soon found their marks. The sea was becoming red with the pirates' blood.

Witnessing the rapid demise of its consociate, the second pirate junk turned and began sailing with the wind. The shrewd General Jiguang turned his flagship to sail across the wind, in a reaching maneuver with the wind abeam. He knew from experience that sailboats can usually move faster when sailing across the wind than in any other direction. "Helmsman, set an interception course to port," he barked to Cheng Chong, the head helmsman. But the enemy captain countered with his own reaching maneuver to starboard, and was clearly making a run for it. Soon Cheng Chong bemoaned "they are moving out of cannon range, Captain." "Then prepare the rockets!" Captain Jiguang shouted back. He had reckoned that the pirates did not have rockets of their own.

By this time the Chinese had made two additional rocket innovations. Rockets with double boosts were produced. At the rear of the rockets, there were four gunpowder boosters. The weapon could fly above water for a few kilometers. Secondly, a large number of "multi–shot rockets" could be positioned in a bucket, which had two layers for orientation and direction purposes. Once the soldier fired the fuse, all of the shots were made at the same time, covering quite a noticeable range. Besides this kind, there were other variations, capable of firing 2 to 100 shots at a time.

The ordnance technicians had packed three multi–shot rocket buckets with gunpowder. "Range now about five li," cried Lu–Pan, observing through the spyglass. "Tighten the mainsails, the wind is shifting!" commanded Captain Jiguang. "Aye, Captain," answered the experienced sailors. They expertly adjusted the ship's spinnakers to gain added speed as the ship inclined slightly with the added wind load. Both the *Pan Gu* and the fleeing pirate junk were now running with the stiffening wind.

"Range holding steady at five li!"Lu-Pan shouted. Qi Jiguangwas mystified. How could the pirates be making such headway? Maybe their junk was more lightly loaded, than his ponderous, six–masted *Pan Gu* flagship, he thought. But his ship was still a far cry from the great Admiral Zheng He's 134 meter flagship which had nine huge masts.

Selected by the Yung–lo Emperor Cheng Zu Zhu Di to be commander–in–chief of the Chinese missions to the "Western Oceans," Zheng He first set sail to southeast Asia and across the Indian Ocean in 1405, commanding 62 ships and 27,800 men. Later voyages went to the Persian Gulf, Arabia, Egypt, the Red Sea, the coasts of India and east Africa. Besides flattering the Emperor's vanity, these expeditions had the effect of extending China's political sway over maritime Asia for half a century, but they did not, like similar voyages of European merchant–adventurers, lead to the establishment of trading empires. Chinese emigration did increase later, resulting in Chinese colonization in Southeast Asia and the accompanying tributary trade into the 19th century. At its peak, the great Chinese fleet comprised 337 ocean going ships, 188 transports, and over 200,000 men. This dwarfed the more famous Spanish Armada of 1588, which on its arrival in the English Channel had some 120 ships with 24,000 men.

General Jiguang brooded. *Why, why, why did the succeeding Hung–hsi Emperor Ren Zong Zhu Gaozhi in 1425 suddenly suspend naval expeditions abroad and disband Zheng He's troops in Nanking? Was the fleet only intended to flatter the Emperor's vanity? And then to have Emperor Xuan Zong Zhu Zhanji not only disband the huge fleet, but order all the ships <u>burned</u> a few years later? What a waste! And for what? China could have extended its hegemony over most of the civilized world,* he lamented. The xenophobic Chinese, even more so than the Japanese, would close their doors to the outside world for centuries.

Back to the task at hand. "Are we closing?" he asked Lu–Pan. "No, still holding steady, Captain," came the reply. If the pirates distanced themselves anymore, they would soon be out of firing range, even for his rockets. "Prepare to fire the first salvo!" he commanded his rocketeers. "And get the second salvo ready to fire immediately!" His ordnance technicians prepared and carefully oriented the upper layer of four–booster rockets for maximum range. "Ready, now. At my command, ready…ready…aim…FIRE!" Jiguang shouted above the whistling wind. "Second barrage, get ready and adjust your aim," he shouted.

He felt exhilarated as the twenty rockets flew smooth trajectories with the wind. Closer…closer…then they flew over and started splashing into the ocean *ahead* of the enemy pirate ship, a distance of six li (three kilometers). One, then two fire rockets ripped through the mainsail of the pirates, splashing harmlessly behind the target. The rocketeers knew from experience how much to adjust the firing angle of the lower tier of twenty booster rockets. They made their best educated guesses to compensate for wind speed and direction. "Steady as she goes," ordered Qi Jiguang. "Second team, fire at my command…now ready… now ready…hold course …hold…aim…aim, I say…and FIRE! Eighteen two–booster rockets suddenly swooshed out of their individual firing cylinders into the ocean air. Two rockets stayed hung up in their holders. Apparently duds. A rocketeer reached in, then boom!, one of the "duds" exploded. It had gotten snagged in its holster and was not a dud after all. The hapless rocketeer fell back, eyes blinded and clothes set afire by the explosion. He held up a bloody stump, all that was left of his right arm. The medics would be busy this afternoon, reckoned Captain Jiguang.

Five and one–half li ahead, the forty some–odd buccaneers watched helplessly as the four–booster rockets flew gently curving trajectories toward them, the crow–shaped wings and rear tail fins keeping them on a deadly course. BAM! BAM! BAM! Their junk shuddered as three of the weapons hit their mark, two towards the rear and one just ahead of amidships. Those alone would have been enough to disable their ship.

But then two more lethal hits. CAROOM! Swoosh... boom! Very soon it was all over for the pirates and their splintered junk—now literally junk.

General Jiguang and his weapons experts had learned through years of experience just how to maximize the lethal effect of each booster rocket, and make it into an explosive rocket. First they made the boosters which would be attached to the rear of the rocket. They packed them with gunpowder and inserted a fuse which would fire the rocket. Then they'd hollow out longitudinal spaces on two sides of the cylindrical rocket shell, towards the front end because the tip would always strike first. Next they would pack explosive powder into both hollowed–out spaces, and securely seal them with bamboo stock and sticky glue. The Chinese rocket technicians did not know aerodynamics, but they learned to make the cylindrical sides smooth and not let any protrusions stick out, or the missiles would lose accuracy. They had also learned over the years how to better cut the balls or crow–like wings and load them with gunpowder, and how to angle the tail feathers or fins for maximum accuracy.

After the month–long battle with Japanese pirates in the Taizhou Prefecture, Qi Jiguang's army and naval forces had inflicted over 5,000 casualties on the pirates. Qi Jiguang's soldiers established a name for themselves among both the people of Zhejiang and its enemies.

Partly as a result of Qi Jiguang's military success in Zhejiang, pirate activities surged in the less well-defended province of Fujian. More than 10,000 pirates had established strongholds along the coast from Fu'an in the north to Zhangzhou in the south. In July 1562, Qi Jiguang led 6,000 elite troops south into Fujian. Within two months, his army eradicated three major lairs of Japanese pirates at Hengyu, Niutian, and Lindun. However, his own army also suffered significant losses to fighting and diseases. Seeing the pirate infestation in Fujian subdued, Jiguang then returned to Zhejiang to regroup his force. The Japanese pirates took the opportunity to invade Fujian again, this time succeeding in conquering Xinghua (present day Putian). In April 1563 Jiguang led 10,000 troops into Fujian and regained Xinghua. Over the next year, a series of victories by his army finally saw the pirate problem in Fujian fully resolved.

A final major battle against Japanese pirates was fought on the island of Nan'ao, which lies near the boundary between the provinces of Fujian and Guangdong, in September 1565. There Jiguang joined arms with his old comrade Yú Dàyóu again to defeat the remnant of the combined Japanese and Chinese pirate force.

With the pirate situation along the coast under control, General Jiguang was called to Beijing in late 1567 to take charge of training troops for the imperial guards. In the next year, he was given command of the troops in Jizhou to defend against the Mongols. He soon began the repair work on the segment of the Great Wall between Shanhai Pass and Juyong Pass. Meanwhile, he also directed the construction of watchtowers along the wall. After two years of hard work, more than a thousand watchtowers were completed, giving the defensive capability in the north a great boost.

Qi Jiguang also conducted a month–long military exercise involving more than 100,000 troops in winter 1572. From the experience of the maneuver he wrote Records of Military Training, which became an invaluable reference for military leaders after him. Over the 16 years when he was in Jizhou, not a single Mongolian raider crossed to the south of the Great Wall.

In early 1583, Jiguang was relieved of his duty on the northern frontier and assigned an idle post in Guangdong. His already ill health worsened in the next two years, forcing him to retire to his hometown. He finally died in 1588, days before the Lunar New Year. His life was probably best summarized by his own poem: *For three hundred sixty days a year, I hold my weapon ready atop my steed.*

WAN–HU

The capital city of Beijing in the Ming period had grown to an even grander scale than under the Mongols. By 1525, plans had been laid to build additional an outer city wall, in order to accommodate the increasing number of inhabitants living outside the city. Unlike the simple wall construction of pounded earth in Mongol times, the city walls of the Ming capital city were to be faced with a layer of bricks to prevent weathering.

"You mean those are the largest diameter rockets you could find?" asked Wan Hu incredulously. His assistant nodded. "Sir, I have scoured the countryside, and could find nothing bigger." "You mean you were too *lazy* to look anywhere outside the confines of Beijing, don't you," retorted Master Hu. "I cannot launch this chair with rockets this small! These puny things are only eight centimeters across, and that simply won't do. Sit down! Not one more word out of you. Yeng, you come here." "Yes, Master," said Yeng–Wang–Yeh ("*Foremost of the Ten Yama Kings of Lords of Death*"), taking a gracious low bow. Returning the bow to his chief servant, Wan Hu looked him straight in the eyes, then breathed in a low voice, "I have trusted you as my head servant for many years, now. You must find the biggest size, and I mean the biggest size rockets you can. My careful calculations show that even with the full complement of 50 rockets, my rocket chair won't get off the ground with anything less than 10 centimeter diameter rocket boosters, and it will be iffy at that." "Yes, Master, I will immediately look into this matter," said the obeisant servant. Wan Hu had hoped and prayed that he would fly in the sky like an eagle with the boosting power from the rockets and the kites. In the Chinese Confucian tradition, the concept of heaven was important. Heaven had nothing to do with the after–life, as Christians would think of the term. Rather, Heaven was a divine ruler, where the ancestors resided and from which emperors as Sons of Heaven drew their mandate to be earthly rulers.

Two months later, Wan Hu's cast of servants were busy installing as many booster rockets as physically possible in every nook and cranny of his large, sturdy "rocket chair." Years before, Hu had amassed a small fortune as a local government official, subject to bribes, in the prosperous city of Beijing. His financial means also allowed him to indulge his favorite passion, which was stargazing and wondering about the heavens.

"Fine, fine, it looks fine," beamed Master Hu as the rocket chair took form. His head servant Yeng–Wang–Yeh had managed to locate a stockpile of 100 double booster rockets in Nanjing, each 12 centimeters in diameter. The local garrison commander could not believe it when Yeh had told him of Hu's hairbrained scheme to fulfill his dream of getting himself closer to the stars by launching himself in his rocket chair. Yeh could only convince him to part with 50 of the rockets, and at that he had to pay a premium price. Rockets were weapons of war, not a means of transportation and a ticket to the stars, the commander had told him.

Hu's servants were tying as many rockets as possible to his sturdy chair and angling them upward for flight, but were coming up short of the 50 desired. "Master, I don't think we can possibly tie any more than 47 of these rockets to your chair," said one of his exasperated servants. Looking over the whole assembly, Hu had to agree. *Oh, well*, he thought. *My calculations assumed only 10–centimeter diameter booster rockets, and we are using 12–centimeter ones. This should be enough, and even give me a small margin.* He had instructed his servants to put the chair up on rocks, so that they could access all areas of the chair, even including two rockets perilously close to where he had to set his feet. He also had them attach two huge kites, one to each arm of the large chair.

Upon completion of the whole affair, Wan Hu gathered his crew of 53 servants. "Turn me around, so that I directly face the wind," he told them. Wan had picked out a tall grassy knoll just outside the capital city

Wan–Hu's Rocket Chair

of Beijing, and had waited for a windy day so that the kites would have maximum lifting effect. Six of the strongest men strained and heaved to lift and turn the heavy chair into the wind, while assistants held the kites in position.

To make an impression Wan had dressed himself in his best imperial finery. He turned to face the crowd, hiding his disappointment at the small crowd of a only a few hundred who had gathered in the wind to watch the launch. He had expected well over a thousand spectators. Doffing his hat, he took a deep bow, and announced in a loud voice, "Today, we are going to make history with this spaceship. I will be China's first yǔháng yuán ("universe navigator"). I will go where no man has gone before. When I return, I will be the Emperor's tàikōng rén ("space person," referring to taikonauts who have actually been in space). My two kites will ensure a soft landing." More than a few in the crowd thought to themselves, *another moronic idea from him. Let's see if he can even get off the ground without blowing himself up.* Wan had the character of a somewhat nutty professor, and a reputation as an eccentric.

> *"Strap me in securely," he instructed several of the servants. "And please untie the kites, I want to hold them in my hands for better control. Just make sure that my forearms are securely attached to the chair." Immediately the servants complied.*

He meditated. *They just don't know. I have been living the Mahayana ideal of bodhisattva, one who delays the full enjoyment of Nirvana for the sake of serving others.*

The bodhisattva ideal is to exercise compassion, based upon the sacrificial example of Gautama ("Buddha, the Enlightened One") and other Buddhas to lead others to Nirvana rather than selfishly enjoying it. Gautama proclaimed the "Middle Path" which lay between the two extremes of self–indulgence and self–mortification. He declared the "Four Noble Truths":

1. Suffering is a universal fact.
2. The origin of suffering is in craving or desire.
3. The cessation of suffering is accomplished by the forsaking of desire.
4. The way leading to the end of desire and the cessation of suffering is the Noble Eightfold Path (or Way).

Two years before, a disciple of the Great One had taught Wan Hu that the consequence of following the Eightfold Path is Nirvana, the state in which desire is extinguished. The holy teacher had emphasized time and again that *Nirvana* is <u>not</u> the negative experience of "nothingness" or "annihilation," but the positive experience of freedom or "emancipation." Nirvana centers upon the concept of energy. Everything in the universe has energy, including the individual human self. But the goal is to be able to <u>control</u> one's energy, even to the extent of appearing and disappearing to other disciples at will. His father Huan Yue, a monk, had reached the experience of enlightenment and entrance into *Nirvana* in this life and had earned the title of *arahat* before his death. All craving is gone. His father knew the impermanence of self, so there was nothing about him to be born again, to live again, or to die again. Full *Nirvana*, even for the Enlightened, comes only at death, when the constituents of the Self are finally dispersed. There is an impersonal Reality into which one can be absorbed.

As a layman, Wan could not hope to reach *Nirvana* in this life, but he could amass merit for a next existence which would be more favorable for attaining the ultimate goal. Buddhism developed considerable strength in India, especially under King Asoka, who was an ardent Buddhist of the third century B.C. By 1525 Mahayana Buddhism had been significantly modified from its Indian origins in 500 B.C. by centuries of association with the other religions of East Asia, especially Chinese and Japanese religions, with which it tended to blend.

Buddhism's Eight–Fold Path

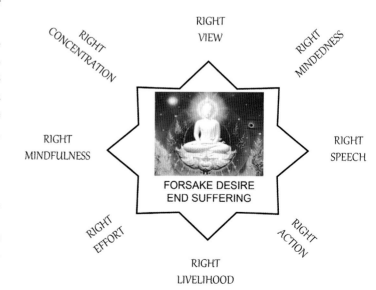

RIGHT CONCENTRATION
RIGHT VIEW
RIGHT MINDEDNESS
RIGHT MINDFULNESS
RIGHT SPEECH
RIGHT EFFORT
RIGHT ACTION
RIGHT LIVELIHOOD

FORSAKE DESIRE END SUFFERING

I have to overcome my bad past karma, the bribes, the lies, the deceits, he continued. I have to earn back good karma. I am giving the ultimate sacrifice, perhaps even my life, for the sake of the advancement of mankind.

"Now light the torches," Wan instructed. Forty seven servants lit their torches at once. "Now fire the fuses of the rockets, simultaneously." Each of the 47 servants approached with a flaming torch, and fired the fuses of all 47 rockets at the same moment. Their job done, the servants speedily retreated to a safe distance… and waited.

All of a sudden, a huge explosion shook the earth. The shock waves reverberated for some seconds, gradually diminishing. Rocket exhaust and smoke everywhere were obscuring what was happening to Wan Hu and his rocket chair. From a safe distance, the spectators saw both kites rise a few meters above the knoll, then they disappeared in the tremendous blast.

Floating above, Wan looked down at what was left of his bloody corpse. Down below, a multitude of servants rapidly drew around in a circle. His favorite, Shen Nung, broke down and sobbed on Wan's chest. *But it does not matter. Why are they staring at my broken body? Here I am, not down there! Can they not see? That is just a lifeless, bloody corpse torn to pieces. I am here. I am here!* He could clearly see sorrowful assistants picking up the shattered pieces of his lower legs. Others soon found his mangled left arm some distance away. Then they all just stood there. He looked up to Heaven, expecting to see the non–corporeal entity *Shangdi*, an omnipotent, just, and monotheistic supreme being.

Like many of his fellows, Wan Hu believed that Chia–ching, the reigning 11th emperor (1521–1566) of the Ming dynasty, had lost his Mandate from Heaven to preside as earthly ruler over the people. Notoriously cruel, Chia–ching caused hundreds of officials who had the temerity to disagree with him to be tortured, demoted, or killed. In his later years, he would selfishly spend much of his time and money patronizing Taoist alchemists in the hopes of finding an elixir to prolong his life. In such cases, many Chinese felt they had the right to rebel against, if not overthrow, the disenfranchised ruler.

It was an insensate but penetrating darkness. I could just feel it.

Moving along a huge tunnel. Where is everybody?

Ahead of the passage, a light.

Closer and closer, then intensely bright.

Now an unearthly, effulgent radiance, everywhere, permeating everything.

Such peace, such a feeling of joy and all–surrounding happiness absolutely beyond human description.

Then, I saw…Confucius! Confucius took me by the hand and gently guided me in.

Love, everywhere an unearthly tremendous love all fulfilling and all encompassing.

There…over there… my father, reaching out his hand. My mother, Juan Na. Many ancestors I had worshipped since childhood, they were all here! But, I did not have to call them by name. I just knew who each and everyone was, with an unfathomable knowledge!

I am so ALIVE now

Whoa, what's happening? THE COSMOS. THE BUDDHA - DIVINE BEING, ALL PENETRATING, ALL KNOWING…

WOW…WOW…WOW!

Though doomed to fail, the Chinese scholar Wan Hu has been universally acknowledged as the first man to try flying to space with the help of rockets. In memory of Wan, NASA named the Wan–Hoo crater on the back of the Moon after him. The Discovery Channel's show "MythBusters" attempted to recreate Wan Hu's flight using materials that would have been available to him. The chair exploded on the launch pad, with the crash test dummy sitting in it suffering severe burns. An attempt was also made using a chair with modern rockets attached, but it barely made it off the ground before going out of control. It was determined that small rockets that can be strapped to a chair cannot provide enough thrust to significantly lift the chair, let alone achieve escape velocity.

China eventually followed the trail blazed by Wan Hu's experiments and turned his centuries–old dream into reality. Astronaut Yang Liwei's history–making flight in October 2003 to space transformed him into an instant hero for millions of Chinese. With his flight aboard the Shenzhou V spacecraft, China joined the elite club of space powers that can put a man into space and return him safely to Earth.

ORVILLE

"It must be remembered that there is nothing more difficult to plan, more doubtful of success, nor more dangerous to manage than the creation of a new system. For the initiator has the enmity of all who would profit by the preservation of the old institution and merely lukewarm defenders in those who would gain by the new ones."

Machiavelli, The Prince 1513

"To design a flying machine is nothing. To build one is something. But to fly is everything."

Ferdinand Ferber, 1898

Modern aerodynamics emerged about the time that the Wright brothers made their first powered flight in 1903. Today and for the foreseeable future we must depend upon the stalwart early airplane pioneers of the late 19th and early 20th centuries. Modern rockets and missiles of all types today depend on our knowledge of aerodynamic effects for their stable flight while flying in the Earth's sensible atmosphere. All missiles must have some sort of aerodynamic control surfaces, like fins and wings. The missile body itself provides some lift which must also be accounted for in the design. In fact, cruise missiles are basically airplanes in cylindrical form.

DECEMBER 17, 1903

The Outer Banks near the village of Kitty Hawk, North Carolina contain 130 miles of salt marshes and desolate sand dunes. Nicknamed the "graveyard of the Atlantic," the outer banks are notoriously treacherous. The Wright Brothers, Wilbur and Orville, both bachelors, had made their home here for the last three months, living in a wooden shed while they conducted test flights and put the finishing touches on, and checked out their first self–powered

Wright Flyer. They had come here to take advantage of the favorable onshore winds, and the sand that would soften the occasional crash landing with Wilbur or Orville on board. Kitty Hawk provided the secretive brothers with wind, and isolated testing grounds for flight experiments. It also gave Wilbur the opportunity to study birds, the masters of the art of flying.

The Wrights had spent the last four years performing wind tunnel tests and constructing and ground testing a succession of heavier–than–air machines. Today was the culmination of literally thousands of hours of testing kites and flying gliders above the sands at Kitty Hawk. They were desperate to see if their new machine would work.

The 1903 Wright Flyer was a braced biplane structure, with two wings spaced seven feet apart. Forward of the wings was a twin–surface horizontal elevator, and to the rear was a twin–surface vertical rudder. Wing spars and other long, straight sections of the craft were constructed of spruce, while the wing ribs and other bent or shaped pieces were built of ash. Aerodynamic surfaces were covered with a finely woven muslin cloth. This Flyer was the Wrights' largest so far, with a wingspan of a little over 40 feet, a surface area of 510 square feet, and a weight of 625 pounds. A series of lightweight struts and wires held the wings together. It was constructed of spruce and ash covered with muslin. The framework "floated" within fabric pockets sewn inside, making the muslin cover an integral part of the structure. This ingenious feature made the aircraft light, strong, and flexible. The Wrights' carefully designed cotton and willow wings, which could warp in flight, together with elevators and rudders, would control the machine in the air. A hand–controlled pair of horizontal elevators mounted in the front was designed to control the craft's ascent and descent. Steering control was provided by a smaller pair of vertical rudders, centered at the rear of the airplane.

Of course, the Flyer had to have an engine. Because no engine manufacturer had a powerful, lightweight engine that met the Wrights' specifications, and no one was willing to develop one, they had decided to design and build their own engine from scratch with the help of their talented mechanic, Charles Taylor. Taylor was employed as a machinist in the Wright brothers' bicycle shop. Their simple four–cylinder gasoline engine developed only a puny 12.5 horsepower after the first few seconds of operation. Moreover, the engine could not be throttled; a hand lever only allowed the pilot to open or close the fuel line. In order to start the engine, a coil box was connected to the spark plugs, and two men pulled the propellers through to turn the engine over and start it.

producing thrust through motion in the air. The Wright's propeller has been tested out to be 81% efficient. It is just mind–boggling that the Wrights could be so close to

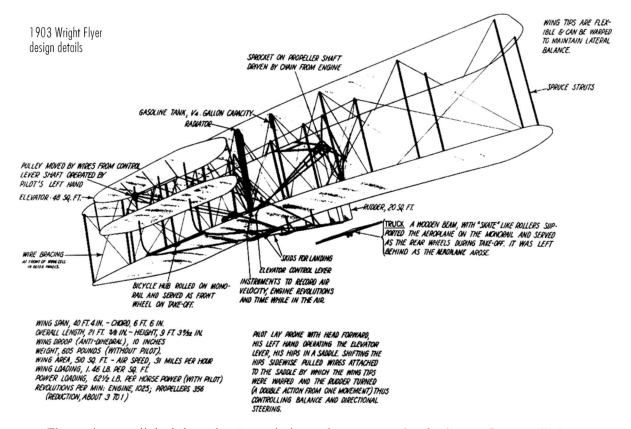

1903 Wright Flyer
design details

SPROCKET ON PROPELLER SHAFT
DRIVEN BY CHAIN FROM ENGINE

WING TIPS ARE FLEX-
IBLE & CAN BE WARPED
TO MAINTAIN LATERAL
BALANCE.

SPRUCE STRUTS

GASOLINE TANK, V₄ GALLON CAPACITY
RADIATOR

PULLEY MOVED BY WIRES FROM CONTROL
LEVER SHAFT OPERATED BY
PILOT'S LEFT HAND
ELEVATOR 48 SQ. FT.

RUDDER, 20 SQ. FT.

TRUCK A WOODEN BEAM, WITH "SKATE" LIKE ROLLERS SUP-
PORTED THE AEROPLANE ON THE MONORAIL AND SERVED
AS THE REAR WHEELS DURING TAKE-OFF. IT WAS LEFT
BEHIND AS THE AEROPLANE AROSE.

WIRE BRACING
AT FRONT OF WING CELL
IN OUTER PANELS.

SKIDS FOR LANDING

BICYCLE HUB ROLLED ON MONO-
RAIL AND SERVED AS FRONT
WHEEL ON TAKE-OFF.

ELEVATOR CONTROL LEVER
INSTRUMENTS TO RECORD AIR
VELOCITY, ENGINE REVOLUTIONS
AND TIME WHILE IN THE AIR.

WING SPAN, 40 FT. 4 IN. – CHORD, 6 FT. 6 IN.
OVERALL LENGTH, 21 FT. ⅜ IN. – HEIGHT, 9 FT. 3⁵⁄₃₂ IN.
WING DROOP (ANTI-DIHEDRAL), 10 INCHES
WEIGHT, 605 POUNDS (WITHOUT PILOT).
WING AREA, 510 SQ. FT. – AIR SPEED, 31 MILES PER HOUR
WING LOADING, 1.46 LB. PER SQ. FT.
POWER LOADING, 62½ LB. PER HORSE POWER (WITH PILOT)
REVOLUTIONS PER MIN: ENGINE, 1025; PROPELLERS 356
    (REDUCTION, ABOUT 3 TO 1)

PILOT LAY PRONE WITH HEAD FORWARD,
HIS LEFT HAND OPERATING THE ELEVATOR
LEVER, HIS HIPS IN A SADDLE. SHIFTING THE
HIPS SIDEWISE PULLED WIRES ATTACHED
TO THE SADDLE BY WHICH THE WING TIPS
WERE WARPED AND THE RUDDER TURNED
(A DOUBLE ACTION FROM ONE MOVEMENT) THUS
CONTROLLING BALANCE AND DIRECTIONAL
STEERING.

The engine was linked through a transmission made of gear–driven sprockets and bicycle chains to twin contra–rotating pusher propellers mounted on the rear of the plane, which it turned at an average speed of 348 rotations per minute. The wooden propellers were carved by hand from laminated layers of spruce, with tips covered in muslin to keep the wood from splitting. They measured 8.5 feet across. Recognizing that propeller blades could be understood as rotary wings, the Wrights were able to design their twin wooden pusher propellers on the basis of their wind–tunnel data. They were the first to realize that a propeller is a wing in rotation. This was invention touched with genius. Everyone else at the time was thinking of propellers as rotary paddles, like the screws on a ship. The revolving wing, on the other hand, would need to be angled for lift. But since the outside of a spinning blade travels faster than the inside, they had to gradually shift the blade's tilt, called angle of attack, along the blade's length to maximize its efficiency. The Wrights quickly figured out what the correct twist should be, and in so doing invented the modern propeller. Modern wooden propellers are 84–85% efficient in

where we are today, for the very first propeller!

Most important of all, the Wright Flyer had the successful control system that the brothers had developed for their gliders. A main feature of this system was their ingenious new invention, a device called "*wing warping*" for twisting the wing tips in order to preserve the plane's balance and control in flight. The device warped the wing tips in opposite directions simultaneously. This increased the angle of attack of the wings on one side of the craft and decreased it on the other, enabling the pilot to raise or lower the wing tips on either side in unison in order to maintain balance or to roll into a turn for lateral control. It worked!

The only problem was that there was no seat. Instead, the pilot had to lay prone in the middle of the lower wing. A wire was strung from each wing to a padded wooden "cradle" that fitted around the pilot's hips. The pilot had to shift his hips from side to side to operate the wing warping mechanism, which also moved the rudder. The rear rudder was directly linked to the wing–warping system in order to counteract problems of yaw produced by the warping of the wings.

While he was performing these acrobatics, the pilot also had to operate a small hand lever to control the forward elevators, which provided pitch control and some extra lift. Quite a bit of extra lift, as the Wright Brothers would soon find out.

The Wright Flyer also had no wheels. Instead of wheels, it had wooden runners. The Wrights knew that it would be difficult to operate a wheeled aircraft from the rough and sandy surface where they planned to fly, so they decided to launch their machine into the air with a smooth run down a 60–foot–long monorail track. The launch rail consisted of four 15–foot two–by–fours, the thin upper edge of which was protected by a metal cap strip.

At the beginning of each flight the airplane was positioned at the head of the rail. A restraining line ran from a clip near the pilot's position at the leading edge of the lower wing to a stake driven into the ground behind the machine. When the pilot was ready, he released the restraining rope with the hand clip, and the machine moved down the rail on two modified bicycle wheel hubs. The engine could not be throttled; another hand lever only allowed the pilot to open or close the fuel line. In effect, each landing was a controlled crash landing.

The Wright brothers had constructed as much of the Flyer as possible in their Dayton, Ohio workshop, then shipped the parts to Kitty Hawk for final assembly onsite at Kitty Hawk. They had made every single piece of it, at a cost of less than $1000. Their only official government support had come from Richard Rathbun, the assistant secretary of the Smithsonian. In 1899 he received a letter from Wilbur Wright asking for background material on "proper construction of a flying machine" that would help the Wrights in their own engineering. Rathbun assembled a packet of materials and references that encouraged and informed the Wrights' progress at a crucial time in their development.

The Wrights had tried to fly three days earlier on December 14. Men of a nearby U.S. Lifesaving Service station, the Kill Devil Hills Life Saving Crew, hauled the flying machine up the slope to where it would take off. Wilbur had won the coin toss, climbed aboard the Flyer, and lay prone on it. After a somewhat difficult start, the craft began to move down the ironclad launch rail and then rose from its track. But Wilbur had miscalculated and turned upward too soon, not realizing the effectiveness of the front elevators. The Flyer climbed a few feet, stalled, and then settled onto the ground near the foot of the hill. Wilbur was unhurt, but he had smashed the elevators. Repairs took two days, with no work on the Sabbath out of respect for their father, Milton Wright, who was a bishop in the Church of United Brethren of Christ. The machine was ready to fly again late on December 16.

December 17 dawned with a strong wind blowing from the north and rain beating down. The rain stopped, but the bitter wind kept blowing, whipping the sand dunes of Kill Devil Hill. Ice coated the rain pools. Impatient to get on with the day, the younger brother Orville was already out on the dunes. "*Curses*," he thought to himself. "*Will these winds ever die down?*" Orville had gotten up early to measure the winds with a hand held anemometer. It was today that they had planned to put their finished aircraft to the test in controlled flight. Over the last few weeks the weather conditions on the Kitty Hawk sand dunes had become unpredictable, and today the weather was horrid—cold and stormy. With winter approaching, the situation would likely deteriorate over the coming months. Orville dejectedly trudged back on the sand dunes to their work shed, where his older brother Wilbur was cooking breakfast, as he had customarily done for the last three months.

"The eggs and ham are almost done," Wilbur pointed out as he gave their breakfast one last stir. A delicious aroma of eggs, ham and bacon filled the shed. "I measured the winds just minutes ago at 27 miles per hour, almost the same as the 28 mile per hour reading I took an hour ago," said Orville, putting the anemometer down on their makeshift dining table. "The winds are still too high for safe flying," responded Wilbur. They both resignedly sat down to eat a hearty breakfast.

Less than two miles away, the men of the local lifeboat station also had their eye on the weather. Orville had asked the men the night before for their help once again as they planned to undertake yet one more series of test flights. John T. Daniels stood on the promontory, scanning the horizon with his binoculars for any sign of a boat in distress. It was not uncommon this time of year for offshore storms to jeopardize seamen close to the coast. Today, six lifeguards would take turns patrolling up and down the sand dunes. He had been watching the Wright brothers with fascination, and was expecting to help them with their new flying machine today. But so far, he saw no sign of activity from the brothers' camp.

Just after 10:00 a.m., Wilbur and Orville went outside to check on their new machine. While Orville was confident in their ingenuity, the brothers were only too aware that this machine had never been tested in flight. They had experienced many frustrating failures with previous versions. Testing any machine is risky, but in a 28 mile per hour wind, they were putting their lives at risk.

At 10:20 a.m. the winds were still too strong, gusting to almost a gale. But time was running out, and the brothers were getting increasingly frustrated. They had made a solemn promise to their father that they would be home in time for Christmas, and the journey from Kitty Hawk to Dayton, Ohio would take three days in the best of times. So today, with less than five days to spare, all they could do was wait, and pray that the wind would die down. Impatient at the weather delay, Wilbur decided to stroll out on the dunes and check conditions for himself. *"How much longer, just how much longer must we wait,"* thought Wilbur to himself, his mounting frustration becoming unbearable.

Orville Wright, 1903

Wilbur Wright, 1903

Wilbur chided his younger brother, "look, Bubs, we have promised father that we would be home by Christmas. Let's just do it, I am tired of waiting. Besides, now its your turn at the controls. We cannot afford any accidents. Let's see if you can keep from crashing it." "O.K, Ullam," responded Orville.

Growing up, the Wright brothers had private nicknames for each other. Wilbur was known as "Ullam," short for Jullam, which is German for William. Orville was known as "Bubbo" or "Bubs," Wilbur's pronunciation of "brother" when Orville was a baby.

Orville felt trepidation. This was a high–risk venture. What if he crashed it so bad that they couldn't make repairs in time? There was just no time left, it had to be today or nothing. Butterflies and a knot were beginning to manifest themselves in his stomach. Looking each other sternly in the eye, the brothers solemnly shook hands and went out into the wind. Reluctantly, Orville raised the flag to summon the Kill Devil Lifeboat Station men, who had generously agreed to help. Upon seeing the signal, four men (John T. Daniels, W.S. Dough, A.D. Etheridge, and W.C. Brinkley) set out from the lifeboat station, along with a boy, Johnny Moore, from the nearby resort of Nags Head. They buttoned up their coats and put up their collars against the stiff, cold north wind and trudged towards the Kitty Hawk sand dunes.

Around 10:35am all six men carefully maneuvered the machine to the top of a dune, with the boy looking on. The men turn the craft to face the wind. The brothers had meticulously taken many photographs of their kite and glider tests. Today was no exception. "John, would you please manage the camera while I assist Bubbo with the machine?" asked Wilbur. "But I have never in my life taken a photograph!" retorted Daniels. "Look, it's very simple. Here, you just hold it like this and look through the lens," said Wilbur, demonstrating. "Then you squeeze this bulb to take a picture," Wilbur added. Orville then impatiently grabbed the camera and tripod, positioning the camera alongside the Flyer's projected flight path, and aimed at a spot where he thought the airplane might rise into the air. "Mr. Daniels, would you *please* snap a good picture in case something interesting happens?" he insisted. "For all we know, the winds are only going to get worse."

Behind Orville, Wilbur and one of the men positioned themselves to hand crank the propellers and draw fuel into the cylinders. Two others carried a dry battery box coil onto the lower wing and connected it to the spark plugs. "All clear," shouted Wilbur. With the battery connected, the two men laboriously turned and cranked the propeller by hand, in a clockwise direction when facing forward. In turn, it simultaneously cranked the other rear propeller counterclockwise through a system of gear–driven sprockets and bicycle chains. The gasoline engine coughed, then sprang to life, with a roaring sound much like a present day driving lawnmower. Daniels stationed himself by the camera. With Wilbur steadying the right wingtip, Orville carefully took a prone position on the lower wing of the biplane with his hips positioned in the padded wooden cradle. He grasped the front control handle and moved it to ensure mechanical connectivity to the set of elevators. He released the line holding the airplane in place. It clattered down the wooden launch rail and into the wind. Wilbur started his stopwatch to time the flight while simultaneously running alongside, steadying the right wing to ensure a successful takeoff. As the craft left the end of the rail, Orville pulled back on the control handle to make the Flyer rise. The craft quickly rose about nine feet into the wind and reached the whopping speed of 30 miles per hour. His heart pounding, Orville struggled with the control handle. The machine pitched up and down under its own power for 12 momentous seconds. The four other men and the boy looked on in amazement—witnesses to the historic event. After it had flown 120 feet from the end of the rail, the Flyer landed in the sand—with the machine and the pilot intact except for one damaged skid. Wilbur was so excited at the sight that he forgot to turn off his stopwatch. Orville was so surprised that he forgot to throw the engine switch to stop the propellers after the Flyer touched the ground again. But Daniels kept his cool. He snapped the historic photograph of the Flyer's first successful flight in the air. The first photograph he ever took would become one of the most famous in the history of aviation. The Wrights' machine had flown, but only for 12 seconds.

The Wrights were so sure of their calculations that they showed no surprise when the machine flew. Five persons besides the Wrights witnessed the flights. For the first time in history, a powered flying machine had taken off from level ground, traveled through the air, and landed under the control of its pilot.

A bit shaken, Orville stood up and wiped the sand off his clothes. Wilbur breathlessly caught up with Orville and the landed craft. "The rear rudder," stammered Orville. "Control of the rudder was quite difficult. As a result, the machine would rise about ten feet, then suddenly dart for the ground. She felt like a bucking bronco." "Do you think we balanced the rudder too near the center?" Wilbur commented. "Yup."

Not satisfied, the brothers wanted to try again. They repaired the cracked skid, and at about 11:20am, Wilbur made the second flight and covered a distance of 195 feet. Taking turns as was their custom, Orville made the third flight, covering just over 200 feet in 15 seconds. Each attempt pushed their machine a little further and the

The Wright Flyer's Historic Flight, 10:41 a.m. December 17, 1903

brothers learned how to control the Flyer a little better. Just before noon Wilbur got ready for the fourth attempt, and what would be the last flight. The flight began like the others, with the Flyer pitching up and down. After about 300 feet he got the wobbly craft under control and began traveling on a fairly level course. But before a minute elapsed, the Flyer began bucking again, suddenly pitching forward headlong into the sand. Time: 59 seconds. Distance: 852 feet. The feat astonished young Johnny Moore, who ran down the beach exuberantly shouting to another native, "They done it, they done it, damned if they ain't flew!"

As had happened to him a few days before on December 14, Wilbur's crash damaged the Flyer. The front elevator frame was badly broken, but the main frame was intact, and Wilbur was unhurt. "Well, it doesn't look too terribly bad," said Orville. "*I can beat him!*" he thought. "*What if I can break 1000 feet?*" "Bubs, you mean you're game to try again with the elevators hanging on by a thread?" Wilbur asked his brother, incredulously. "Why not?"

After carrying the machine back to the starting point, the group was discussing the events of the day when a gust of wind slowly tumbled the aircraft backward. The cast aluminum feet that attached the engine to the propeller snapped. The crankcase shattered when it struck the sand, the chain guides and propeller supports twisted, and the ends of virtually all the ribs snapped off. John Daniels, who held on too long, went fell into the machine. "Ahhh!" shouted Orville in panicked surprise, unable to rescue the hapless man. Fortunately, John was not seriously hurt, but this time the plane was destroyed beyond repair. It had a accomplished a milestone, but would never fly again. The brothers would ship the remains to Dayton. Orville later wrote in his journal: "His escape was miraculous, as he was in with the engine and chains." Daniels would later comment that he not only had taken the first photograph of an airplane in the air, but had also been the first victim of an accident with a powered flying machine! The 1903 Wright Flyer would never fly again.

> *Nevertheless, with Wilbur's final long, sustained effort there was no question the Wrights had flown. After four years of painstaking experiment, they had solved the riddle of flight, learning a secret man had coveted since he first watched the birds.*

Around 1:00pm the brothers took a break. They thanked the Lifesaving crew and ate lunch—eggs, ham and biscuits—again. They were eager to tell their father that they would be home in time for Christmas. And they wanted to tell the world of their triumph.

Ninety–six miles away in the seaport town of Norfolk, Virginia, H.P. Moore was at work in the circulation department of the Virginian Pilot. He was desperate to break into reporting, but he had never been able to convince his editor that he had what it takes. This afternoon, one of the biggest stories of the century was about to land in his lap. The brothers set off for the Kitty Hawk weather bureau, to send a telegram to their father. Orville dictated to the telegraph operator, who concisely tapped out:

> *"Success four flights Thursday morning all against twenty–one mile wind started from level with engine power alone average speed through the air thirty–one miles longest 57 seconds Inform press home Christmas Orville"*

The astounded telegraph operator in Norfolk asked if he could inform his friend, H.P. Moore, who worked for the local paper. The Wright brothers said, "Absolutely Not!" They wanted the news of their amazing flights to come first from their hometown newspaper in Dayton, Ohio. But the telegraph operator ignored them. 3:35pm. Ring, ring. When he got the call from his friend in the telegram office, H.P. Moore couldn't

believe his luck. "What?" he exclaimed. This was the story that would make his career, he realized. But he needed more facts from the eyewitnesses to get the story straight. He wanted to get a quote from the Mr. Wright mentioned by his friend. H.P. picked up the phone and called to Kitty Hawk, but the line to Kitty Hawk was dead. Nevertheless, the ambitious hack pressed ahead with writing up an inventive story, based upon what his friend at the telegram office just told him. With a few embellishments of his own.

> *"The machine flew above the sea for three miles, and gracefully descended to the earth. It had one six–blade propeller beneath it to elevate it, and another propeller at the rear to shove it forward."* Moore concludes his story by writing that Wilbur and Orville celebrated their success by running around shouting, *"Eureka!"* Unfortunately, Moore's story was 99% inaccurate.

The Wrights seemed unmoved by their epochal achievements. Having flown gliders hundreds of times, they were completely confident of success. The five bystanders hardly understood the drama of the occasion. Most newspapers refused to carry the story; later they picked up an exaggerated and inaccurate account. It was years before the American public realized that a new dimension had been added to travel and gave to the Wrights the credit due them.

Seven hundred and twenty one miles away, Milton Wright held in his shaking hands his son Orville's telegram. "Success! Four flights Thursday morning all against 21 mile wind…." The telegram has spelled his son's name wrong——as "Orevelle Wright"—nevertheless Bishop Wright is overjoyed. He immediately sends news of his sons' flights to the office of the local newspaper in Dayton. Frank Toonison is on duty in the newsroom of the *Dayton Journal*. As a journalist, Toonison prides himself on his exacting professionalism. Bishop Wright's two sons may well have made a flying machine. But a 57–second flight (a two–second error by Orville) is just not big enough news for Toonison. If they had flown 57 minutes, they might have made it into the Dayton Journal. Toonison tosses the missive aside and discards it. Meanwhile, the Virginian Pilot in Norfolk splashes the headline, **"Flying Machine soars 3 miles in teeth of high wind over sand hills and waves at Kitty Hawk on Carolina Coast."** "No Balloon Attached to Aid It" inaccurately said the subheadline. The role of the four lifeboat men who made the events of that cold December day at Kitty Hawk possible is not even mentioned! But the Wright Brothers reputation as the fathers of modern flight was sealed on December 17, 1903. Little did they foresee that 66 years later, a piece of their historic flyer would accompany the first men to land on the moon.

One would think that the newspapers and telegraph companies would be chomping at the bit to announce this great feat. Wrong. Only three or four newspapers reported the event the next morning, and their accounts were inaccurate. The Journal in the Wright's home city of Dayton, Ohio did not even mention it! Other reports contained factual inaccuracies with misattributions of distances, times, and even misspellings of the Wrights' names. Although the Wrights attempted to correct the mistakes, the general public would misunderstand the full significance of their efforts for several years. The Wrights' efforts received a more positive reception from the scientific community. The first scientific description of the machine and its flights appeared the following March in a magazine entitled, of all things, Gleanings in Bee Culture. The Wrights did not foresee how greatly the airplane would change civilization. They also did not believe at first that it would ever be possible to fly at night, because you can't see anything.

Seldom has such an auspicious event gone by so unheralded. The conventional wisdom of the day held that if God had meant man to fly, He'd have given him wings. The leading scientists of the day and the loud chorus of the press volubly spouted conventional wisdom that human flight was impossible. They laughed at eccentrics who believed otherwise. An editorial in the New York Times written after Langley's second failure on December 8, 1903, predicted that manned flight was achievable only if "scientists and mathematicians worked on it around the clock for the next one to two million years." It wouldn't take a million years. It wouldn't even take ten days.

SAMUEL PIERPONT LANGLEY

Among the community of experimenters in aviation at the turn of the last century, one stands out from the crowd in terms of reputation and funding—Samuel Pierpont Langley, the third secretary of Smithsonian Institution, otherwise known as "the Castle," in Washington D.C. At first glance, it is almost inconceivable that the Wrights could succeed where Langley could not. Langley considered himself the leading figure in heavier than air flight. During the 1890s, he had built a steam–powered model airplane, which he termed an aerodrome. Langley was influenced by an "aerial steam carriage" which had been patented by William S. Henson of Great Britain in 1843. It had fixed wings, a steam engine, two propellers behind the wings, and a passenger cabin. But Henson's "airliner" was never built. After building one unsuccessful model, Henley gave up on the project. Now Langley saw a golden opportunity.

In 1896 a Langley subscale 13–foot unmanned model aerodrome, propelled by a one–horsepower steam engine, flew in a circular path for half a mile in about 1½ minutes. Langley's successful experiment caught the attention of powerful friends and acquaintances in Washington, including inventor Alexander Graham Bell (the second president of the National Geographic Society) and Theodore Roosevelt, at that time assistant secretary of the Navy. Spurred by the beginning of the Spanish and American War in 1898, the Navy endorsed further experimentation by Langley for military purposes, and the U.S. War Department gave Langley two grants totaling $50,000 (a princely sum at that time) to build a passenger–carrying "aerodrome." Additional support came from the Smithsonian Institution and Alexander Graham Bell. Thus enriched, Langley purchased materials and hired a young Cornell University engineering student, Charles Matthews Manly, as his chief assistant. Eventually Langley's team grew to 10 people, as they set busily to work in space located behind the Smithsonian.

Langley, Manly, and the rest of the team built several subscale models of Langley's aerodrome, but they had problems demonstrating powered flight even by models. Ominously, a quarter–scale, powered model of the full–size aerodrome was unable to maintain even level flight when flight–tested in August, 1901.

No wonder. Langley's efforts were fraught with problems. In fact, it seems he was over his head in dealing with the complex technologies underlying powered human flight. He never accepted the curved–wing theories of previous experimenters such as Lilienthal. He assumed that one could simply "scale up" a full–size machine from his smaller models without drastically changing its basic structure. Chief among the flaws in Langley's full–size aerodrome designs was a drastically weak and understrength structure. Langley also had no concept of the controllability a machine would need to maintain stable flight. His craft lacked any effective control once they took to the air, because they possessed but a single large pivoting tail and a tiny rudder underneath the fuselage.

Langley designed a larger aircraft, the Great Aerodrome, fitted with an internal combustion engine putting out a robust 52 horsepower, over four times that of the Wright Flyer. That's great, but incredible as it seems, the Great Aerodrome was afflicted not only by a weak structure and no effective controllability, it also lacked any landing gear whatsoever. In Langley's case, this would definitely be hazardous to a hapless pilot's health since the Great Aerodrome would be launched off a track atop a houseboat in the Potomac River and splash down in the frigid water after each flight! Nevertheless, by the fall of 1902 Langley, Manly, and the rest of the team had finished building a full–size Great Aerodrome. It was too late to fly before winter set in, so they had to wait months until the warm summer of 1903.

Just to check things out, they built a subscale model of the Great Aerodrome, and it somehow managed to make a 1,000–foot flight on August 8, 1903, several months before the Wrights' own successful flight. Encouraged, Langley made preparations for a manned flight of the full–scale aerodrome. It underwent a series of ground tests and engine runs in preparation for its first flight. During the course of these, a propeller disintegrated, damaging the machine. But finally all was ready. On the morning of October 7, 1903, Charles Manly set forth from the Washington waterfront with two tugs to assist in flight operations, bound for the lower Potomac. They towed a huge houseboat along behind them, atop of which sat an impressive launching catapult and the Great Aerodrome. Professor Langley had suddenly decided to remain ashore, claiming pressing business kept him in town.

It had taken the Langley team five years of hard effort to reach this point, and Langley's own research with test rigs and flying models went back over a decade further, to 17 years. They had exhausted $73,000 in research funds, equivalent to approximately $1,700,000 today. Langley's assistants had labored for weeks to construct a suitable launching catapult almost 100 feet long, painstakingly assembling and installing it piece by piece atop a large rented houseboat— at 75 feet long, one of largest in the region. In fact, it resembled more of a small barge than a houseboat. Because of his considerable fame, Professor Langley had negotiated favorable terms on the lease for the huge thing. The bespectacled Charles Manly had been Langley's trusted chief mechanic and pilot for years, not to mention his invaluable "assistant in aerodromics." Manly would be the prospective pilot. The Aerodrome A, as Langley called it, was his pride and joy. It had wide gossamer–like tandem wings, four in all. The mid–chord of the forward pair was separated by a good 15 feet from the rearward pair. Perched on a launching track, Aerodrome A made an impressive sight, like two huge condors mating. But looks can deceive. The contraption was not airworthy and had at least one fatal weakness. The framework that Langley had designed for the tandem wings was far too weak to sustain the flight loads the aircraft would soon experience.

Presently, the whole assemblage reached the designated launch point in the lower Potomac and anchored the houseboat. Charlie Manly clambered up the wooden framework as two other men made the final preparations for launch. The houseboat rocked gently in the waters of the Potomac. "Ahh, don't touch it!" shouted Manly. One of the crewmen had inadvertently knocked his elbow against the aerodrome. It was frail beyond imagining, and a wire plucked or a wing spar jostled would set the entire structure aquiver like a bowlful of jelly. Charlie had to wait almost 30 seconds for it to stop. He had set forth from the Washington waterfront full of hope and expectation. His boss would be impressed today! He settled himself into the aerodrome's cockpit and ran up the engine. Then he waved to a crewman on the houseboat, who immediately fired two signal rockets into the air, announcing the imminent takeoff. Tugboat horns sounded in reply. Many spectators had gathered around on the two tugboats and other small craft.

Charlie fired the big spring catapult under the watchful eyes of the crewmen. He braced himself. Please God, please God, he prayed. Since Manly's "cockpit" was little more than an open shell, he knew it was doubtful he could have survived an impact with any sort of velocity without sustaining serious injury. The Great Aerodrome made a loud, roaring grinding noise as it raced down the launch rail. Nobody noticed the pin sticking out of the houseboat's launching mechanism. It snagged the front bracing strut, hooking the Great Aerodrome as expertly as a fisherman gaffing a flopping flounder.

Fortunately for Manly, the plane *"simply slid into the water like a handful of mortar,"* as a Washington Post reporter, George Rothwell, bluntly wrote. Manly, wet but otherwise unhurt, clambered aboard a boat. Critics had a field day. The *Post*'s headline the next morning read,

BUZZARD A WRECK

LANGLEY'S HOPES DASHED

Two days later subscribers to the New York Times awoke to read, *"The ridiculous fiasco which attended the attempt at aerial navigation in the Langley flying machines was not unexpected.... It might be assumed that the flying machine which will really fly might be evolved by the combined and continuous efforts of mathematicians and mechanicians in from one million to two million years ...No doubt the problem has attractions for those it interests, but to the ordinary man it would seem as if effort might be employed more profitably* [elsewhere]."

First failure of the manned Great Aerodrome, October 7. 1903

Undaunted, Langley and his team made immediate preparations for another try amid this rising criticism. They returned the water–soaked remains of the Great Aerodrome to the Smithsonian and set about repairing it.

December 8, 1903 had dawned grey and cold. Weather conditions had fluctuated during the day but improved in the early afternoon. Shortly after 2:30 P.M., the two tugs departed the Washington waterfront and pulled the houseboat and its precious cargo down the Potomac. This time they headed towards the relatively nearby Arsenal Point, a journey of at most 20 minutes. As soon as the houseboat was anchored, the weather turned for the worse. Scudding clouds and occasional squalling gulls rode a shifting and bitter wind that whipped the river's surface under an evil–looking, ever–darkening sky.

"The gusting air is shifting its direction most abruptly and disconcertingly," remarked Professor Langley. His neatly trimmed white beard fluttered in the breeze. He had to hold onto his black–banded white hat to keep it from blowing off his head into the water. Winds rocked the ungainly and bluff–sided houseboat, ruffling the hair of observers, chilling them to the bone, accentuating the bleakness of the ice–filled river and warning of a cold winter to come. More onlookers had gathered to watch the spectacle than before. Reporters from the *Washington Post* and Washington's *Evening Star* began taking notes.

"Should we attempt to fly in these conditions?" queried Manly. He had pulled down his derby hat to keep his dark brown hair, parted in the middle, from blowing all over the place. Langley was carefully surveying the situation with his piercing gaze. Already the houseboat rocked uneasily, its anchor chains alternately tautening and creaking as the variable gusting winds pushed it about. "Any delay now will mean our flight will push off well into next year, past winter. I don't want a repetition of the several–month delay we had to endure last year because of wintry conditions, do you?" Langley rhetorically asked his assistant. Manly nodded, "I agree." "This dwindling day seems to be our last, best chance," surmised the Professor.

Determinedly, Manly mounted the little open cab of the *Great Aerodrome*. It was 4:45 P.M. The darkening sky was already so dim as to prevent most photography. Manly cranked up the engine which would drive the two pusher propellers positioned behind the forward pair of wings. The engine roared as Manly gave it more throttle. Shafts whirred, propellers beat the air. For a few seconds, the Great Aerodrome sat on its perch as the propellers gained speed. The Great Aerodrome quivered expectantly, and then the catapult released. Again, Manly breathed a prayer. This time, unimpeded by snagging bits on the catapult's wooden framework, the contraption raced down the short 70–foot track. It quickly gained more and more speed, the wings producing more and more lift. Finally, the all–too–frail structure could withstand no more, twisting violently under rapidly fluctuating aeroelastic loads. With a loud crack, the aft wing's spar abruptly failed, and the aft wings folded and buckled upward like a flapping pigeon's. Manly felt an extreme swaying motion immediately followed by a tremendous jerk which caused the machine to quiver all over. The lift from the remaining front wings pitched the mortally wounded Great Aerodrome nose up as its propellers desperately thrashed the air and its pilot vainly deflected the tail to try to restore it to level flight. It paused momentarily in the air, with Manly precariously caught between the deadly props behind him and yet dangling in front of the clattering engine. Then it fell tail–first before the horrified yet fascinated eyes of the spectators and reporters, sinking beneath the ice–filled Potomac waters with a barely heard splash, like an Olympic diver.

Plunging into the icy water, Manly felt an immediate shock from the cold. He fought off the urge to gasp for air under the water. Then his cork–lined canvas jacket got caught in a protuberance. With a surge of adrenalin, he exerted all the strength he could muster to rip his jacket in two. He swam clear of the remaining mess of wires, struts, and cloth–covered wings, any one of which could have snagged and trapped him again like a Chesapeake crab in a pot. Coming up for air under blocks of ice, he had to dive again to swim to open water. Mr. Hewitt, one of the houseboat workers, heroically plunged in to help him. Manly clambered aboard the houseboat, where crewmen cut off his frozen clothes from his shivering body as others plied the shaking, cursing, disheartened man with whiskey. Upon learning of the disaster, the *New York Times* chided Langley, urging him to "stop wasting time and money on further air experiments…Life is short," they added, "and he is capable of services to humanity incomparably greater than can be expected to result from trying to fly."

The *Great Aerodrome*'s poorly braced wings undoubtedly experienced asymmetric changes in center of pressure, resulting in wing flutter and flexing. The wings twisting in both in angle of attack and dihedral angle (the angle between upwardly inclining wings and a horizontal line) produced the swaying motion that Manly experienced immediately before the final catastrophe. This kind of motion would produce exactly the diverging kind of structural loading necessary to destroy the *Great Aerodrome*, effectively replicating the loads pattern experienced by a modern aircraft during a rolling dive pullout—the most demanding structural test that an airplane can undergo.

A cold midnight had struck on the Potomac before the houseboat and its mangled cargo of wreckage, tied to the stern of the tugboat, reached dockside in Washington, D.C.

The next day, December 9th, the wreckage returned to the Smithsonian. Samuel Langley's failures had been profound, his larger machine doomed to meet a watery end before it even flew. Popular and professional opinion alike, and a wayward press, consigned Langley to history's dustbin. Sadly, the aging Langley did not outlast the failure, dying in 1906. The great man deserved a far kinder fate. A letter that Wilbur Wright wrote to Octave Chanute in November 1906 best sums up Langley's place in aviation history.

"The knowledge that the head of the most prominent scientific institution [Smithsonian] of America believed in the possibility of human flight was one of the influences that led us to undertake the preliminary investigation that preceded our active work. He recommended to us the books which enabled us to form sane ideas at the outset. It was a helping hand at a critical time and we shall always be grateful. Of his actual work, his successes and his failures, it is perhaps too soon to make an accurate estimate, but entirely aside from this he advanced the art greatly by his missionary work and the inspiration of his example. He possessed mental and moral qualities of the kind that influence history. When scientists in general considered it discreditable to work in the field of aeronautics he possessed both the discernment to discover possibilities there and the moral courage to subject himself to the ridicule of the public and the apologies of his friends. He deserves more credit for this than he has yet received. I think his treatment by the newspapers and many of his professed friends most shameful. His work deserved neither abuse nor apology."

Although he failed at manned, powered flight, having someone of his stature involved in flight gave credibility to the whole field. Langley is still acknowledged as a significant aeronautical experimenter. The first aircraft carrier ever, an Air Force base, and a NASA research center were named after him, yet now he is a footnote in the history books, eclipsed by two bicycle makers from Ohio.

The Wrights worked tirelessly at the base of the Kill Devil Hills, near Kitty Hawk, preparing for Wilbur's December 14, 1903 attempt with the Wright Flyer. The Wright Brothers and competing Langley team seemed unaware of each other's almost simultaneous efforts to conquer the skies. A succession of bad storms and minor defects would delay the Wrights' next flight experiments at Kill Devil Hill until December 17, just *nine days* after Langley's failed attempt!

Milton Wright

Susan Wright

Reuchlin Wright, the eldest
Wright child, at age 17

Lorin Wright, the
second son at age 16

Orville Wright at age 8

Wilbur Wright at age 12

Katharine Wright, the
youngest Wright child, at
age 4

THE LEARNING YEARS

Born in 1828, Milton Wright was a minister in the United Brethren Church, a professor of theology, editor of his church newspaper, and eventually an elected bishop in his church. In 1888, he broke with the liberal leadership of the United Brethren Church and started his own conservative sect, the Church of United Brethren, Old Constitution. He married Susan Catherine Koerner in 1859. Wilbur (born 1867) and Orville (born 1871) were the third and sixth of seven children. Only four sons and one daughter survived past infancy—Reuchlin, Lorin, Wilbur, Orville, and Katharine.

None of the Wright children had middle names. Instead, their father tried hard to give them distinctive first names. Wilbur was named for Wilbur Fiske and Orville for Orville Dewey, both clergymen that Milton Wright admired. They were "Will" and "Orv" to their friends, and "Ullam" and "Bubs" to each other. Katharine was "Swes". In Dayton, their neighbors simply knew them as the "Bishop's kids."

The five Wright children grew up in a home where, as Orville later explained, "there was always much encouragement to pursue intellectual interests; to investigate whatever aroused curiosity. In a less nourishing environment," Orville believed, "our

The Wright home in Dayton, Ohio 1895

curiosity might have been nipped long before it could have borne fruit."

Interestingly, neither Wilbur Wright nor his younger brother Orville Wright had bothered to finish high school. The boys went through high school, Wilbur in Richmond, Indiana and Orville in Dayton, Ohio, but neither received a diploma. Wilbur did not bother to go to the commencement exercises, and Orville took special subjects rather than a prescribed course curriculum in his final year. Mechanics fascinated them even in childhood. To earn pocket money they would sell homemade mechanical toys in Dayton.

When Wilbur Wright was about 21, Lorin—his older brother by five years—wrote a letter home from Kansas where he had gone to find his fortune. He worried that Wilbur was aimless and would never amount to anything much. This would later haunt Wilbur.

The Wrights didn't drink or smoke. The two boys had tempers, but no matter how angry they ever got, neither was ever heard to utter a profane word. Neither Wilbur nor Orville ever married—despite the fact that their older brothers, Reuchlin and Lorin, and their sister, Katharine, did. "Ullam, it is up to you to marry first, because you are older," Orville commented one day. "Bubs, I may be older than you, but not necessarily wiser," Wilbur replied. "I just don't have time for a wife, you know that." "Me neither," said Orville. "Remember the last time we were riding the traction car at Simms Station? Some older woman sat down next to you, and you both began talking up a storm. Then when she got off at our stop, you offered to carry her packages. You'd think that you had known her all your life! But then when a younger woman sits next to you, you sooner or later begin to fidget. Then you get up and go stand on the platform until it is time to leave the car." "So?" Wilbur retorted. "You're woman–shy yourself. Young women, at least."

By 1889 Orville had started a printing business, building his own press, and encouraged Wilbur to join in. As soon as they had mastered the customary printing services, they moved on to building their own printing presses and launched a weekly paper, the West Side News, with Wilbur as editor.

Wilbur was 25 and Orville 21 when the brothers went into the bicycle business in Dayton in 1892. They began by selling, renting, and repairing bicycles. In 1896 they began to hand–build them, assembling the machines in a room above their shop. Their top–of–the–line *Van*

The Wright Cycle Shop

Cleve and the less expensive *St. Clair* were customized to their customers' specifications. The brothers added a few original improvements to the customary components, including an oil–retaining wheel hub and coaster brakes, which still find use today. Nevertheless a few years later, Wilbur would complain that although he had built a successful bicycle business with Orville, his success was merely modest. Like Lorin had told him years before, he too worried that he "would never amount to much."

Meanwhile, Otto Lilienthal, a German civil engineer, became the first person to pilot a glider in flight and was known as the "bird man of Europe." Lilienthal had covered distances of up to 800 feet in elegant gliders that looked like huge birds, but made of wax cotton stretched over willow frameworks. Lilienthal controlled and steered the glider by swinging his body from side to side. From 1891 to 1896, he made about 2500 glider flights. One of the principal innovations of Lilienthal's gliders was his translation of the shapes of birds' wings—based on his own years of study—into glider wings.

In 1897, a Scottish engineer named Percy S. Pilcher first used a towing technique to launch a glider. Both Lilienthal and Pilcher died in glider crashes. After reading about the death of pioneer glider Otto Lilienthal due to a diving crash in 1896 where he plunged head first into the ground and broke his neck, the Wright brothers became increasingly intrigued with flying.

Wilbur did not seriously begin to pursue his interest in aeronautics until 1899, when he happened to read a book on birds. Wilbur would later say,

"For some years I have been afflicted by the belief that human–powered flight is possible. My disease has lately increased in severity, and I fear that it will cost me an increasing amount of money, if not my life. We could not understand why birds had any flying capabilities that could not be made by man on a larger scale, and like them for flight. If birds could be effortlessly sustained in flight, we did not see why man couldn't be sustained by the same means for stable flight."

He sent off to the Smithsonian Institution for a list of recommended readings about man's attempts at flight, and managed to get Orville interested in the then–outlandish idea that men could indeed fly. The Wrights had plenty of reason to doubt it. Some of the most prominent scientists of the era had declared heavier–than–air machines impossible. All previous attempts by well–capitalized, well–educated and much–celebrated aviation pioneers like Samuel Langley had failed. And Orville and Wilbur Wright were mere bike shop partners, sons of an ordained minister of the Church of the United Brethren in Christ. They began without much capital, without much formal education, and, to all appearances, without much chance at all of succeeding.

That same year of 1899 they experimented for a day or two with a 5–foot biplane kite. Employing a five–foot, double–foiled kite, they tested their own theories on lift and their "wing–warping" method for the control of altitude and direction. This required the differential twisting of the curved wings in order to shift the kite's direction, which the Wright brothers eventually called "helical twisting". Combined with a rudder, this principle of warping or twisting eventually became the key to the Wright brothers' success, allowing the pilot to lean into turns while maintaining aerial stability. Throughout 1899 and 1900 the Wright brothers applied their experimental efforts toward creating a glider capable of carrying the weight of a man. They tested different materials, further

analyzed the flight of birds and the shape of birds' wings, and further developed the notion of wing–warping. A final kite test in the summer of 1899 achieved only limited results. Undeterred, the Wright brothers advanced their designs and the understanding of the processes at work. They had pretty much amassed all the scientific knowledge of aeronautics then available.

The Wrights also relied on the experiences and glider flights of a French–born, Chicago–based amateur scientist named Octave Chanute. From 1896 to 1897, Chanute and his assistants built five different glider types and completed over 1,000 experimental flights, many 200 to 300 feet in length. After 1900, the Wrights began a correspondence with Chanute, who in time advised them on aspects of their experimentation.

On the advice of the US Weather Bureau (later the National Weather Service) in Washington, DC the Wrights selected for their experiments a narrow strip of sand called Kill Devil Hill, near the settlement of Kitty Hawk, North Carolina. This area was known for its steady winds and high sand dunes. In 1900, they tested their first glider that could carry a person. This glider measured 17 feet from wing tip to wing tip, and cost a mere $15 to build. They returned to Kitty Hawk in 1901 with a larger glider. They showed that they could control sidewise balance by presenting the right and left wings at different angles to the wind. But neither the 1900 nor the 1901 glider had the lifting power that they had counted on.

The Wrights set up a six–foot wind tunnel in their shop and began experiments with model wings. They tested more than 200 wing models in the tunnel. After further study and comparison with their experimental results, the Wrights concluded that all published tables of air pressures on curved surfaces up to that time must be wrong.

From the results of their tests, the brothers made the first reliable tables of air pressures on curved

Dan Tate and Edward Huffaker launch Wilbur aboard the second Wright glider in 1901. Tate was a "Banker" (a native of the Outerbanks) and Huffaker was a friend of Octave Chanute. On right, Wilbur after landing the 1901 glider.

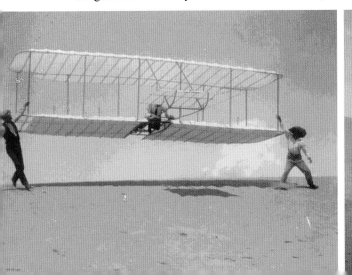
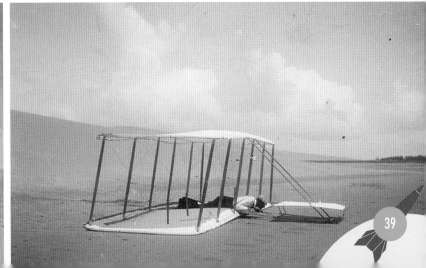

surfaces. These tables made it possible for them to eventually design a heavier–than–air machine that could actually fly. The Wrights experimented with gliders to experimentally solve many problems of flight control. They built a third glider and took it to Kitty Hawk in the summer of 1902. This glider, based on their new calculations, had aerodynamic qualities far in advance of any tried before. With it, they solved most of the problems of balance in flight. They made nearly 1000 glides in this model, and on some flights covered distances of over 600 feet. Their basic patent, applied for in 1903, relates to the 1902 glider.

Before leaving Kitty Hawk in 1902, Wilbur and Orville started planning a powered airplane, the historic Wright

the first successful airplane. From a modern viewpoint their reputation is so strong that it's hard to believe that there were so many other experimenters close to being first in flight. The Wrights' success was not assured—even they had moments of doubt that piloted, powered, heavier–than–air flight would exist in their own lifetime. Yet it is undeniable that Wilbur and Orville conquered the problem that had long bedeviled their competitors. Why did these brothers succeed where so many others who were so much better equipped for success had failed?

The home–schooled Wright brothers turned their deficiencies into assets. They did not have extensive formal education, but they read broadly and incessantly, absorbing intellectual insights from a broad spectrum.

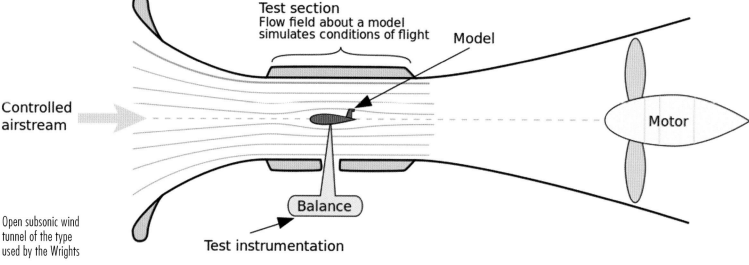

Open subsonic wind tunnel of the type used by the Wrights

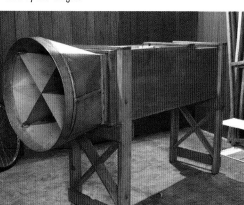

The Wright's six–foot wind tunnel

Drift balance in Wright wind tunnel

Lift balance in Wright wind tunnel

Flyer. They completed it by the fall of 1903, and made history on December 17. No matter how you define flight, Orville and Wilbur Wright are credited with having invented

They *did not accept conventional wisdom* as unalterably true. When things did not work as the books said they ought to, they did not hesitate to question the books and the 'experts'. They took nothing for granted. The Wrights thought through, invented, redesigned, or tested every element of their planes.

They invented the first wind tunnel in order to test the wind resistance formulas of Otto Lilienthal, the German mathematician and civil engineer who became known as the 'bird–man of Europe' after thousands of glider flights. Like the Wrights, Lilienthal believed that man must master control of a flying machine if he was to conquer the skies. After years of studying birds' wings, Lilienthal believed the "gentle parabolic curve across the wings' upper surface" gave his gliders the necessary lift to glide long distances.

Lilienthal had published meticulous accounts of his work, complete with diagrams and photographs, which the Wright brothers used. But they also found errors in Lilienthal's research, particularly with wing surface–to–weight ratios and his search for a suitable steering method. The Wrights realized that the movable wings of Lilienthal's gliders caused instability as did the wings' relatively short length and the pilot's overall lack of control.

Otto Lilienthal preparing for flight with his second "glider," August 1884.

Rather than attaching bird–like wings to their arms, the Wrights conducted many experiments with kites, made in subscale like the Wright Flyer. They found that by twisting the wings of their kites in flight, they could control their movement through the air. The Wrights thus were the first in the world to develop the concept of "wing warping," a device for twisting the wing tips to enable their kites to make commanded turns in flight. Never mind that today's aircraft make commanded turns by actuating ailerons on the wings in opposite directions

instead of wing–warping, the Wrights did not have the benefit of this hindsight. They took the next step after kites by constructing a series of gliders. Throughout the autumn of 1902, gliders were a regular site at Kitty Hawk, the wind providing the power for the willow and waxed cotton machines. Wilbur wrote in his journal, "in two days, we made 250 flights in these new machines, so that we were able to take it out in any kind of weather." And the sand would soften the occasional crash landing with Wilbur or Orville on board.

The Wright brothers did not often agree with each other. They turned this problem into an asset also. They didn't know what was right when they began to study the problem of aviation. But they knew they wouldn't accept what was wrong. Their father Milton Wright had always encouraged the boys to argue and debate, even allowing them to raise their voices to each other at the dinner table, so long as they remained respectful. Orville and Wilbur argued with each other constantly, and liked arguing so much that they'd switch sides so each could make an argument for the other side! Their conflict was constructive. When they were through, they knew where they were and could go ahead with the job. Their habit of argument and enquiry kept them disciplined, and prevented them from accepting easy, wrong answers and from giving up when problems seemed insuperable.

The Wright brothers were not engineers. They were bicycle mechanics. They lacked many of the advantages that previous experimenters had lacked. However, they understood what it took to keep a bicycle upright and rolling. At a time when some theorists of flight supposed that flying would be mainly a matter of developing an engine powerful enough to keep a craft in the air, and when others thought that the key issue in flying would be to build a very stable aircraft, the Wrights understood the advantages of deficiency. A powerful engine would be a heavy engine. Better to have less power and lighter weight. A firm and stable aircraft would be one that stayed on the ground. They knew from their bicycling experience that balance and control allowed a rider to make good use of instability. In fact, some of today's most modern aircraft are inherently unstable, kept in flight by rapid computer–driven flight control responses beyond human capability.

"Ullam, are you sure that playing around with the curvature on the upper side of the wing will have such a great effect on lift? Don't you have to curve the lower side, increasing the camber as well?" Orville once asked his brother. "Well, my research shows that it is very tricky. You're right, you have to curve the lower side also. But you can't just make it symmetrical, I don't think. I have been studying George Cayley's pioneering work," the older sibling responded. "It is tricky because a great deal also depends on what you do to the bottom side of the wing. I also looked at Penaud's rubber band–powered toy airplane. Alphonse Penaud introduced his toy in 1871. He was an enthusiastic student of Cayley who furthered his scientific work. His plaything actually flies. I couldn't believe his craft also has a retractable undercarriage and glass–enclosed cockpit."

"Well, can you just flatten the lower side of the wing?" inquired Orville. "No. We need to test this in our wind tunnel with a variety of shapes. I think that flattening the lower side too much increases the drag. You know, it's all a question of making the airflow go faster above the wing, and if at all possible retard it below the wing. I can't think of any way to retard the airflow speed under the wing's lower surface, other than putting some kind of obstacle like a series of small flat plates. But all this would do is increase the overall drag and slow the whole aircraft down. I'll have to look some more at Lilienthal's work. Octave Chanute has also done a lot of work recently."

"I can't think of a clever way to slow the slipstream down below the wing either," Orville said. "Both Cayley and his student Alphonse Penaud said to just make the lower surface more or less flat," Wilbur added. *Of course, neither brother could foresee that in 1964–1974 Richard T. Whitcomb and his staff would invent and test a "supercritical airfoil" at NASA's Langley Research Center with just the opposite, a flat upper side that would fly more efficiently at transonic speeds.*

Wilbur pondered the problem and eventually came up with an ingenious solution. He effectively turned the wing on its side! He took a bicycle wheel, turned it sideways, and securely attached it to Orville's front handlebars. He then affixed two metal plates on opposite sides of the wheel. The right plate was flat and created drag, producing a clockwise moment on the wheel as one rode the bike. The left plate was larger and curved, like an airfoil. In other words, the airfoil would act like a wing in flight. Angled against the wind, it would produce lift sideways and create a countervailing moment in the counterclockwise direction.

Wilbur wanted to experiment with different airfoil shapes and different angles of attack, and experimentally prove his theory about airfoil design by measuring the different points of balance where equilibrium was achieved. With another touch of genius, he and Orville would do this by measuring how far off of its centerline the wheel was displaced when the bike was ridden at speed.

"O.K., that's fast enough! Hold steady!" exclaimed Wilbur as Orville pedaled furiously past him. "The scale still reads 15 degrees, and that's it," Orville shouted back. An experienced bicyclist like his brother, he was just getting used to riding at speed while balancing the strange contraption strapped to his handlebars. Needless to say, Orville had to maintain a very steady speed and ride a straight trajectory for these measurements.

The Wrights' ingenious "testing machine"

Fifteen degrees was not what Wilbur expected from previous calculations. The brothers had carefully fabricated many airfoil sections in their shop and then tested them in their handmade wind tunnel. Based upon these tests and their incipient knowledge of the new field of aerodynamics, Wilbur calculated that the airfoil shape should maintain a displacement angle of about five degrees off of centerline, so that its lift would exactly counterbalance the drag produced by the flat plate.

Wilbur and Orville returned to their makeshift wind tunnel with yet more airfoil shapes. Along with other researchers, they had discovered that wings produce lift by increasing the airflow speed above the wing compared to below it, and the difference in dynamic pressure was what produced a lift in the upward vertical direction to maintain flight. They had meticulously handcrafted a series of airfoil wing shapes of different cross sections and tested them in the wind tunnel. Wilbur had to correct his aerodynamic force measurement equation to match the new test data. Additional tests with the horizontal bike wheel in motion, acting like an aircraft on its side, corroborated the new equation.

Next page: Airfoil and wing notation. If a horizontal wing is cut by a vertical plane parallel to the centerline of the vehicle, like cutting a loaf of bread with a bread knife, the resultant section is called an *airfoil section.*

The generated lift and the stall characteristics of the wing depend strongly on the geometry of the airfoil sections that make up the wing. A wing's *chord* is the distance between its leading and trailing edges, and the "chord line" is a straight line connecting the two. The geometric "angle of attack" is the angle between the chord line and the direction of the undisturbed "free stream" flow in front of the wing.

The "mean camber line" is the locus of points midway between the wing's upper and lower surfaces, measured along the chord line. A wing's *camber* is the distance between the chord line and the mean camber line as shown. In other words, "camber" measures how "curvy" a wing is. Most wings are not rectangular and everything except the span varies: chord, chord line, camber, camber line, and thickness. The "mean chord" of a wing is its total planform area (the yellow area shown in the top view below) divided by its span. For unsymmetrical wings, the planform is the silhouette of a wing when viewed directly from above or below, perpendicular to the direction of flight. Since the leading and trailing edges are curvy, calculus has to be used to determine the planform area.

Pioneering aeronauts learned that wings had to be "unsymmetrical" in order to optimize their lift characteristics. Based upon the research of aeronautical pioneers like Sir George Cayley, Otto Lilienthal, Samuel Langley, Octave Chanute, and especially the Wright Brothers, knowledgeable aeronauts knew that a wing's camber had to somehow be varied between the wing's leading edge and trailing edge across a wing's chord to ensure the highest performance in flight.

Geometric parameters that have an important effect on the aerodynamic characteristics of an airfoil section include: (1) the leading edge radius, (2) the mean camber line, (3) the maximum thickness and the thickness distribution of the profile, and (4) the trailing edge angle.

By comparison, symmetrical airfoils would have poor performance, generating less lift, and they would stall at relatively low angles of attack. Symmetrical airfoils generate zero lift at zero angles of attack. In contrast, early wind tunnel tests showed that cambered airfoils in a subsonic flow generated lift even when the airfoil section angle of attack was zero. The question was, how much to vary the camber? This, along with many other airfoil design characteristics, would remain a subject of aeronautical research for many decades to come.

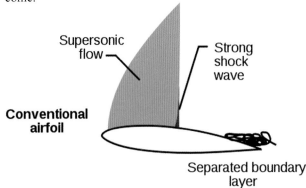

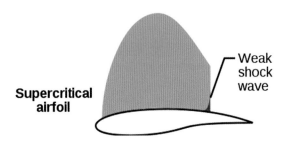

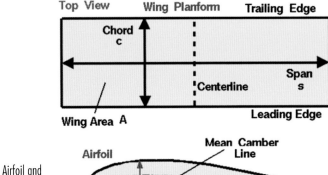

Top View

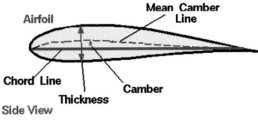

Airfoil and wing notation

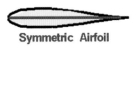

Symmetric Airfoil

But be careful! A symmetrically–shaped wing can also generate lift (at non–zero angles of attack), simply by using a positive angle of attack to deflect air downward. Symmetrical airfoils are, in general, less efficient and lack the lift provided by cambered wings at zero angle of attack. Nevertheless, they are used in aerobatics, as they provide practical performance both upright and inverted. For flight speeds near the speed of sound (transonic flight) or above the speed of sound (supersonic flight), airfoils with complex asymmetrical shapes are used to minimize the drastic increase in drag associated with airflow near the speed of sound. Such airfoils are called "supercritical airfoils".

testing, and repairing their powered machine and conducting new flight tests with the 1902 glider. Wilbur made the first attempt at powered flight on December 14, but he stalled the aircraft on take–off and damaged the forward section of the machine. Three days were spent making repairs and waiting for the return of good weather. Then the historic first flight took place with Orville on board December 17.

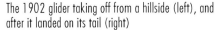

The 1902 glider taking off from a hillside (left), and after it landed on its tail (right)

Previous page: The supercritical airfoil maintains a lower Mach number over its upper surface than the conventional airfoil, which induces a weaker shock. "Shock waves" form when an aircraft goes faster than the speed of sound (Mach greater than 1). At the region where this occurs sound waves traveling against the flow reach a point where they cannot travel any further upstream. The pressure progressively builds in that region, and a high pressure shock wave rapidly forms. *Think of a traffic jam on a highway. Cars and trucks begin to pile up, then a crash happens when people follow too close, and a pileup occurs.*

With the major aerodynamic and control problems behind them, the brothers pressed forward with the design and construction of their first powered machine. They returned to their camp near the Kill Devil Hills in September 1903. They spent the next seven weeks assembling,

The 1902 Wright glider as shown here initially had a fixed tail. It was only a partial solution to the turning problem.

THE ART OF FLYING

The Wrights knew that the only plausible customer who would invest the $100,000 it would take to move embryonic airplane technology forward to practical use was the military arm of a national power. But following their historic flight, Orville and Wilbur faced unexpected indifference from the military in the United States. In the early 1900s there was no threat of war, no one was planning ahead, and the Army was not interested in evolving what they still thought was an unproven technology forward to implementation as a warfighting instrument.

Determined to move from the marginal success of 1903 to a practical airplane, the Wrights in 1904 and 1905 built and flew two more aircraft from *Huffman Prairie*, a pasture near Dayton, and presently in Area A of Wright Patterson AFB. They continued to improve the design of their machine during these years, gaining skill and confidence in the air. In 1904, they made 105 flights but totaled only 45 minutes in the air. Two flights lasted five minutes each. On Oct. 5, 1905 their machine flew 24 miles in 38 minutes and 3 seconds, performing circles and other maneuvers. Then, no longer able to hide the extent of their success from the press, and concerned that the essential features of their machine would be understood and copied by knowledgeable observers, the Wrights decided to cease flying and remain on the ground until their invention was protected by patents and they had negotiated a contract for its sale. They used this downtime to refine their thoughts, make contacts in Europe, and pursue the business of aviation sales. Yet they were careful to keep their most important innovations in airplane stability and control to themselves.

Early fliers in this period exhibited remarkable courage. Usually a flight ended in a crackup. As the pilot crawled out of the wreckage (assuming he was still alive), two or three mechanics would rush out to the plane with their tools and apparatus and repair the broken wires and struts. The dauntless pilot, if he was still able, often would take off again as if nothing had happened. In 1908 the Wrights felt safe enough to resume experimental flights near Kitty Hawk which this time newspapers reported at great length. Immediately after these trials, Wilbur went to France where he aroused the admiration and enthusiasm of thousands. The French were astonished when he made flights to altitudes of 300 feet and more, flew in circles, did figure–eights, and stayed aloft at will. In 1908, becoming airborne was still a challenge, but staying airborne was a singular triumph. The first French biplanes had only been able to do hops in a straight line, their turns were out–of–control, and their biplanes were unable to sustain flight. Wilbur arranged with a French company for the construction of his machine in France. After he returned to the United States in May 1909, he made demonstration flights in the fall of 1909 from Governors Island, New York, around the Statue of Liberty, up to Grant's Tomb, and back.

While Wilbur was in France, Orville made successful flights in the United States. On August 20, 1908, Orville arrived at Fort Myer, Virginia (near Washington, DC), with the disassembled parts of a new airplane and engine, and with mechanics Charlie Taylor and Charles Furnas in tow to assist in its assembly. As Orville and his crew went to work each day in the old balloon shed chosen for the assembly of their Flyer, they could hear their competitors Glenn Curtiss and Scott Baldwin chugging along overhead aboard the SC–1, the two–man airship that would soon be the first powered dirigible to enter the U.S. Army inventory.

On September 3, Orville made his first test flight, 1½ circles around the artillery practice field at Fort Myer in one minute, 11 seconds. By September 9 he was making flights of up to 11 minutes in duration with Army Lt. Frank Lahm as a passenger. That morning he made 57 complete circles at an altitude of 120 feet over the training field. He remained in the air one hour and two minutes, and set several records the same day. Orville had joined his brother in the headlines of the world's newspapers. The Wrights were the first in the world to put together the airfoil shape, the propeller, and a three–axis control system for stability in flight. Three–axis control allowed them to stop an airplane from pitching forward unexpectedly, or rolling over and going into a spin, or yawing—slipping sideways into an aerial skid. By controlling these dangerous characteristics, the Wright Brothers proved that sustained flight was truly achievable. They were the first to take control of the sky. But then tragedy struck.

On September 17, Charlie Taylor, the Wrights' long–time trustworthy mechanic, was slated to fly as Orville's passenger in his continuing series of historic flights. Early that morning, Orville and Charlie had gotten together to discuss the day's events. "Orv, I just finished fabricating two larger propellers a couple of days ago. I don't know—do you think we should install them to give us better performance today?" Charlie suggested. "Well, let's see. How much do you weigh?" responded Orville. "'Bout 195 pounds, give or take." "OK, then, I think it would help for you to put 'em on, to compensate for our heavier weight. Plus, I would like to see how much these bigger props help. The crowd of spectators has been growing. Let's impress them," said Orville. Charlie installed the new props, as ordered.

An Army lieutenant came walking up to Orville, an expectant look on his face. "Sir, could I give it a try as your passenger today?" he asked. "I have been working without salary for the AEA on Curtiss's tetrahedral aerodrome *Cygnet*." Orville pondered. *This is my chance to beat him,* he thought. *I can steal this bright young Army lieutenant away from Curtiss and get him on our side.* Glenn Curtiss, another aviation pioneer, was becoming more and more his arch–nemesis of late. "Sure. Just wait here a minute… on the other hand, come 'ere and help me check out the Flyer before we start her up," Orville replied. "What's your name?" They struck up a conversation.

Lt. Thomas Etholen Selfridge, a 20–year–old West Point graduate (class of 1903), had served with artillery regiments and distinguished himself as a troop commander during the San Francisco earthquake of 1906. Selfridge had committed himself to a career in aeronautics, with the notion "that sometime or other the U.S. government would require flying machines in the Army and … when that time came, the services of an officer who had made an expert study of the subject would be in demand, and he would be sure of promotion into a field of great usefulness." Selfridge soon earned a place of honor in the annals of flight as a result of his work with the fledgling Aerial Experiment Association (AEA) headed by Alexander Graham Bell. He worked tirelessly with the little band of AEA enthusiasts to build, perfect, and test a series of new experimental flying machines.

Charlie was not too happy to give up his turn, but Orville convinced him that replacing him at the last minute with Lt. Selfridge was a good idea. Orville and the army lieutenant as his passenger were making another series of demonstration flights at Ft. Myer on September 17. As usual, hundreds of spectators had gathered to witness the historic flights. Just after takeoff, they were flying at an altitude of 125 feet when Orville heard a sudden thumping coming from the rear of the airplane. He looked around, but saw nothing.

Tom Selfridge smoking pipe

Crashed Wright Flyer that took the life of Lt. Selfridge and seriously injured Orville Wright. It was the first passenger fatality in an airplane accident.

Suddenly, they felt two large thumps, and the airplane shook violently as Orville tried to control its descent to the ground. One of the two pusher propellers had split, cutting a guy wire. Out of the corner of his eye, Lt. Selfridge saw the wire coil about the blade and snap it across the middle with a kapow! Then the rudder collapsed. The machine became more difficult to manage, wobbling violently. About 20 feet from the ground, the airplane started to correct itself, but it was too late. The airplane hit the ground in full view of the horrified spectators.

After the accident, Charlie Taylor investigated the crash scene and found that the new propellers that they installed before the flight had delaminated. He reported his findings to Orville, who was in the hospital recovering from his injuries. Charles was the first person to investigate a powered fatal accident flight. This accident would be the most serious in the Wright brothers' career. Orville suffered a broken thigh and two broken ribs. He would carry the pain of the injuries received that day for the rest of his life. Nevertheless, Orville reappeared at Fort Myer the next year (1909), fully recovered. He completed official tests for the War Department with no evidence of nervousness. However, Tom Selfridge suffered a fractured skull, and died a few days after the crash. He would earn one final, though unwanted, distinction. Selfridge was the first man in the world to die in the crash of a powered airplane.

ERA OF PATENT LAWSUITS

As the Wright's fame grew in the early years of the 20th century, aviation pioneers and innovators quickly popped up all over the place. Having convinced the world, they finally managed to convince their own tentative government. In 1908, the Wrights closed a contract with the US Department of War for the first military airplane, and in 1909 the Army took delivery of a Wright Flyer, the world's first warplane. In August, 1909 they closed another contract with some wealthy men in Germany for the formation of a German–Wright Company. Later that year, they formed the Wright Company in New York City to manufacture airplanes. The French followed suit, and orders for aircraft poured in. The Wrights set up airplane factories and flight schools on both sides of the Atlantic. The European powers and Russia jockeyed for position to outdo each other with innovative airplane designs.

While the Wright brothers earned some money, they were troubled with imitators, infringements on their patents, conflicting claims, and lawsuits. Unfortunately, once they had demonstrated their aircraft in public, it was easy for others to copy them — and many did. The Wrights were dragged into time–consuming, energy–draining patent fights in Europe and America.

The era of the patent suits began on August 18–19, 1909, when the Wrights filed a bill of complaint enjoining Glenn Curtiss and the Herring–Curtiss Company from making, selling, or exhibiting airplanes that infringed on the Wright patents. The following day, they filed suit prohibiting the Aeronautic Society of New York from exhibiting a Curtiss airplane. Glenn Curtiss and Augustus Herring had incorporated the Herring–Curtiss Company on March 3, 1909.

Glenn Curtiss

Glenn Hammond Curtiss was born and raised in Hammondsport, New York, on the shores of Lake Keuka in the Finger Lakes wine country. He left school at age 15 to accept a job as a bicycle delivery boy for Western Union. Curtiss earned an early reputation as a bicycle racer, then moved on to motorcycles. It was the beginning of a career devoted to speed. He made national headlines in 1907 when he set a new world speed record of 136.3 mph with a Curtiss motorcycle at Ormond Beach, Florida. But it was the engines, not the motorcycles, that would bring him real fame. Curtiss started by pursuing the small market for flying–machine engines. He began to offer his services to aeronautical experimenters, whose names appeared in local newspapers —including the Wrights.

Glenn Curtiss was the quintessential Yankee, with a rugged face and hard lines etched across his forehead by years of concentration. A bristling mustache completely hid his upper lip, so that he always appeared to be frowning into the camera recording his image. In fact, those who knew him well found it difficult to remember the last time they had seen him smile. Frivolity was simply not in his nature. Neither was he given to idle

Wright brothers
at the Belmont
Park Aviation
Meet in 1910

chatter, hard drink, or profanity, although he was known to use harsh language for effect. Of medium height, his lanky build and aloof manner led a surprising number of casual acquaintances to remember him as being tall.

Wright brothers at the Belmont Park Aviation Meet in 1910

Curtiss met the Wrights in the fall of 1906, when he traveled to Dayton to repair the engine of an airship that Frederick Walker (Casey) Baldwin, another of the many pioneers in the burgeoning aviation industry, was operating at the local fairgrounds. Curtiss and Baldwin found the Wright brothers to be very friendly, exhibiting the "frankness of schoolboys in it all," and having a "rare confidence in us." Curtiss wrote to the Wrights on December 30, 1907, bringing them up–to–date on AEA activities and offering a "gratis" 50 horsepower Curtiss engine for their continuing experiments. Also active with the AEA, Selfridge followed up with a letter to the Wrights on January 15, 1908. His questions to the brothers were straightforward and practical. What

had been their experience with the travel of the center of pressure on the wings? The location of the center of pressure (cp), or aerodynamic center (ac), compared to a craft's center of gravity (cg), was known to be crucial in ensuring the stability of aircraft in air. Both were located on the longitudinal axis. For stability, it was known that the ac had to be located in front of the cg, as any kid today knows who attaches a paper clip to the front of his paper airplane to make it go farther. But if too far forward, the cg would render an airplane difficult to control because of a tendency to pitch down. On the other hand, if the ac migrated a bit too far in front of the cg during flight, the aircraft would pitch up, tend to stall, and plunge down to the earth. What was a "good efficient method" of constructing light, strong ribs that would maintain their camber? How should the fabric be applied? Could the Wrights offer any general structural advice?

The Wrights provided specific answers to these questions and directed the newcomers to their patents and published papers for additional details. Why were the Wrights so careful in their dealings with the Europeans and yet so apparently open in these early contacts with the AEA? They probably assumed that the new AEA and competing domestic aviation enthusiasts would not pose a threat for a very long time.

Within a few months of their August 18–20, 1909 patent suits, the Wrights sought injunctions restraining visiting aviators Louis Paulhan and Claude Grahame–White from operating aircraft that infringed on their U.S. Patent No. 891,393, granted on May 23, 1906. Interestingly, this basic patent incorporated engineering drawings from the 1902 Wright Glider, not the 1903 Wright Flyer. The patent had nothing to do with power and everything to do with control. Additional Wright patent applications had been approved in the United States, France, Belgium, and Germany. But it is the 1902 glider patent that is the great–granddaddy of everything that flies.

The patent suits consumed Wilbur and Orville between 1910 and 1912. In 1910, the Wrights and their licensee, the *Compagne Generale de Navigation Aerienne* (CGNA), sued six rival French aircraft manufacturers (Blériot, Farman, Esnault–Pelterie, Clement–Bayard, Antoinette, and Santos–Dumont) for infringing on the Wrights' French patents. The following year, a consortium of German aircraft builders, in turn, sued the German Wright Company in an effort to overturn the Wright patents in that nation.

The Wrights simply rolled over independents like Paulhan and Grahame–White. The cases involving the Herring–Curtiss Company and European firms proved more difficult, expensive, and time–consuming, however, and seldom produced a clear–cut resolution. Courts invalidated the Wrights' German patent, arguing that prior disclosure and the publication of information on basic elements of the Wright airplane before patent approval had compromised their claims. The French suit was complicated by the peculiarities of the legal system. Lacking a spirited prosecution by the CGNA, the case remained unresolved until the Wrights' French patents expired in 1917.

The situation in the United States was just as complex. The Curtiss–Wright controversy involved the technique for lateral control of aircraft in flight. In January 1910, a U.S. Circuit Court enjoined Glenn Curtiss from the manufacture or sale of aircraft. Curtiss posted a $10,000 bond and appealed. He could legally continue flying until the appellate court reached a decision, but he took a terrible risk in doing so. If Curtiss lost, financial ruin would follow. A new federal agency, the *Advisory Committee for Aeronautics*, mediated the rancorous and complex dispute between Glenn Curtiss and what was also known as the Wright–Martin Company. Two years later, the federal government approved the creation of a patent pool composed of leading manufacturers who would purchase the rights to all aeronautical patents and share access with other members of the group. This cross–licensing arrangement consolidated patent rights and cleared the way for volume production of aircraft during WWI as well as during the postwar era.

In January 1914, the U.S. Court of Appeals ruled for the Wrights. Rather than clobber their principal rival, Wright Company management sensed the opportunity for monopolistic profits. They announced a schedule of rates that they would charge anyone who wished to exhibit an airplane in the United States. Curtiss, represented by the best lawyers that money could buy, announced that he would immediately alter his control system so that it no longer infringed on the Wright patent. Few aviation people believed that to be possible, but it was enough to muddy the waters and set the legal process in motion once again. As part of his defense, Curtiss borrowed Langley's unsuccessful aircraft from the Smithsonian Institution and rebuilt it to prove that the *Aerodrome* could have flown before the *Wright Flyer*, although Langley hadn't succeeded in flying it. This time, after significant modifications from the original model, the Aerodrome flew. But the ruse didn't work — Curtiss had

made too many modifications to get Langley's aircraft in the air and the courts ruled in favor of the Wrights. Yet although the case resolved the Wright/Curtiss dispute, it left an enduring resentment between the Wrights and the Smithsonian, who persisted in backing Langley, who had served as the third Secretary of the Institution. The Smithsonian's Annual Report of 1915 stated, "The tests thus far made have shown that former Secretary Langley had succeeded in building the first aeroplane capable of sustained free flight with a man." The Smithsonian officially disregarded the Wrights as the inventors of the first successful airplane!

Outside the courtroom, the world seemed no friendlier to Wilbur and Orville. As their airplane went from experimental to developmental, so too had the brothers evolved, from experimenters to entrepreneurs. They would find that role infinitely more challenging, and far less rewarding. The aircraft business was uncertain and dangerous. Most of the money to be made was in exhibition flying, where the audiences wanted to see death–defying feats or airmanship. The Wrights sent out teams of pilots who had to fly increasingly higher, faster, and more recklessly to satisfy the crowds. Inevitably, the pilots began to die in accidents and the stress began to take its toll on the Wrights. Additionally, their legal troubles distracted them from what they were best at — invention and innovation.

By 1911, Wright aircraft were no longer the best machines flying. Although the Wright brothers invented and flew the world's first airplane in 1903, the Europeans took their work and expanded it greatly. Alberto Santos–Dumont of France ranks as the third person to fly a powered aircraft. An experienced airship pilot, he heard of the Wrights' progress during a 1904 visit to the U.S. In 1906 Santos–Dumont made Europe's first heavier–than–air flight in a plane modeled after a box kite. Another Frenchman, Henri Farman, completed Europe's first circular flight in 1908 and went on to win prizes in his biplane. Farman's aircraft construction firm contributed to the vast growth in airplane numbers during World War I. German, British French and even Russian airplanes in that war were far superior to the ones America produced. In fact, after America entered World War I, the "Great War," in 1917 its pilots all flew British and French machines. From 1914 through 1918, manufacturers in France, Germany, the U.K., Italy, and the U.S. produced some 200,000 aircraft for the burgeoning industry.

Wright family members believed that the tension and exhaustion of the patent and business battles weakened Wilbur and increased his vulnerability to the typhoid that took his life on May 30, 1912. Nor could they mistake the fact that the patent wars were joined, not only before the bar of justice, but also in the court of public opinion. The public assumed that the patent suits had retarded the growth of American aeronautics, and many in government and industry agreed.

After Wilbur's untimely death, Orville decided to work on alone, and in 1913 won the Collier Trophy for a device to balance airplanes in flight automatically. However, he felt less enthusiastic about working on alone with the Wright Company. Ultimately, Orville did profit from the patent suits. He sold his interest in the Wright Company to a group of New York financiers in 1915, at a time when a Wright monopoly still seemed possible, a perfect time to sell. The amount of the sale was reputed to be $1.5 million, enough to enable Orville to live comfortably for the rest of his life. This was a blessing, as he was sick and tired of troubles with imitators, infringements on the Wright patents, conflicting claims, and lawsuits.

Yet Orville didn't want to just get out of the business after winning the Collier Trophy. He was still interested in continuing to work with new aeronautical innovations, so he went back to inventing. He built a small laboratory in his old West Dayton neighborhood. He wanted to stay closer to home in Dayton, instead of having to live in New York all the time, or travel abroad. World War I was still going on, and it had put a damper on things. Orville didn't want to straddle both sides of the fence with both the French and the Germans who were on opposite sides in the war.

Logically, he named his own shop the Wright Aeronautical Laboratory. Here, he worked on anything that caught his interest. He did some aeronautical work, helping to develop a racing airplane, guided missile, and "split flaps" to help slow an aircraft in a dive. But he also worked on an automatic record changer, a toaster, and children's toys. In short, he tinkered. His fame as the co–inventor of the airplane endured, however, and he put it to good use. Orville's laboratory grew and became known as the Wright Aeronautical Laboratory. His and Wilbur's legacy would be permanently enshrined in the expansive Wright–Patterson Air Force Base, located in the Wright's hometown of Dayton, Ohio.

ABOUT TECHNOLOGY

The 20th Century spawned wave after wave of ingenious scientific discoveries and technology development. It is in this context that the development of today's rocket science needs to be told.

Technology is not inherently good. No technology is absolutely, by its very nature, "good." And none is bad. But neither is technology ever neutral. Depending on how we design the technology, and even more on how we use—or misuse—the technology, it will affect us, it will change us, in some way.

Dr. Wernher von Braun, one of the greatest of rocket scientists with whom we shall become acquainted later, once said, "All of man's scientific and engineering efforts will be in vain unless they are performed and utilized within a framework of ethical standards commensurate with the magnitude of the scope of the technological revolution. The more technology advances, the more fateful will be its impact on humanity…If the world's ethical standards fail to rise with the advances of our technological revolution, the world will go to hell. Let us remember that in the horse–and–buggy days nobody got hurt if the coachman had a drink too many. In our times of high–powered automobiles, however, that same drink may be fatal…."

Even the most horrific of technology advances can be turned to good. The "mutually assured destruction" strategy of the Cold War between the United States and the Soviet Union worked for decades, and saved the world's civilizations from destruction. In this case, the outcome of the horribly destructive technology of the H–bomb became benevolent when it was applied to peaceful uses like in nuclear power plants. Similarly, rockets and missiles can be used for good or bad, for destroying cities or for taking humans to space, the ultimate frontier.

In 1914, Dr. Robert Goddard began serious experiments in rocketry at Clark University, and the Panama Canal had opened. 1915 was a banner year for technological achievements. Albert Einstein postulated his general theory of relativity and Margaret Sanger was jailed as the author of *Family Limitation*, the first popular book on birth control. Frederick Winslow Taylor, father of "Scientific Management," died, while disciples like Henry Ford were applying his ideas in the process of achieving prodigies of production. Ford produced his one–millionth automobile the same year. In 1915, Alexander Graham Bell made the first transcontinental call, from New York to San Francisco, with his trusted colleague, Dr. Thomas A. Watson, on the other end of the line. Motion pictures began to reshape American entertainment habits, and New Orleans jazz began to make its indelible imprint on American music. At Sheepshead Bay, New York, a new speed record for automobiles was set, at 102.6 miles per hour, a figure that many fliers of the era would have been happy to match.

Orville Wright summed up our loftiest ambitions for aviation when he said that it had been his hope, and that of his brother Wilbur, that they were giving the world "an invention which would make further wars practically impossible." The Wrights thought that by being able to see everything, surprise attacks would be impossible. War would become inconceivable unless a government were willing to go into a grinding war of attrition and of mutual annihilation. They thought no sane government would do this. But as history soon proved, the Wright brothers' thinking was naïve. In Europe tensions soon boiled over into World War I.

By late 1914 the state of deadlock on the Western Front had become clear to the governments of the warring countries and to many members of their general staffs. Each side sought a solution to this deadlock, and the solutions varied in form and manner. By the end of 1914 the casualties the French had so far sustained in the war totaled about 380,000 killed and 600,000 wounded; the Germans had lost a slightly smaller number. With the repulse of the German attempt to break through at the Battle of Ypres, the strained and exhausted armies of both sides finally settled down into trench warfare, a stalemate that would last for years. The trench barrier was consolidated from the Swiss frontier to the Atlantic. The power of modern defense had triumphed over the attack, and stalemate ensued. The military history of the Western Front during the next three years was to be a story of the Allies' attempts to break this deadlock. Repeated French attacks in February–March 1915 on the Germans' trench barrier in Champagne won only 500 yards of ground at a cost of 50,000 men. An even worse military failure was the joint offensive launched by the Allies on Sept. 25, 1915. In all, for a little ground, the Allies paid with 242,000 men, against the defenders' loss of 141,000. All of these attacks were disappointing failures, partly because they were preceded by prolonged bombardments that gave away any chance of surprise and allowed time for German reserves to be sent forward to close up the gaps that had been opened in the trench defenders' ranks by the artillery bombardment.

The American inventor Richard Jordan Gatling demonstrated a similarly dashed hope for the benefits of technology. He is best known for inventing the Gatling gun, the world's first practical machine gun, after he noticed the majority of dead returning from the American Civil War died of illness, rather than gunshots. In 1877, he wrote: "It occurred to me that if I could invent a machine – a gun – which could by its rapidity of fire, enable one man to do as much battle duty as a hundred, that it would, to a large extent supersede the necessity of large armies, and consequently, exposure to battle and disease would be greatly diminished."

Alfred Bernhard Nobel was a Swedish chemist, engineer, and industrialist who invented dynamite and other, more powerful explosives in the mid–1800s. His hope was to save miners' lives by replacing black powder, a form of gunpowder, with a dependable, much more powerful explosive that would reduce the number of mining explosions and men required for their placement. The founder of the Nobel Prize was also a pacifist who hoped that the destructive power of his inventions would help bring an end to war.

Whether the effects and changes of the Wrights' aviation technologies, or modern rocket science, turn out to be good or bad, or both inseparably together, is not predestined in the inherent qualities of the technology itself. Rather, the human consequences of the airplane or rocket depend on a much broader context on the values within which we live our lives, the so–called "soft" and "mushy" things like politics and culture. Unfortunately, for many space enthusiasts logic does not determine the history of technology, and technologically "sweet" solutions do not always triumph over political and social forces.

Every technical innovation seems to require additional technical advances in order to make it fully effective. In this sense, *invention is the mother of necessity*, in contradistinction to the conventional view that *necessity is the mother of invention*. Once the Wrights invented the airplane, all sorts of things then really needed to happen. Over the course of the next 30 years, the airplane was in a sense reinvented as the Wrights' achievement was completely rethought and reworked by emerging groups of professionals dedicated to the airplane's improvement and greater practicality. The invention quickly necessitated all sorts of auxiliary technologies: advanced structures and materials, new wing shapes, streamlined aerodynamics, retractable landing gear, efficient low–drag engine cowlings, variable–pitch propellers, and much more. Synchronization gear was invented so that fuselage-mounted machine guns could fire bullets through the arc of a spinning propeller without striking the blades. But perhaps even more importantly, the airplane also necessitated new social forms and organizations, like military air services, airlines, airports, government bureaus, research laboratories, engineering curricula, and much else. While it might be said that each of these other developments occurred in response to a specific need, it was the original invention that mothered the necessity.

Likewise the invention of the rocket required new advances in gyroscopic control, guidance, gimbal steering, power driven fuel pumps, valves and other devices, lightweight structures, advances in electronics and avionics, the development of geosynchronous communications satellites, and on and on.

It is axiomatic that technologies incessantly improve over time to improve our daily lives. They almost never stand still. The laptop computer of yesterday is obsolete. The 20–year–old storm windows in homes become obsolete, replaced by double–paned glass windows that sandwich inert argon gas, which serves as an excellent insulant. Today's windows are also easier to open and clean. In the 1950s, predictions were made that the automobile piston–driven gasoline engine would reach a plateau in improvements by 1977. And yet gasoline engines have continued witnessing a steady technological improvement well into the 21st century, including computer–driven sensors, diagnostics, and instrumentation that would not have been dreamed of in 1977. Today's automobiles "adjust" themselves to the individual driver, change drivers and the car will drive differently. As a modern car ages, embedded computers can actually make up for wear and tear on parts, so that it drives like it was almost new. The rockets driving the space vehicles of today will soon be driven to obsolescence as new advances render engines more robust, reliable, and less costly. Engineers, technicians and myriad specialists continue pouring forth a steady stream of innovations and technological improvements everywhere, impacting our daily lives. These incremental improvements, like good news, almost never make the headlines.

NACA

The genesis of what came to be known as the National Advisory Committee for Aeronautics (NACA) occurred at a time of accelerating cultural and technological change. Amidst the gathering whirlwind of the World War I, social change and technological transformation persisted.

American flying not only lagged behind automotive progress, but also lagged behind European aviation. This was particularly galling to many aviation enthusiasts in the United States, the home of the Wright brothers. Impressed by the Wrights, the Europeans began a rapid development of aviation, and their growing record of achievements underscored the lack of organized research in the United States.

Sentiment for some sort of center of aeronautical research had been building for several years. At the inaugural meeting of the American Aeronautical Society, in 1911, some of its members discussed a national laboratory with federal patronage. However, the American Aeronautical Society's dreams were frustrated by continued in–fighting among other organizations which were beginning to see aviation as a promising research frontier, including universities like the Massachusetts Institute of Technology, as well as government agencies like the U.S. Navy and the National Bureau of Standards.

The difficulties of defining a research facility were compounded by the ambivalent attitude of the American public toward the airplane. While some saw it as a mechanical triumph with a significant future, others saw it as a mechanical fad, and a dangerous one at that. If anything, the antics of the "birdmen" and "aviatrixes" of the era tended to underscore the foolhardiness of aviation and airplanes. Fliers might set a record one month and fatally crash the next. There were fatalities in Europe as well, but the Europeans also took a different view of aviation as a technological phenomenon. Governments, as well as industrial firms, tended to be more supportive of what might be called "applied research."

As early as 1909, the internationally known British physicist, Lord Rayleigh (John William Strutt, 3rd Baron Rayleigh), was appointed head of the Advisory Committee for Aeronautics; in Germany, Ludwig Prandtl and others were beginning the sort of investigations that soon made the University of Göttingen a center of theoretical aerodynamics. Additional programs were soon under way in France and elsewhere on the continent. Similar progress in the United States was still slow in coming. Aware of European activity, Charles D. Walcott, secretary of the Smithsonian Institution, was able to find funds to dispatch two Americans on a fact–finding tour overseas. Their report, issued in 1914, emphasized the galling disparity between European progress and American inertia. Walcott was flabbergasted. The visit also established European contacts that later proved valuable to NACA.

The outbreak of war in Europe in 1914 helped serve as a catalyst for the creation of an American aeronautical research agency. The use of German dirigibles for long–range bombing of British cities and the rapid evolution of airplanes for reconnaissance and for pursuit underscored the shortcomings of American aviation. Against this background, Walcott pushed for legislative action to provide for aeronautical research allowing the United States to match progress overseas. He received support from Progressive Party leaders in the country, who viewed government agencies for research as consistent with Progressive ideals such as scientific inquiry and technological progress. By the spring of 1915, the drive for an aeronautical research organization finally succeeded.

The enabling legislation for NACA slipped through almost unnoticed as a rider attached to the Naval Appropriation Bill, on March 3, 1915. It was a traditional example of American political compromise. As before, the move had been prompted by the Smithsonian. The legislation did not call for a national laboratory, since President Woodrow Wilson apparently felt that such a move, taken during wartime conditions in Europe, might compromise America's formal commitment to strict nonintervention and neutrality. Although supported by the Smithsonian, the proposal emphasized a collective responsibility through a committee that would coordinate work already under way. The committee was an unpaid panel of 12 people, including two members from the War Department, two from the Navy Department, one each from the Smithsonian, the Weather Bureau, and the Bureau of Standards, and five more members acquainted with aeronautics. Despite concerns about appearing neutral, the proposal was tacked on as a rider to the naval appropriation bill as a ploy to clear the way for quick endorsement.

For fiscal 1915, the fledgling organization, called the *Advisory Committee for Aeronautics* (ACA), received a budget of $5000, an annual appropriation that remained constant for the next five years. This was not much even by standards of that time, but it must be remembered that this was an advisory committee only, *"to supervise and direct the scientific study of the problems of flight, with a view to their practical solution."* Once the ACA isolated a problem, its study and solution was generally done by a government agency or university laboratory, often on an ad hoc basis within limited funding. The main committee of 12 members met semiannually in Washington. An Executive Committee of seven members, characteristically chosen from the main committee living in the Washington area, supervised its activities and kept track of aeronautical problems to be considered for action. It was a clubby arrangement, but it seemed to work.

The prefix "National" soon became customary, was officially adopted, and the NACA emerged as a widely recognized term among the aeronautics community in America. NACA was pronounced as individual letters, rather than as an acronym. The agency's immediate task was to survey the state of the art and reinsure a lead role for the United States in aeronautics. In 1916 NACA called the first joint meeting of the U.S. aircraft industry and government agencies. This conference was instrumental in the creation of the Manufacturers Aircraft Association and in the recommendation for the formation of the Aircraft Production Board for WWI aircraft production. In a wartime environment, NACA was soon busy. It evaluated aeronautical queries from the Army and conducted experiments at the Navy yard; the Bureau of Standards ran engine tests; Stanford University ran propeller tests. NACA also mediated the complex legal dispute between Glenn Curtiss and the Wright–Martin Company involving a technique for lateral control of aircraft in flight, as described previously.

The authors of NACA's charter had written it to leave open the possibility of an independent laboratory. Although several facilities for military research continued to function, NACA pointed out in its first *Annual Report for 1915* that civil aviation research in its own laboratory would be in order when the "Great War", as World War I was called, ended. Under wartime pressures, the Army had already relocated its own research center to McCook Field, near Dayton, Ohio (the site of the present Wright–Patterson Air Force Base).

The best option for a NACA laboratory seemed to be collaboration with the more bountiful resources of the military in the development of a new U.S. Army airfield, across the river from Norfolk in Hampton, Virginia. Construction of the airfield got underway in 1917, but was hampered by the confusion following America's declaration of war on Germany and by the wet weather and marshy terrain of the Virginia tidewater region. The declaration of the WWI Armistice on November 11, 1918 cleared the way for completion of the joint–use military facility. Named after Samuel Pierpont Langley, the famed aeronautical pioneer and former secretary of the Smithsonian, Langley Field was formally dedicated on June 11, 1920. The inaugural ceremonies included various aerial exhibitions and a fly–over of a large formation of planes led by the dashing Brigadier General William "Billy" Mitchell. Visitors found that NACA's corner of Langley Field was comparatively modest: an atmospheric wind tunnel, a dynamometer lab, an administration building, and a small warehouse. The NACA aeronautical research facility was renamed the Langley Memorial Aeronautical Laboratory, soon shortened to the familiar, cryptic "Langley" There was a staff of 11 people—plenty of room to grow. Langley was the first facility of the US government to coordinate research in the civil and military sectors. It should not be confused with the military's collocated Langley Field, a large base whose military influence remained strong. Today Langley Air Force Base is headquarters of the huge Air Combat Command (ACC).

Engineers came to Langley from all over the country. Early employees often had degrees in civil or mechanical engineering, since so few universities offered a degree in aeronautical engineering alone. However, a number of factors converged in the 1920s to metamorphose NACA from an underfunded advisory institution to a huge driver for aeronautical innovation in the United States. On January 29, 1920, President Wilson appointed Orville Wright to NACA's board. He would serve longer than any other board member. By the early 1920s, it had adopted a new and more ambitious mission: to promote military and civilian aviation through applied research that looked beyond current needs. NACA acquired three more major facilities across the country: the Ames Aeronautical Laboratory (collocated with the Navy's Moffett Field in California), the Aircraft Engine Research Laboratory (Lewis Research Center in Cleveland, Ohio), and the Muroc Flight Test Unit (collocated with Edwards AFB in California). Over the years NACA researchers pursued this mission through the agency's impressive collection of in–house wind tunnels, engine test stands, and flight

test facilities. Commercial and military clients were also permitted to use NACA facilities on a contract basis.

In the 1920s the fledgling domestic airline industry began a period of burgeoning growth as privately owned companies forged transcontinental routes across America, flying passengers as well as delivering the U.S. mail. Pan American Airways inaugurated international services between Florida and Cuba, as well as between Texas and Central America. NACA's aviation research budget grew thanks to increased funds for both civil and military (U.S. Army and U.S. Navy) aviation. The Air Commerce Act of 1926 improved lighted runways, radio communications, and established federal guidelines for pilot proficiency as well as aircraft design and construction, and the study and application of aviation technology became more structured.

The Army Air Service became the Air Corps. Gradually, universities became centers for the theory of flight. From a handful of prewar courses dealing with aeronautical engineering, universities like the Massachusetts Institute of Technology evolved a plan of professional course work leading to both undergraduate and graduate degrees in the subject. In 1929, a survey by an aviation magazine reported that 1400 aero engineering students were enrolled in more than a dozen schools across the United States. The California Institute of Technology became a major beneficiary of the Guggenheim Fund's foresight.

The largesse of the Guggenheims is legendary. Daniel Guggenheim represented the second generation of the dynasty founded by copper–mining tycoon Simon Guggenheim. Daniel became interested in aviation after World War I, mainly because his son Harry was a naval flier. He founded the Daniel Guggenheim Fund for the Promotion of Aeronautics, which financed America's first scheduled airline, supported Jimmy Doolittle's famous first "blind flying" test, provided money for aeronautical engineering programs at several universities, and financed Charles Lindbergh's tour of the United States after his historic 1927 solo flight to Paris. Harry F. Guggenheim once told Professor Theodore von Kármán jokingly, "You know, when my grandfather was active, anybody who did not make a great fortune was considered a failure. In the time of my father, if one did not expand the empire, he was also considered a failure. But all they expect of me now is to spend the money graciously and usefully." This he did, among other things heading up the Daniel and Florence Guggenheim

Foundation, which for decades provided funds for the promotion of aeronautical research.

During the late 1920s and into the 1930s, NACA made substantial contributions to the aeronautical sciences. One stellar example was the evolution of modern, retractable–geared aircraft. As means to increase speed, retractable landing gear was not unknown, since this approach had been tried on various airplanes before World War I. However, retractable gear required additional equipment for raising and lowering and appeared to lack the ruggedness and reliability of conventional, fixed gear outside the plane. On the other hand, people suspected that fixed gear was a major drag factor, but nobody had accurately assessed aerodynamic liability. NACA engineers set up a series of tests in the then–new Langley propeller research tunnel to get an accurate estimate of a fixed landing gear's drag using a Sperry Messenger airplane. The results were astonishing.

They estimated that fixed gear created nearly 40 percent of the total drag acting on the plane! This eye–opening news dramatically demonstrated the performance penalty incurred by fixed gear, and prompted rapid development of retractable gear for a variety of airplanes.

Another of many examples was the breakthrough NACA design of a low–drag cowling. Most American planes of the post–WWI decade mounted air–cooled radial engines, with the cylinders exposed to the air stream to maximize cooling. However, the exposed cylinders also caused high drag. Because of this, the Army Air Corps had adopted several aircraft with liquid–cooled engines, in which the cylinders were arranged in a line parallel to the crankshaft. This reduced the frontal area of the aircraft and also allowed an aerodynamically contoured covering, or nacelle, over the nose of the plane. But the liquid–cooled designs carried weight

NACA's 20–foot propeller research wind tunnel, completed in 1927

penalties in terms of the myriad cooling chambers around the cylinders, gallons of coolant, pumps, and radiator. The U.S. Navy decided not to use such a design because its added maintenance requirements cut into the limited space aboard aircraft carriers. Moreover, the jarring contact of airplanes with carrier decks created all sorts of cracked joints and leaks in liquid–cooled engines. Although air–cooled radial engines simplified this issue, their inherent drag meant reduced performance.

The NACA cowling, as fitted on the Curtiss Hawk

In 1926, the Navy's Bureau of Aeronautics approached the NACA to see if a circular cowling could be devised in such a way as to reduce the drag of exposed cylinders without creating too much of a cooling problem. Investigators at Langley conducted hundreds of tests using the new 20–foot propeller research tunnel which could accommodate full–sized airplanes. NACA acquired a Curtiss Hawk AT–5A biplane fighter from the Army Air Corps and fitted a cowling around its blunt radial engine. The results were exhilarating. With little additional weight, the Hawk's speed jumped from 118 to 137 MPH, an increase of 16 percent. The virtues of the NACA cowling received public acclaim the next year, when Frank Hawks, a highly publicized stunt flier and air racer, added the NACA cowling to a Lockheed Air Express monoplane and racked up a new Los Angeles/New York nonstop record of 18 hours and 13 minutes. The cowling had raised the plane's speed from 157 to 177 MPH. After the

flight, Lockheed Aircraft sent a telegram to the NACA committee: "Record impossible without new cowling. All credit due NACA for painstaking and accurate research." By using the cowling, the NACA estimated savings to the industry of over $5 million—more than all the money appropriated for NACA from its inception through 1928. Talk about a payoff!

By the 1930s, NACA had matured as an entity. NACA and the National Bureau of Standards contributed basic research; the aeronautical industry designed and developed aircraft, and the armed forces tested them for conformity to specifications, often in collocated facilities. Not a bad deal for NACA.

NACA reports began to emerge from an impressive variety of tunnels that went into operation during the 1930s. NACA developed a series of thoroughly tested airfoils and devised a numerical designation for each airfoil: a four–digit number that represented the airfoil section's critical geometric properties. By 1929, Langley had developed this system to the point where the numbering system was complemented by an airfoil cross–section, and the complete catalog of 78 airfoils appeared in NACA's annual report for 1933. Engineers could quickly see the peculiarities of each airfoil shape, and the numerical designator ("NACA 2415," for instance) specified camber lines, maximum thickness, and special nose features. These figures and shapes transmitted the sort of information to engineers that allowed them to select specific airfoils for desired performance characteristics of specific aircraft.

Open test section of the full–scale wind tunnel (completed in 1931) at NACA in Hampton, Virginia

Langley and NACA experienced a spurt in growth during the World War II wartime period. In 1938, the total Langley staff came to 426. Just seven years later, in 1945, Langley numbered 3000 personnel. There were some areas of flight technology, such as rocketry, in which NACA did not become involved. Nevertheless, when NACA was transformed into the well–known National Aeronautics and Space Administration on October 1, 1958, the new space agency could reach back into some forty years of American and European writing and research on rocketry and the possibilities of space flight.

THE WRIGHT BROTHERS' LEGACY

Looking back upon the Wright brothers' accomplishments, it is amazing that they were able to get their airplane off the ground on December 17, 1903, much less fly it and control it. They had rapidly mastered the art of flying, in spite of an unforgiving design by today's standards and a rudimentary control system that was very complex to operate. Most importantly, there has been a general lack of success of those trying to strictly emulate their design and fly it even many years later. A group of student test pilots and flight–test engineers at the Air Force Flight Test Center designed and tried to fly a replica of the 1903 Flyer on the hundredth anniversary of Orville's first flight. First, they had to install several safety–enhancing features different from the original: a larger canard elevator located further forward to increase pitch damping; ballast to improve stability by shifting the center of gravity further forward; a different airfoil cross section with less camber, to reduce pitching moment; reduced anhedral and perhaps even dihedral to eliminate spiral mode instability; and a larger vertical tail located further aft for added directional stability. After complex ground analyses and in–flight simulations, the team suggested additional changes: a pitch stability augmentation system, no flying in crosswinds, landings over a wide area to avoid the necessity of making large–input rolls, and finally a ground–based simulator to prepare for the first flight of the "replica." Clearly the Wright brothers were superlative pilots by the standards of any era, then or now.

Charles Taylor worked with Orville in the lab, helping out with some of his inventions and experiments, and kept his car in good running order. But there was less and less work to do, so finally he got restless and took a job with the Dayton–Wright Company, the successor to Orville's Wright Company, in 1919. Orville and Charles continued to see each other frequently. Orville would bring odd jobs to Charlie at the plant where he was working in downtown Dayton, and Charlie would visit Orville at his laboratory. Then in 1928 Charlie moved to California, taking a job in a machine shop in Los Angeles.

The 1920s brought the Wrights some level of recognition for their overall contribution to the development of heavier–than–air flight. By that period, bad feelings resulting from the Wrights' seemingly self–serving patent litigation had subsided, and Americans placed the Wrights' and the country's contributions in perspective. Air flight developed more systematically by the mid–1920s as centralized research operations introduced innovations and encouraged development through interest–grabbing demonstrations and flight records. Charles Lindbergh, a friend and supporter of the Wrights, made the world's first nonstop solo flight across the Atlantic, from New York to Paris, on May 20–21, 1927. The litany of names of the pioneers who followed the trail blazed by the Wrights should sound familiar: the American timber merchant William E. Boeing who founded the Aero Products Company in 1916; James S. McDonnell, founder of the McDonnell Aircraft Corporation in 1939; Donald W. Douglas who established the Douglas Aircraft Company in 1921; John K. (Jack) Northrop who formed the Northrop Corporation in 1932; Allan Loughead who established the Lockheed Aircraft Company in 1926 (the spelling of Loughead was changed to match its pronunciation); Leroy R. Grumman who founded the Grumman Aeronautical Engineering Company in 1929; Igor Sikorsky of helicopter fame who formed the Sikorsky Aero Engineering Corporation in the early 1920s; Juan Terry Trippe who incorporated Pan American Airways in 1927; and abroad, the British Aircraft Corporation; Hawker Siddeley, a group of British manufacturing companies engaged in aircraft production; France's Aerospatialé, Germany's Daimler–Benz Aerospace; and Europe's Airbus Group just to name a few.

In 1929 Orville received the first Daniel Guggenheim Medal for his and Wilbur's contributions to the advancement of aeronautics. Orville helped oversee the Guggenheim Fund for the Promotion of Aeronautics, an effort that helped America recapture the technological lead in aviation during the 1930s. He also worked tirelessly to help unknown inventors bring their ideas to market, working from his lab in Dayton.

Until the end of his life, Orville Wright continued a long, running battle with the Smithsonian that had begun with their duplicity in the Curtiss patent suit. After the First World War, the Smithsonian exaggerated Langley's contributions to aeronautics while seeming to belittle the Wrights. Friends of Orville set the record straight, but the Smithsonian kept on. In retaliation, Orville sent the 1903 Wright Flyer, the airplane in which he and Wilbur had made the first powered flights at Kitty Hawk, to the Science Museum of London in England in 1928 to exhibit there in perpetuity. In the 1930s, Charles Lindbergh, the first aviator to fly from New York to Paris nonstop, attempted to mediate the feud, but to no avail. It wasn't until 1942 that Orville Wright's friend and biographer, Fred Kelly, convinced the Smithsonian to back down and publish the truth. That done, Orville changed his will and bequeathed the Flyer to the Smithsonian for display in the "National Capital only." He sent word to England that the Flyer was to be brought home to America. Its return was delayed by the Second World War, but it was finally returned in 1948. The first successful airplane, the Wright Flyer of 1903, finally graced the North Hall of the Smithsonian's Arts and Industries Building in 1948, 45 years to the day after its historic flight at Kitty Hawk. Orville Wright's estate attached a qualifying label to the display of their 1903 Flyer at the Smithsonian: "The original Wright brothers' aeroplane: the world's first power–driven, heavier–than–air machine in which man made free, controlled, and sustained flight."

Orville's Wright last big project was, fittingly, an aircraft. He helped to rebuild the 1905 Flyer III, the first practical airplane, which he and Wilbur had perfected at Huffman Prairie. This was put on display at Deeds Carillon Park in Dayton, Ohio in 1950, but Orville did not live to see the ceremony. He died of a heart attack on January 30, 1948 while fixing the doorbell at his home. Shortly before he died, he wrote a note to his long–time friend Charles Taylor: "I hope you are well and enjoying life, but that's hard to imagine when you haven't much work to do." It was signed, "Orv." Wilbur was elected to the NACA Hall of Fame in 1955, followed by Orville in 1965.

Orville Wright in 1928

ROBERT

"Every vision is a joke until the first man accomplishes it; once realized, it becomes commonplace."

Robert Goddard, 1920

"Don't you know about your own rocket pioneer? Dr. Goddard was ahead of us all."

Wernher von Braun, when asked about Goddard's work following World War II

During the 1920s, the subject of space flight more often than not seemed to be the province of cranks and science fiction writers spinning wildly improbable tales. But visionary researchers in the United States, as well as Great Britain, Germany, Russia, and elsewhere were taking the first hesitant steps toward actual space travel. In America, Robert Hutchings Goddard is remembered as one of the foremost pioneers. Goddard had a vision for the age of space, but the world was too slow to make it happen before his death. Goddard was the first to:

- mathematically prove that a rocket can propel in a vacuum (1907)
- patent the oscillator tube, which was later used in the radio industry (1912)
- patent the concept of the multi-stage rocket (1914)
- prove by testing that rockets can propel in a vacuum (1915)
- invent the prototype of the bazooka (1918)
- develop a rocket using liquid fuels (between 1921-1926)
- have a liquid-fueled rocket lift its own weight (Dec. 6, 1925)
- successfully fire a liquid fuel rocket (March 16, 1926)
- use a De Laval (expanding cone) nozzle in a rocket
- launch a rocket with a scientific payload (a barometer and a camera in 1929)
- develop turbopumps for a liquid propellant rocket
- develop a liquid propellant rocket cluster
- use a pulse-jet engine
- develop gyro stabilization apparatus as an internal guidance system for rockets (1932)
- use vanes in rocket motor blast for guidance (1932)
- fire a liquid fuel rocket that traveled faster than the speed of sound (March 8, 1935)
- use a rocket engine pivoted on gimbals controlled by a gyro mechanism (1937)

MARCH 16, 1926

"Henry, help me load these crates into the coupe," he said matter-of-factly. For a time, Robert Hutchings Goddard had missed his gifted instrument maker from Stockholm, Nils Riffolt. Riffolt had been serving as a part-time instrument maker in the physics department at Clark University in Worcester, Massachusetts, while taking graduate work. He received his Master's degree in June 1924 and left Worcester to go with Louis Thompson to the Naval Proving Ground at Dahlgren, Virginia. But Henry Sachs, his new machinist, was working out well. Henry was also a highly skilled instrument maker, and had entered Dr. Goddard's employ from the U.S. Bureau of Standards in August, 1924. None too soon. The son of a German blacksmith, he had accepted Goddard's offer of "six months' work, maybe longer." Remarkably, he would end up staying with Robert until February, 1931.

Once his car was loaded, Dr. Goddard carefully locked the door leading to his physics workshop and climbed into his coupe besides Sachs. "I like your predilection for coupes," said Henry. "You bet," answered Robert. "You don't have to take an excess of people." Like many inventors, the close-mouthed Goddard had always been secretive, and had scrupulously covered his pioneering rocket technology advancements with patents, courtesy of his patent attorney Charles T. Hawley (during his lifetime Goddard filed 83 patents). Goddard had earlier worked on solid propellant cartridge rockets, but when he glommed onto the superior liquid-fueled rockets, he told Riffolt, "let's keep this under our hat." On August 2, 1924 he had asked his bemused assistant to sign a pledge of secrecy, with Riffolt's signature notarized. *"I, Nils Riffolt, of Worcester, Massachusetts, hereby agree not to divulge the nature of any of the work which has been done by me, or has come to my knowledge, in the rocket investigation upon which I have been working under Dr. R.H. Goddard, of Worcester, Massachusetts."* Then he turned around and had Sachs sign a similar pledge of secrecy on August 21.

Between Goddard and Sachs were two fresh liters of liquid oxygen in spherical double-wall glass Dewar flasks. Linde Air Products Company would not guarantee more oxygen for several weeks, so Goddard knew that if their March flight failed, it would be April, at least, before they could try again. The liquid cryogen cost them $1 per liter, and small supplies had to be sent by truck from Linde's plants in Texas and Oklahoma.

Robert Goddard bundled himself against the Massachusetts chill. The balding professor wore his habitual cold weather gear: warm woolen cap, a muffler, a greatcoat buttoned up to his neck, and high–buckled galoshes. In this costume he and Henry left the Clark campus, driving along Freeland and Cambridge streets to Southbridge, then up Pakachoag Hill to the municipality of Auburn and Aunt Effie's farm, some 8 kilometers south of Worcester in east central Massachusetts.

The Ward farm in Auburn, locally known for its fine strawberries, was owned by Miss Effie Ward, a spinster and distant relative of Goddard's, who ran it with the help of a hired man. Tall, spare, and chipper, "Aunt" Effie wore her iron gray hair drawn back in a bun and kept a large family of cats in her kitchen. She accepted her unusual kinsman without hindrance or questions, allowed him to put up his testing tower in the back acres and offered him an unused henhouse to store his equipment. She amicably tolerated the occasional blasts from the field. On sighting Goddard, she would say proudly, "Here comes the rocket man!" Never mind the neighbors' occasional complaints at the noisy goings–on.

They parked at a ravine some distance from Effie's farmhouse. "Whew, these things are heavy," exclaimed Henry. Goddard nodded. "Lordy, it feels like this crate is loaded with Napoleon's brass plates," continued Henry. Goddard and Sachs carried, slid, and shoved the wooden crates containing tools, propellant tanks, metal piping, and the rocket's motor to a secluded spot near a cabbage patch. First they erected a pipe launching frame, then started carefully attaching the fragile rocket inside it. The two men worked through the cold morning hours, rigging their gear. His invention, Goddard was finding, demanded novelty throughout—in its design, its fuels, its components, and makeshift fabrication.

Early rocket test run at Effie Ward farm

Historic photo of Robert H. Goddard beside his 1926 liquid– fueled rocket. The rocket is on top, receiving its fuel by two lines from the tank at the bottom.

Shortly after noon the entire ungainly contraption was ready for launch. It stood over 11 feet (3.4m) high. The tanks contained 10.4 pounds (4.7 kg) of propellants, and Robert calculated his rocket would generate over 9 pounds (40 newtons) of thrust. The term "newton" derives from Sir Isaac Newton (1642–1727), the famous English scientist. A newton is slightly less than a quarter–pound in English units. As the oxidizer and fuel burned off and the rocket maintained a steady thrust, Robert expected it to lift not only itself, but its inner frame into the air as well. The igniter was on top, then the combustor/nozzle combination, then over a meter of asbestos–wrapped propellant tubing, then a metal swing set–like outer framework some 8 feet (2.4m) high, then the coated asbestos cone/fuel tanks assembly about in the center of the framework, and finally a short metal tube at the bottom to hold the rocket steady during launch. The base of the outer frame formed a 5.6 by 5.6 foot (1.7m by 1.7m) square in the snow.

Percy Roope, assistant professor of physics at Clark, arrived at the Ward farm with Mrs. Goddard. Robert's face lit up at the sight of his slender blonde wife Esther. "Hi!" he exclaimed. "Let me help you with your camera and gear," he added as he offered his arm to her. Esther carried her husband's latest purchase, a French "Sept" motion picture camera, so named because it ran for seven seconds without rewinding. On many occasions, she had chatted briefly with Aunt Effie in her warm, commodious kitchen, with its familiar window box of bright red, pink, and white geraniums. Miss Ward, never quite sure about rockets, once again offered Esther and her fellow adventurers a cup of hot malted milk as her specific against chills. Esther took moving pictures of Goddard's rocket setup with the Sept. Finally, she took a still of Robert holding the outer supporting frame and facing the camera.

The big moment had arrived. As crew chief, Sachs held a blowtorch attached to a long stick so that he could reach the igniter 11 feet (3.4m) above. The simple igniter held a match–head filled stub. When black smoke issued from the igniter, he turned a valve and lit an alcohol stove beneath the engine. For an instant, the rocket did nothing at all while the propellants flowed up the tubes towards the combustion chamber. Then the engine ignited with a flash. Then it stayed put, burning off the propellant load so that it could lift itself and the attached inner frame. Goddard waited for 90 seconds, then released his rocket. He and Roope edged behind a sheet–iron barricade, Sachs ran toward them, and the rocket roared off as oxygen and gasoline combusted. In Goddard's own words:

"Loaded with liquids, the lower part of the nozzle burned through and dropped off, leaving, however, the upper part intact. After about 20 seconds, the rocket rose without perceptible jar, with no smoke and with no apparent increase in the rather small flame, increased rapidly in speed, and after describing a semicircle landed 184 feet (56.1m) from the starting point — the curved path being due to the fact that the nozzle had burned through unevenly, and one side was longer than the other. The average speed, from the time of the flight measured by a stop watch, was 60 miles per hour (96.5 km/hr). This test was very significant, as it was the first time that a rocket operated by liquid propellants traveled under its own power."

An understatement. This seemingly unimpressive flight of 2½ seconds and 41 feet (12.5m) altitude heralded a whole new era in rocket flight. This was an achievement of the same worth as the Wright Brothers first heavier–than–air flight of December 1903. Unfortunately, no photographic record of this historic flight exists, as Esther's camera had run out of film. While the flight may not sound all that impressive today, many people consider it to mark the beginning of the space age. Doggedly and almost single–handedly, Goddard had put together and flown the world's first liquid propellant rocket.

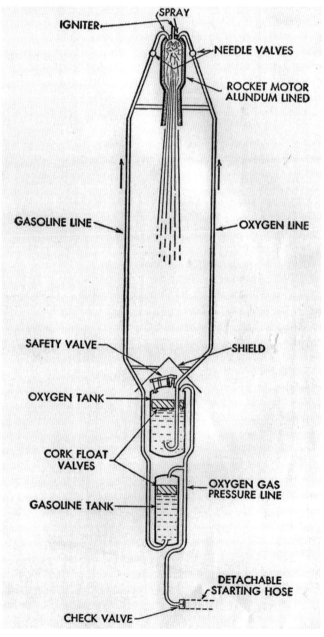

Cross section of the 1926 liquid–fuel rocket.

THE CHERRY TREE

Robert Hutchings Goddard came into this world on October 5, 1882. His father, Nahum Danford Goddard, was a great believer in education and encouraged his son to experiment with things. Father and son would spend many hours hiking through the woods studying and communing with nature. The boys in the neighborhood called him "Robbie," regarding him as a lively, although somewhat sickly, companion. Robert suffered a variety of ailments diagnosed as colds, pleurisy, and bronchitis. Later in his youth, he would suffer from bouts of tuberculosis and would become prematurely bald as a consequence. At any rate, he became motivated by reading science fiction as a boy. Armed with a telescope, Goddard became excited by the possibility of exploring space while still in primary school. Goddard's interest in rockets began in 1898 when, as a 16–year–old, he read the latest publication, *The War of the Worlds*, of that early science fiction writer, English novelist Herbert George "H.G." Wells. The book which so excited Goddard was later made into a 1938 radio program that nearly panicked our entire nation when it was broadcast. Orson Welles' (no relation) too realistic rendition of H.G. Wells' work still causes many to shudder. The young Goddard's interest would soon become a life–long passion. He would later write, "How many more years I shall be able to work on the problem, I do not know; I hope, as long as I live. There can be no thought of finishing, for 'aiming at the stars,' both literally and figuratively, is a problem to occupy generations, so that no matter how much progress one makes, there is always the thrill of just beginning…"

Like other autumn days on Maple Hill, October 19, 1899 was quiet, with the trees dressed in beautiful flaming yellow, orange, and red leaves. After his long rest–time in bed, the sickly Robert was glad to be outdoors. He stood for a minute on the back porch. In the yard, enclosed by chicken wire, near the tool shed, was "Uncle" George Boswell's collection of cocker spaniels, two dozen of them. They yawped and sprawled over one another, some nuzzling up to their mothers who lay contentedly sunning themselves beside the fence. Robert strolled past the shed where the sight of Boswell's immaculately arranged tools was "always a feast to my eyes." The tools were oiled, sharpened, and each took its proper place. Robert picked up a bench saw, hatchet, and a small ladder George had nailed together, and headed down to the fruit orchard to do some pruning.

He stopped before a gnarled cherry tree at the edge of a narrow brook and propped his ladder against it. Hooking his saw onto a branch, Robert paused, enjoying the smell of meadow and woods below, hearing the soft burbling sound of the brook and the varied birds' songs. Lying in the tree, as if in a womb of branches and leaves, he lost all sense of time and space and immediacy. While suspended there, a fantasy shape took hold in the mirrors of his mind, an image sharp enough to shut out the surrounding scene. *A mechanical device materialized from nowhere, functioning perfectly. Faster and faster it whirled until it began to lift, twirling and spinning above Worcester and sickness and fruit trees, upwards into space!* The 17 year old teenager was transfixed. When he collected himself, he began to prune one of the dead branches. But after a few cuts with his saw, he climbed down. He walked back to the tool shed and the house.

Robert Goddard, Age 5

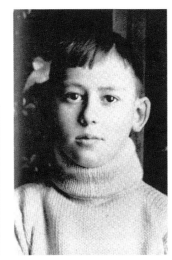
Age 11

Age 25

Goddard's photograph of the cherry tree, taken in 1900. His rough homemade ladder is just visible leaning against the trunk.

Goddard marked the day, October 19, in his dairy every year thereafter, referring to it as "Anniversary Day." That evening, however, he laconically wrote, "…trimmed large cherry tree…" Later he amplified the experience in his autobiographical notes. "On this day I climbed a tall cherry tree at the back of the barn. It was one of the quiet, colorful afternoons of sheer beauty which we have in October in New England, and as I looked toward the fields at the east, I imagined how wonderful it would be to make some device which had even the *possibility* of ascending to Mars, and how it would look on a small scale, if sent up from the meadow at my feet. I have several photographs of the tree, taken since, with the little ladder I made to climb it, leaning against it," he wrote. "It seemed to me then that a weight whirling around a horizontal shaft, moving more rapidly above than below, could furnish lift by virtue of the greater centrifugal force at the top of the path…I was a different boy when I descended the tree from when I ascended. Existence at last seemed very purposive."

Although in 1899 young Goddard still had no clear concept of the rocket vehicle through which space flight would come to pass, his first idea from the cherry tree experience was that centrifugal force might somehow be used was a breakthrough in his thinking. A propulsion device could work just fine *in vacuo*. There would be no need for an atmosphere, simply because a rocket carries all of its own fuel with it in airtight chambers. In fact, the gifted Goddard soon realized that rocket performance becomes better and better at increasing altitudes, as both its aerodynamic drag and the retarding back atmospheric pressure aft of the nozzle steadily diminish, until both become zero for all intents and purposes.

In September 1901, shortly before he was 19, Robert enrolled as a sophomore at Worcester's new South High School. Bouts with various ailments had delayed his formal schooling, but the ambitious Robert used these periods to intensely study physics, chemistry, and science. The eager, hardworking student was received as a minor phenomenon by his younger classmates, who described him as a "shark" and twice elected him as their class president, an exalted post for which he equipped himself with a volume on parliamentary law. He excelled in his studies, particularly in technical subjects: mathematics, mechanics, astronomy, and geometry. He graduated with honors in June, 1904.

THE UNIVERSITY YEARS

During his undergraduate years at "Tech" from 1904–1909, fellow students considered Goddard a brilliant but acceptably congenial class leader. They found fewer eccentricities in him than in their faculty, one of whom, absent–mindedly making his way to a campus meeting, stepped on the tail of a wayward cat and absently apologized to the animal. Worcester Polytechnic Institute had been founded in 1865 by Ichabod Washburn, the Worcester wire manufacturer, and John Boynton, one–time farmer, peddler of tinware and manufacturer, to meet the city's emerging need for skilled engineers, mechanics, and technicians. In the view of Boynton, education was meant to be usefully and profitably applied.

Professor Alexander Wilmer Duff, head of the department of physics, was accustomed to the attitudes of Worcester's young students of engineering. Alert and witty, the brush–mustached Dr. Duff had earned his doctorate at the University of Edinburgh and brought to the Tech the stern requirements of European scholarship. Students looked to him for crisp, matter–of–fact answers, readily translated for the uses of industry.

Goddard, however, was always raising questions. Duff was fascinated by Goddard's ferret–like curiosity, took him on as his laboratory assistant, and recommended him as a tutor to help him pay his way. Goddard's drive for knowledge, Duff suspected, concealed more than it revealed. It was at least a match for the Professor's own astringencies.

A. Wilmer Duff sitting at his desk in 1907

As part of his undergraduate studies for a B.S. in physics, Goddard conducted research on the possibility of navigating interplanetary space. Writing a lengthy theme paper for an English class on October 3, 1907, he wrote: "The problem of navigating interplanetary space is a complex one, but with whatever solution there may be, certain definite physical laws must be followed… The discussion falls naturally into three divisions: the sustaining of life in space, the protection against accident during transit, and the means of propulsion." Regarding the latter, he stated: "The third and most fundamental consideration is that of propulsion…we can apply Newton's law of action and reaction, and the law of conservation of energy, and in consequence…the first law simply means that it is impossible to produce motion in one body without pushing against some other body, i.e., against matter or ether." Unaware that the 19th century concept of ether had been discredited by this time, Goddard said, "The ether does exert a resistance to the motion of a charged body, and one form of this resistance is manifest in radiation pressure—the pressure exerted by all ether radiations on the surfaces upon which they impinge."

Goddard also studied the deviation from Ohm's law in certain conductors. In November 1908 he wrote an article for the Journal of the Worcester Polytechnic Institute entitled "On Some Anomalous Electrical Conductors." He stated: "When an electromotive force, *emf*, is applied to the ends of a conductor, a current of electricity flows between the points of application, and under all ordinary conditions this current is directly proportional to the emf applied. This fact is known as Ohm's Law, and the ratio of the emf to the current is called the resistance of the conductor, the reciprocal of the resistance being the conductivity." Goddard observed some interesting exceptions to Ohm's Law, wherein the emf and the current are <u>not</u> proportional. In addition to the electric arc and certain heated substances called *pyroelectrolyte*s, Goddard experimented with dissimilar solid substances in a powdered state as anomalous electrical conductors which deviated from Ohm's Law. He wrote: "Three chief anomalies have been noticed: 1) a change of resistance with time under the application of a constant emf; 2) a lack of proportionality between emf and current at any time; and 3) a development of a difference between the resistance of the conductor in the direction of the original current and in the opposite direction, or an asymmetric conductivity." In other words, he discovered that rapid changes in applied voltage (emf) caused deviations from Ohm's Law in powdered conductors. "If, however, the powder itself was stirred or jarred, the conductivity, after the disturbance, was found to have risen. This showed that the effect was not a result of changes at the surface of the electrodes… It is difficult to explain the action of a vacuum unless the chains of particles are broken down in some way, as the conductivity of a coherer has been shown actually to increase with a vacuum."

Jumping ahead to 1912, then–Dr. Goddard conducted research with variable dielectrics and James Maxwell's so–called electrical "displacement current." On April 27, 1912, he read his paper, "On Mechanical Force from the Magnetic Field of a Displacement Current" to the spring meeting of the American Physical Society in Cambridge, Massachusetts. He continued his displacement current work for some months.

In a sense, Professor Goddard was a century ahead of his time. His research covered three phenomena: 1) deviations from Ohm's Law in conductors, 2) the effects of varying emf upon electrical conductivity in solid dielectrics, and 3) the electrical displacement currents of Maxwell. It is these very types of phenomena that some scientists and physicists the world over would study in the 21st century with hopes of uncovering how to do "field to field interactions" with Einstein's mysterious space–time continuum (one could call it a "21st century ether") and produce some kind of propulsive force without propellants. This topic will be covered in *The Never–Ending Frontier*.

At any rate, Goddard foresaw the possibilities of really fast space propulsion, propellantless or not. He presciently observed some years later that: "If a device that could exert a constant propelling force were to be employed, and the speed were increased during the first half of the journey [to Mars] and decreased during the other half (by reversing the rocket), the time could be considerably reduced. Thus, if the average velocity were 7 miles/sec (11.3 km/sec), the time would be reduced to seven weeks. The limit to the reduction of time is imposed by the nature of the device used to secure this reduction."

In June, 1909 Robert graduated from Tech. His graduation thesis was called, "On Some Peculiarities of Electrical Conductivity Exhibited by Powders and a Few Solid Substances." Professor Duff had encouraged the subject, seeing in it a sound approach to the development of radio. His student was apparently preoccupied with something, but whatever it was, he was unwilling to unburden himself. Perhaps Duff wondered about this as he considered his student's thesis. It was a well–informed paper. The lad really understood the principles of physics and how to apply them. He ought to be successful in radio engineering. *In radio,* Duff thought, *there was promise enough for any ordinary lifetime.*

Before his graduation at Tech, the thoughtful Goddard was already planning his post–graduate education. On March 29, 1909 he met with Dr. Webster, professor of physics at Clark University in Worcester, Massachusetts. Webster thought he could get a fellowship and a Master's degree in one year. Clark, founded in 1887, was among America's first graduate schools for science. The Worcester university was a bold experiment: to match foreign graduate schools by offering high academic standards. Goddard's association with the school had begun as an undergraduate in the fall of 1908, and would last many years as he pursued his dreams of manned flight with rockets. In the early 1900s, Americans were still streaming to the greater and lesser universities of Europe for advanced degrees. In fact, Dr. Duff had encouraged his bright student to go abroad for advanced studies after two years at Clark. At the time, Germany was preeminent in physics research and discoveries.

Arthur Gordon Webster had succeeded Professor Albert Abraham Michelson as the head of Clark University's physics department. Michelson had conducted fundamental research into the velocity of light and, through his studies, was able to determine distances to the stars. In 1887 he and Edward Williams Morley had conducted a groundbreaking experiment

The Magnetic Laboratory where Goddard conducted experiments. Photos inside show some of his experimental equipment. It was renamed the "Skull Tomb" in Goddard's honor in 1965.

which proved the constancy of the speed of light in all directions. In so doing they also disproved the "ether" theory. The ether was thought at the time to be a diaphanous all–encompassing medium permeating the entire universe and everything in it. Physicists theorized that the ether must exert a drag on the earth as it moved through it in opposite directions during its orbit around the sun. No drag. Therefore no ether. Michelson was the first American scientist to win the Nobel prize, in 1907. His work also formed a starting point for Einstein's later studies in relativity.

Professor Arthur Webster

Dr. Webster was an outstanding mathematical physicist in his own right, born of New England stock and trained in the classical German method under von Helmholtz at the University of Berlin. He authored basic textbooks on dynamics and electromagnetism. In the three decades that he taught at Clark, Webster saw fit to accept and prepare only 27 candidates for their doctorates. Among them was Goddard, who would someday succeed him.

The professor and his singular scholar were opposite numbers in both personality and approach. Webster was flamboyant in his views, wore an expansive white mustache, and was outspoken on virtually all subjects in all fields that attracted him. He successfully urged his accomplished students to publish their findings on the dire threat that if they didn't, he would do so himself. Unlike the reticent and introspective Goddard, he set little value on practical or applied science, regarding it as a low order of activity, not much better than engineering.

Robert matriculated into Clark University in September, 1909. As before, he wisely planned his doctoral work at the same institution while studying for a Master's degree under Dr. Webster. One day in April 1910, Dr. Webster concluded his formal discourse on classical electromagnetism via relativity with a nod. He went at once to his office, where students, if they wished, could seek out his assistance. Goddard sheepishly trailed behind a few students to Webster's office, and awaited his turn with his mentor. If there was any important flaw in his thinking, he looked to Webster to help him dispose of it. He softly knocked. "Enter, take a seat," said Webster peremptorily without looking up. The last student had put him into an ill humor, with his preoccupation about commerce and the value of his degree, instead of the ideal of comprehensive and exacting theoretical scholarship as in Europe. Goddard timidly took the closest seat across from Professor Webster's desk. Dr. Webster glanced up at his stellar student. His gaze softened. Goddard began, "Professor, I have tried and tried, but conventional rockets are too inefficient. If the rocket's energy could be more fully utilized, stepped up from 5 to 50 percent in efficiency, then it might really go places. I think…a liquid–fuel rocket should employ the perfect blend of liquid hydrogen, the ideal fuel, and liquid oxygen, the flawless oxidizer." He paused, then tentatively offered Dr. Webster a piece of paper. "In my efforts to create a better rocket and find adequate fuels to propel it, I have written down some suggestions for future experiments."

Webster stopped, reading intently. *"Try, if possible, an arrangement of hydrogen and oxygen explosive jets, with compressed gas in small tanks – which are subsequently shot off – giving perhaps 40 percent or 50 percent efficiency. To get even 50 percent efficiency, it will probably be necessary to have small explosive chambers and jets, into which the (not too violent) explosive is fed. They should also be small in number. Otherwise a large mass of metal will be needed for the large high–pressure chambers, which will cut down the efficiency per pound greatly…"* There followed a set of clearly delineated assumptions, and complex rocket equations. As usual, Goddard's calculations appeared flawless.

He put the paper down. "But neither liquid is available today that I am aware of," he retorted, his eyebrows furrowed. "Liquid hydrogen, which exists as a liquid at –253°C [–423°F] is definitely unmanageable and unobtainable today. And I don't know, isn't liquid oxygen available from Linde's air liquefaction process?" "Yes, sir," Goddard replied. "I did a search in the US Patent Office archives at our library, and found that U.S. Patent No. 727,650 was issued to Carl von Linde of Munich, Germany in 1903. It specifies how to liquefy air, obtaining liquid oxygen in the process." Professor Doctor Carl Paul Gottfried von Linde (1842–1934) was a German engineer who developed refrigeration and gas separation technologies. "Is it available in commercial quantities, and how much does it cost?" asked Dr. Webster. Goddard thought a minute. "Well, I sent a letter in December to Herr Doctor Linde in Munich. He replied that he had formed the Linde Air Products Company in the US in 1907. "Where is the US Linde company now?" queried Webster. "It is headquartered in Cleveland, Ohio. "You need to find out how or if liquid oxygen is available from their process, and most importantly how much it costs. I don't have a very ample research budget for rockets, you know." "I guess I'll have to go to Cleveland and find out," replied Robert.

Robert knocked softly on the unprepossessing front door. Rudolf Wucherer, Linde's son–in–law, answered the door, eyeing Goddard intently. "Helloo. Vhat may I do fur you?" "May I speak to Herr Carl Linde?" answered Goddard. "Hey is not here. He ist still in München. But cume in, cume in und make yourself confortable" responded Rudolf in heavily accented German. The cluttered room contained some chairs, two large desks stacked with valve parts and piles of papers, and one had to step over or around various half–built valve head assemblies and metal cylinders. Linde's two sons Friedrich and Richard stood up from behind their desks and approached to shake hands with their visitor.

"My name is Friedrich, and this is my brother Richard," offered Friedrich in much better understandable English. "Ve hav come here to expand our father's operations in the US. Sit se down and may I offer you some coffee?" "Why, yes, thank you," said Robert. Over coffee and tea, the brothers explained how Dr. Linde had in 1906 negotiated a stake in the Brin Oxygen Company in exchange for rights to his patents in the UK and other countries. Dr. Linde now held a board position with Brin. "Would you explain your new process for air liquefaction?" asked Robert. Friedrich answered, "Vell, my father's apparatus for the liquefaction of air combines the cooling effect achieved by allowing a compressed gas to expand with a counter–current heat exchange technique. Ve use the cold air produced by expansion to chill ambient air entering the apparatus. Over a period of time, deese combined effects gradually cool the apparatus and the air witin eet to the point of liquefaction."

"Ah, yes, you combine the Joule–Thomson effect with your new counter–current heat exchange invention," observed Dr. Goddard. James Prescott Joule and Lord William Thompson Kelvin had discovered that atmospheric air when discharged through a valve from a space under high pressure into a space maintained at lower pressure by causing the gas to pass off will have a lower temperature, the decrease in temperature being directly proportional to the decrease in pressure. "How do you contain this supercold liquefied air?" asked Goddard. "Vell, szee dees dewar flask here?" responded Richard. His proficiency in English seemed to hover midway between his brother Friedrich's and brother–in–law Rudolf's. Goddard took hold of the flask. It seemed a bit heavy. "You see, ve uuse dees dewar vacuum flasks to keep ze stuff cold. Zhey can hold cold für several hours," interjected Rudolf. Years before, Sir James Dewar (1824–1923) had invented a double–walled silvered glass or metal flask with a vacuum between the walls to provide thermal insulation. It worked quite well for the Linde Company.

Robert started to get the idea. "But, how can I get just pure oxygen from your process?" he asked. "Friedrich, could jew explain dhees?" queried Rudolf of his brother–in–law Friedrich. Friedrich responded, "in 1895 my fater succeeded in liquefying air by first compressing it and then letting it expand rapidly, thereby cooling it. He then obtained oxygen and nitrogen from the liquid air by slow warming. Ve plan to commercialize our product to metal welders here ünd in other countries who need the highest temperature possible to melt and join metals. Dis ees dzee new oxyacetylene process." In the early days of oxygen production the biggest use by far for the gas was the oxyacetylene torch, invented in France in 1904, which revolutionized metal cutting and welding in the construction of ships, skyscrapers, and other iron and steel structures. "But, how could I obtain liquid oxygen, not in gaseous form?" asked Robert. "I need cold liquid oxygen, by itself and as pure from impurities as possible." Friedrich stopped and pondered a minute. He turned his gaze to Rudolf, then Richard. "Vell, perhaps ve can see if we can bleed off zome of zhee gaseous oxygen during our production process, ünd reduce it to liquid form. I theenk ve can obtain a modest supply of oxygen in dees vhay forr you," said Richard. "And what might that cost me?" asked Robert. "Vell…" Richard thought. "Ve would propose a price of, say, $1.00

per liter. But zhou would have to provide orr obtain your own dewar flask, or thermos bottle, to keep ze stuff cold," responded Friedrich.

From Dr. Webster's highly polished lectures, Goddard eagerly mastered the fundamentals of physics, the nature of motion and matter, the principles of electricity and mechanics, of heat and light and sound. He noted in his diary, "I have learned the poetry of mathematics and science. The most wonderful thing in nature is the uniformity, that 2 + 2 = 4, and that all the atoms of the same element are alike, and the constant development towards higher perfection, even if temporal and at times uncertain…"

Goddard was appointed Senior Fellow at Clark University on May 7, 1910, and on June 16 he received a Master of Arts degree for his thesis, "Theory of Diffraction." It bore little relationship to his consuming interest in obtaining great altitudes, nor did his doctoral dissertation, again under Webster, which he began that fall of 1910 at Clark. It was entitled "On the Conduction of Electricity at Contacts of Dissimilar Solids."

A quick study in complex technical subjects, Goddard completed his doctoral thesis on May 30, 1911. His performance on the written examination on June 9 elicited accolades from his professors. Dr. John C. Hubbard telephoned, "very brilliant—one of the best they ever had." In his diary, Robert wrote "loved the way Webster questioned on Maxwell's equations. The next day, Robert passed his oral examination with flying colors. Webster and the doctoral committee were especially pleased. Under a barrage of questions, Goddard was lucid on every point. Dr. Webster told him, "You did yourself proud, and everybody connected with you. Was afraid you couldn't stand it, but you did, all right. We asked you a good deal we usually don't get around to, but it didn't take much time." Dr. Taber said "I was proud to know that one of my students could have such a grasp of his subject." Webster told Dr. Thomas L. Porter, "very spectacular examination." On June 15, 1911 Dr. Robert Goddard graduated from Clark, *cum laude* (the only one). He soon became a notable, if occasionally notorious, professor of physics at the University.

THE BUILDING YEARS

Madame Gram Goddard, Robert's grandmother, now in her 83rd year, lay dying of old age ills in her bedroom in the Goddard farmhouse on Maple Hill in Worcester. *She was not of the usual mold, in New England or elsewhere*, her grandson reflected as he watched her life ebbing on his daily visits to Maple Hill. Out of the clan of prosperous Yankee Goddards, she had happily elected to marry one of the least provident members. She had raised their only child, Nahum, and Nahum's only child, Robert, to carry on the struggle, blessing each for his special brand of courage. To the end she was especially proud of her grandson. "You know," she often said to anyone who would listen, "Robert is really superior. The world will hear more of him."\

Mary P. (Upham) Goddard, Robert's grandmother, 1910

She was alert to the end. Robert talked with her often as she rested, covered with quilts, while the housekeeper hovered nearby. He felt impelled to tell her all he believed in her fading days, how his rocket would one day offer mankind great danger but also enormous possibility. She listened gladly, accepting his hopes while helping him to dismiss his doubts, saying, "Well, Rob, we can just stop where we are." She died on October 17, one year to the day before the Communist October revolution in Russia, and was buried on October 19, 1916. On that same day in 1899, 17 years before, Robert sadly recollected how he had gone down through the unkempt field, past the empty barn to the brook, and climbed the cherry tree to dream of the future. Now he attended her funeral services in the farmhouse, and was the last to see her before the burial.

Robert recorded in his meticulously maintained dairy, "went over to Gram's and saw Gram in parlor. Walked up on hill to leaning oaks, pretty, oak deep red, maples red and orange….Saw cherry tree…." Robert meditated on his life–changing experience that afternoon of October 19, 17 years earlier.

Over the ensuring years after earning his doctorate, Robert Goddard relentlessly continued his rocket research, between classes, advising his students, and overseeing graduate research at Clark University. In 1914 he received two patents. One was the first for a rocket using solid and liquid fuel, and the other was for a multi–stage rocket. America's involvement in the World War I (called the "Great War" to end all wars) caused some disruptions in Goddard's tenure at Clark. He long suspected that his rocket concept had potentialities as an antiaircraft and bazooka–like antitank weapon, as a mobile substitute for artillery, or as a naval torpedo. But he also realized that specific and perhaps lengthy development would be required. After the Armistice was signed on November 11, 1918 (famously known as the 11th hour of the 11th day of the 11th month), Goddard returned to Worcester and his classes at Clark.

He appeared the model of a Massachusetts pedagogue. Now at 36, he wore clothes respectably aged and drab in color, casually well–fitted, and with the stiff white collar and conservative silk cravat of the period. As if to strike a balance with his baldness, he had grown a mustache which he kept close–cropped. In lecturing his students, he was enthusiastic, with occasional interjections of whimsy. He especially enjoyed demonstrating the principles of physics with apparatus he carefully arranged. After a particularly successful classroom demonstration, he sometimes underscored his effects by stepping to the window, staring out at the campus clock tower and whistling, "it's a long way to Tipperary."

On a bright spring day in March, 1919, Arthur Webster gently stroked his expansive, bushy white moustache. "Through your incisive lectures, you cleared my blurred impression of Newton's reaction principle," said Goddard demurely. Dr. Webster added, "According to the Third Law, nothing, absolutely nothing is needed by a reacting device to react against. Not gravity, nor magnetic fields, no even the air." Concurring, Goddard said, "that's why I think that my reaction devices, which

carry all their propellant, should be able to rise above the earth's atmosphere, and create thrust even in the vacuum of space." Both men then turned to the task at hand, and the enterprising Goddard began affixing an igniter to a cartridge filled with a mixture of charcoal and potassium nitrate. Then Webster held Goddard's rocket still while he drilled holes for four screws to secure the cartridge in place.

At last the homemade–looking solid–fueled rocket was ready for testing. Goddard cautioned, "Professor, I think you should back off a bit. You just never know about these things." Webster took a few steps back. Goddard pushed down the ignition switch. A loud bang ensued as the test rocket sputtered to life, producing a lively pyrotechnic display. The physics laboratory filled with acrid smoke. "Can't you turn the stupid thing off?" said the Professor, coughing. "No," replied Goddard, apologetically.

Professor Robert Hutchings Goddard

Robert had ingeniously worked out the kinematics of multiple stages, which makes it possible to loft heavier payloads. The purpose of staging rocket engines—where separate rocket engines fire in a sequence one right after another as each has depleted its fuel—is to throw away each stage as its fuel runs out and the stage becomes mere useless dead weight. After each stage separation, the rocket becomes lighter and it takes less fuel to move it. But Robert's work on multiple charge rockets continued to be plagued by failures. Goddard's assistant Nils Riffolt was impressed by Goddard's gift of resilience. "One thing would go wrong after another", he observed. "He'd come in with pencil sketches drawn on the backs of envelopes and we'd talk over ways to handle the problems. After a while, we knew we were running down rabbit trails, that a workable cartridge rocket was at least a long way off." Goddard's diary would faithfully record the results of each try. "Got apparatus set up on June 23: tried it on June 26…nozzle split. July 14: chamber cracked…July 17: cartridge jammed. August 7: tried multiple, fired four shots…August 8: tried multiple with parachute…it fell and magazine blew up." After many such frustrating failures, the indomitable Goddard would say, "These were not failures, but valuable negative information."

Goddard lecturing at Clark University, 1924

On September 11, 1919, Robert continued writing furiously, polishing his cover letter to Dr. Charles D. Walcott, Secretary of the Smithsonian Institution in Washington, DC. Pausing circumspectly, Goddard eyeballed his earlier treatise of August 21, 1914. "The Problem of Raising a Body to a Great Altitude above the Surface of the Earth." It read,

"The problem is concerned with the practicality of doing two things, namely: the raising of

apparatus, such as recording instruments, to a great altitude, and letting it fall back to the ground by suitable parachutes; and second, the sending of apparatus to such great distances from the earth that the apparatus comes under the influence of the gravitational attraction of some other heavenly body." Such as the moon, Goddard mused to himself.

Now he daydreamed. *How would it be to explore the solar system with rockets? At a scale of miles, 240,000 to the moon. Then onward, about 50 million to the planet Mars, to view the network of long straight canals discovered by [the Italian astronomer Giovanni] Schiaparelli during the opposition of 1877. Then on to Neptune, the last planet in our solar system... But I must expand my thinking to orbits around a different star. Change the scale of miles to one of light years—9 light years to the Dog Star, Sirius, then 900,000 to Andromeda's "nebula." Even at their best, my rocket ships could only travel about 250 miles a minute. At such speed, man would take less than a day to reach the moon, several months for a trip to Mars, and over 100,000 years to approach the nearest star. But suppose I could build a "spaceship" to travel at speeds approaching the speed of light? Even at the velocity of light, which flashes from the earth to the moon while two human steps are taken, and which scientists claim will remain unattainable by man, the extent of a human lifetime would limit us to nearby areas of space.*

Thus did Robert dream of impossible future conquests, as men have dreamed throughout the ages. *Will new, scientifically established obstacles, unlike the angry gods, storm waves, and airplane flight of generations past, place a final limit to the achievement of mankind? Or may we eventually discover that limits are relative to years, that dreams and reality transpose with time, as do Einstein's energy and matter? If only I could break the speed of light barrier. Curving space, and slowing time, perhaps?* Goddard's imagination carried man through galaxies. His life would keep us aware of the human challenge to all limits.

Back to the task at hand. *The rocket must expel jets of gas*, he thought. *In order to maximize the kinetic energy of the propellants exhausted out of the rocket nozzle, and thereby maximize the necessary reaction forward—by Newton's axiom that for every action there is an equal and opposite reaction—I must use as light a weight as possible of gas. Of course, hydrogen as the rocket fuel.* Robert set down his pen, pondering, his brow furrowed. *The greatest energy available per pound is from chemical propell*ants. *Hydrogen molecules come in pairs, as H_2. For combustion, this requires an oxidizer which is lacking two electrons in its outer shell, thereby combusting with the hydrogen molecules which have two donor electrons in their outer shell. A glance at Mendeleev's periodic table of the elements tells me that a single molecule of oxygen is the ideal oxidizing element.* He recalled his April 1910 discussion with Dr. Webster to this effect.

He quickly picked up his pen. "By the application of several new principles, the writer is certain that an efficiency of 60 percent or more may be attained, the estimation being made as though the entire rocket consisted of the burned, or exploded, material. Inasmuch as these principles are of some practical value for military purposes, the writer has protected himself by patents covering these points, specifications of which are appended to the present discussion."

Goddard eventually achieved an overall efficiency of over 63 percent in his static rockets, "the highest ever obtained from a heat engine," he wrote. The best reciprocating steam engine, he observed, gave only 21 percent efficiency and the best internal combustion engine some 40 percent. His rockets reached a "velocity of ejection" of 7000 feet/sec (2134 m/sec) or better. With this velocity, he added, the rocket motor could lift a one–pound payload to 35 miles with a starting weight of only 89.6 pounds. Goddard's computations were theoretically precise. They merely assumed an ideal rocket, which decades of development to date have never attained. But his theory, as such, was impressive and sound, and his optimism was irrepressible.

He wrote on: "The principles concerning efficiency, described in the specifications, are essentially three in number. The first concerns efficiency, and is the use of a nozzle, of proper length and taper, through which the gaseous products of combustion are discharged. By this means the work of expansion of gases is converted into kinetic energy, and also complete combustion is ensured."

"The other two principles concern the increase of efficiency which results from making the ratio of the mass of the burned substance to the total mass as large as possible. The second principle is embodied in a reloading device, whereby a large mass of the burned material is used, a little at a time, in a small, strong combustion chamber. The third principle consists in the employment of a primary and secondary rocket apparatus, the secondary—a copy in miniature of the primary—being fired when the primary has reached the upper limit of its flight."

Robert meticulously spent the rest of the afternoon and well into the evening meticulously going over every equation, the wording, the justification behind every assumption or simplification in his treatise. He unwittingly set the standard for all future rocket equation calculations in the 20th Century, down to endowing the letter R as representing the force due to air resistance, and the letter c as the velocity of ejection of the mass expelled. Goddard correctly determined that the burned propellant exhaust velocity c is almost a constant for an ideal rocket and nozzle, because well–designed rocket nozzles are almost 'ideal.' He wrote, "Also, in the following calculations, it will be assumed the thermodynamic efficiency is 100 percent; i.e., all the 50 percent of the heat energy available for propulsion is transformed into velocity of the mass ejected. This assumption will not lead to serious error, for the reason that the temperature of burning of both propellants is very high as compared with the temperature of the atmosphere." At very high temperatures the chemical combustion process approaches 100 percent thermodynamic efficiency (almost 50% overall efficiency) in transforming the total heat energy of the propellants into kinetic energy to move the rocket forward.

Finally, carefully and painstakingly he carefully bound his manuscript, a revision of the statement originally written on August 21, 1914. He titled it, "A Method of Reaching Extreme Altitudes." His new, solid propellant device would find use for daily observations at moderate altitudes. He also discussed the possible attainment of the most extreme altitudes. Moreover, he correctly suggested that with a velocity of 6.95 miles/second, without air resistance, an object could escape Earth's gravity and head into infinity, or to other celestial bodies. This would become known as the Earth's 'escape velocity.' The impeccable manuscript was bound in leather, with marbleized end papers, and its title embossed in gold.

SMITHSONIAN MISCELLANEOUS COLLECTIONS
VOLUME 71, NUMBER 2

A METHOD OF REACHING EXTREME
ALTITUDES

(WITH 10 PLATES)

BY
ROBERT H. GODDARD
Clark College, Worcester, Mass.

(PUBLICATION 2540)

CITY OF WASHINGTON
PUBLISHED BY THE SMITHSONIAN INSTITUTION
1919

On the manuscript's arrival at the Smithsonian, Secretary Walcott turned it over to Dr. Charles G. Abbot, the Assistant Secretary and Walcott's later successor. Dr. Abbot was a tall, mustached, distinguished–looking astrophysicist who was engaged in exploring solar radiation. "Within an hour," Abbott said later, "I enthusiastically recommended it as the best presentation of a research project I had ever seen." Goddard had established a firm theoretical base that rockets could be used to explore the upper atmosphere. Abbot advised his superior on October 2, 1919, "Goddard claims the possibility of reaching several hundred miles altitude!" His funding solicitation would lead to a stipend of $5000 from the Smithsonian, spread out over five payments of $1000 each. A princely sum in those days.

On January 3, 1920, Goddard received his author's copies of the 69–page monograph. The first edition of 1,750 copies had been issued in the Smithsonian Miscellaneous Collections, Volume 71, Number 2, 1919. For eight days, there was no mention of Goddard's report in the press. But on Monday, January 12, 1920, the Worcester professor and his theory made front page news across the nation. The headlines all registered the same astounded note:

Modern Jules Verne Invents Rocket to Reach Moon
Aim to Reach Moon with New Rocket
Claim Moon May Soon Be Reached
Savant Invents Rocket Which Will Hit Moon

Not only did these New Year's headlines invade the privacy of his dream, but henceforth he would often be greeted obnoxiously across campus with, "Dr. Goddard, how goes your Moon rocket?" The general public ignored the scientific merit of the paper — latching instead onto Goddard's Moon rocket proposal. At the time, such an endeavor was absurd and most dismissed Goddard as a "crank." The experience taught Goddard a hard lesson— one which caused him to shy away from future opportunities to publicize his work.

In late August, 1920, a solid–fueled rocket became dislodged and fell, naturally exploding the magazine, which contained nine cartridges. Goddard and Riffolt gathered up the pieces and a few days later tried again. This time they supported the rocket in a vertical position by three threads fastened to posts just outside the launching frame. On firing, four of the cartridges successfully entered the combustion chamber, exploded and carried the rocket to its highest altitude—60 feet (18.3m) off the ground. When the fifth cartridge failed to fire, the rocket crashed beyond repair. Goddard's commentary was as bland as ever. "The fifth tappet had broken away from

Charles G. Abbot

the corresponding feeding springs," he wrote. The parachute failed to open as well. Characteristically, he pointed out improvements to be considered. But after January, 1921, he made no further attempts with the cartridge rocket.

One advantage of using smokeless powder was that the fuel contained its own source of oxygen, indispensable in the emptiness of space. But there were notable disadvantages. The system's intermittent and explosive thrust made the rocket almost impossible to control. Ideally, he needed a smooth, continuous thrust of power; a combustion, rather than a jarring explosion, of the fuels; and a very high velocity of the escaping gases. It would be decades before solid fuels would be developed to meet such requirements.

After years of research and experimentation with dogged determination, Goddard had come down to earth as resoundingly as his ill–starred solid rockets. Goddard returned to his theorizing. His new approach, after so many failures, led to liquid–fueled rockets as he had suggested to Dr. Webster as a graduate student back in 1910. Though he had made vast improvements over the old Chinese solid rockets, he realized that liquid propellants could provide more energy for propulsion than an equal weight of gunpowder or other available solid fuels. Besides producing greater thrust per kilogram of fuel, one of the great advantages of liquid fuel is that it can be controlled, whereas solid fuel burns to completion once ignited. The primary source of fuels was hydrogen; of oxidizers, oxygen. Goddard's calculations showed that a tremendous thrust was possible by combusting the two elements.

"Oxygen," he wrote, "is the most obvious liquid to use with a liquid fuel because it gives complete combustion." He added dryly that there were "difficult problems involved in using the very cold oxygen." It boiled at $-183°C$ ($-297°F$), had to be kept below this temperature, and was subject to other severe difficulties of handling. Increasing the pressure increased its boiling point, but not appreciably for the modest pressure vessels available for containing cryogens at the time. Goddard's venture with liquid oxygen was a step of faith into a wilderness. Liquid oxygen was a key to future rocket development, an indispensable substance for the famous (or infamous for the Allies) German V–2 missile in World War II and subsequent high–altitude missiles. One day rocket men would refer familiarly to "LOX," but in 1920 there was nothing familiar about it.

The situation with liquid hydrogen was even worse. It was still unmanageable and unobtainable in the 1920s, but there were substitute fuels such as propane, ether, alcohol,

gasoline—a whole series of hydrocarbons. Goddard experimented with most of these fuels before settling on gasoline as cheap, easily obtainable and dependable.

What Goddard called the waist–high "rocket engine" was just a combustion chamber with a rocket nozzle underneath it. The liquid oxygen (LOX) and fuel (gasoline) flowed up to the top of the combustion chamber from tanks at the bottom of the rocket through two asbestos–wrapped aluminum tubes. An asbestos–coated cone protected the fuel tanks, located *underneath* the rocket engine. A horizontal metal tube held the nozzle in place and the two propellant tubes at its ends. The two propellant tubes then continued vertically downward for the distance of the height of a normal man, at which point on the downward side they proceeded inwards towards the fuel tank (a larger cylinder about 61 centimeters in length) and the oxidizer tank (a thinner cylinder of the same length). Liquid oxygen flowed from the bottom of the LOX tank and gasoline from the fuel tank. Finally at the top they were attached to the combustion chamber. A big part of knowing what makes rocket engines tick is to learn about the propellants required for their operation, especially since 80%-90% of a rocket vehicle's weight is taken up by propellants.

Preliminary Tests

Dr. Goddard worked with his technician Henry Sachs to tighten the fittings on the 12 pound (5.4 kg) rocket's static test fixture. "Would you hand me the camera, please, so that I can take pictures of the rocket in its final form?" he asked. This model included pumps, and operating and governing devices, as he had mentioned in the letter to the Smithsonian of October 5, 1925. "Too bad," said Henry as Dr. Goddard steadied the camera. "The two separate engine system arrangement actually involved less work of construction." "But we found them next to impossible to work together. No dice," Goddard retorted. *This double–acting engine has to work*, he reticently thought.

After Goddard took several pictures, Henry quickly checked over the gauges and pipes to ensure everything was ready for the static test. Slowly at first, then Shhh– WHOOSH! the sound built up and held steady as the rocket strained against its constraints. It was actually lifting itself, its entire weight off the ground! Not quite a foot, the limit of its holding frame. It slowly bounced up and down in a tug–of–war against gravity. "Hooray!" both men exclaimed. This model operated perfectly for 27 seconds, until the liquid propellants were used up. Dr.

Goddard had been worried that they were reusing the same thrust chamber and nozzle as in a number of previous tests, with various governing devices. Within a minute, Henry and the Professor examined the chamber and nozzle. They were uniformly and only moderately heated, a good sign. In fact, the rocket was in as good a condition after the run as before.

This December 6, 1925 test was an important one, in that it demonstrated that the problems of pumping, governing, and control of heating were solved, but it also showed that a rocket on so small a scale as this model would not lift itself sufficiently to give a flight, including, as it did, devices which would not be necessary in a larger rocket.

Below: Static test stand that Goddard and Sachs used on December 6, 1925. Goddard noted that "This was the first test in which a liquid–propelled rocket operated satisfactorily and lifted its own weight."

Although the problem of satisfactory operation was solved, it was not possible to have a flight unless a larger model were made, or a different principle permitting greater lightness were employed. Nevertheless, to save money Goddard and Sachs did not undertake the construction of a larger model. They used another plan of construction, which involved back pressure in the supply tanks, and eliminated the pumps, engines, and most of the moving parts, but did not permit such accurate control of the combustion, nor the use of the lightest form of supply tanks, as did the December model. Goddard was worried that the lack of control of combustion manifested itself in the frequent burning out of the nozzles.

He was understandably impatient. Fifteen years of research was long enough. The combustion chamber would be on top, with the nozzle attached below. The motor at the top would have to pull the rocket. The fuel tanks directly below the nozzle would have to be protected by an asbestos

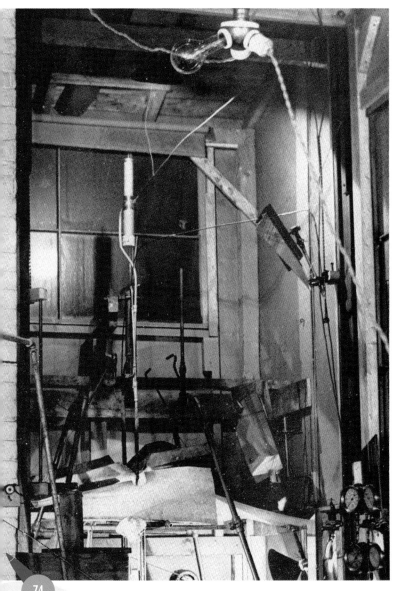

Goddard and larger liquid–fuel rocket, December 1926.

cone. Asbestos–wrapped aluminum tubes connect the motor to the tanks, providing both support and fuel transport. Using the same static test stand in a structure adjoining the Clark University physics laboratory, Goddard and Sachs conducted a test of the revised configuration on January 20. They measured a pull throughout the test of more than 9 pounds (40 Newtons). "OK, I am encouraged," Robert told Henry with a grin on his face. "Let's make a similar model for flight, of as light a construction as possible. Maybe we can have it ready for a flight test by mid–March."

PRIMER ON ROCKET BASICS AND TERMINOLOGY

- The very word "rocket" is a misnomer because it doesn't distinguish between the entire vehicle and the engine by itself. It is better to use the words "missile", "vehicle", "spacecraft" for the entire system; and to associate the word "rocket" with the engine or motor that makes it move.

- A rocket engine is driven by combustion of liquid propellants. These are classed as monopropellants (like hydrazine) or bipropellants (like LOX and gasoline). The term "fuel" is also a misnomer and confusing. Fuel is properly used for monopropellant engines only. This type of engine is smaller (i.e., used as a thruster or an orbital maneuvering engine) and never used where big thrust is needed (like launching boosters off the ground). It takes larger bipropellant engines to produce the high thrust levels needed to launch vehicles, and we'll mostly discuss this type in the book. Using the term "fuel" for bipropellants is also confusing, because you don't know which propellant you're talking about, the fuel, oxidizer, or both? In order to provide combustion, you have to have both an oxidizer (i.e., LOX or nitrogen tetroxide) and a fuel (i.e., liquid hydrogen, kerosene, hydrazine, or hydrazine compound). They are totally different, in regards to everything: different tanks, pumps, characteristics, pressurization systems, density, energy content, chemical composition, look, feel, smell, and taste. Taste? You better not.

  - There's a big difference between storable and cryogenic liquid propellants. As their name implies, the former can be stored for long periods of time (months, years) at 'ambient temperatures' (say –12°C or 11°F) while the latter has to be kept supercold (say –240°C or –400°F)

  - The term "mixture ratio" signifies the ratio by mass of oxidizer to fuel as both are sprayed into the rocket combustion chamber. At an optimum oxidizer to fuel mixture ratio of 2.56, a LOX/kerosene rocket of today generates a temperature of combustion of 3,670K° (6146°F), much hotter than a blowtorch.

- A rocket motor is driven by combustion of solid propellants. Solid propellants are blends of flammable compounds such as ammonium perchlorate. We'll address solid propellants in other chapters.

Oxidizers

Cryogenic liquid oxygen (LO_2) was the earliest, cheapest, safest, and eventually the preferred oxidizer for large space launchers. High performance cryogenic bipropellant rockets commonly use LOX as oxidizer. Liquid oxygen, as normally supplied, is of 99.5 percent purity and is covered in the United States by Military Specification MIL–P–25508. High purity liquid oxygen has a light blue color and is transparent. It has no characteristic odor. Liquid oxygen does not burn, but will support combustion vigorously. The liquid is stable; however, mixtures of fuel and liquid oxygen are shock–sensitive. Gaseous oxygen is dangerous on the test stand and elsewhere, because it can form mixtures with fuel vapors that can be exploded by static electricity, electric spark, or flame. Liquid oxygen is obtained from air by fractional distillation. As an aside, the cost of liquid oxygen in large quantities varies according to the energy cost of liquefying the air, in 2013 a typical number was 20 cents/kg or 84 cents/gallon.

Another common oxidizer is the storable nitrogen tetroxide (N_2O_4), nicknamed "NTO." NTO by itself is actually not a very good oxidizer for military applications which have to withstand temperature extremes. It freezes at –11.2°C (12°F). At sea-level pressure it boils at 22°C (71°F). If confined in a propellant tank, it can stay liquid but begins to exert increasing vapor pressure as the temperature rises. For this reason, propellant chemists often add varying proportions of nitric oxide, and the resulting oxidizer is called "mixed oxides of nitrogen" (MON). The number included in the description indicates the percentage of nitric oxide by weight, e.g., MON–3, MON–10, MON–25 contain 3%, 10%, 25% nitric oxide. Chemists often add small amounts of other additives like iron (Fe) as well to improve its properties as an oxidizer.

The term nitrogen tetroxide (N_2O_4) is used herein, but it actually refers to N_2O_4 mixed with varying proportions of nitric oxides and other additives by propellant chemists. The chemistry can get pretty complicated, so that even rocket engineers simply call it "nitrogen tetroxide (N_2O_4)" and we'll adopt that nomenclature here.

The family of nitric acids are often used as oxidizers. In the United States, the nitric acid formulation most commonly used is type III-A, called inhibited red-fuming nitric acid (IRFNA), which consists of HNO_3 + 14% N_2O_4 + 1.5-2.5% H_2O + 0.6% HF (added as a corrosion inhibitor). There are many other kinds of oxidizers, but these are the main ones.

Fuels

At first, experimental rocket engines after Goddard used any number of gasoline–like rocket fuels derived from petroleum, such as kerosene, diesel oil, or even paint thinner. However, as rocket engine technology improved, additional restrictions had to be placed on the exact composition and purity of fuels that would work. A 75% ethyl alcohol–25% water mix was common in early missiles like the V–2 and Redstone MRBMs. The most commonly used fuels today include cryogenic liquid hydrogen (LH_2), kerosene, and hydrazines. Liquid hydrogen is fairly straightforward. Like oxygen, it has to be kept in liquid form to feed through injectors into the combustion chamber. Hydrogen vapors are even more flammable and explosive than gaseous oxygen. Both have to be safely vented away from rocket operations at all times. Costs of LH_2 are much higher than liquid oxygen. Like LOX it varies according to the energy cost of producing it (while LOX is produced through an air liquefaction process, LH_2 is produced from the steam reforming of natural gas and must be compressed and cooled to much lower temperatures). In 2013 a typical number was $4.60/kg or $1.22 per gallon.

The kerosene that rockets use is not the same as the kind you get for camping. Rockets have to use a special grade of kerosene suitable for rocket engines. Petroleum–based fuels vary greatly in physical properties, so stringent, tightly controlled specifications had to be developed to govern kerosene's use as a rocket fuel. By 1954 a standard U.S. kerosene rocket fuel—Rocket Propellant grade RP–1—had emerged and was defined in Military Specification MIL–R–25576.

Rocket propellant RP–1 is a straight–run kerosene fraction, which is subjected to further treatment, including acid washing and sulphur dioxide extraction. In 2013 RP–1 cost about 55 cents/kg, or about $1.92/gallon. Jets use a different formulation of kerosene; common types are Jet–A (U.S. only), Jet A–1, JP–5, JP–7 and JP–8.

Also common are storable fuels derived from hydrazine. The problem with hydrazine by itself is that it freezes at 2°C (35°F). It doesn't boil until 114°C (237°F), so vaporization is not a problem. Variants of hydrazine like MonoMethyl Hydrazine (MMH), $H_2NN(CH_3)_2$ or Unsymmetrical DiMethyl Hydrazine (UDMH), $CH_3(NH)NH_2$ are used in rockets to overcome the freezing problem. Also commonly used are blended fuels, such as Aerozine 50 (or "50-50"), which is a mixture of 50% UDMH and 50% hydrazine. Propellant chemists often blend additives of various sorts to these blends to suit particular purposes. Aside from rocket chemistry specialists, propellant terminology is usually shortened to the dominant species: "NTO", "MMH", "UDMH", etc. We'll adopt that practice as well, and use the terms "kerosene" and "RP-1" interchangeably.

Goddard's combustion chamber was welded to a "de Laval" nozzle. Goddard realized that the products of combustion of gaseous, liquid, or solid propellants in rockets must be ejected from the combustion chamber at a high velocity in order to produce thrust, much like a kid's blown–up balloon flies through the air when you release it. Borrowing from the Swedish scientist, engineer, and inventor Carl Gustaf Patrik de Laval (1845–1913), Goddard had installed de Laval's patented nozzle to the combustion chamber so that the exhaust gases would exit at high speed, producing greater thrust. De Laval pioneered the development of high speed turbines. He built his first impulse steam turbine in 1882. Further advances followed, and in 1893 he built and operated a reversible turbine for marine use. A Laval reaction turbine (patented in 1883) attained a speed of 42,000 revolutions per minute. Laval continued improving his turbine until by 1896 he was operating a complete power plant using an initial steam pressure of 23.44 megapascals (3,400 psi). He invented and developed the divergent nozzle used to deliver steam to the turbine blades. His flexible shaft, used to eliminate wobbling which can be dangerous at high speeds, and his special double–helical gear formed the foundation for most of the steam–turbine development that followed.

Goddard's knowledge of the kinetics of chemical processes in the nozzle was essential to determine the thrust required. Rocket thrust decreases with the increasing mean molecular weight of the combustion products. Mixtures of low molecular weight and high heat of combustion, therefore, are used for rockets. The outer part of the nozzle was covered with a large number of small metal pipes, through which cold liquid oxygen, or oxidizer, would flow on its way to the combustion chamber. The high temperature of combustion in pure oxygen required heat resistant materials, and to help overcome this Goddard pioneered the technique of having the LOX itself cool the combustion chamber on its way from the oxidizer tank. A liquid fuel rocket also needed fuel in order to combust with the oxygen in an exothermic chemical reaction, that is one that produces heat. But this was no ordinary heat, rather a temperature of combustion of thousands of degrees, in fact about half the heat on the surface of the sun. Goddard's rockets, and other rockets of today, required two lines running into the combustion chamber, one feeding fuel, the other oxygen, similar to the way a steel–cutting blowtorch operated, except here both lines carried liquids, not gases. Goddard's design used gasoline as fuel, and liquid oxygen as oxidizer.

de Laval Nozzle

Most rockets today depend on the de Laval nozzle. Hot gases under extremely high pressure (about the same as a scuba tank) enter the converging nozzle section. They reach the speed of sound at the throat and then exit the nozzle at supersonic speeds. It sounds counter–intuitive that the gases' velocity increases as they are squeezed through the throat and then keeps increasing in speed as the nozzle cross section gets bigger, but that is how nature works.

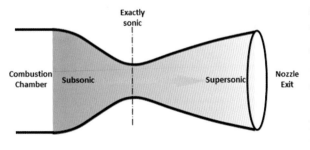

Exactly sonic

Combustion Chamber — Subsonic — Supersonic — Nozzle Exit

Chemical Combustion Physics

The energy source most useful to rocket propulsion is chemical combustion. For liquid propellants, the energy from a high–pressure combustion reaction of propellant chemicals, usually a fuel and an oxidizing chemical, permits the heating of reaction product gases to very high temperatures (2500–4100 °C, or 4500–7400°F). Solid propellants also use an oxidizer and a fuel for combustion, but in a different form. In both cases the exhaust gases are subsequently expanded in a nozzle and accelerated to high velocities (1800–4300 m/sec or 5900–14,000 ft/sec). These velocities are very high, from 5 to 13 times the speed of sound, or around 3 times the speed of a high–powered rifle bullet. A typical rocket is only able to use around half of the chemical combustion energy for propulsion. First off, there is about a 1% loss in the combustion chamber itself due to incomplete burning and mixing of the propellants. Then another 2% or so of the heat is lost to the chamber walls. A well–designed de Laval nozzle can turn almost all of the remaining combustion energy into kinetic energy in the rocket's exhaust jet. However, a rocket can only use a portion of the energy in the exhaust jet for propulsion, and this is where the biggest loss comes in. This is the propulsion efficiency, and it is a bugaboo. Only if the rocket happens to be moving forward at exactly the velocity of the gases coming out of the nozzle can it use most of the kinetic energy in the exhaust jet for propulsion. In this case the residual kinetic energy and absolute velocity of the exhaust jet are zero and the exhaust gases stand still in space. But that is almost never the case. This is just the nature of the beast. Here's why.

When a launch vehicle like the Space Shuttle is still sitting on the ground with all of its engines going, its propulsion efficiency is zero! It is not moving forward (upward) at all. As it takes off, the relative difference in speed between its nozzle exhaust gases and the ambient atmosphere starts to decrease. So more and more of the energy is useful for propulsion. Then as the Shuttle approaches hypersonic speeds around 8 times the speed of sound, the difference is practically zero and its propulsion efficiency approaches 97% or so. If you were somehow suspended in the air right next to the Shuttle as it flew by, you would not feel any wind from the exhaust gases coming out of the nozzle, but they would still be quite hot. This residual energy accounts for most of the 3% loss, the rest is heat lost to the walls in the chamber and nozzle, and a little combustion loss from poor mixing and incomplete combustion. The exhaust jet heat quickly dissipates in the thin atmosphere. Then as the Shuttle gains speed on its way to orbit, it is going too fast to use all of the kinetic energy in the exhaust jet stream because in effect it is having to drag the jet behind it at the difference in speed between its own velocity and the exhaust jet's velocity. So the propulsion efficiency varies from a little over zero to about 97% and back down again. If you average them out over the whole trajectory they amount to maybe 55% or so.

But not all this energy is useful energy for propulsion. The propellants have to drive the turbines to turn the turbopumps, provide cooling fluids to the chamber and nozzle walls, provide electrical power, and so on. So a typical rocket only uses around 50% (half) of the energy available for propulsion to move it forward. Nevertheless, this is way more efficient than your car, which only manages to use about 10% of the available energy in the gasoline/air mix to move itself forward, by the time you account for losses in the pistons, drive train, wheels, road friction, and so on.

OF ROMANCE AND MARRIAGE

Miriam P. Olmstead was, like Robert, an honor student at South High School in Worcester. As graduating seniors, both were selected by Miss Marrietta Knight, Rob's English teacher, to give speeches at the commencement ceremony on June 24, 1904. The students rehearsed and were well prepared for their talks, but not in the least for each other. Miriam's talk, entitled, "The Wealth of One Field," described the wealth of nature to be found within the compass of a single New England field. She declaimed first. She was a shy, luminous young lady, with the beauty of some creature from the Italian Renaissance. Or so Robert thought as he listened. He almost forgot his speech.

When Robert arose, Miriam grew equally fervent. She listened breathlessly to his discourse, "On Taking Things for Granted." It seemed to him an abysmal folly to take anything for granted in an age of magnificent possibilities. He concluded his oration with a summary of his future life's work: "Each must remember that no one can predict to what heights of wealth, fame, or usefulness he may rise until he has honestly endeavored, and he should derive courage

from the fact that all sciences have been, at some time, in the same condition as he, and that it has often proved true that *the dream of yesterday is the hope of today and the reality of tomorrow.*" One wonders what the rocket pioneer was dreaming and hoping. Through his longstanding dedication to rocket research, the dream of yesterday has indeed become the reality of tomorrow.

On meeting Miriam, he felt stirrings of young manhood, unfamiliar and long delayed. Science had ill prepared him for this encounter with Miriam. He wondered why he had never noticed her before. There were lost years to make up. For a time, his dream of space was blurred.

It began like a spring idyll. On a warm June evening in 1904, when South High's senior class held its graduation ceremonies, they had their first date, the valedictorian and the salutatorian of the class. He escorted her by trolley to the embellished auditorium, stopping in the town square to buy a bouquet of red roses. The classmates, Miriam recalled, were "surprised and pleased to see them together." That night Robert made a most unscientific entry in his diary: "Miri…Magnificent time!!!" He transformed Miriam into "Miri," the name of his love.

Courtship followed that summer, grave and measured in New England style. Robert and Miri walked in Hadwen Park, an expanse of greenery enclosing trails, a pond and the bandstand where his grandfather had played, overlooking the city's cemeteries. On these adventures, they were capably chaperoned by Miriam's mother, who sat sewing under a tree while he led Miri to a cloistered spot where South High's biology class had gone bird watching. There, holding her hand, Robert said he hoped they would become "more than friends." It was, in the accustomed ritual, "an understanding."

The understanding led to a curious and tentative romance. Like Robert, Miriam had pronounced ideas of her own. She, too, had grown up near sickness. Her late father, Dr. James Olmstead, Jr., had been superintendent of the Connecticut Hospital for the Insane and had settled his family in a house on the hospital grounds.

While Goddard was drawn to the hard science of physics, Miriam was fascinated by biology and metaphysics. She introduced him to the hidden labyrinths of Browning and Wordsworth, Thoreau and Whitman. The genii of physics and metaphysics looked on dubiously. While Robert considered his sparkling Miriam, his vision of space lost some of its sheen. "My own dream did not look very rosy," he recalled at the end of his high school years. "I had on hand a set of models which would not work, and a set of suggestions which I had learned enough physics to know were erroneous." But Robert was incessantly persevering. He would go on to say, "For even though I reasoned with myself that the thing was impossible, there was something inside me which simply would not stop working."

Miriam posed new and unexpected hazards to Robert. She urged on him such works as Ralph Waldo Trine's *In Tune with the Infinite*; while he preferred to reread H.G. Well's *The War of the Worlds*, which he considered less elusive. She also demonstrated an unfamiliar quality, a mystical outlook he could detect neither in his mother nor in his grandmother. Miriam's unfamiliarity made him apprehensive. Nevertheless, Robert accepted that Miriam expected marriage to follow college, although he tried to persuade her that first he must earn his doctorate, then put aside $1,000 as a reserve, and then get on with his research. "After all," he observed, "a man who would hold himself back from marriage is in the same class with him who runs away from battle."

At the moment, he had no intention of running away from battle. That fall of 1904 Miriam enrolled as a biology major at the well–respected Smith College for women in Northampton. Concerned that he would have to eventually provide for a family, Robert decided on a "practical education, from an economic standpoint," at the Worcester Polytechnic Institute.

Even for an inventive man, the affair with Miriam was traveling at frightening speed. In almost daily courtship by mail and in visits between Worcester and Smith College, they exchanged such tender gifts as the poems of Browning for a gold–plated locket. Her mother and his parents had quietly approved of an eventual marriage, after a proper courtship. But the courtship ran on and Miriam grew restless. "Five years is a long time," she told him, a distressing comment which he recorded in his diary. During the Thanksgiving holidays of November 1905, in the parlor of his grandmother's house, Robert gave Miriam a small diamond ring. They were engaged. His family, after a respectful wait, came in to welcome her and admire her trophy, which they had already examined. Apparently discomfited, he would not resume his daily record for almost a month.

The following summer witnessed the high tide of their romance. Miriam arranged to study German and board with an erudite pastor in Chesterfield, New Hampshire. Robert took quarters down the road from the parsonage. After Miriam's lessons, they spent long afternoons walking, talking, and riding through the countryside in the pastor's surrey. "Happiest two weeks at a stretch I ever had," he wrote that summer, remembering her "in evening near yellow lamp, her dress blue and white lace, with the fire–light shining on her…" One day, Miriam spoke of a short story she was reading in German. "Alles Hat Ein Ende." Perhaps she was warning him. At the time he could see no reason why love, too, might end.

But end it did. Four years had passed since they met at Mechanic's Hall in South High. Absence from each other for extended periods put a strain on their relationship. There was little hope left for the affair when Robert and Miriam were graduated from college. After graduation from Tech, Robert set off for Northampton and Miriam. There, they talked through earnest and agonizing evenings before he added this entry: "Our minds have grown apart in the last four years, and it will take time to bring them together." However, Miriam guessed, perhaps correctly, that she had found a master at delaying tactics. She left that summer of 1909 for a chaperoned tour of Europe. When she returned that autumn, Robert sensed a new conviction about her, a sureness and independence of spirit. She was pursuing her own dreams, in comparative anatomy and biology. She didn't much care for Worcester and was only mildly interested in Robert's rocket activities and his dream. She accepted a position in New York City at the research laboratory of the Department of Health. When Miriam took a train for New York, to pursue her own career, Robert knew that it was over. "The years forever fashion new dreams when old ones go," he had written in his diary. It was more a question than a statement. The space pioneer, feeling his old isolation, then observed, "God pity a one–dream man!"

Almost a decade later, Goddard met a tall, blue–eyed young secretary in the president's office at Clark University, Esther Christine Kisk. He found her shy, intelligent, and attractive. Esther Kisk agreed to type his papers after school hours and he began to call at her home more frequently than the stenographic work required. Miss Kisk, when Goddard met her, was much younger at 17 years old, the forthright daughter of Swedish immigrants. In her quiet determination, she resembled him strikingly. She had frugally saved her Clark salary to enroll in Bates College at Lewiston, Maine, where she was repeatedly elected to class offices. Goddard sent her a succession of letters and elaborate boxes of candy, called her his "golden girl" and fondly told her she was "puppy–like, enjoying everybody and everything." On vacations and holidays the persistent professor called at the Kisks', taking the fair–haired young student on countryside drives in his Oakland coupé. Occasionally he invited her to dinner at his home. Nahum Goddard would study her quizzically, wondering whether she could deal with the unpredictable problems his son presented.

The lovely Esther Christine Kisk, 1917

"Hi, Dr. Webster," said Carrie Dolby, a sophomore engineering student at a time when very few women were entering that field. "Isn't this just a splendid spring day!" she exclaimed. "Hmmm…yes," absentmindedly grunted the Professor as he hurried on. The head of the physics department at Clark, Arthur G. Webster, was on his way to his cluttered and comfortable office in the physics building at Clark. His white hair and flamboyant mustache were combed, his collar immaculate behind a black four–in–hand tie, his rimless pince–nez magnifying the intensity of his demeanor. In his 60th year, Webster was an acknowledged master of fundamental theory, a pioneer in dynamics, an expert in ballistics.

But there was a dark side. For years he had kept up a good front, but he was all torn up inside and clinically depressed. His wife had berated him for making an unsuccessful run as a Democrat for the U.S. House of Representatives in March, 1920. He was also aghast at the curtailment of academic freedom at Clark at the time. Other Clark professors sought employment elsewhere, among them Dr. Harry Elmer Barnes, the eminent professor of history.

Dr. Wallace W. Atwood, a geographer from Harvard University, was the new president of Clark when a socialist, Scott Nearing, was invited to address the students' Liberal Club in the school auditorium. Nearing was rightfully denouncing the control of universities by wealthy trustees, when Dr. Atwood twice attempted to dismiss the audience. Failing that, he ordered the auditorium lights turned off and adjourned the meeting.

But worse of all, Webster deplored the fact that he was losing his edge academically. He had sought to encompass the whole field of physics, but modern discoveries had made it impossible for him, or indeed for anyone. And then for his wife to ridicule him that morning for not being called to a larger university. *Why don't I just end it all*, he thought in deep, dark despair. *Bob Goddard can fill my shoes just fine. Of what use am I to anyone? I've had a good run, it's time to say goodbye.* He stoically reached into the lower left–hand drawer and pulled out the bottle of Johnny Walker Red, taking a long draught. Putting it back, he opened the lower right–hand drawer. He pulled the pistol out. Opening his mouth, he stuck the barrel tightly into the roof of his mouth. He pulled the trigger.

Shortly after Webster's death, Goddard was appointed as his successor to head Clark's department of physics. Then he proposed to Esther Kisk, persuading her to leave college to marry him. Her parents had all along been dubious about her marrying a man so much her senior and apparently in frail health with bouts of tuberculosis. He assured them that his tuberculosis was under control, submitting himself to the Kisk family doctor, who said that he and his bride might expect a reasonably lengthy life together if he took care of himself.

They were married in a small, quiet ceremony at St. John's Episcopal Church in Worcester on Saturday morning, June 21, 1924. The Reverend Charles Lancaster Short officiated. There were no attendants. Esther was resplendent in a gown of white georgette crepe made over Canton crepe, and a corsage bouquet. After the ceremony Esther changed into a roshanara crepe gown with a gray wrap and a hat to match. They left in a blue coupe for a week's honeymoon in the White Mountains of New Hampshire. Goddard's colleagues at Clark, considering him an inveterate bachelor, were surprised at the match. His former nurse, Miss Doyle, upon noting a resemblance, exclaimed, "Heavens preserve us, Robert! You've married your own mother!"

WESTWARD HO!

After his successful first flight on March 16, 1926, Dr. Goddard was thrilled with his triumph but resolved to say little about it. If people thought him daft when he was merely designing rockets, who knew what they'd say when the things actually started to fly? When word nonetheless leaked out about the launch and inquiries poured into Clark, Goddard answered each with a pinched, "Work is in progress; there is nothing to report." When he finished each new round of research, he'd file it under a deliberately misleading title — "Formulae for Silvering Mirrors," for example — lest it fall into the wrong hands.

But rockets are hard to hide, and as Goddard's grew steadily bigger, the town of Worcester caught on. On July 17, 1929, an 11–ft. missile caused such a stir the police were called. Where there are police there is inevitably the press, and next day the local paper ran the horse–laughing headline:

MOON ROCKET MISSES TARGET BY 238,799 ½ MILES

The reporters said the professor was trying again to fly up to the moon, which he wasn't, and that one of his rockets had blown up, which it hadn't. The local fire marshal had simply called the rocket a fire hazard, which it was, and ruled it out of the sovereign state of Massachusetts.

The reticent Goddard was mortified. At the post–mortem the next morning, Larry Mansur (then a student in physics at Clark University) tried to sum up their feelings with a wheeze. "They ain't doing right by our Nell," he said. Nell was a line from an old play. Thus "Nell" for many years became an affectionate name for the rocket, although Esther never cared much for it.

For Goddard, the East Coast was clearly becoming a cramped place to be. He was ridiculed by the press, and this caused him to continue most of his later experiments in secret. In 1930, with the promise of a $100,000 grant from financier Harry Guggenheim and support from Charles A. Lindbergh, world famous aviator, Goddard and his wife Esther headed west to Roswell, New Mexico, where the land was vast and the launch weather good, and where the locals, they were told, minded their own business. Goddard could conduct experiments without the humiliation of the news media.

One morning on December 30, 1930, the doctor and his men were absorbed in checking and installing their flight equipment. Long rehearsed, they had little need for instructions. Henry Sachs, the crew chief and instrument maker, and Al Kisk, Goddard's brother–in–law, had climbed the tower. They fastened cables to the striped rocket to hold her steady until the hoped–for moment of release. Below them, the brothers Lawrence and Charles Mansur paid out wire from the tower to the control shelter they had recently built 1,000 feet away and buttressed with sandbags. When—and if—the rocket flew, Larry would be farther out on the prairie, measuring its approximate speed and altitude with the aid of his recording telescope and stop watch.

Goddard, perhaps alone, concealed anxiety. There would be hundreds of experimental tests ahead, and more than could be accomplished in a lifetime. After each small advance had come failures and breakdowns. Although acclaim accompanies scientific success, the scientist accepts failure as a means of learning.

Esther drove into the valley alone, her motion picture camera ready. She had declined to be picked up by one of

The July 1929 fiasco at the Ward farm. From left, the crude test stand, photo sequence of rocket in flight, and Goddard's flight crew with shattered rocket.

Mescalero Ranch, near Roswell, New Mexico

Goddard towing Nell for the December 30, 1930 launch in Roswell, New Mexico.

the crew, deeming it a waste of time. As happened over four years before, the professor helped his slender blonde wife out of the car with her camera and gear. Feeling the hot sun on his back, Robert made a final check of the rocket, examining its controls, its connections, its pressure tanks. With his wife, he then joined the rest of his crew at the control shack. Esther adjusted her movie camera on its tripod, pointing it toward the tower from a hole cut in the shelter wall.

The launch sequence began. When the pressure–generating tanks built up to 200 pounds per square inch, Goddard quietly ordered: "Ignition!" Al Kisk fired the igniter. Through the open slot in the shack, Esther recorded the flame shooting out from the base of the rocket. Goddard brought the second pressure gauge on the fuel tank up to 225 psi. His hands clammy, he could almost feel the rocket straining at its cables. Watching the lift indicator, Sachs moved the release lever. Al Kisk, seeing that the lever had moved, pulled the releases.

This was the magic moment. The slender rocket worked up out of the tower, slowly at first, like a heavy man pushing himself up out of his chair. As the rocket neared the top of the tower, it climbed faster, gaining speed. Its jet was

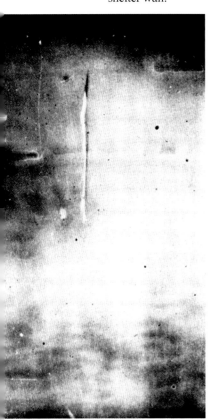

December 30, 1930 launch as recorded by Esther Goddard

still pointed and strong. Its nose came out of the tower, and then the rocket was out completely. It was leaning a bit to the southwest, but it was already higher than on any earlier flights. Probably 1,000 feet up and still climbing, though slanting off more. It was still making altitude and barely turning. Its red stripe was visible. But now it was slanting over.

At this point, the apex of its flight, Goddard's rocket was 2,000 feet above the prairie floor, by far his best altitude yet. The doctor apprehensively watched for its parachute to emerge from the nose. Something in the rocket had jammed. He heard Esther's hand–wound camera run down and stop. She had her movie film of the ascent. With his crew, Goddard followed the rocket's jagged descent. It came down shrill and whistling, banging into the dry lands a half mile away. Near the launching tower, Sachs was shouting at Kisk, who was already in the touring sedan, heading toward the fallen machine. Nell had achieved a maximum speed of 500 miles per hour (805 km/hr), perhaps the greatest obtained by a man–made contrivance up to that time.

Robert recorded, "It is believed that the present flight proved two things: first, that a light, very high speed rocket can be made for obtaining records, and for other uses, at a few thousand feet, without the employment of gyroscopic stabilizers; secondly, that the attainment of great heights cannot be made without the use of automatic stabilization. It seems desirable, therefore, to proceed at once with automatic stabilized flight…" The flight of December 30 was a promise of future flights.

In the open, roasted stretches of the Western scrub, the fiercely private Goddard thrived. Over the next nine years, his Nells grew from 12 ft. to 16 ft. to 18 ft., and their altitude climbed from 2,000 ft. to 7,500 ft. to 9,000 ft. He built a rocket that exceeded the speed of sound and another with fin–stabilized steering, and he filed dozens of patents for everything from gyroscopic guidance systems to multistage rockets. Astronautics was no longer a fancy; men would learn to navigate in space as they had learned to navigate on the seas and in the earth's atmosphere.

In 1963 Dr. Wernher von Braun, by then a world famous rocket scientist in the United States, reflected on the history of rocketry, and said of Goddard's work: "His rockets ... may have been rather crude by present–day standards, but they blazed the trail and incorporated many features used in our most modern rockets and space vehicles." The first successful Sputnik would be launched in October 1957, and the race into space quickly became irreversible with the competition between the U.S. and the Soviets to land a man on the moon in the 1960s. Flight into space was becoming the reality Goddard had foreseen.

Picking up the pieces after the successful December 30, 1930 launch

Above: Professor Goddard and assistants working on a rocket in his Roswell, New Mexico facility in 1940,
Left: Goddard's 1934-1935 rocket and 1939–1941 rocket as exhibited in the Smithsonian Institution's Aircraft Building until August 1959.

THE COMPETITION

Konstantin Tsiolkovsky

As is frequently the case with scientific theory and invention, developments proceeded in parallel with Goddard's in various parts of the world. While Goddard was engaged in building models of a space–bound vehicle, he was unaware that an obscure schoolteacher in a remote village of Russia was equally fascinated by the potential for space flight. Konstantin Eduardovich Tsiolkovsky began to devote more attention to space problems. In 1895 his book Gryozyo zemle i nebe "*Dreams of Earth and Sky*" was published, and in 1896 he published an article on communication with inhabitants of other planets. That same year he also began to write his largest and most serious work on astronautics, "*Exploration of Cosmic Space by Means of Reaction Devices*," which dealt with theoretical problems of using rocket engines in space, including heat transfer, a navigating mechanism, heating resulting from air friction, and maintenance of fuel supply.

Konstantin Eduardovich Tsiolkovsky

The genesis of the very word "astronautics" is interesting. A science writer named J.H. Rosny, Sr. was the first to coin the word in 1909, while a member of a French astronomical committee. In 1912, while Robert Goddard was beginning his serious research with rockets, Robert Esnault–Peltierie of France published a historically important paper in which he described the basics of a trip to the moon and the planets. To present his paper before the august French Physical Society, comprising the leading physicists of the day, Esnault–Peltierie had to give it a bland and obscure title: "*Consideration of the Results on the Indefinite Decrease in the Weight of Engines.*" Even so, the Society would only publish a summary. In 1930 Esnault–Peltierie gave the results of his studies of the possible exploration of space by rockets in a book entitled L'Astronautique, after which "astronautics" became an accepted term. Borrowing the term, the United States has coined the term astronauts, which strictly speaking means navigation among the stars, even though they have been restricted to voyages in the general vicinity of the earth. The Russians regularly refer to their space pilots as "cosmonauts," where cosmonautics refers to the science of travel in the cosmos, which is the 'universe' bounded by one star, our sun.

Although he did not experiment with rocket engines, Tsiolkovsky showed why rockets would be necessary for space travel and proposed liquid hydrogen and liquid oxygen as the most efficient propellants. He conducted the first studies that demonstrated that space travel was, at least theoretically, possible and advanced the concept of multistaging. Tsiolkovsky suggested that the staging should be set up on a principle similar to an aircraft jettisoning its empty wing tanks, but he was not the first to do so. Nearly 400 years before Tsiolkovsky advocated the use of multistage rockets (also called "step" rockets in those days), Conrad Haas, an artillery officer, had proposed the same idea with black–powder rockets. The French mathematician Jean d'Alembert and philosopher Denis Diderot also discussed the concept of a multistage "rocket train" in their *Encyclopédie, Ou Dictionnaire Raisonné Des Sciences, Des Arts Et Des Métiers* which was published between 1751 and 1772.

By 1898, Tsiolkovsky had already asked the fundamental question upon whose answer all space flight depended: could one build a rocket that could fly faster than its own exhaust gases? To remain in orbit a spacecraft must achieve a speed of 8000 meters/sec. To completely escape Earth's gravity and go into endless space, 11,000 m/s are required. Even using hydrogen as a fuel, a rocket's exhaust could not exceed about 3600 m/s. However, Tsiolkovsky's equation showed that up to the speed of light, 300 million m/s, there is no limit to the speed rockets can reach. The one major limiting factor is that a rocket must lift its own weight off Earth's surface. And it must continue thrusting throughout the ascent trajectory, or else it will fall back to Earth. Tsiolkovsky's theoretical computations showed that to do this, at least 75 percent of a rocket's weight had to be its fuel. But missiles have to include massive structures, heavy rocket engines, all the propellants and other fluids, and hopefully a payload. Even today, many rocket designs require that over 90 percent of a missile's weight be devoted to structure, engines, control systems, avionics/electrical, and fuel. Principally fuel. This leaves less than 10 percent for payload. According to the equations, the paradox is that if the Earth's gravity were significantly bigger, even rockets could not escape the Earth's gravity and we would have to rely upon some other heretofore unknown means of space propulsion.

By 1903, Tsiolkovsky had also given careful thought to manned–spacecraft design and his description of such a ship demonstrates how completely realistic fact and fantasy had by this time emerged. In his own words:

> *"Let's imagine the following configuration: a metal elongated chamber (having forms of least resistance), provided with its own light, oxygen, with absorbers of carbon dioxide, noxious effluvia and other animal excretions, intended not only for the maintenance of various physical devices, but also to provide life support to the men controlling the chamber. The chamber contains a large supply of liquid hydrocarbon and liquid oxygen which, when combined, immediately form an explosive mass."*

Tsiolkovsky proposed dividing his spaceship into three bays. The top bay in the rocket's nose housed the crew. Here would be the control panels, automatic instruments, decompression chambers, and comfortable couches. The second bay contained the oxygen supply and the 'bathtubs' filled with water in which the crew members would immerse themselves to ease the powerful G–forces experienced during the rocket's launch and re–entry. The third bay housed the pumps necessary to move the propellants from the fuel tanks to the engine. The fuel tanks were below this third bay and would contain, Tsiolkovsky suggested, a liquid hydrocarbon and liquid oxygen.

Also in 1903 Tsiolkovsky published his results in *"Investigations of Space by Means of Rockets,"* the very year the Wright Flyer stumbled into the air under its own power. Many years later this groundbreaking work was hailed by the Soviet Union as the forerunner of space flight. Tsiolkovsky died in 1935, an honored hero of the Soviet Union and was given a state funeral. One is impressed even today by Tsiolkovsky's vision of space exploration as an inevitable process that would transform and spread human life throughout the solar system. "Earth is the cradle of the mind," Tsiolkovsky wrote, "but one cannot live in the cradle forever."

Tsiolkovsky later in life

A young Hermann Oberth, 1929

Oberth later in life

Hermann Oberth

The other giant of the pioneer space trio (besides Tsiolkovsky and Goddard) was Hermann Julius Oberth of Germany. He published, largely at his own expense, a 92–page space–flight treatise, *"Die Rakete zu den Planetenräumen"* (The Rocket into Interplanetary Space) in 1923, four years after the appearance of Goddard's groundbreaking work, *"A Method of Reaching Extreme Altitudes."* The book describes Oberth's mathematical theories of rocket flight, his research in solid–fuel and liquid–fuel rockets, and the possibilities he saw of exploring the earth and beyond. Like Goddard, Oberth identified liquid hydrogen and liquid oxygen as the most efficient propellant combination. Oberth's book is regarded as one of the pioneering works of the science of rocketry, but Goddard correctly foresaw that the bellicose Germans would eventually create weapons of war with their rockets. In a 1923 memorandum to the Smithsonian Institution, he summed up his sentiments. "I am not surprised that Germany has awakened to the importance and the development possibilities of the work, and I would not be surprised if it were only a matter of time before the research would become something in the nature of a race."

Oberth knew his mathematics. What dismayed the Yankee Goddard was not only Oberth's boldness, but more likely his brilliance. Oberth was not content with a mere moon–striking rocket, but carried his calculations, beyond Goddard's, to the mathematics of spatial trajectories. Oberth discussed staging and described a three–stage launch vehicle. There is an advantage to dropping off spent booster stages, engines, and tanks and continuing on with a lighter load, he said. While we take it for granted that rockets can escape Earth's orbit, in Oberth's time the general public did not. In his treatise, Oberth demonstrated the theoretical possibility of space flight and formulated its basic mathematics. In addition to proposing designs for man–carrying spacecraft and high–altitude research rockets, Oberth advanced the concept of orbital rendezvous for refueling and resupply by reviving the idea of orbiting a space station or large satellite—an idea first suggested in an 1870 Atlantic Monthly magazine fiction serial, "The Brick Moon" written by Boston clergyman Edward Everett Hale.

A second book by Oberth, "Wege zur Raumshiffahrt (Path to Space Travel), countered arguments which had been brought up by the scientific and technical community. It talked about earth satellites, space stations, solar collectors, space suits, and practically all the things being done today in space.

Oberth's books excited the imagination of many young men, creating great interest in rocketry all over Europe. Some rocket enthusiasts worked alone, but others banded together to form amateur rocket societies with friends and coworkers in America, Germany, and the Soviet Union. Rocket enthusiasts attempted to demonstrate the feasibility and the practicality of reaction systems, and installed rockets on automobiles, on ice–sleds, on sail–planes and motor–planes. They also designed and launched regular skyrockets, in some instances as a means of mail delivery. None of these early pioneers had ever even hinted at the use of rockets for military purposes, although solid propellant missiles—often in traditional black powder rocket form— had been used for many years by the armies of many nations. All these amateur efforts were greatly hampered by funding shortages; this situation forced the groups to 'pass the hat around' at public "shows," talks, demonstrations, static firing performances, and oftentimes unsuccessful rocket launches. But there was never enough money around to do an adequate job of designing, manufacturing and testing the rockets before flight attempts. Progress was therefore extremely slow.

To those seriously interested in reaching other planets within their lifetimes, it seemed that development of the liquid rocket engine was necessary to build space ships. This would require the collaborative effort of thousands of engineers and technicians. In June 1927 Johannes Winkler called to order in Breslau the first meeting of the world's first Society for Spaceflight, *Verein für Raumschiffahrt* or VfR (later, it became *Verein für Raketenforschung*, Society for Rockets). The membership grew from three to 500 within the year, including most of the European space pioneers. Willy Ley was instrumental in publicizing the Society and editing its newsletter. Oberth joined this group. Max Valier, an Austrian writer and self–proclaimed astronomer residing in Munich, took up the cause. Based upon the publicity efforts by rocket amateurs; and Valier's articles, books and speeches, the German general public also became interested. In 1928 a German movie company in Berlin, Ufa Studios, decided to film the space movie, "*Die Frau im Mond*" (The Lady on the Moon). This was to be done under the direction of the famous producer Fritz Lang. He in turn hired the 35 year old Hermann Oberth as scientific advisor for this venture. Oberth objected when scientific accuracy was sacrificed for romance. How could one present a love story taking place on the Moon and have the lead characters talk to each other and hold hands through space suits? So the airless Moon was endowed with an atmosphere (as well as water). Other compromises were made for the sake of drama. At least the script included a more realistic rollout of the vehicle to the launch pad, and the now–famous countdown sequence before ignition and lift–off.

Kegeldüse

Lang provided Hermann Oberth with a small amount of funds to design, construct, test, and launch a liquid–propellant rocket to publicize the film. All VfR members were greatly elated by support that was gained by this association. Oberth built a rocket engine for combustion of liquid oxygen and gasoline, inventing a conical combustion chamber to mix the propellants. His Kegeldüse ("cone" jet nozzle) was made of steel and lined with copper. The assembled rocket was 1.8 m tall and was to have been launched to an altitude of 64 km over the Baltic Sea from Greifswalder Oie. Oberth's mathematical equations and trajectory projections were meticulous and flawless. The great American rocket pioneer Dr. Wernher von Braun–whose life will be discussed in a subsequent chapter––years later said of him,

> *"Hermann Oberth was the first, who when thinking about the possibility of spaceships grabbed a slide-rule and presented mathematically analyzed concepts and designs... I, myself, owe to him not only the guiding-star of my life, but also my first contact with the theoretical and practical aspects of rocketry and space travel. A place of honor should be reserved in the history of science and technology for his ground-breaking contributions in the field of astronautics."*

Kegeldüse Rocket

However, Oberth's engineering acumen—the nuts and bolts of rocketry—was not quite so flawless. One of the assistants hired by Oberth to fabricate the rocket was Rudolph Nebel, a World War I fighter pilot with (unfortunately) little actual engineering experience. Oberth also had no practical engineering or organizational ability, and was unable to produce the liquid rocket in the four months allotted.

Oberth then turned to a much larger (and impractical) 11–m tall hybrid rocket that was to burn a to–be–determined carbon compound with liquid oxygen. This also proved impossible. A disappointed Oberth simply gave up and returned to his native Romania to resume his teaching career in Schaessburg – returning, however, for the film's premiere on October 15, 1929. Ufa studios took ownership of the unfinished rockets. Among the design team was an eager 18–year–old student named Wernher von Braun, whose enthusiasm for space flight never waned.

In December 1929, after the failure of Oberth to produce a liquid propellant rocket for Fritz Lang, the VfR was in shambles. Winkler had resigned as president. The VfR held a meeting with Rudolph Nebel, Karl Wurm, Hermann Oberth, Klaus Riedel, Johannes Winkler, and Willy Ley in attendance. They decided to try and get the Oberth rocket materials back from Ufa and press on to demonstrate flight of a liquid propellant rocket. For this purpose the Oberth rocket was much too ambitious and probably wouldn't work anyway. Nebel proposed building a new *'Minimum Rakete'* or *'Mirak'* to demonstrate that it could be done.

Since the mid–1800s; rifled, breech–loaded artillery guns had greatly improved their accuracy and range. Likewise, rifles and guns used during the American Civil War had vastly improved accuracy and lethality over those used in the Revolutionary War of the late 1700s. Black powder rockets, difficult to handle and not as accurate, had fallen out of favor and could not effectively compete with conventional artillery. Military experts and lay people alike had come to regard rockets, which burned a gunpowder–like fuel in a metal or paper casing, as little better than interesting fireworks.

However, by coincidence, Lieutenant Colonel Karl Emil Becker of the German Army Ordnance Command had just begun to investigate the revival of the rocket as a weapon. Becker realized the limitations of German (or anyone else's) artillery. The Paris Gun used by Germany to shell the French capital in 1918 wore out its main and reserve barrels after lobbing only 320 shells from 130 kilometers away. Granted, 130km is quite a distance for a 21–cm howitzer, but its accuracy on an unguided, unpowered ballistic trajectory at that range was at the mercy of prevailing winds and other uncertainties.

Becker believed that by replacing conventional gunnery with liquid–fueled rocket engines, one could eliminate not only barrels and their massive supporting equipment, but also all limitations on range and payload, and with guidance improve the accuracy as well. Moreover, the surprise deployment of stunning new weapons could have a dramatic psychological effect. A rain of fairly accurate long–distance projectiles stood a good chance of collapsing an enemy's morale.

new rocket–car engine at the Heylandt Company in Berlin in March/April. Tragically, on May 17 Valier was killed by a terrific rocket car explosion in a laboratory experiment, due in part to his cavalier attitude towards safety. But the work had to go on. By July of that year the *Kegelduese* was being run for 90 seconds, generating 7 kgf (kilogram-force) and consuming 6 kg of liquid oxygen and 1 kg of gasoline in that time, with a specific impulse of 90 seconds.

Max Valier's rocket–propelled racecar

Mirak

The Mirak was basically a head and a "guiding stick," similar to the designs for powder rockets. The cast aluminum head looked like an artillery shell. You had to remove it to put in the liquid oxygen inside. Inside the top of the head, above the LOX, was a safety valve. Oberth's steel and copper–lined *Kegeldüse* was inserted, nozzle down, into the bottom of the head, so that the chamber would be surrounded by LOX in hopes of cooling the combustion chamber. The guiding stick was an aluminum tube which held the fuel (gasoline) and compressed CO_2 gas for pressurization, and was attached off to one side, akin to how the Space Shuttle was strapped to the large External Tank. For the LOX tank, the combustion chamber's heat was supposed to evaporate some of the LOX, creating pressure to force LOX into the chamber. In January 1930 Nebel and Riedel, the two brains behind German rocket technology at the time, began a series

The VfR group obtained funding from two private benefactors and began to ground test and perfect Oberth's 'Kegelduese' conical rocket engine in the summer of 1930. During this time, Max Valier adapted a rocket engine to propel automobiles. He tested his

Historic view of the Verein für Raumschiffahrt, 1930. Wernher von Braun is standing behind Klaus Riedel who is holding an early version of the minimum rocket, or "Mirak"

of tests of the Mirak rocket at the farm of Riedel's grandparents near Bernstadt, Saxony.

American Interplanetary Society

At about the same time (in April 1930), Fletcher Pratt, husband and wife G. Edward and Lee Pendray, science fiction writers David Lasser and Laurence Manning, together with seven others founded the American Interplanetary Society (AIS) in New York City, for the "promotion of interest in and experimentation toward interplanetary expeditions and travel." The AIS approached Goddard to publicize his work. He refused. The general public had ignored the scientific merit of his 1919 paper, "A Method of Reaching Extreme Altitudes." As we have seen, in his paper he outlined his ideas on rocketry and suggested, none too seriously, that a demonstration rocket should be flown to the Moon. The newspapers and capricious public media latched onto Goddard's Moon rocket proposal. At the time, such an endeavor was absurd and most dismissed Goddard as a "crank." The experience taught Goddard a hard lesson — one which caused him to shy away from future opportunities to publicize his work.

The AIS, rebuffed and learning that no one in the United States aside from Goddard was working with rockets, turned its attention to rocket research underway in Europe, where rocketry was beginning to attract a following. In April 1931, founding members Edward and Lee Pendray traveled on vacation to Germany where they made contact with the German VfR. The visiting Americans were given a preview of the future when a member of the VfR, Professor Willy Ley, took the pair to *Raketenflugplatz* where they witnessed a successful Mirak static rocket firing. Unfortunately, the timing was such that the Pendrays never met Hermann Oberth.

Nevertheless, upon returning home in mid–1931, the Pendrays filed an enthusiastic report of their visit, prompting the AIS to build its first rocket. AIS members conducted their own rocket experiments in New York and New Jersey and did pioneering work in designing and testing liquid–fueled rockets. Goddard occasionally corresponded with the AIS's *Bulletin*, but as always he remained aloof from other American researchers, cautious about his results, and concerned about patent infringements. AIS enthusiasts attempted to make improvements to the German Mirak design. A sister organization, the Institute of Aeronautical (later Aerospace) Sciences (IAS), was founded in 1932. The first test flight of the so–called ARS–1 in November that year ended with the American design firmly on the ground. Unassisted by the reticent Goddard, the American rocketeers improved on their design. They successfully launched the ARS–2 on May 14, 1933. However, the rocket veered after takeoff, reaching only 75m in altitude. Nevertheless, they kept trying to refine the design.

Following VfR's example, in April 1934 the AIS changed its name to the American Rocket Society (ARS) and published the *Journal of the American Rocket Society*. With burgeoning membership over the next three decades, the ARS and IAS merged in early 1963 to form today's American Institute of Aeronautics and Astronautics (AIAA).

Repulsor

By the spring of 1931, Nebel, Klaus Riedel, and the other German Mirak designers realized that using liquid oxygen to cool the combustion chamber simply would not work for two reasons. The heat of combustion through the steel chamber wall was enough to cause it to burn through, creating an explosion. Secondly, the heat turned the LOX into gas, and the excessive pressure eventually burst the oxygen tank to create an explosion. These two factors were related, because gaseous oxygen cannot transfer heat off the chamber walls near as well as LOX.

To solve this problem and other problems, the Germans made four innovations. First, they had to cool the heat of combustion in the engine chamber itself. In line with Oberth's original suggestion, they substituted alcohol for the gasoline. It took 3.5 kilos of LOX to burn 1 kilo of gasoline, but a kilo of alcohol required only 2 kilos of the expensive and hard–to–handle cryogenic liquid oxygen. In addition, they could add water to the alcohol to lower the combustion temperature without suffering too much in performance. They eventually settled on a mixture of 60% alcohol/40% water as a compromise (later, the German V–2 rocket would adopt a mixture of 75% alcohol and 25% water).

Second, the VfR group invented a cooling jacket around the engine and nozzle. At first they used plain water, but after much experimentation they found they could circulate the alcohol/water fuel mix through the cooling jacket before injecting it into the combustion chamber. This worked, and versions of this technology, called *regenerative cooling*, are still being used today.

Third, they addressed the problem of pressuring the propellants fed into the combustion chamber. You have to pressurize the propellants entering the chamber, for the same reason that you have to blow really hard in order to inflate a balloon (but imagine the "balloon" being made out of steel or aluminum instead of rubber!). Since the infancy of rocket science, combustion under pressure has been vital to rocket performance, because thrust is directly related to the pressure inside the combustion chamber walls. After trying other approaches, the VfR group used a bottle of compressed nitrogen to build up pressure in the propellant tanks and expel both fuel and oxidizer into the combustion chamber. During World War II the Germans improved on this technology by using the violent decomposition of hydrogen peroxide (H_2O_2) in the presence of a catalyst like potassium permanganate to generate high pressure gas. Today's rocket engines are much more complicated and more sophisticated arrangements are used to pressurize the propellants.

Fourth, the experimenters changed the configuration. They made the combustion chamber and nozzle out of aluminum instead of steel (of course, today sophisticated alloys are used instead). They also adopted Goddard's odd-ball looking "nose drive" configuration. So instead of having the engine and tail fins of the rocket at the rear, they had the engine/nozzle at the front and the propellant tanks trailing behind like sausage links! The modern concept makes much more sense today, but remember, in those days rocketry was in its infancy. *We would not be where we are today in rocket science were it not for the persevering rocket enthusiasts of the past.*

Repulsor rocket nozzle

At least the Germans had the sense to move the trailing propellant tanks closer to the centerline and added a slender "guide stick" for stability. Behind the forward-mounted engine and nozzle was the liquid oxygen tank, then the watered alcohol tank, then the pressurization system—practically the opposite of the order used today. Willy Ley christened their new creation the 'Repulsor' to avoid using the word "rocket" which implied a powdered propellant at the time. He took the name from the science fiction novel by Kurt Lasswitz, Auf Zwei Planeten (On Two Planets), in which Lasswitz invented "repulsit" as his substance for propulsion. On May 14, 1931 the Repulsor failed its first flight test. It got into a looping trajectory, sending the VfR experimenters running for cover, but reached 60m in the process. Like Goddard, the Germans did not give up but put their heads down and plowed ahead. Eventually they developed a new Repulsor that was 3.5m long, 10cm in diameter, weighed 10kg empty, 20kg fully fueled, and had a thrust of 60 kgf. However, in September 1931, von Braun was recruited away by Lt. Col. Karl Becker and assigned to work under Capt. Walter Dornberger of the Army Ordnance Command at Kummersdorf, an Army proving ground south of Berlin. After his departure, a Repulsor launched by the German rocketeers reached a whole kilometer in range, demonstrated recovery by parachute, and was reused. A few months later this improved to 1500m and 3000m downrange. By March 1932 a Repulsor reached an altitude of four kilometers. In fact, the Germans had to resort to launching them only partially fueled to prevent them from landing outside the perimeter of the Raketenflugplatz.

Specific Impulse

A basic measure of any rocket's performance, whether it uses liquid or solid propellants (or both for a 'hybrid' rocket), is the term "specific impulse" (I_{sp}). The specific impulse a rocket can deliver is simply the ratio of its thrust (the force it produces) to the weight flow rate of the propellants (through its combustion chamber and exit nozzle) that it takes to produce that thrust. The units are in seconds, because specific impulse represents the kilograms-thrust (kgf) or pounds-thrust (lbf) – an engine or motor produces per kilogram of rocket fuel (or solid propellant for motors) per second of operation. This makes perfect sense, at least to rocket scientists. The more bang you get out of a rocket for the same amount of propellant, the more efficient it is—at producing thrust. Using the Repulsor above as an example, suppose the VfR experimenters filled it full of propellant (10 kg), and someone visually timed how long the rocket burned on a long test flight to propellant depletion (40 seconds). Then its propellant flowrate is 10kg/40sec = 0.25 kg/sec. Its specific impulse is its thrust (60 kgf) divided by 0.25 kg/sec, or 240 seconds (disregarding the kg force to kg mass units). It doesn't matter if the Repulsor ran for 10 seconds, or 120, or 55 minutes. Its delivered Isp was still 240 seconds, during that time as long as its thrust and propellant weight flow rate stayed the same.

Rocket Aerodynamics

Meanwhile in the United States, Goddard was rapidly learning how aerodynamic forces act on a rocket in flight through the "school of hard knocks" in New Mexico. In those days the principles of aerodynamics about bodies of revolution (in other words, cylindrical missiles) were not well understood by the engineering community, and no one knew much about guidance and control in the early 1930s.

Aerodynamic forces are used differently on a rocket than on an airplane. On an airplane, lift is used to overcome the weight of the aircraft, but on a rocket, thrust is used in opposition to weight. Because the center of pressure is not normally located at the center of gravity of the rocket, aerodynamic forces can cause the rocket to rotate in flight. The lift of a rocket is a side force used to stabilize and control the direction of flight. While most aircraft have a high lift to drag ratio, the drag of a rocket is usually much greater than the lift.

British Interplanetary Society

In yet another parallel development, Mr. P.E. Cleator founded the British Interplanetary Society (BIS) in London in January 1933. The BIS aimed to not only promote and raise the public profile of astronautics and space exploration, but also to undertake practical experimentation into rocketry

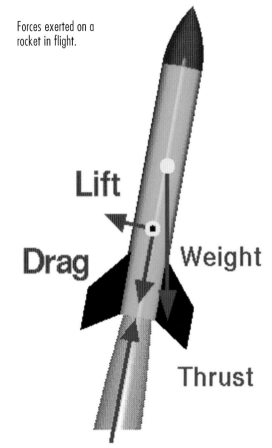

Forces exerted on a rocket in flight.

along similar lines to the German VfR and the American ARS. However, early in 1936 the Society found this ambition thwarted by the Explosives Act of 1875, which prevented any private testing of liquid–fuel rockets in the United Kingdom. Not to be discouraged, in the late 1930s the BIS devised a project of landing people on the moon by a multistage rocket, each stage of which would have many narrow solid–fuel rockets. The BIS lander was gumdrop–shaped but otherwise quite like the Lunar Excursion Module (LEM) developed by the United States for the Apollo moon landings in the 1960s. The British considered that the lander crew cabin would have to rotate to provide artificial gravity by centrifugal force, so claiming to have invented the first instrument for space travel – a navigation mechanism that would cancel out the rotating view.

For amateur rocket enthusiasts in Germany, the die was cast. Rocket development was, and still is, complex and expensive. The cost and the difficulties of planning and organization meant that, sooner or later, the major work in rocket development would have to occur under the aegis of permanent government agencies and government–funded research bodies.

Sadly, soon the VfR disintegrated in political disputes and withdrawal of funding by its wealthiest backers. Rudolph Nebel was unable even to pay the water bill for *Raketenflugplatz*, where old pipes and faucets were constantly leaking water. The government cancelled the lease and took the property back. Dornberger and the German Army took over rocket development. The remaining private experimenters – Nebel, Winkler, and Puellenberg – were ordered by the Gestapo to cease and desist. Henceforth only military applications of rocket technology were to be pursued. Adolph Hitler and the Nazis came into power in 1933. Goddard's prescient prediction about a "bellicose Germany" would soon come true.

GALCIT

Similarly in America, significant team research began in 1936 at the Guggenheim Aeronautical Laboratory, California Institute of Technology (GALCIT). In 1939, this group received the first federal funding for rocket research, achieving special success in rockets to assist aircraft takeoff. The project was known as JATO, for jet–assisted takeoff, since the word "rocket" still carried negative overtones in many bureaucratic circles. JATO research led to substantial progress in a variety of rocket techniques, including both liquid and solid propellants. Work in solid propellants proved especially fortuitous for the United States during WWII. American armed forces made wide use of the bazooka (an antitank rocket) as well as barrage rockets (launched from ground batteries or from ships) and high velocity air–to–surface missiles.

Nazis

By the late 1930s Goddard grew troubled. He had noticed long before that of all the countries that showed an interest in rocketry, Germany showed the most. Although Goddard was not recognized as a scientist of any significance in the U.S., scientists in Germany saw his work as very important as their country prepared for war in Europe. Goddard's experiments included fuel feeding devices, propellant pumps, gyroscopic stabilizers, and instruments for monitoring the flight of rockets. During his life he accumulated 83 patents, for everything from multistage rockets to fuel pumps and clustered engines. Now and then, German engineers

would contact Goddard with a technical question or two, and he would casually respond. But in 1939 the Germans suddenly fell silent.

With a growing concern over what might be afoot in the Third Reich, Goddard paid a call on Army officials in Washington and brought along some films of his various Nells. He let the generals watch a few of the launches in silence, then turned to them. "We could slant it a little," he said simply, "and do some damage." The officers smiled benignly at the missile man, thanked him for his time and sent him on his way. For some unexplained reason, the U.S. military appeared to have remained underwhelmed with the potential of rockets. As before with the Wrights, the United States seemed willing to let other countries take the lead in trailblazing

a disruptive new technology. The missile man Goddard, however, apparently knew what he was talking about. Five years later, the first of Germany's murderous V–2 rockets blasted off for London. By 1945, more than 1,100 of them had rained down on the ruined city.

How much technology did Nazi Germany steal from Goddard ? Not a whole lot it turns out. As explained below, the patent infringement finger should be pointed at the U.S. government instead. First of all, given Goddard's reticence, the VfR and other German rocket enthusiasts made their own innovations before the war. Much later, in early 1975, a West Point cadet wrote a letter to Dr. Wernher von Braun, the chief wartime German rocket scientist, asking about the persistent accusations that he and his team in Germany

The U.S. Army "bazooka" was an outgrowth of Goddard's World War I research.

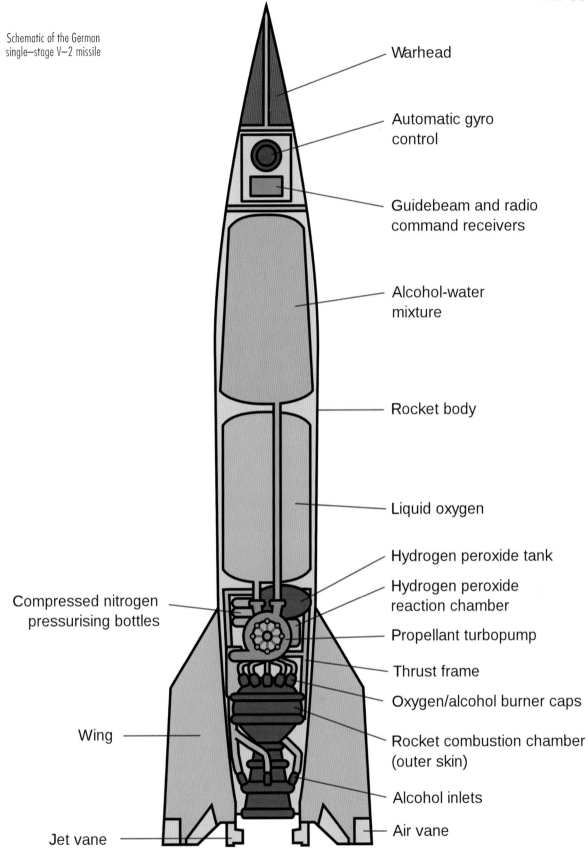

Schematic of the German single–stage V–2 missile

Warhead

Automatic gyro control

Guidebeam and radio command receivers

Alcohol-water mixture

Rocket body

Liquid oxygen

Hydrogen peroxide tank

Hydrogen peroxide reaction chamber

Compressed nitrogen pressurising bottles

Propellant turbopump

Thrust frame

Oxygen/alcohol burner caps

Wing

Rocket combustion chamber (outer skin)

Alcohol inlets

Air vane

Jet vane

basically copied Goddard's patents and performed little original engineering in developing the V–2 missile. Von Braun, who was not in good health by that time, chose to respond in detail to the young engineering student's apparently sincere question.

"Around 1930, when I was eighteen years

old and a member of the German Society for Spaceflight ... Dr. Robert Goddard was one of the great international names in the concept of flight through space, and I counted him among my boyhood heroes. I had read his booklet "A Method of Reaching Extreme Altitudes" which described the multistage principle and presented some advanced ideas on how to improve the performance of solid fuel rockets. In subsequent years, when I developed liquid rockets for the German Ordnance Corps which ultimately led to the V–2, I occasionally saw illustrations (e.g., a Goddard proposed aerial torpedo), or statements (e.g., Goddard says: "Man can reach the moon") in aviation journals. However, at no time in Germany did I or any of my associates ever see a Goddard patent. I was not even aware of the fact that Goddard worked in the field so dear to my own heart, namely liquid propellant rockets, let alone that even as early as 1926 he had successfully launched the world's first liquid–propellant rocket."

Von Braun, who had arrived in America only weeks after Goddard's death, stated in his letter how, in 1950, the Army had asked him to review and analyze "a large stack of Goddard patents" related to a lawsuit filed against the government by the Goddard estate and the Guggenheim Foundation, an early sponsor of Goddard's work (Goddard is credited with 214 patents, of which 131 were filed after his death). Interestingly, the Army backed the plaintiff (Goddard's estate) in this case, contending that the V–2s brought to America and some of the new postwar U.S. rockets infringed on the reclusive Goddard's patents, and so the government should pay royalties to his estate. Von Braun further noted in his letter to the cadet:

"The government lawyers apparently took the position that at least the V–2 design could not possibly have violated Goddard patents, because they had been highly classified and had, therefore, not been available to the German engineers. Moreover, even if any infringement existed, the United States would automatically have acquired all rights to the V–2 technology by virtue of the captured V–2 missile falling in the category of war booty under international law."

Von Braun wrote that he was asked to do "a detailed written assessment" for use by the court, regarding whether any V–2 engineering designs had infringed on Goddard patents. He stated that he could certify "there were indeed infringements all over the place"—from

the application concepts of jet vanes to turbo–pumps to guidance–and–control gyroscopes. Von Braun added in his letter:

"All the Goddard patents I saw were classified and had never been published, even as late as 1950. I was fully unaware of them while in Germany, and even in the United States I saw them for the first time only five years after my arrival, and upon receipt of a secret clearance.

All the patents I reviewed in 1950 rendered impressive proof that Dr. Goddard had indeed a brilliant and imaginative mind. The patents covered not only design features actually (but unwittingly) used in the V–2, but numerous alternate options."

The ailing von Braun concluded to the West Pointer:

"It might be of interest that (maybe in part as a result of my affirmative report) the lawsuit led to an amicable settlement, under which the U.S. Government paid a generous sum ($1,000,000, at that time the largest patent settlement that the government had ever given) to the Goddard estate and Guggenheim Foundation."

Moreover, von Braun went out of his way to acknowledge Goddard's achievements. In March 1957 he was invited to address a banquet of the Worcester Engineering Society in the Massachusetts. He asked to be taken by Esther Goddard to the launch site of history's first liquid–propelled rocket by her husband 31 years earlier (March 16, 1926). Finding no historical marker there, von Braun began a campaign that evening to erect a proper one. His efforts, through the ARS, succeeded with the 1960 dedication of a memorial at the site. Other honors in Goddard's memory followed, often with von Braun's involvement.

Because of Goddard's reticence, in contrast to the more visible personalities in the VfR, and because of the publicity given the German V–2 of World War II, the work of British, American, and other groups during the 1930s has been overshadowed. Their work, if not as spectacular as the V–2 project, nevertheless contributed to the growth of rocket technology in the prewar era and to the successful use of a variety of Allied rocket weapons during the war. Although groups like the AIS, ARS, and BIS succeeded in building and launching several small chemical rockets, much of their significance lay in their role as the source of a growing number of technical papers on rocket technologies. With a number of successful test launches they pointed the way to space programs in the United States and Europe.

Goddard went to his grave suspecting that the Nazis stole most of his trade secrets. In 1945 he was found to have throat cancer, and before the year was out, he would be dead. Shortly after nine o'clock on the morning of August 10, 1945, Goddard lay dying. Only four days before, the *Enola Gay* B–29 bomber had dropped the first atomic bomb on Hiroshima, Japan.

THE DEMISE OF THE "LONE RANGER"

Goddard was probably the last American to try to carry out large–scale engineering ventures in astronautics by himself. He could be said to be the last Lone Ranger—like the jack–of–all–trades "Renaissance Man" Leonardo da Vinci—in this field. There were two central reasons for this change. The first is relatively easy to comprehend and it has already been hinted at here—the development of something as complex as a rocket and spacecraft capable of operating outside the earth's atmosphere is too large for any one individual, regardless of how much mastery of however large a body of knowledge, to oversee. The breadth and depth of engineering and scientific information is simply too large for a single person to fully comprehend. It must be parceled out and managed through a team approach.

The second reason is more complex, and ultimately more interesting. Before WWII, by all accounts, engineering education in the United States was overwhelmingly oriented toward training young engineers in a very practical "shop culture." Instructors in engineering focused on practical application, not towards research and theory. Where research was conducted, it usually emerged naturally from the implementation of engineering projects and focused on the narrow questions informing such work.

This began to change in the first part of the 20th century as an influx of European engineers came to the United States and brought their educational ethos to the nation's academies. In the aerospace engineering community this was personified by such men as the brilliant Hungarian aerodynamicist Theodore von Kármán, whom we will meet in the next chapter. Dr. von Kármán came to the California Institute of Technology (Caltech) in the 1930s and was one of the founders of the Jet Propulsion Laboratory (JPL). He was not only a brilliant, hard–edged aeronautical engineer, but also a leading theorist who contributed important concepts to aerodynamics. At the same time, the requirements of complex high–technology artifacts required for war prompted the United States to expend for the first time massive amounts of money for technology projects. Those with broad–based theoretical implementation were most readily funded.

By the end of World War II, however, most engineering in the United States had become so theoretical that much of its practical application was lost on working technicians. Increasingly, it became difficult to distinguish between engineering projects and purely scientific explorations without immediate practical application. The results of this change soon became manifest in the engineering disciplines. American engineering faculty were no longer necessarily experienced in industry's practical needs, and had instead made their careers as theoretically oriented researchers who published scholarly papers in journals but did not design and build artifacts for public use. These two emerging subcultures were sometimes contradictory and often combative.

The complex theoretical foundations and components of huge aerospace projects have made it less and less likely that a single individual, or even a single genius with some assistants like Thomas Edison, could carry a path–breaking new development through to successful completion. Aerospace technology work since WWII has of necessity been a group effort with individuals in charge of certain segments of the work. There might be an overall project manager, but the demands of a multi–billion dollar project always forced more breadth and depth of knowledge than even the genius of a Thomas Edison, Leonardo da Vinci or Robert Goddard could master. Indeed, it might be said that the Lone Ranger syndrome was a chimera all along, for complete success was always beyond even the most creative genius' grasp.

THEODORE

"When the going gets tough, the tough get aeronautical."

H. M. Murdock, The A–Team

Vehicles in flight, be they airplanes, gliders, missiles, or rockets are subject to tremendous forces at high speeds. In fact, our atmosphere presents an almost impenetrable shield against all but the largest meteorites.. The Earth's sensible atmosphere extends up to 400 kilofeet (122 km, or 75.7 mi) or so. But it doesn't end there. A tenuous atmosphere extends to much higher altitudes, even to low Earth orbit (LEO). This is why satellites in LEO must occasionally boost themselves, to make up for atmospheric drag which inexorably slows them down due to the very thin atmosphere in their orbit. Space stations in LEO like the International Space Station and the Soviet Mir ("Peace" in Russian) must especially account for thin atmospheric drag, otherwise they will eventually decay from their orbit, reenter the Earth's atmosphere, and burn up. This happened to the U.S. Skylab in 1976, parts of which landed on the eastern Australian coast. Fortunately, nobody was killed.

The construction of large towers, buildings, and bridges also requires predictions of aerodynamic forces and the creation of an optimum design to minimize these forces. The consideration of aerodynamic forces of flat surfaces such as the side of a building or superstructure is not new. In 1910 Alexandre–Gustave Eiffel achieved remarkable experimental results measuring the wind resistance of a flat plate, using the Eiffel Tower as a test platform.

The science of aerodynamics is, and always has been, critical to the design and operation of rockets and space vehicles that have to traverse the atmosphere. They require accurate navigation, guidance, and control—all of which depend on aerodynamics. Among others who played a prominent role in the development of modern aerodynamics was the Hungarian–born engineer Theodore von Kármán, whose contributions led to major advances in such areas as solid–propelled rockets, turbulence theory, and supersonic flight.

TACOMA NARROWS BRIDGE

The morning of November 7, 1940 was unusually blustery in the Narrows of Puget Sound, named for Peter Puget, a 2nd lieutenant in the British navigator George Vancouver's expedition of 1792 to explore the main channel. Puget Sound stretches for 100 miles from north to south, and there was no way to get across it from the major population centers of Seattle, Tacoma, Everett, and Port Townsend except by boat. Washingtonians had long lamented having to take ferries across the sound which connects the Olympic Peninsula with the mainland of Washington state. Puget Sound's excellent deepwater harbors had attracted increasing population growth and the need for outposts to serve the rich farmlands along the river estuaries. The Puget Sound Naval Shipyard in Bremerton added military shipping to the sound's already large volume of local and international trade, making matters worse.

Tacoma Narrows is the single point in the 20,000 square mile Puget Sound where the mainland and the Olympic Peninsula are close. Work on a suspension bridge began and it was opened on July 1, 1940. Suspension bridges provide an economical solution to the problem of long spans over navigable waterways like Puget Sound where it is difficult to sink piers into the bottom. However, early suspension bridges in the 18th and early 19th centuries were plagued by serious problems of stability and strength against wind forces and heavy loads, failures resulting from storms, heavy snows, and droves of cattle. The German–born engineer John Augustus Roebling solved the problem by adding a web truss to either side of his roadways. He produced a structure so rigid that he successfully bridged the Niagara Gorge at Niagara Falls, New York, the Ohio River at Cincinnati, and, finally, in his masterpiece, the East River between Brooklyn and Manhattan at New York City. The Tacoma Narrows bridge was an experiment by American engineers to replace the traditional deep stiffening web trusses with shallow plate girders to stiffen the roadway, justified because only automobile traffic would be carried by the bridge. Its design represented the pinnacle of suspension bridge lightness, grace, and flexibility. The bridge's designer, L.S. Moisseiff, was rich, successful, and famous in bridge–building circles. The mile–long bridge was the third longest of its kind, and it made a pretty sight as its thin ribbon of steel spanned the narrowest section of Puget Sound.

Yet from the day of its opening, and even during its construction, something was noticeably wrong with the bridge. Moisseiff used a new deflection theory method of calculating stresses, whereby shear and bending loads are partly carried in the cables, rather than relying on stiffening trusses. The vertical and torsional rigidity of the bridge were much lower than contemporary bridges, and the bridge relied on its dead load for rigidity with little inherent structural damping. The highly flexible main span undulated even in relatively mild breezes, rising up and down as much as four feet in winds of only three to four mph. Engineers installed hydraulic buffers at the towers and added heavy tie down cables to the side spans to take some of the sway out by stabilizing the structure, but these measures were not successful in controlling the oscillations. In fact, the bridge's motion became so spectacular at times that it was soon nicknamed "Galloping Gertie." Sensation seekers would come from distant areas for a chance to ride the galloping roller coaster. By November 7 Galloping Gertie had served as a tourist attraction—swaying and flexing—for four hectic months, in addition to its practical use for transportation. The authorities kept a watchful eye on her behavior, but they had become increasingly complacent when no changes were observed from week to week. Everyone was confident that the bridge was structurally safe despite the swaying.

The night before—November 6, 1940—a storm had arisen, but the next morning nothing unusual was observed other than the bridge bouncing in the usual manner. By 10:00 A.M. the wind had picked up to 35–46 mph, the severest the bridge had yet encountered. The center span was undulating 3–5 feet. The alarmed authorities stopped all traffic crossing the bridge. Suddenly, at a few minutes past ten, the motion took on an ominous character. The rhythmic up–and–down action abruptly gave way to a violent twisting spiral motion, much like a snake writhing in the grass after its head is cut off.

In the next few minutes, the writhing corkscrew motion got out of control with increasing violence. Professor F.B. Farquharson of the University of Washington took a position on the side span and began taking motion pictures of the whole ordeal. "My God!" he shouted. "One edge is some 28 feet higher than the other edge, then in the next moment it is 28 feet lower." Nearby, newspaper reporters continued snapping pictures and taking notes. "Look, that guy's car is stalled, and he's fallen out of his vehicle!" one of them exclaimed. "Hey, that's Wendell Jones, one of our reporters!" Wendell had been hurled from his vehicle, his hat falling to the frigid waters of the Sound below. The cables in the main span, instead of rising and falling together as in the usual bouncing movement, pulled and wrenched in opposite directions, tilting the deck from side to side as much as 45 degrees from horizontal. Unbelievably, lamp poles on the bridge tilted almost to the horizontal. Jones hung on for dear life to the curb. Finally, after a few minutes there was a brief lull in the motion, and he crawled on hands and knees to safety. "My dog, he's still in the car!" he screamed. Two men started to make a dash for the car, thinking maybe they could rescue the dog and retrieve the car. But they immediately realized that the risk was too high. They decided that discretion was the better part of valor.

Wendell Jones' car on Tacoma Narrows Bridge

Amazingly, the steel girders, suspenders, and concrete roadway endured the tortuous stresses for a half hour without falling apart. Wendell Jones's car miraculously stayed on the roadway, with his panic–stricken dog inside. A center span floor panel dropped 195 feet into the water below. Finally at 11:02 a.m. the structure could stand no more. Lamp posts started to give way and fall. Then the 2,800–foot main center span buckled, and a 600 foot section of girders and roadway tore loose and went crashing into the Sound with a deafening roar. Seven minutes later the rest of the center span sections fell to their doom. The 1,100–foot side spans, losing support, twitched and sagged down 30 feet, forcing the two towers to rebound and flex back towards shore. So violent was the shock generated in the side spans as the main span fell that

Professor Farquharson was thrown to the concrete floor. He managed to scramble to his feet, and quickly ran out of harm's way back to the shore. Fortunately, there were no human casualties that day. Just one stalled car, and one pet dog—the only life lost in the disaster.

Tacoma Narrows Bridge collapse, looking west

INQUIRY

On November 8, the day after the Tacoma Narrows bridge disaster, Dr. von Kármán (only his family and closest associates called him 'Theodore') was dumbfounded when he read a news item that reported the Governor of Washington state, Clarence D. Martin, as saying the bridge was built correctly and that a new one would be built according to the same basic design. Von Kármán, a noteworthy authority in aerodynamics, had one of his mechanics from the California Institute of Technology (Caltech) build a small rubber model of the bridge to determine what went wrong. He suspected that the fault lay with the shedding of the so–called "Kármán Vortex Street" *(we'll explain what this mysterious term means later)*. Kármán set the model on the living–room table of his home in Pasadena and turned on an electric fan. The model wavered in the breeze. He varied the setting of the fan. At a certain wind speed, the model started to oscillate, showing instability, which grew greater when the oscillation coincided with the rhythm of the air movement from the fan. "As I suspected," he told his sister Josephine in fluent Hungarian. Josephine (Pipö) von Kármán was her brother's closest confidant,

adviser, and friend. "The villain was the Kármán vortices, amplified when the wind swayed the flexible bridge at just the right frequency to resonate with the frequency of the vortex shedding." She nodded in agreement, not having the foggiest idea of what he was talking about.

Von Kármán wrote up his conclusions in a paper which he sent to the *Engineering News Record*, copied Dr. Farquharson at the University of Washington, and wired a summary with his reasoning to Governor Martin in Olympia. He said in effect that if the Governor were to permit the new bridge to be built in the same way as the old bridge, it would fall in the same way. Von Kármán explained the cause of the failure and compared it to the structural instability of a badly designed airplane wing under certain flying conditions. This instability occurs when an accidental twisting of the wing produces forces which increase the twist, he said. Dr. Farquharson, who had been conducting studies of the bridge motion, agreed.

A month or so later, von Kármán was invited to join a federally appointed accident investigation team in order to investigate the Tacoma Narrows bridge collapse. The team's chairman was O.H. Ammann, a Swiss–born engineer, who was the chief engineer for the Port Authority of New York. He was also the designer of the Triborough Bridge and other bridges in the New York area. Two other experienced civil engineers represented bridge trusses and beams. Von Kármán said jokingly that he was there "only to represent the wind" as an aerodynamics expert.

In preparation, von Kármán and his colleagues tested a scaled model of the bridge under varying wind conditions in the Caltech wind tunnel. He and his assistant Louis Dunn found that the model was quiet with no oscillations until they reached a specific wind velocity. Then the model vibrated violently. Duncan Rannie, a Caltech professor, made the necessary calculations and documented a theory of the vibration of suspension bridges in wind.

There were three possibilities for the collapse: (1) Periodic wind gusting 'in tune' with a natural frequency of the bridge. Rannie's calculations showed that this requires precise pressure variation which is unlikely to happen in turbulent wind flow. (2) Von Kármán vortex shedding off the bridge's 'blunt body'. If the frequency of vortex shedding caused by the wind matches a natural frequency of the structure, the driving force for vortex formation feeds off the motion of the structure in a self–reinforcing loop. (3) Self–excitation – here the driving force for oscillation is a function of bridge twist and rate of change of twist and involves interaction between structure and wind. Hence the wind provided the power and the motion supplied the power–tapping mechanism. Essentially, the bridge experienced flutter which excited torsional response modes, to which the structure had little resistance.

Von Kármán maintained that vortex shedding caused the bridge to fall, but Dunn and Professor Rannie argued for the third explanation—self–excitation. To settle the argument, von Kármán directed that they measure the frequency of the bridge's twisting from motion pictures while he calculated the bridge's wind–induced frequency of vortex shedding. The answers came back as 0.2 Hertz for the bridge twist, and about one Hertz for the vortex shedding in a 42 mph wind. Presented with the mismatch, von Kármán conceded defeat, but still maintained that vortex shedding could have contributed to the fall. They all came to Seattle well–armed with data to bolster their case for what kind of aerodynamic forces caused the bridge to fall.

"I told you that Moisseiff properly calculated for a design load with 50% margin that accounted for all the forces and moments acting on the bridge," Mr. Ammann triumphantly huffed at Dr. von Kármán. "He correctly designed the bridge in terms of failure criteria—where the stresses exceed the yield strength of the steel and the tensile strength of the concrete roadway. Why do you insist that we have to build a bridge and stick it into a wind tunnel?" Conventional wisdom was on Ammann's side. Bridge engineers had many decades of experience in designing bridges for all possible loads. Ancillary wind loads against such a heavy immovable object could not possibly cause a collapse when compared to the many tons of steel, girders, stone, and concrete of bridge construction. Or so he thought.

"But your thinking is dominated by the consideration of static forces, like the weight of water pressure—which creates no motion—instead of dynamic forces which produce motion or changes of the motion, ja?" von Kármán hammered back at him. *Von Kármán's voice was his most unusual characteristic—low, slow, and heavy with the accents of his native Budapest—which left the uninitiated baffled.* Bridges had been observed to oscillate in wind before, but nobody had thought such motion was important. Bridge failures were usually blamed on other things. On top of all this, the bridge engineers, excellent though they were, couldn't see how a science applied to a small, unstable thing like an airplane wing could also be applied to a huge, solid, nonflying structure like a bridge.

"I'm telling you," von Kármán insisted, "you're only talking about the static loads. My calculations, backed up by test data gathered in the Caltech wind tunnel, show that what you call small loads can be very dangerous indeed when they cause repeated, amplified oscillations. The applied dynamic forces from the winds caused the bridge to sway up and down as everyone has observed. But never until November 7 were the winds strong enough and steady enough to produce a deadly interaction between the bridge structure and wind. It so happened that the November 7 winds caused twisting and amplified the bridge's torsional response modes."

"Mr. Moisseiff ignored, or was unaware of, the volume of evidence for wind–induced vibration of bridges stemming back to 1818. He extended the slender span concept too far, despite notable gaps in our understanding of aerodynamic loading, to a depth–to–span ratio of 1:350, twice that of the recently completed Golden Gate bridge with a span of 4,200 feet. I think that vortex shedding also contributed to the fall," von Kármán concluded. "What's vortex shedding?" queried Ammann.

"Under just the right conditions, the frequency of vortex shedding matches the bridge's periodic swaying motion, such that on each sway, a wind–induced vortex develops on the leeward side of the bridge and is alternately shed from both sides, like this, ja?" Von Kármán picked up a flat piece of cardboard to demonstrate. "It is just as if you pushed on the back of a girl swinging at just the right time, and she swings higher and higher. Now pretend that my hand represents a vortex shed behind the bridge, like this…" Professor Rannie added, "if you don't stop pushing, the oscillations are not damped, but rather continue amplifying out of control. Eventually, the little girl falls out of her swing seat. If the bridge had swayed five times faster, one oscillation per second, we could have laid the blame entirely on the Kármán vortices shedding. Look at this graph. When scaled, it shows that the original Tacoma Narrows bridge design had a natural frequency of vibration, or what we scientists call "resonant frequency," of about one Hertz. The motion pictures show the bridge vibrated at only one–fifth Hertz. At an average wind speed of 42 mph, Dr. von Kármán calculated the frequency of wind–induced vortex shedding as about one Hertz. So the winds would have had to blow a hell of a lot faster to entirely lay the blame on vortex shedding." "But I still think that vortex shedding was a contributing factor," von Kármán added.

Eventually, von Kármán and his colleagues carried the day. The Seattle meetings concluded that the Tacoma Narrows bridge failed for two reasons: weak structural and torsional rigidity resulting from faulty design and lack of stiffening web trusses; and disregard of self–exciting aerodynamic effects which introduced a new failure mode. The bridge had a solid sidewall and roadway, and it was evident that an interaction between structure and wind caused flutter and torsional response modes, hitherto ignored in bridge design.

Further wind tunnel testing of a 3–D scale model showed the same aerodynamic self–excitation failure mode and proved the validity of similitude between model and a prototype replacement bridge design. The new design was modified considerably to achieve stability. It incorporated a greater depth of stiffening truss to span, double laterals, and hydraulic damping devices. But most importantly for aerodynamicists, it had open steel grid slots between traffic lanes, to equalize the air pressures above and below the deck; and replaced the solid side plates with ones containing a lot of big holes for the winds to go through. These ideas were incorporated into a new bridge across the Tacoma Narrows which was built a couple of years later. Eventually, the plate girder was abandoned altogether in suspension bridge design, and the Tacoma Narrows Bridge was replaced in 1950 by a new span stiffened with web trusses.

Aerodynamic considerations would be taken into account in the design of all extant and future bridges after the 1940 bridge collapse. The Federal government also worked up a plan for examining all the major suspension bridges, like the Golden Gate, Triborough, and Oakland Bay, from this dynamic point of view. All these bridges checked out safely. At the Golden Gate, for instance, the ocean winds meet an open network of stiffening trusses. To develop dangerous oscillations, the wind would have to appear at a certain time and at a velocity of 110 mph over a long period of time.. No such velocity has ever been recorded in that area, so the Golden Gate is relatively safe. Every tall bridge faces unique aerodynamic engineering challenges; for example, the bottom of the roadway in the Millau Viaduct in France is shaped like an upside down wing, so that cross winds generate a downward force against the foundations, stabilizing the structure. Interestingly, when the state of Washington tried to collect the $6,000,000 insurance policy on the Tacoma Narrows bridge, they found that the wily insurance agent had pocketed the premium and had not obtained a policy! He never figured that something as big and durable as a bridge would ever collapse. He ended up in jail, one of the unluckiest men in the world.

THE EARLY YEARS

Theodore von Kármán's unusual amalgam of scientific genius and personal magnetism stemmed from parental influence and the context of his childhood. He grew up in Budapest, Hungary before the turn of the century, in a middle–class Jewish family. His mother, Helen Kohn, descended from scientists, scholars, and men of practical affairs. For example, her grandfather— the author Moses Kunitz —wrote an important book on commentaries on the *Sefer ha–zoar,* the 13th century Hebrew "Book of Splendors," which speculated on the nature of creation, the problem of evil, and the significance of good deeds. Her father enjoyed considerable success as a tenant farmer near Budapest, raising cattle and bees and growing tobacco and wheat. Helen enjoyed the advantages of an English governess thanks to her family's wealth. Kármán's father, Maurice, also numbered learned men in his family, but none like those of his wife. He owned a tailor shop in the provincial city of Szeged and made clothing for the local nobility. Maurice von Kármán achieved eminence not through family position, but by drive and brains. He became a professor at the University of Budapest and commissioner of the Ministry of Education, reformed the secondary–school system of the country and founded an elementary school system called the "Minta." The Minta –aka *Model Gymnasium*—was an open educational laboratory, a "nursery for the elite." Its graduates included such renowned scientists as Edward Teller, Leo Szilard, George de Hevesi, George Polya, and John von Neumann.

The Kármáns worked hard to make a stable home and had a devoted marriage. The eldest offspring, Elemer, born in 1874 was followed a year later by Feri. Then the Kármáns lost one son, and to celebrate the healthy birth of a third boy on May 11, 1881, they named him Todor, or "gift of God." His full name in Hungarian—von Sköllöskislaki Kármán Todor—became known in English as Theodore von Kármán. In 1885 the Kármáns had a third son, Miklos, followed by the youngest, a daughter they named Josephine. Theodore would establish an especially strong emotional and mutually supportive —and chaste—bond with his "darling little sister," whom everybody called "Pipö," later in life. He was also profoundly influenced by his father.

The elder von Kármán not only wrote his children's primers and fairy tales (which he later published), but oversaw their upbringing down to the slightest detail.

He sought openly to influence their values and goals. Skeptical of the efficacy of public kindergarten and grammar school, he hired a former student to tutor them at home. He also ensured that each child not only study, but teach. Hence, Todor learned from Feri, and taught Miklos. Young Todor also participated in several required pastimes: fencing, hiking, skating, music, and trips to the countryside to visit his maternal grandfather. The mostly happy family also entertained great crowds of relatives with parties, and Maurice entertained eminent Hungarian intellectuals, artists, and politicians in their well–appointed family compound.

Theodore (Todor) von Kármán, aged four or five

Child Prodigy

It was from his mother that von Kármán must have inherited a great talent. A marvelous woman of keen humor and intelligence, she was gifted with an unbelievable memory. She could accurately recall details of past conversations with the family's guests to whom she had been casually introduced, and could even recite exactly what they had worn on their previous meeting! One day when he was six, Todor's elder brothers Elemer (aged 13) and Feri (12) were slowly working their way through some arithmetic problems. Little Todor walked over to look at their work. He suddenly began reciting the answers to the problems. To their amazement, every figure proved correct. Even more astounding, in the coming weeks they discovered he could multiply five– and six–digit numbers in his head, yielding sums in the millions! Their mother eventually became aware of this gift and asked how he did it. Todor could not say, and remained just as puzzled about it later in life. Nevertheless, she sensed her husband's disapproving reaction and managed to dampen her son's enthusiasm for a little while.

Inevitably, the secret got out. At one of the family parties, Todor stood by himself watching the relatives drink and enjoy themselves. Suddenly, on cue, Elemer and Feri took him by the hand and led him to the center of the room. Then one uncle, in collusion with the older boys, called for silence. He began to ask Todor multiplication problems of increasing difficulty. All eyes widened as the child's answers matched exactly the hand calculations that followed. The boy enjoyed the attention and, like an entertainer, knew instinctively how to please his audience. He paused expectantly before giving his answer and responded well to the applause and laughter that followed. But among the faces he saw was his father's, which looked not the least bit pleased. It bore a stern and dark expression.

After the guests left, Todor was ushered into the study for a talk. His father, whom he considered "a very wise man," asked him to do an extraordinary thing: banish mathematics from his mind for the next few years. The boy could hardly believe it. His father had always helped him do his sums and encouraged his work. Why should he be prevented from studying a subject for which he had such obvious aptitude? Maurice based his decision on traditional pedagogy. He had known boys of similar talent who squandered their abilities on mathematical tricks rather than serious study. To save Todor from this fate, he made him concentrate on a wide, liberal curriculum. He banned all math (but one algebra text) from the home and presented his son with books on geography, history, and literature. Significantly, the boy meekly accepted the harsh decree. Even though he disliked these subjects, for three years he carefully avoided mathematics in order to please his father. In fact, he obeyed Maurice's dictum to such an extent that he lost the knack of rapid mental multiplication and became a slow multiplier. Moreover, although Todor eventually spoke six languages, he could perform such calculations solely in Hungarian. Thus, he not only repressed this unique mathematical capacity; he reverted to the level of competence that existed prior to the incident with his father. Here it remained for the rest of his life.

Maurice nurtured "a general humanistic interest" in all his children. He taught that science offered only one way of seeing the world. A sunrise might be viewed as God's will; as the basis of artistic expression; or as the fulfillment of Kepler's laws of planetary motion. Scientific knowledge, said the elder Karman, merely organized sensory experiences in logical patterns, while art and religion interpreted the same phenomena emotionally. Thus, all three—religious, poetic, and scientific ways of knowing—had equal validity and significance to mankind.

From this principle Todor grasped the idea of a fundamental harmony in the world, not understood by purely rational or purely instinctive means. In this framework, God and science coexisted peacefully. Maurice von Kármán also imparted a breadth of thought which transcended not only epistemological categories, but narrow, parochial interests. He emphasized that discoveries should be shared and disseminated freely, not safeguarded like personal possessions. Nothing less than the "expansion of man's intellectual heritage" was the objective passed from father to son.

The elder Kármán inculcated Todor not just with the importance of the basic laws of nature and the value of open discussion. He also believed that productive inquiry began with a mind receptive to new ideas. The fostering of intellectual curiosity affected the young Kármán more than any other of his father's precepts. Maurice piqued his interest in the physical world by posing simple questions: Why are the planets fixed in their orbits? Why are some raindrops large and some small? What are the principles of telegraphy? Observation alone would not suffice; the boy had to **understand** how these worked and the father went

to great lengths to teach with examples and experiments. Indeed, when Todor left home for his first day of school, he could not have been more thoroughly prepared to meet paternal expectations.

At age 9 Todor enrolled in his father's Minta. Under the direction of University of Budapest scholars, the Minta method stressed experiential learning: Latin from reading public statues, and advanced math (which Kármán now studied voraciously) from statistical analysis. Memorization was forbidden and rules were arrived at by the student's own inductive logic. This was a skill which the younger Kármán would practice the rest of his life.

Inductive logic involves reasoning from particulars to a generalization: working out particular instances, reasoning to other instances, and then up to generalizations or principles. Deductive logic is its opposite: using a general scientific principle or rule to arrive at (deduce) a particular instance. The former is more powerful, as fundamental scientific discoveries are usually arrived at using inductive means.

A respectable Budapest family in the mid–1890s. Surrounding the elder Kármáns —Helen and Maurice— are (left to right): Josephine, Feri, Elemer, Todor, and Miklos

Maurice von Kármán also defied tradition by assigning graduate students to classroom teaching and prompting free discussion among students and instructors. He came to espouse this philosophy, and in his words:

"My father taught me to believe in God, saying that science deals only with consistency, not with truth. It is man's way of organizing his experiences, and nothing more, my father would say. I never saw any conflict between science and religion, and I think it is because of these early discussions with my father.

We cannot say with certainty that there are no miracles, because the world may be much more complicated than we see it. This limitation of science was always ingrained in my way of thinking. Science is not something whose truths we can absolutely believe. The moment we cannot explain some phenomenon with the laws we have obtained up to now, we have to change these laws and find some new ones that fit."

The young Todor succeeded brilliantly at the Minta, passing through the lower grades with ease. His high school career ended with great promise. For example, along with a number of superior students, he entered a national competition for the Eötvös prize, awarded annually to the most deserving science or math student in Hungary. A nationwide committee screened all applicants and a select few were chosen to apply their imaginations to some highly complex mathematical problems. In 1897, young Kármán won the contest. The intellectually gifted genius planned to perhaps enter a great foreign university.

However, that very year his father suffered a mental collapse. Years of feuding with colleagues and rivals, and debilitating migraine headaches had taken their toll. Suffering a nervous breakdown, Maurice would have to enter a Budapest sanitarium, where he would be isolated from his wife and children for four years. This episode devastated the Kármáns. Brothers Elemer and Feri took jobs to support the family. Deeply troubled by his father's illness, Todor stayed in Budapest to save money and assist his parents. Maurice had been his closest companion, his mentor, and the center of his life. He enrolled at the Royal Joseph University of Polytechnics and Economics, a former trade academy that had become a technical school through royal patronage.

Despite Maurice's release from the sanitarium in 1901 and the resumption of close ties between father and son, by now the younger Kármán had discovered his own identity and began to orient his career according to his own tastes—towards engineering and away from pure science. On completing his undergraduate studies with distinction in 1902, Theodore decided to pursue his engineering career in the academic world, which would enable him to fulfill his wide scientific interests and to practice the art of teaching, a trait which his father had inspired in him. He retained

a brilliant memory and discovered he could learn varied subjects with ease. He also found his thought processes slow but deep, a characteristic which enabled him to make key advancements in the field of aerodynamics while relatively young. In later years, he would be delighted when engineers to whom he had imparted his scientific attitude and methodological approach acknowledged him as their teacher.

The Romantic

As a student, and early on in his career, Theodore von Kármán had acquired quite a reputation as a ladies man. He cut a dashing figure, with a mop of curly brown hair atop an elegantly shaped head. He was well known to pursue dalliances with women, and was a great party-goer. Uniquely attractive, he was five feet eight inches tall, had a sizable aquiline nose, sparkling grey eyes, and a half–smiling, lively expression. There was an aspect of shyness in his manner, characterized by a gentle shrugging of his shoulders as he spoke. His Hungaro–English accent made him habitually accent the first syllable of words and he was sometimes difficult to understand. But he turned this weakness into a strength and a point of charm. His melodious baritone accent was pleasing to the ear. His looks and manner complemented his voice. His hands, which habitually sliced and swept the air, always accompanied his discourse.

von Kármán as an undergraduate at the Royal Joseph University, Budapest, around 1900.

Between 1903 and 1906 von Kármán served on the faculty of the Polytechnic University and as consultant to Ganz and Company, the principal Hungarian engine manufacturer. The research that von Kármán conducted on the strength of materials prepared the way for important later contributions to the design of aircraft structures. He attracted attention, and was soon awarded a prestigious two–year fellowship to the University of Göttingen, Germany, in order to obtain a doctor's degree. By that time Göttingen had become a training ground for the most elite scientists, researchers, and engineers in Germany; and Germany itself was an acknowledged world leader in the sciences. In due course von Kármán finished the fellowship, but decided to interrupt his doctoral studies. His father's words of wisdom came back to him: *Don't specialize. Don't content yourself with study in only one nation.* In other words, be an internationalist. Von Kármán took it one step further. He resolved to <u>never</u> be a nerd, whether of the engineering, technical, or scientific variety. The best vaccine against nerditis, he decided, was waiting for him in Paris.

In March, 1908 he arrived in the City of Lights with Julius A. Vészi, his friend from Göttingen. They settled into a strategically located apartment on the Left Bank, overlooking the river Seine. Von Kármán did not choose to go to Paris with Julius Vészi by accident. Vészi had three beautiful daughters, Hungarian and Jewish, of course. Von Kármán took a fancy to Margit, the oldest one. Conveniently, she had just divorced her husband, the Hungarian dramatist and playwright Ferenc Molnár. By day he would drop in on classes at the Sorbonne, and listen to lectures by notables such as Madame Curie (born Maria Salomea Skłodowska), famous for groundbreaking research into radioactivity, and the first person to be awarded two Nobel prizes—in physics and chemistry. By night he would prowl the Parisian night life, sometimes accompanied by Margit, at other times by another beautiful woman.

Margit—his favorite—was studying sculpture and writing for a Parisian newspaper. Early one morning von Kármán was sipping coffee at a student hangout on the Boulevard St. Michel, trying to recover from a hard night of partying. Margit found him and came rushing up with a request. "Sweetie, would you give me a ride to an historic event? The Englishman Henry Farman is attempting the first two–kilometer airplane flight in Europe, and my newspaper has assigned me the story! Can you take me to the airfield—please, please, please?"

Margit Vészi

"What airfield?" von Kármán asked. "I didn't know there was one." "Well, they have a makeshift one at Issy–les–Moulineaux, a small Army parade ground southwest of Paris, on the Left Bank," she said. "Well, ok. When do you need to go?" "Five a.m."

"Five–what?" von Kármán retorted, not sure he heard her right in the din of the café. "Five in the afternoon, why I suppose. Sure."

"No honey, five in the *a.m.* But this is the world's first! Well, o.k., the first on the Continent. We can't miss it, don't you see!" "I am not interested in historical occasions at 5 a.m.," he told her bluntly. "Besides, I'm working off a hangover." But she was persuasive, and they ended up driving to witness this "historic event."

They arrived just before five and found a small, odd assortment of literati and street urchins watching in the twilight as the Voisin was pushed out of a hangar at the south end of the field. Farman then threaded himself through the tangle of wire braces onto a seat on the lower wing. The engine turned over with a bang and as the propeller started to turn, von Kármán watched expectantly as the pilot wheeled the clattering machine 150 feet down the field. When the aircraft reached a distance of 500 yards from a marker on the far side of the parade ground, the Englishman gave the beast full throttle and up it went, making lazy circles from the starting line to the turning point and back again. At the end of the second pass, he maneuvered the fragile steering rudders so that the plane sailed safely between two poles planted at the finish line.

In von Kármán's own words:
"We parked in an open space and joined the small excited crowd gathered to see the event.

A short distance away was a Voisin biplane, a fragile box kite made of sticks, paper, and wire. I found myself watching with great interest as the bearded pilot, Henry Farman, got in front between the wings and started up the motor. There was a clatter, and up went the odd contraption. Manipulating it against the strong wind he flew to a one–thousand–meter marker, turned, and flew back again. He maneuvered the fragile steering rudders so that the plane sailed safely between two poles planted at the finish line. The event was finished quickly amid cheers, and I learned it was somewhat of a record."

A record indeed. Farman had traveled 2,005 meters in 3½ minutes, setting a world record for an officially witnessed flight. The crowd was awestruck by the event. This watershed event provided the inspiration for the young man who was to became a founder of the aeronautical and astronautical sciences. Three things about the display excited Kármán's scientific imagination:

1. How had the Voisin been able to overcome Sir Isaac Newton's theory of air resistance and particles, which seemed to pose an insurmountable barrier to flight?
2. How had the aircraft's wings supplied enough lift to overcome the plane's considerable drag?
3. Finally, how had the fifty-horsepower Renault engine produced enough power to levitate its own weight, as well as that of the airframe and Farman himself?

Karman began to investigate. He would spent his career seeking solutions to these three problems. Shortly thereafter, Professor Ludwig Prandtl whom he had gotten to know during his fellowship, invited von Kármán to return to Göttingen as his assistant on dirigible research and to complete his doctoral studies. The environment at the university was admirably suited to develop von Kármán's talents. He would meet and work with the world's best mathematicians and scientists. He responded, in particular, to the school of the eminent mathematician Felix Klein, which stressed the fullest use of mathematics and of the basic sciences in engineering to increase technological efficiency.

However, there was a catch. Prandtl's offer to Kármán came with the title "Privat Dozent," the purgatory of German academic life. Professors normally headed their own institutes; associate professors acted as paid faculty and researchers. Privat dozents, however, were temporary workers awaiting permanent appointment elsewhere, with no tenure, no faculty voting rights, and no salary except what students were willing to pay! Until an associate vacancy opened, men like Kármán could do nothing but wait. As a friend observed, the best way to hasten promotion would be to sink the ship that recently carried Göttingen's mathematicians to a conference in New York. Kármán suggested a more devious, if less deadly, plan that better suited his persona. A privat dozent, he joked, had the *right* to teach, and the *duty* to marry the daughter of his superior.

In the meantime, Prandtl kept the Hungarian busy with a dirigible project. Count Ferdinand von Zeppelin had conceived the idea of rigid, nonmotorized airships at the turn of the century. The German army expressed interest, and Felix Klein persuaded the government to underwrite Prandtl's investigations of its flight characteristics. "In my mind the handsome gas–filled bags of the skies were one of the great products of early aeronautical engineering," von Kármán once said. Kármán not only collaborated in building the airship wind tunnel, but undertook some of Göttingen's first aeronautics experiments and taught courses in mechanics.

USS Macon mooring at Sunnyvale, California, after flight from Lakehurst, New Jersey, 1933. A year and a half later she crashed.

The German airship Hindenburg begins to fall seconds after catching fire during its attempt to dock with its mooring mast at Lakehurst, May, 1937. Thirty-six people were killed in the 1937 Hindenburg disaster. It shattered public confidence in giant, passenger–carrying rigid airships and marked the end of the airship era.

LUDWIG

The German physicist Ludwig Prandtl was an interesting personality, with his distinguished–looking, formal Prussian mien. Prandtl was precise of speech, and rather aloof in manner. He cast a formidable, no–nonsense, dignified appearance with thick black mustache and beard, wire–rimmed glasses, and well–groomed dark brown hair. But he was a man of strange contrasts. Sophisticated and gifted in science, and a great teacher on a one–to–one basis, he was nonetheless terribly naïve about life and childlike in behavior. He had a well–known love of toys, and was somewhat impulsive. One day around 1914, a young Dr. von Kármán came to visit Prandtl, his supervising professor and mentor, at the University of Göttingen in Germany. "Hello, Herr Doctor" von Kármán cheerily announced in perfect German as he entered Prandtl's office. Prandtl, of course, could speak no English. He looked up from his work, and the two men exchanged brief pleasantries. Finally, Prandtl got to the point. "I have been thinking about getting married," he announced. "Oh, that's nice. Who with?" asked von Kármán. Prandtl just stared at the wall in silence. "Uhh. Umm…" "Yes?" queried von Kármán again. "Well, you see, that is what I'd like to talk to you about…I just don't know how to go about finding a wife." "You, what?" asked von Kármán incredulously. "You're nearing 40 years of age. You are secure financially. You have a full professorship, a permanent position at this prestigious university, and you don't know how you can find a wife?" "No, I am afraid I do not," replied Professor Prandtl, somewhat sheepishly.

Professor
Ludwig Prandtl

and asked which one of them wanted to marry Prandtl. Both daughters replied that Prandtl had been nice to them. He had made formal calls and had shown them funny mechanical toys, but he had never spoken of love. There was a further family discussion, and it was decided to give Gertrude, the elder daughter, and to save the younger daughter for another candidate. Prandtl didn't object. The marriage turned out quite satisfactory, and their union produced two daughters!

It should come as no surprise that great scientists like Prandtl and the brilliant 18th century Englishman Isaac Newton, and great inventors like the Wright Brothers, should be so inept in the art of love. All except Prandtl were bachelors. Though adept at lovemaking, Theodore von Kármán, too, would remain a bachelor all of his life. It is a dichotomy that great brains can be so focused on their passions for discoveries and science that they exclude everything else, including marriage. For example, Newton's only close contacts with women were his unfulfilled relationship with his mother, who had seemed to abandon him, and his later guardianship of a niece. He also cloistered himself for some months while fanatically working out his priceless masterpiece, *Philosophiae Naturalis Principia Mathematica*, which is acknowledged as a fundamental work for the whole of modern science. Alas, such is the price of brilliance.

Notwithstanding his social awkwardness, Ludwig Prandtl was a pioneer in aerodynamics and the associated field of fluid mechanics, without which the science of rocketry would not be where it is today. In fact, Prandtl is commonly regarded as the father of modern aerodynamics. He made a profound discovery on the motion of fluids around stationary objects. Through careful analysis and experimentation, he found that "boundary layers" of viscosity existed along the borders of such solid forms, interacting with adjacent free–moving fluids. The minute friction that developed along the surfaces of such shapes, he determined, could have a far–reaching influence on the overall flow pattern around an object, thus retarding its efficiency of motion. Today we understand that a very thin layer of fluid— the boundary layer—attaches itself to a body—no matter how smooth——as it moves in a fluid, such as the hull of a boat crossing a lake. But back then, this mysterious phenomenon had eluded the mathematical as well as the experimental factions of fluid mechanics. Prandt used approximate mathematical calculations and water canal experiments to unveil this mystery. Then he

Dr. Prandtl was well-acquainted with von Kármán's savvy as a romantic and ladies' man. Hence his embarrassed request for assistance. After some verbal exchanges, von Kármán gave him his best advice.

"Why don't you write to Mrs. August Föppl, the wife of your mechanics teacher in Munich? The Föppls have two eligible daughters, one in her early 20s, the other in her late 20s."

Taking Kármán's advice, Prandtl asked Mrs. Föppl for the hand on one of her daughters; but he didn't specify which one! Flattered but puzzled, Mrs. Föppl called her daughters into a kind of emergency meeting made a historic scientific leap by deducing that the same boundary layer that existed in fluids must also act in the air, which today's aerodynamic engineers also recognize as a "fluid." Prandtl introduced the world to the idea of "drag" in aeronautical design.

AACHEN

After four years as a privat dozent Theodore von Kármán despaired his prospects as a junior researcher in the inflexible, rigid German academic system. In 1912 he applied for and won a professorship in applied mechanics at the College of Mining Engineering in Selmeczhanya, Hungary. But after arriving there in the fall of 1912, within weeks he decided he'd made a mistake coming to such an academic backwater. Von Kármán packed his bags and returned once again to Göttingen, leaving the job temporarily in the hands of a friend.

Luckily, his old friend Felix Klein soon obtained for him a position as professor of aeronautics and mechanics at the Technical University in Aachen, Germany. Not only that, but Klein also obtained for Kármán the directorship of the Aachen Aerodynamics Institute, a weighty title for someone so young at the age of 31. Young though he may have been, von Kármán liked everything about Aachen. Centered in the prosperous Rhineland at the corner of the Belgium, Dutch, and German borders, and only a short train ride from France, the town had the bustle and cosmopolitan flair he missed in Göttingen. After more than six years among the stolid Protestant burghers, Kármán relished the worldly, leisurely atmosphere of Aachen, former seat of medieval emperors and bastion of Roman Catholic influence.

Theodore von Kármán would remain as director at Aachen until 1930, with an interruption during World War I when he was called into military service. While at the Military Aircraft Factory at Fischamend in Austria, he led the development of the first helicopter tethered to the ground that was able to maintain hovering flight.

After the war, as his international reputation grew, so did that of the Aachen Institute. Students came from many countries, attracted by the intellectual and social atmosphere he had created. To help reestablish contacts and friendships broken by the war, he was instrumental in calling an international congress on aerodynamics and hydrodynamics at Innsbruck, Austria, in 1922. This meeting resulted in the formation of the International Applied Mechanics Congress Committee, which continues to organize quadrennial congresses, and gave birth, in 1946, to the International Union of Theoretical and Applied Mechanics, with von Kármán as honorary president.

Lieutenant Theodore von Kármán of the Austro–Hungarian Air Force in World War I. "I was always comfortable with the military," he maintained.

Von Karman teaching at Aachen Technical University. "Some of the faculty and students regarded me suspiciously…Aachen was used to elderly white–bearded professors," he remarked.

Early wind tunnel at the Aachen Aerodynamics Institute, with trumpet–shaped opening to draw air in from outside. it was later modified to a closed–circuit wind tunnel patterned after the one at Gottingen .

The world's first "captive helicopter," developed by von Kármán, Wilhelm Žurovec, Oszkár von Asboth, and Oberstlt Stephan von Petróczy for the Austro–Hungarian Air Force in World War I, taking off in test flight.

Close–up of engines and counter–rotating propellers of the von Kármán, et. al. helicopter.

Von Kármán (center) with his staff at the Aachen Aerodynamics Institute, 1920s. Many German aircraft manufacturers supported the Institute's research.

Josephine (Pipö) von Kármán, her brother's closest confidant, advisor, and friend, mid–1930s.

Von Kármán never married. His mother Helen, and his sister, Josephine, lived with him from 1923 onward in the Netherlands near Aachen and later in Pasadena, California. His sister was his manager and hostess until her death in 1951 in America. Brother and sister were devoted to each other, and her death plunged von Kármán into deep depression for several months, during which he was unable to work.

Von Kármán had begun traveling widely in the 1920s as a lecturer and consultant to industry, with wide–ranging trips to Japan, Europe, and the United States. During his first visit to the U.S. in October 1926, he met Orville Wright. Wright invited him to lunch at the Engineers' Club in Dayton. Afterwards, they went together to his private laboratory. Von Kármán later remarked that this visit was the highlight of his trip. He found the famous aviator both sincere and unassuming, and knowledgeable with the fundamentals of aerodynamic theory. Orville told him that before their historic flight at Kitty Hawk, he and his brother Wilbur spent almost 2000 hours with their small wind tunnel, studying the relative merits of various wing shapes. Von Kármán remarked, "scientists study the world as it is; engineers create the world that never has been."

As he passed his 40th year, von Kármán and his colleagues succeeded in reformulating the boundary layer equations so that comparisons between theory and experiments could be undertaken. First, he integrated Prandtl's work into a single momentum equation. Then, rather than rely on his mentor's partial differential equations, Kármán worked with his colleague Karl Polhausen to develop an integral relation in which a general solution resulted from assumptions on the distribution of flow velocity. It enabled Kármán to describe the boundary layer as a complete mechanism, using a method that came to be known as the Kármán–Pohlhausen approximation. This new application of a well–known mathematical technique had crucial practical implications, allowing an accurate prediction of drag on surfaces moving through air or fluid. It affected not only the design of aircraft and future rockets, but was of critical importance to petroleum and hydraulics–related industries, which were deeply concerned with the frictional motion of liquids through pipes.

On the strength of discoveries such as these, Dr. von Kármán gradually transformed the University of Aachen into an aeronautics powerhouse, the rival of Göttingen. Students the world over learned of the institute in Aachen and its atmosphere of free inquiry and sought admission. Von Kármán's fame and reputation grew as he lectured to audiences over the length of Europe, explaining with exuberance the unfolding of aeronautical knowledge.

Although America possessed the facilities to train engineers and NACA offered superb facilities for practical research, the country lacked a nerve center for advanced studies in theoretical aerodynamics. Germany, with its great universities and research facilities, led the world in this respect. So it is no wonder that on one of his trips to the United States he was invited to assume the direction of the Guggenheim Aeronautical Laboratory at the California Institute of Technology (GALCIT) in Pasadena, California and of the Guggenheim Airship Institute at Akron, Ohio. His love for Aachen made him hesitate, but the darkening shadow of German Nazism caused him to accept in 1930. He would never regret his decision.

The following pages present a brief overview of aerodynamics. Disciplinary specialists will find this treatment very basic and elementary, whereas to the lay person, the subject of aeronautical engineering is practically incomprehensible. Hopefully the material will strike a balance in between. In any event, aerothermodynamic effects critically affect rocket and missile performance in flight.

The brilliant English scientist Isaac Newton's work in setting forth the laws of mechanics marked the beginning of the classical theories of aerodynamics. After years of experiments and hard work he formulated the three laws that bear his name:

1. *An object at rest tends to stay at rest and an object in uniform motion tends to stay in uniform motion unless acted upon by an external force.*

2. *An applied force on an object equals the rate of change of its momentum with time. An object's momentum is simply its mass times its velocity. This simple–sounding law would form the foundation of engineering mechanics, forces, and moments for centuries to come.*

3. *For every action there is an equal and opposite reaction.*

EXTERNAL AERODYNAMICS

An aircraft or rocket in flight must balance and control the four principal forces acting upon it: (1) weight pulling it downward towards the center of the Earth; (2) lift on the wings, fins, and body surfaces pulling it upwards to counteract gravity; (3) drag, the air resistance pushing it backwards against the direction of motion, and most importantly (4) propulsion which balances against drag and induces forward motion. Unfortunately, these four effects are closely intertwined. What you do to one affects the others, and both rocket engineers and airplane designers have to be well–versed in how they all interact. In addition, it takes many specialists in each of these fields in order to design and produce a modern rocket vehicle or spacecraft, such as aerodynamic engineers to maximize lift and minimize drag. These two figures of merit are summarized as a ratio of an aircraft's overall lift to its overall drag. This "L over D" or L/D is a measure of overall aerodynamic efficiency.

Dynamic Pressure

The concept of dynamic pressure is important because it greatly affects rockets (or for that matter, any vehicle) in flight. In very simplistic terms, you can consider dynamic pressure—a force per unit area—as a giant "hand" trying to push back on a vehicle in flight.

the density of air þ times velocity squared. A fine point: velocity is different than speed. Speed is how fast you're going directly straight against the air (like a car on a level road), velocity accounts for the angle at which you're going in free flight, like an airplane. We'll use the term velocity since we're talking aerodynamics here.

Newton considered the pressure of the air acting on a moving inclined plate as arising from the impingement of particles on the side of the plate that faces the airstream. Common sense tells us that this pressure is greater the more you incline the plate to the air flow. But Newton went a lot further. Through experiments bolstered by deep thinking, he figured out that the pressure acting on the plate was proportional to the product of the density of the air, the area of the plate, the square of the velocity, and the sine of its angle of inclination as shown in the figure.

However, as we'll see below Newton's dynamic pressure formulation failed to account for the effects of the flow on the upper surface of the plate where low pressures exist and from which a major portion of the lift of a wing is produced. The idea of air as a continuum with a pressure field extending over great distances from the plate was to come much later.

Static Pressure

The term "pressure" usually refers to "static pressure," which is simply the force per unit area exerted

Comparison of Aerodynamic Forces

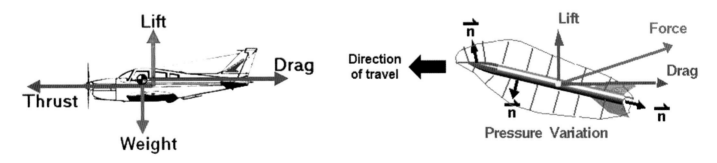

Aircraft

Missile

Dynamic pressure on a body in flight underpins our understanding of aerodymanics today and is central to the determination of lift and drag coefficients as explained below. In engineer–speak dynamic pressure, denoted as q, represents the kinetic energy per unit volume of a fluid particle. Engineers define it mathematically as one–half

by a fluid that is not in motion. For example, the pressure of air at sea level is 101 kilopascals, or 14.7 psi (technically speaking, psia for pounds per square inch absolute). The pressure in your car tires is 240 kpc (35 psi), and so on. In the aerodynamic context, static pressure, commonly called "freestream static pressure",

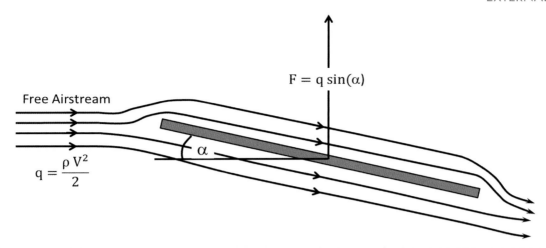

$$F = q \sin(\alpha)$$

Free Airstream

$$q = \frac{\rho V^2}{2}$$

α

is the pressure of the atmosphere at the altitude at which an aircraft is flying, and is generally measured using static ports on the sides of the fuselage.

Stagnation Pressure

Stagnation pressure is the sum of the static and dynamic presssure acting on a body or surface (for non–moving bodies the stagnation and static pressure are the same). Aircraft in flight can't determine the dynamic pressure directly. They use pitot tubes to measure the stagnation pressure in flight first, then subtract off the freestream static pressure to determine the dynamic pressure of the air in motion around the craft, then apply Newton's formulation for dynamic pressure to determine their airspeed. By the way, instead of separate pitot and static ports, the two can be combined in a pitot–static tube, but this is less common.

Lift

Even before World War I, the practical elements of lift were pretty well understood, as we have seen in the work of Otto Lilienthal and the Wright Brothers. However, while they were great experimentalists and aeronauts, the Wrights did not fully understand the scientific underpinnings of aerodynamics. With the relatively poor communications in those days, researchers and scientists like Robert Goddard would often make great discoveries and advances in science independently; whereas in today's Internet world such developments are collaborative endeavors, relying upon the work of many, one building upon the other.

Unbeknownst to the Wright brothers, the discovery of the scientific nature of lift is due to Martin Wilhelm Kutta, a German mathematician, and to Nikolai.E. Joukowski, a professor of mathematics at the University of Moscow early

in the 20th century. Joukowski showed by mathematics that when you subject a body in motion to circulating air, or in other words when you spin it in a wind, its motion changes due to lift. The Kutta-Joukowski theorem is well-known to aerodynamicists. It relates lift to an airfoil's velocity, and the air density and circulation. In 1913 the German mathematician Otto Blumenthal also made important (but little-acknowledged) contributions to Joukowski and Kutta's work.

In 1897, unknown to any of these men, Frederick William Lanchester—an amazing British engineer and inventor—had independently proposed a circulation theory of lift of an airfoil of infinite span and a vortex theory of the lift of a wing of finite span. Ludwig Prandtl was yet a fifth person that arrived independently at the same hypotheses as Lanchester and developed the mathematical treatment some years later. Prandtl's work, refined and expanded by subsequent investigators, formed the theoretical foundation of the field.

Magnus Effect

But even before Kutta, Joukowski, Blumenthal, Lanchester, or Prandtl; the German physicist and chemist Heinrich Gustav Magnus was the first to experimentally investigate the effect in 1853, so it is often called the Magnus Effect. A cylinder spinning in a fluid tends to speed up the flow near its surface, in the direction of its spin, and slow it down where its spin is contrary to the flow direction. This creates lift perpendicular to its axis on the side where the flow is flowing fastest. In aircraft design, the Magnus Effect became the basis for the theory of lift. This is because the wing is so shaped as to cause the same type of circulation as obtained by spinning. When the airplane wing is propelled in straight flight the circulation of air around the wing creates the pressures that result in lift.

The Magnus
Effect

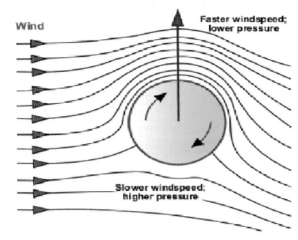

Spinning balls in a thin medium like air are also subjected to lift and transverse aerodynamic forces that cause them to swerve in flight. In this case the forces are due to the asymmetrical nature of the twin swirling eddies that develop at the rear of a spinning sphere, and to a lesser extent the Magnus effect. In everyday life, twin swirling eddies are readily evident as they form behind both sides of a motorboat moving leisurely through the water.

As another common example, consider the motion of a tennis ball under spin. As the ball is hit by the racquet with a downward undercutting chop, the ball rotates as it moves forward. The air passes over the top at a higher velocity and lower pressure, and passes underneath at a lower velocity and higher pressure than is the case without spin. As a result, there is an upward force on the ball and it tends to rise. This represents lift. Conversely, if the ball is 'topped,' that is given a top spin, it sinks at a faster rate than if hit without such spin, making it harder to return.

Curving Baseball

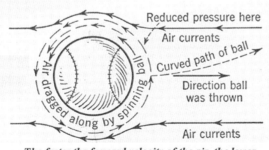

The faster the forward velocity of the air, the lower the pressure exerted sideways.

Other examples are when baseball pitchers pitch a 'curve ball' to the batter, and when golfers hit a 'slice' off the tee, where the ball enters a curved trajectory due to side–spin, rather than going straight. Cricket balls— unlike the balls used for baseball, tennis, and golf— have a raised equatorial seam that plays an important part in making the eddies asymmetric. A bowler in cricket who wants to make

the ball swerve imparts spin to it, but he does so chiefly to ensure that the orientation of this seam remains steady as the ball moves toward the batsman.

Explanation of Lift

There are two alternative explanations of lift. Both are based upon sound scientific principles and give the same result. In 1738 the Swiss mathematician and physicist Daniel Bernoulli published the simpler of the two in his book *Hydrodynamica.* In a nutshell, the "Bernoulli Principle" that bears his name states that if you decrease the speed of a flow, its pressure decreases. Conversely, by slowing a fluid down you increase its pressure. Fluid mechanics and civil engineers apply this principle all the time in the design of dams, water channels, airplane wings, or wind tunnels. The Orville chapter described how the Wright Brothers took advantage of Bernoulli's Principle in their wind tunnel design.

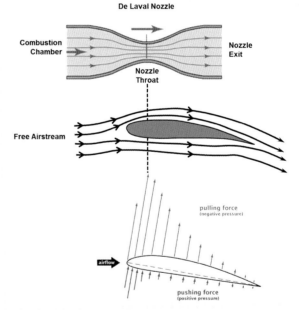

The preceding Robert chapter described the de Laval rocket nozzle shown above which is also based upon Bernoulli's pioneering work. Bernoulli formulated the equations that describe how fast the flow increases when you decrease the cross–sectional area of the flow. When the flow gets past the nozzle, Bernoulli's Principle applies in reverse. The fluid's static pressure decreases, so guess what — its speed increases. The airplane wing shown below the de Laval nozzle operates the same way. Engineers illustrate the direction of the airflow with the streamlines shown. Notice how the air has to flow over the top surface of the wing. In essence, it has farther to travel than the air flowing beneath the wing. So basically it has to speed up. When

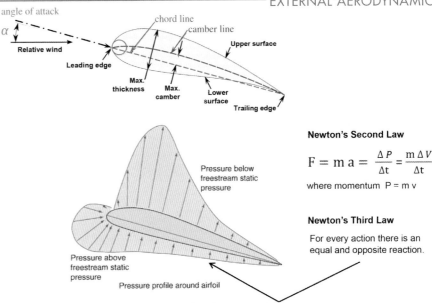

Newton's Second Law

$$F = m\,a = \frac{\Delta P}{\Delta t} = \frac{m\,\Delta V}{\Delta t}$$

where momentum $P = m\,v$

Newton's Third Law

For every action there is an equal and opposite reaction.

a flow speeds up, its pressure decreases in accordance with Bernoulli's equations. Notice how the nozzle throat and areas where the flow streamlines are squeezed together perfectly line up with where the pressure is lowest on the wing's upper surface.

But be careful. Don't fall into the "Equal Transit–Time Fallacy." When the air splits into two, the air going around the top DOES NOT meet the same air that went around the bottom. In fact, measurements have shown that it goes much faster and gets to the wing's trailing edge before the air that took a shortcut, even though it has farther to go! In this case, nature's generosity means that wings generate lots more lift than if the same two air flows met behind the wing.

The second explanation of lift is a bit more complicated and involves Newton's Laws of Motion, conservation of mass, and conservation of energy.

Consider air flowing over a wing. *Airfoils and their notation were introduced in the Orville chapter.* The pressure is higher in front of the wing as it comes against the airstream as shown. This acts like a force F on the mass of the airstream m. According to Newton's Second Law this produces an acceleration a of the airstream flowing over the wing. In other words, the force F induces an equivalent change in the air's momentum, P. But momentum P is the air's mass times its velocity. By the law of conservation of mass, the mass of the air, m, stays the same because air is not being added or taken away, so the only way to increase its momentum is to increase its velocity V (remember, ΔV is a change in velocity). Since it's going faster, its kinetic energy is much higher (kinetic energy is one–half mass times velocity squared). Now we apply the law of conservation of energy, which holds that the total energy of the airstream is constant, since no energy is being added to it or taken away. Since the total energy of the air is equal to its potential energy plus its kinetic energy, the potential energy of the flow has to go down since its kinetic energy went up. What does this mean? Think in terms of a bottle of compressed air. Its potential energy is the pressure in the bottle times the area of the top of the bottle where the air escapes times the length of the bottle neck through which the escaping air has

to travel, roughly speaking. Since the physical geometry of the wing isn't changing, then the only thing that can change is pressure, and it has to go down (to make the potential energy go down).

Look at the bottom of the wing. Just like with the flat plate above, the air is hitting it from below, the "action". If you push on something, there has to be an equal and opposite reaction in accordance with Newton's Third Law. In this case, the opposite reaction is that the wing is forced upwards, in the same general direction as the Second Law effect.

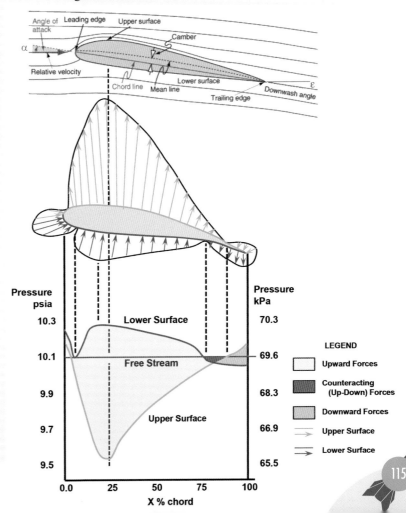

Whew! That was complicated but we're not quite done yet. The interaction of a cambered airfoil with the air creates an additional pressure pattern towards its aft end as shown on the previous page. Suppose you're cruising along at 10,000 feet (3,050m) in a typical subsonic airliner. The ambient static air pressure—shown as the red free stream line— is around 10.1 pounds per square inch absolute (psia), or 69.6 kilopascals (kPa). This what a properly positioned static port on the plane's fuselage would measure. The "angle of attack" is the angle between the chord line and the free air stream. The green and purple colors illustrate what's going on the upper and lower surfaces of the wing, respectively. The arrows show the direction of aerodynamic forces, up for positive force (lift), down for negative force, and right in the direction of drag. Notice how the dotted vertical lines line up with different points on the airfoil, indicating regions of maximum and minimum static pressures on the wing surface. The "downwash angle" shown near the trailing edge is associated with both the trailing edge and spinning vortices generated by the wing tips. These are associated with "induced drag due to lift" as will be explained below.

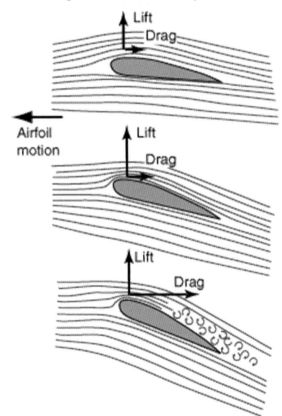

Cambered airfoils produce lift even at zero angles of attack, defined as the angle between the free air stream and the chord line. Sport planes can even fly upside down by using a negative angle of attack to produce "lift" upwards!

When the angle of attack gets too big the airflow separates from the top surface and forms a turbulent pattern. Lift decreases, drag greatly increases, and the wing 'stalls'. Pilots have to recover from the ensuing spin or other wild gyrations before it is too late.

The "lift coefficient" is a non–dimensional number that relates the lift generated by an airfoil, the dynamic pressure of the fluid flow around the airfoil and the planform area of the airfoil. It may also be described as the ratio of lift pressure to dynamic pressure. The lift coefficient also relates the total lift generated by an aircraft to the total area of the wing of the aircraft. In this application it is called the aircraft lift coefficient (C_L). Of course, the concept of lift coefficient also applies to rockets and missiles in flight, where it is also known as C_L. Because of the "real world" complexities involved, C_L often has to be determined experimentally, usually with scale models in wind tunnels.

Drag

In early times, it was thought that the impelling force of a projectile was associated with forces exerted on its base by the closure of the flow of air around the body. This conception of air as an assisting medium rather than a resisting force persisted for centuries, even though by the 16th century it was recognized that the energy of motion of a projectile was imparted to it by the catapulting device. Newton's pioneering work on dynamic pressure led to the conclusion that the aerodynamic drag on a body is directly proportional to the density of the air (or other gas), the cross–sectional area of a body perpendicular to the direction of motion, and the velocity squared.

Drag Coefficient

By the Wright brothers' time, a dimensionless quantity called the "'drag coefficient" had been coined to quantify the drag or resistance of a body in a fluid environment such as air or water. Instead of Newton's cross–sectional area engineers used the term "reference area," whose meaning changed depending upon the type of body. For rockets without fins, automobiles, and many other objects, the reference area is the frontal (i.e., cross–sectional) area of the vehicle when viewed from ahead. For airfoils or rocket fins, the reference area is the chord of the airfoil multiplied by the length of the span, which can be easily related to wing area. Since this tends to be a rather large area compared to the projected frontal area, the resulting drag coefficients tend to be lower than for cars or bluff bodies. Submerged streamlined bodies like submarines use the wetted surface area.

Today, we know that there are four types of drag: (1) friction drag, (2) induced drag due to lift, (3) form drag, and, for supersonic flight, (4) wave drag. Sophisticated computer programs are used to derive the lift and drag on complex bodies like the Space Shuttle

1. Friction Drag

The biggest source of drag is obviously the first type—air friction—the rubbing of molecules of air against the flying object. In the early 1900s, the mathematics used to describe this friction were so complicated that it was quite hopeless to apply them to the design of an airplane wing.

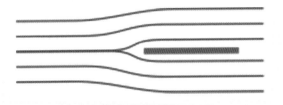

Skin friction drag

Ludwig Prandtl set his mind to this problem and came up with a brilliant method of attack. His *boundary layer theory* afforded a clue to unraveling the mystery of drag. Prandtl ingeniously assumed that the total effect of friction on any part of the airplane can be estimated by the trick of restricting the effects of aerodynamic forces to a thin sheet of air close to the surface, which he termed "the boundary layer." The boundary layer is like a thin film that coats every streamlined object flying through the atmosphere. Within it, air particles change from a smooth laminar flow from the leading edge to a more turbulent state towards the trailing edge. In the process, drag increases. After observing tests in a smoke tunnel and evaluating other data, aerodynamicists concluded that the prime culprits in disrupting laminar flow were traceable to the wing's surface (rivet heads and other rough areas) and to pressure distribution over the wing's surface. Prandtl assumed that despite the wide array of turbulence in the atmosphere, the laminar flow of air over

an aircraft wing remains largely unaffected by the turmoil surrounding it. The rest of the surrounding atmosphere is not affected by the friction of the moving airplane, and its motion could still be explained by Lanchester's air circulation theory. Prandtl's simple concept of the *boundary layer* marked one of the most important breakthroughs in the theory of flight.

2. Induced Drag Due to Lift

While he was pondering how to marry a girl in 1914, Prandtl was also struggling with a second, even less well understood source of drag known as 'induced drag', or 'drag due to lift. Building upon the research of others, Prandtl knew that this drag arises because air curls over the tips of the wing from the bottom to the top, and gives rise to two trails of vortices. These vortices, known as tip or 'horseshoe vortices' because they have a shape vaguely reminiscent of a horseshoe, form behind the wingtips and extend backwards as the plane flies. As planes pull up so that the wings increase their tilt—the angle of attack against the air—drag is induced by the rotation of air around the wing. One can actually see these vortices form in the air during such activities as crop dusting.

What Prandtl was struggling with was how to develop a mathematical theory which explained the tip vortices. He soon succeeded by developing a more realistic model in which the vortex strength reduces along an aircraft's wingspan, and the loss in vortex strength is shed as a vortex–sheet from the wing's

Wingtip vortex behind a crop duster airplane

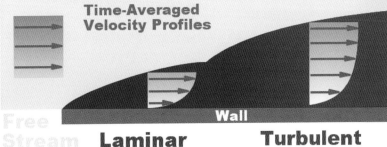

Boundary layer visualization, showing transition from laminar to turbulent condition

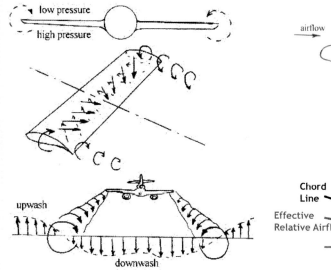

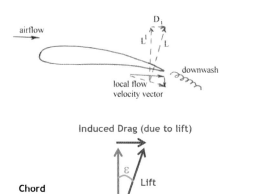

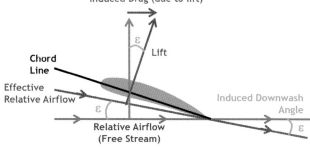

trailing edge, rather than just at the wingtips. Prandtl also showed that there was enough energy in the vortex–sheet to resist a lifting plane's motion in flight, and went on to indicate how to treat it in design.

The curving downward motion of wingtip vortices reduces lift by exerting a downward force on the wing as shown above. The vortices influence the airflow over the entire wing surface, not just the wingtips. Notice how the flowstream has a sideways component to it along the span of the wing, effectively reducing the speed of the air in the chordwise direction, which also lessens lift. The resulting downwash angle ε in the diagram on the right varies along the span. Its effect is to slant the lift vector

In the foreground, a red blended winglet extends up from the port wingtip of a Boeing 737–800. In the background, a yellow wingtip fence on the starboard wing of an Airbus A319

to the right, producing induced drag due to lift.

Engineers came up with what seems now as an obvious solution to reducing wingtip vortices and their

bad effects on lift and drag. Why not break up the vortex at the source? Modern aircraft use wingtip devices to capture some of the energy in the wingtip vortex. They increase the lift generated at the wingtip by smoothing the airflow across the upper wing near the tip, and they reduce the lift–induced drag caused by wingtip vortices, improving the L/D lift–to–drag ratio.

Ludwig Prandtl's pioneering efforts have been enshrined for posterity in the so–called "Prandtl number." The Prandtl number (Pr) is a dimensionless number used in the study of diffusion in flowing systems. It is equal to the kinematic viscosity (also known as momentum diffusivity) of a fluid divided by its thermal (also called molecular) diffusivity. Another way of putting it is that Pr represents the ratio of two diffusion rates: viscous diffusion rate to thermal diffusion rate. In heat transfer problems, the Prandtl number controls the relative thickness of the momentum and thermal boundary layers. When Pr is small, it means that the heat diffuses very quickly compared to the velocity (momentum). This means that for liquid materials the thickness of the thermal boundary layer is much bigger than the velocity boundary layer. The mass transfer analog of the Prandtl number is the Schmidt number (Sc). The dimensionless Schmidt number is equal to the dynamic viscosity times the specific heat of a fluid at constant pressure divided by its thermal conductivity. It physically relates the relative thickness of the hydrodynamic and mass–transfer boundary layers, and is used by scientists and engineers in the study of forced and free convection.

Thanks to Prandtl, great aeronautical progress was made during the decades following the first powered flights. His work led to the design of wing configurations which would minimize the loss of energy due to the tip vortices. In fact, closer to the present time even further improvements have been made in wing design to reduce drag due to the combination of the spanwise flow of air towards the wingtips and the wingtip vortices themselves. Today one can see fixed winglets rising straight up out of the wingtips of modern commercial aircraft. In fact, today's Airbus A380 jumbo jet has winglets both above and below the wingtip.

3. Form Drag

Theodore von Kármán could not resist the old instinct to compete with—and possibly surpass—his mentor. In 1911 von Kármán was able to mathematically explain the third source of drag, known as form drag, that occurs when the airstream fails to stick to the shape of the body, but rather breaks off behind it into a wake.

The science of aerodynamics continues to evolve today, with new applications. Tractor trailers started out with vertical windshields (upper left below). Makes sense, right? A vertical glass pane to protect drivers from the wind. Then truck owners figured out (what took them so long?) that they could decrease form drag—thereby improving mileage—by streamlining the cab and sleeping compartment (upper right). But improvements didn't stop

there. Recently, truck owners have taken to installing shields below the trailer, and "trailer tails" behind it to further reduce form drag and get better gas mileage.

Form drag behind a flat plate. Turbulent wake flow and eddies form behind the plate.

Another manifestation of form drag is seen in the design of golf balls today. Golf balls originally were smooth, then people tried putting bumps on them. Finally, the best performance was found when the surface of a golf ball was dimpled. Most balls on sale today have between 300 to 450 dimples.

The air that slides past the golf ball very close to its surface forms Prandtl's boundary layer. Generally speaking, the boundary layer grows in thickness as it moves over a surface. As air flows around the ball, the place where the turbulent wake starts is called "separation of the boundary layer". This is where the smoothly flowing air departs from the ball and does not close up behind the ball nicely but rather swirls around in small vortices or eddies behind it. This area of separation behind the ball—the wake, much like a wake behind a motorboat—increases drag. In fact, what happens is that the flow within the boundary layer actually reverses in sign at a certain point, generally about ¾ of the way around the sphere. The cause is a negative pressure gradient that builds up behind the ball. As the negative pressure gradient builds up behind the ball, the backflow of fluid at the separation point causes an

accumulation of fluid that causes the oncoming boundary layer to separate. The fluid behind the sphere circulates slowly within the boundary layer and forms eddies.

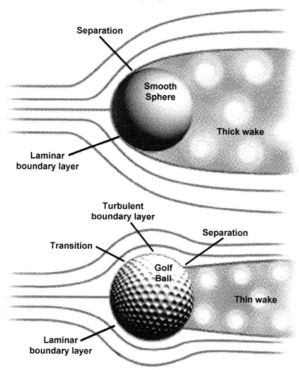

If the boundary layer can be encouraged to stick to the ball a little longer, then the turbulent part of the wake can be reduced. It turns out that adding a little extra turbulence in the boundary layer itself all over the ball allows the main smoothly–flowing air currents to stay closer to the ball and delays the separation of the boundary layer. Indentations, or dimples, on the surface are better at stirring up the placid boundary layer on the ball than bumps. There is also an increase in the Magnus force, giving the ball some lift when it is spinning in the correct direction. When given some backspin, this force helps keep the ball in the air longer. Its range depends on the loft of the club and how it was hit.

In addition to the Magnus force that gives a ball some lift when it is spinning in the correct direction, von Kármán discovered a very important aerodynamic effect having to do with vortex shedding behind a body that produces a wake behind it. For any bluff body (an airplane wing, fuselage, or the Tacoma Narrows bridge) the air gets separated behind it and the phenomenon of form drag manifests itself. One Friday afternoon in 1911, von Kármán became intrigued when a doctoral candidate of Prandtl's named Heimenz was trying to measure the pressure at different points on the surface of a solid cylinder when it was placed crosswise in

a steady stream of water. Prandtl wanted him to study the turbulence formed in the wake of the cylinder by observing the point at which the eddies separated themselves from the viscous boundary layer surrounding the circular cylinder. What was intriguing to Kármán was that no matter what the hapless Herr Heimenz and his professor tried, they could not mitigate some annoying persistent, violent oscillations of the cylinder at certain rates of flow. Von Kármán had a hunch that the solution to this mysterious phenomenon was mathematical.

"I assumed that the water rolled up into vortices as it passed over and under the cylinder and broke off into two trails, one from the top and one from the bottom. Then I considered two possibilities. First I assumed that the vortices are all symmetrical, that is, the top and bottom vortices form at the same time and start out together. But the mathematics of this motion soon told me that it was unstable. A very small deviation in the position of the vortices would gradually grow larger until the flow pattern was destroyed."

"Then I made a second assumption in which the vortices are shed alternately from the top and bottom of the cylinder. One vortex is formed at the top, then one at the bottom, one again on the top, one again on the bottom, and so on. As I examined this motion, the whole solution suddenly leaped into my mind and I saw clearly that the configuration becomes stable when there is a definite geometric arrangement of the vortices. This arrangement occurs only at a certain relationship of two distances—the distance between two single consecutive vortices and the distance between two rows of vortices. Or to put it another way, instead of marching two by two, the vortices are staggered like lampposts on both sides of a street."

With this stroke of genius, von Kármán explained the important phenomenon of vortex shedding behind

Von Kármán's "Vortex Street" in action

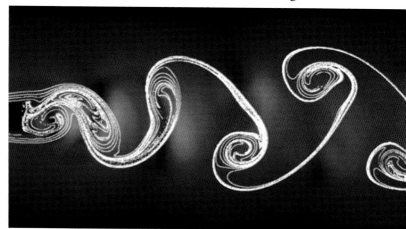

bluff bodies in a flow, which quickly became famous in aerodynamic circles as the "Kármán Vortex Street" after his lampposts analogy. The shed vortices break off into two trails behind the body, one from the top and one from the bottom. The fact that the vortices alternate will shake the cylinder back and forth, whereas if the vortices were shed in pairs at the same time from the top and the bottom, they would probably cancel each other out. But this is not how nature works, and von Kármán proved it mathematically. In the case of the Tacoma Narrows bridge, the ratio of distances that he worked out came out just right at winds of about 42 mph. Von Kármán was convinced that this and the fact that the winds held steady at this velocity contributed to the terrible collapse.

The vortex theory explained many phenomena that were not clearly understood before, including the "singing" wires on biplanes and the problem of the "singing propeller" in torpedoes and submarines. The high tone emitted by its propeller when it oscillated at a certain frequency underwater made submarines vulnerable to detection by enemy sonar. For example, engineers had to redesign the propeller on the USS Nautilus, the first U.S. submarine to surface at the North Pole, in order to eliminate the unwanted tone.

Many other practical applications resulted from von Kármán's rotating eddies. With this information and knowledge of flow dynamics engineers had a clearer understanding of how to minimize drag. Aircraft, dirigibles, and ships underwent significant external modifications to inhibit the drag induced by Kármán Streets. For instance, by forming an airship hull so that the turbulent trails followed the downstream contours of the vessel—rather than separating from the solid body almost on contact, as in the case of a smooth cylinder—the cigar–shaped vehicle would move through the air with more efficiency than a disk of the same diameter. Hence, the well–known term, streamlining: an object designed so that its vortices flowed as far down its own streamline as possible. Streamlining is a basic technique in the design of planes, ships, rockets, racing cars, and even trucks. Modern tractor–trailers have a streamlined "hood" over the cab to minimize drag. Retractable landing gears, slender airfoil profiles, and smooth fillets joining upper wings to fuselages all contributed to the "clean" lines of modern aircraft. Streamlining has embodied a revolution in the contours of objects traveling through air or water.

Even stationary structures such as bridges, radio towers, and power lines may exhibit vortex instability during blustery conditions. The oscillations even in a moderate wind of radio towers, chimneys, submarine periscopes, and other tall thin bodies are now known to be caused by the forces generated by alternating vortices. All such objects must be designed to reduce dangerous oscillations and buffeting. Periodic forcing set up in this way can be highly undesirable and hence it is important for engineers to account for the possible effects of vortex shedding when designing a wide range of structures, from submarine periscopes to industrial chimneys.

In 1957 Christopher Scruton and D.E.J. Walshe, two scientists working at the National Physics Laboratory in Great Britain, invented helical strakes as a way to suppress vortex-induced vibration on cylindrical structures. Helical strakes consist of one or more oval-shaped fins that spiral along the structure's length, like large screw threads. If the fins are effectively sized and designed, then the fins cause the vortices along the structure's length to be broken up into shorter and weaker segments. Furthermore, the breaking up of the vortices prevents eddies from interacting along the span, resulting in a series of vortices that are randomly phased in time and remain attached instead of shedding. While a structure may still experience strong local forces in a strong wind, the randomness of the forcing frequency along the structure's length produces only a small net vibration of the structure. Twisting together of wires in high power transmission lines, or spacing balls along the length of high voltage transmission lines (you can sometimes see them as orange balls) where wind gusts are prevalent also works to discourage the alternate shedding of vortices.

Incidentally, a vortex street will only be observed over a given range of "Reynolds numbers" (Re), typically above a limiting Re value of about 90. *The Reynolds number(after Osborne Reynolds who popularized its use in 1883) is essentially a measure of the ratio of inertial to viscous forces in the flow and is often used by aerodynamicists. It is given by the product of the mean fluid velocity (recall that air is a "fluid") times a characteristic length (usually a dimension crosswise to the oncoming flow, like diameter or width) of an object, divided by the fluid's kinematic viscosity. The range of Re values will vary with the size and shape of the body from which the eddies are being shed, as well as with the kinematic viscosity of the fluid.*

4. Wave Drag

The last type of drag, "wave drag", manifests itself when an aircraft breaks the sound barrier and becomes supersonic. The phenomenon of wave drag is analogous to a speed boat riding in the water. At low speeds the boat's keel rides deep in the water. As the operator increases speed the boat encounters greater resistance, which arises from the waves created by the boat. Similarly, in the case of flight, the resistance of the air rises appreciably when the airplane begins to approach the speed of the air waves created by it—this is the so-called "sound barrier" which was cracked by Chuck Yeager flying the Bell X–1 in 1947. The reason for the sound barrier is that when an aircraft approaches the speed of sound, it approaches the speed of the pressure waves it generates. The pressure waves in air cannot exceed the speed of sound. The pressure waves "bunch up" in front of the aircraft to form a condition known as a shock wave. Shock waves on an aircraft create the sudden increase of resistance often described in newspaper stories as the "sonic wall" or barrier. Once the aircraft exceeds the speed of sound, the flow around it smooths out to some extent and the drag decreases.

An X–15 free–flight model being fired into a wind tunnel vividly details the shock–wave patterns for airflow at Mach 3.

The difference between a motorboat and a supersonic airplane lies in what one can do to overcome this new resistance. A speedboat can rise on its "steps" and move along the water with most of the boat in the air—like a hydrofoil—so a great part of the water drag is lessened. But an airplane cannot move on the step; it is already operating in three dimensions, so it would have to enter a fourth dimension or some such nonsense. Other means have to be found to reduce the new resistance arising out of high speeds.

> *If you take nothing else away from this discourse on aerodynamics, please remember that there are four forces acting on a rocket in flight in the Earth's atmosphere: thrust forward, drag backward, gravity downward, and lift upward.*

UNVEILING THE MYSTERY OF TURBULENCE

One of the biggest impediments to improvements in aeronautical design is turbulent air motion, elusive theoretically. Though defined as the irregular motion of gases or fluids past stationary objects, it occurs more commonly in nature than smooth flow which produces less friction. Turbulence increases frictional resistance, heat transfer, and the diffusion of fluids. Prandtl's boundary layer coats an aircraft's wings in flight. Starting from the leading edge, air particles change from a smooth laminar flow towards a more turbulent state towards the trailing edge. In the process, drag increases. As a case in point, around 1930 aerodynamicists observed tests in a smoke tunnel, evaluated the data, and concluded that the prime culprits in disrupting laminar flow were traceable to the wing's surface (i.e., protuberances like rivet heads and other rough areas) and to the pressure distribution over the wing's surface. In other words, a negative pressure gradient over a wing's surface tends to disrupt the smooth flow of air and creates the onset of turbulence. In the early 1900s, however, the mathematics used to describe this motion was so complicated that no one could apply it to the design of an airplane wing, a complicated geometric object in itself. Obviously, the Wrights did not fully understand turbulence, they just knew how to apply the principles as then known to make airplanes fly.

If laminar flow could be represented as a pair of parallel columns of soldiers, turbulent flow could be represented by an equal number marching at angles to the double rows. The friction encountered as the soldiers clash would slow the progress of the forward–moving columns. But if all the men moved at equal intervals over equal distances, elemental laws of probability could be brought to bear on this apparent chaos. Just as soldiers collide in this example, layers of gases likewise pass over one another, causing friction at the molecular level. Ludwig Prandtl announced this principle at the 1926 Congress of Applied Mechanics held at Zurich. His "Mixing Length Concept" accurately predicted the distance molecules traveled before impacting other molecules, thus losing momentum. His discovery marked an important step toward a fuller understanding of chaotic motion.

Shown below is a flow visualization of a turbulent jet. The jet exhibits a wide range of length scales, an important characteristic of turbulent flows.

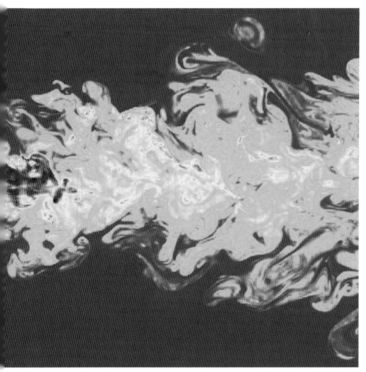

Back in 1822 the Frenchman Claude–Louis Navier and the British George Gabriel Stokes formulated the so–called "Navier–Stokes equations." It is simply amazing that these complex equations were formulated so long ago, even today supercomputers have a hard time solving them without making a number of simplifying assumptions. These beautiful differential equations—yes, they are beautiful once comprehended!—form the theoretical basis for the science of fluid mechanics, which in turn is the foundation of aerodynamics, and the latter influences every rocket or spaceship that has to traverse the earth's atmosphere. These equations, unlike algebraic equations, do not explicitly establish a relationship among the variables of interest—like velocity and pressure—but rather they establish relations among their rates of change. Contrary to what is normally seen in solid mechanics, the Navier–Stokes equations dictate not position but rather velocity. Their solution is called a velocity field or flow field, which is a description of the velocity of a fluid (air is a "fluid") at a given point in space and time. Once the velocity field is solved for, other quantities of interest such as flow rate, drag force, or the path a "particle" of fluid will take may be found. Unfortunately, the Navier–Stokes equations are nonlinear, and a number of assumptions are often made to make their solution tractable, particularly where air turbulence is involved. Even very small perturbations in a flow field grow parasitically on those that are first established, until the flow pattern is so grossly disturbed that it is no longer useful to define a fluid velocity for each point in space. The description of the flow has to be a statistical one in terms of mean values and of correlated fluctuations about the mean. The flow is then said to be turbulent.

Turbulence arises not only behind bluff bodies in air, but also within boundary layers around solid obstacles when the rate of shear within the boundary layer becomes large enough. Curiously enough, the onset of turbulence in the boundary layer can paradoxically reduce the drag force on very bluff objects. In the case of a spherical obstacle, the point at which the boundary layer separates from the rear surface of the sphere shifts backward when the boundary layer is "energized" and becomes turbulent. The turbulent eddies that circulate behind the sphere therefore become smaller. It is turbulence in the boundary layer that is responsible for the dramatic drop in the drag coefficient for both spheres and cylinders that occurs when the Reynolds number is about 3×10^5. This drop enables golf balls to travel farther than they would do otherwise, and the dimples on the surface of golf balls are meant to encourage turbulence in the boundary layer. If swimsuits with rough surfaces help swimmers to move faster, as has been claimed, the same explanation may apply. The transition between laminar and turbulent flow is often indicated by a critical Reynolds number (Re_{crit}), which depends on the exact flow configuration and must be determined experimentally.

Oftentimes, satisfactory flowfield solutions to the Navier–Stokes equations are obtained with simplifying assumptions. For example, in the simple case of an "ideal" inviscid fluid, its acceleration is proportional to the gradient of pressure. This makes sense, as pressure increases in the direction of a fluid flow it should go faster. A flow pattern that is steady, such that the fluid velocity at each given point in space is not changing with time, may be seen in terms of a set of streamlines, the trajectories of imaginary particles suspended in the fluid and carried along with it. In steady flow, the fluid is in motion but the streamlines are fixed. Where the streamlines crowd together, the fluid velocity is relatively high; where they open out, the fluid becomes relatively stagnant. What is counterintuitive is the governing principle that the velocity of inviscid fluids (this is not a big assumption, water and particularly air are fairly inviscid) flowing through a constriction, such as along a pipe of varying cross section, increases while the pressure decreases. Think of it in terms of a bunch of people trying to crowd out through a single exit door after someone yells FIRE! People get munched and crushed, right? But this is not how nature works. The pressure of fluids flowing through a constriction, such as in a channel, actually decreases! It is this pressure decrease, called a positive pressure gradient, that causes the fluid to flow faster through the constriction. These effects are well-known to an experienced rafter, as he threads his kayak through the churning waters of a river to take advantage of faster flows through narrow channels, and stagnant water in broader places. The same thing can be said about lift, the pressure decrease above a curved wing is what increases the airflow velocity over it. Again, this is counterintuitive. One would think that as a curved wing moves forward, it has to "scoot" the air out of the way and the pressure above it should actually increase. Moreover, early in the 20th century, it was counter-intuitively found through experiments that thick wings can actually have less drag than thin ones which have more of a tendency to cause the downstream air to separate. Thicker wings lead to the benefits of retractable landing gear, being able to store munitions and much of the fuel inside the wings, among other benefits. In fact, a man could stand inside the wing root of the Convair B–36, a strategic bomber of 1950s vintage.

Sad to say, until the advent of supercomputers a number of simplifying assumptions had to be made in order to solve for the complex flowfields that really occur in flight. Even with supercomputers, wind tunnels and prototype flight tests are still necessary because of the uncertainties involved. For example, the high Mach number aerodynamics of the first Space Shuttle, Columbia, that was launched on April 12, 1981 were found to differ significantly in some respects from those estimated in pre–flight testing. A misprediction of the location of the center of pressure (due to using an ideal gas model instead of a real gas model) caused the computer to extend the body flap in flight by 16 degrees rather than the expected eight or nine, and side–slip during the first bank reversal maneuver was twice as high as predicted. After the Columbia reached orbit, inspection by astronauts Bob Crippen and John Young showed significant damage to the thermal protection tiles on the Orbital Maneuvering System/Reaction Control System pods at the orbiter aft end. Young reported that two tiles on the nose looked like someone took 'big bites out of them'. Post–flight inspection of Columbia's heat shield revealed that an overpressure wave from the Solid Rocket Booster ignition resulted in the loss of 16 tiles and damage to 148 others. The same overpressure wave pushed the body flap below the main engines at the rear of the shuttle well past the point where damage to the hydraulic system would be expected, which would have made a safe reentry impossible. The crew were unaware of this until after the flight, and John Young reportedly said that if they had been aware of the potential damage at the time, they would have flown the shuttle up to a safe altitude and ejected. Columbia would have been lost on the first flight!

THE BOSS

Shortly after his arrival at the California Institute of Technology in 1930, Theodore von Kármán's laboratory again became a mecca of the world in the aeronautical sciences. Two years later he became a founder of the U.S. Institute of Aeronautical Sciences, consultant to various American industries and to the government. On July 24, 1936, he took the oath of citizenship in the U.S. District Court of Los Angeles. Von Kármán's personal scientific work continued unabated. By 1940, he was an acknowledged world expert in aerodynamics, with important contributions to fluid mechanics, turbulence theory, supersonic flight, mathematics in engineering, aircraft structures, and even wind erosion of soil!

He was an extremely social animal, holding frequent parties at his fine house on South Marengo Avenue in Pasadena, California. Von Kármán held court there from 1930 to 1944. During frequent parties, his home was equally jammed with friends, colleagues, and students. During parties, his gestures often punctuated earthy jokes. He was much interested in poetry and literature and could always supply a story appropriate to any occasion. When the atmosphere became charged with tension in a scientific meeting, he was able to restore balance by drawing on his collection of anecdotes. His memory for anecdotes equaled his childhood recall of numbers, and he could quickly call up stories to whatever the situation demanded. Kármán's attraction also sprang up from the intense focus of his mind. He communicated as if nothing mattered but the person before him. The host of Marengo Avenue effortlessly adjusted his conversation to the tastes of his listeners, never leaving the impression of a great man uttering lofty thoughts. Unimpressed by credentials, or the lack thereof, Kármán felt that station in life mattered little, shared experience a great deal.

Von Kármán meets Jayne Mansfield, one of the early Playboy playmates, in Greece, 1960.

Theodore von Kármán, 1962.

Most of all, Kármán adored women, especially beautiful ones. At all parties, he would amble toward them with the full determination of a Hungarian bachelor, embracing them, speaking to them in intimate confidence. Much more often than not, they loved his attentions as much as he loved their company.

The Kármán home held a wonderful attraction to his many friends, yet it represented work as well as play. "The Boss," as his pupils and associates liked to call him, often told them to bring along their scientific

problems to the Marengo Avenue parties. He would greet them loudly at the door, with a shout of their name and an expression of delight. Welcomed in, he would usher them into a living room teeming with men and women of a dozen professions, from a dozen countries. After depositing these innocents into the midst of a hard–drinking, good–humored crowd, Kármán would often temporarily vanish with several others out the kitchen door and into a small study located in a cottage at the rear of his home. Choked with cigar smoke, dominated by a large carved mahogany desk and a very old, rumpled semi–wing desk chair, it was the scene of thousands of rushed conferences between pupils and teacher. Across a wicker table topped by glass they would joke, quickly exchange papers, scrawl out new equations, and discuss the solutions to especially vexing problems.

Before an hour elapsed, the Boss would adjourn the meeting to greet new guests and see others off. Free to roam, the students ate heartily and drank whiskey or scotch for the rest of the night. Most of them and the other visitors said their goodbyes around midnight. A few who required more of the master's time would retire back to the study. He had a fantastic capacity for work and left behind him wherever he went a trail of bits of paper covered with calculations. "Would you like to work now?" the indefatigable, good–humored professor would ask. For the next two hours he gave then his undivided attention, kindled their enthusiasm, and filled sheets of scratch paper with equations. Sometimes they awoke at 2 a.m. to hear the Boss throw his pencil on the desk after realizing some special mathematical triumph. "His teaching," remarked a former student, "was very personal, by no means confined to the classroom or restricted to scientific subjects." Thus, Marengo Avenue provided the spiritual center of Kármán's life, a place where he conveyed important scientific and personal life lessons to many academic disciples.

Von Kármán held that "there is no such thing as an entirely correct answer to any question in engineering. It is the way the problem is treated and developed. A student who has completed an intelligent analysis, with the proper emphasis and approach, but who comes out with a wrong answer because of a mechanical slip in multiplication, would receive a much higher rating from me than a student with the correct answer but no imagination in his approach." Von Kármán also flavored his lectures with pictorial examples to prove theoretical points. He once likened a vortex to water and soap powder going down a bathtub drain. He described Isaac Newton's concept of inertia as two monkeys hanging at either end of a rope, suspended from a pulley. Through fast climbing, one might overcome his own weight, as well as the "dead" pounds of his lazier partner. His students came to refer to themselves as the "Kármán circus."

Kármán's fine house at 1501 South Marengo Avenue in Pasadena, California. The Boss held court here from 1930 to 1944. Although he later shifted his life to Washington, DC (1945 – 1950) and Paris (1951 to 1963), the home remained dear to him and he maintained it even though he lived there only part of the year.

THE ROCKETEERS

One day in March 1935, three young men, Frank J. Malina, John W. Parsons, and Edward S. Forman approached von Kármán in his office at Caltech with a proposal. Frank Malina was an outstanding graduate student at Caltech in aeronautics, Jack Parsons was a self–taught chemist with considerable innate ability, and Forman was an ardent rocketeer who was as familiar with rocket engines as automobile engines are to auto hobbyists. Parsons was a delightful screwball. He loved to recite pagan poetry to the sky while stamping his feet. He headed a local chapter of a strange religious sect called the Thelemites, which von Kármán would learn years later also practiced sex rituals. Only in California. The son of a one–time business tycoon, Parsons stood six–foot–one, with dark wavy hair, a small mustache, and penetrating dark eyes which appealed to the ladies. Both Parsons and Forman confided to Dr. Kármán that their backyards in Pasadena were pockmarked from the effects of rocket explosions. The three men wished to build and test liquid and solid rockets that could propel themselves from 20 to perhaps 50 miles into space. Von Kármán's open-mindedness was well demonstrated when he granted his approval -- in spite of the general disbelief in academic circles in the possibilities of rocket propulsion and its applications— under two conditions:

1. He would fund none of their activities, whatever equipment they needed would have to be bought at scrap or purchased from the trio's own meager resources; and
2. They could use the Guggenheim lab only during nonoperating hours. Otherwise, Kármán offered to provide advice, allowed Malina to write his Ph.D. dissertation on rocket performance, and let Parsons and Forman assist Malina despite their lack of official ties to Caltech.

Robert Millikan, an influential professor and provost, gave moral support to the project, perhaps because of its applicability to his own cosmic ray research. Millikan looked the part of a distinguished professor, with a serious demeanor and a mane of white hair atop his head. His son Clark Millikan followed in his stead. So did Professor Irving Krick, a GALCIT meteorologist who realized the value of atmospheric probes for weather prediction. However, Malina realized that the real thanks went to Theodore von Kármán, who approved the nation's first university–affiliated rocket program and lent his prodigious prestige to an endeavor of doubtful scientific repute at the time. Von Kármán not only rewarded three rocketeers with important moral support, he now took the gamble of letting them occupy the basement of the GALCIT building. Von Kármán liked the practicality of their goals. They avoided the visionary objectives of "rocketeers" who often dreamed of flying to the moon. Instead, Malina and his friends concentrated on penetrating the edge of space, where they hoped to obtain data on cosmic rays and weather.

Unbeknownst to the three rocketeers, von Kármán had more than a passing interest in rocket technology. He knew of the writings of Konstantin Tsiolkovsky, the relatively obscure Russian teacher and rocket pioneer, who had proposed multistage rockets in 1903 (that year he published *The Exploration of Cosmic Space by Means of Reaction Devices* (in Russian), the first serious scientific work on space travel). As early as the 1920, von Kármán attended a conference of the German WGL (the equivalent of the American Institute of the Aeronautical Sciences) in Danzig at which Professor Hermann Oberth, one of the first pioneers of German rocketry, gave an enthusiastic lecture on the possibilities of escaping from the earth. Von Kármán was intrigued. He defended Oberth against criticism by the distinguished Dutch physicist Hendrik Antoon Lorentz, winner of the Nobel Prize in 1902, who read a long paper at the conference "showing" why it was impossible to escape the earth's gravitational pull. Lorentz was well known in physics circles for proposing the Lorentz–Fitzgerald contraction and the Lorentz transformation which contributed to Einstein's Theory of Relativity. But Lorentz was shortsighted in concluding that Oberth's spaceship could never be built because to acquire an escape velocity from the earth a rocket would need an enormous amount of energy. He argued that even if it used the best fuel of the time the rocket would have to contain so much fuel that it would weigh 34 times as much fueled as empty! Lorentz concluded that this was beyond engineering capability and should be forgotten. Standing up, von Kármán objected, "If we calculate the energy in a pound of kerosene, or other hydrocarbon, and transform this into mechanical work, we would find that there is more than enough energy available to us to send that pound into space." Von Kármán believed in theory and in possibilities. He thought that if the theory is right, technology would catch up, and one should not reject the entire idea because engineers could not implement it with the machinery of the time.

In those early days Frank Malina and his fellows needed every bit of available help, whether money, facilities, equipment, parts, or ideas. As ardent rocket enthusiasts, they had heard of Robert Goddard's pioneering experiments in the 1920s and 30s. When Goddard and his wife chanced to take a vacation in California in August 1936, Harry Guggenheim—Goddard's principal benefactor—pressed him to confer during the trip with Robert Millikan and the GALCIT team. For his part, von Kármán encouraged Malina to talk with Goddard and gather what information he could for his dissertation. Malina introduced himself to Goddard and arranged to confer with the famous rocket scientist on his home turf in New Mexico, where Goddard had set up his operations. Unfortunately, over the years Goddard had learned to rely on the U.S. patent office, rather than open scholarly discourse, for his survival. For decades, the reticent professor had endured countless cruel press clippings characterizing him as an eccentric and a dreamer.

So naturally, the September 1936 meeting between Frank Malina—the eager rocketeering graduate student—and the skeptical, secretive Goddard proved fruitless. Goddard showed his visitor a *New York Times* clipping which made fun of rockets. Goddard would not provide any substantive information, and on the other hand Malina found his host in New Mexico exasperatingly uncooperative. As Dr. von Kármán would later write, "if he (Goddard) had taken others into his confidence, I think he would have developed workable high–altitude rockets and his achievements would have been greater…But not listening to, or communicating with, other qualified people hindered his accomplishments." Malina had already learned the lesson in the weekly GALCIT seminars: "the trouble with secrecy," Kármán told his students and colleagues, "is that one can easily go in the wrong direction and not even know it…Many fine inventors lose the benefits of their inventions because they think they must secretively work out every last detail themselves." Amen.

Undaunted, in the following months Malina and Parsons went scrounging all over Los Angeles for such simple items as pressure tanks, fittings, and meters for their initial rocket experiments. The trio—Malina, Parsons, and Forman—were soon joined by two other rocket enthusiasts: Apollo Milton Olin (Amo) Smith, a talented master's student, and Hsue–Shen Tsien, a brilliant Chinese doctoral candidate. Sometime later a young assistant from the Caltech Astrophysics Laboratory named Weld Arnold joined the team, and became so enthused that he offered to donate $1000 to the cause. Arnold spread his donations out over a period of some months. The now six rocketeers spent most of their time learning the behavior of various rocket propellants. Since there was nothing in the literature that would enable the group to build a solid or liquid rocket to specifications, they had to learn it all from scratch; by analysis, trial, and error. For liquid rockets they tried a chemical propellant composed of methyl alcohol as fuel and gaseous oxygen as oxidizer. For solid rocket experiments, they used a smokeless powder technically called a "double–base propellant" — a mixture of nitroglycerin and nitrocellulose. In granular form, this dangerous solid propellant acts like dynamite, forming a high explosive force in a hundredth of a second after ignition.

As Goddard had found out, the first rocket experiments did not always go as planned. One day the young rocketeers set up a small liquid rocket engine with nitrogen dioxide as an oxidizer, and suspended it like a plumb bob at the end of a 50–foot pendulum which hung from the third floor ceiling to the basement of the Guggenheim lab. They planned to measure the amount of thrust developed by the engine by measuring the amount of pendulum swing after ignition. However, the darn thing misfired, enveloping the entire laboratory building in toxic and odorous fumes. Because of oxidation, a thin layer of rust formed on virtually every piece of equipment. After this accident, some staff members grumbled about the danger of rocket work in the lab. Von Kármán moved the group outside to an unused concrete platform attached to a corner of the building. Sometime later, the campus witnessed two loud blasts from the east side of the GALCIT lab, one right after the other. Students running to the site found Parsons and Forman dazed but unhurt. The force of the second explosion was so great that it blew a piece of a gauge deep into the wall past the spot where Frank Malina's head would have been, had he been sitting upon his stool reading test results. By good fortune, Dr. von Kármán had asked Frank to drop a typewriter off at his home on South Marengo. The loss of Malina would have been a crushing blow to the project, as well as to Kármán. After this incident, students at Caltech irreverently called the rocketeering group the "Suicide Club."

After the fiasco, von Kármán decided to move the adventuresome group as far away as possible from any buildings. After considerable scouting, they found a likely area in the Arroyo Seco canyon, tucked behind Devil's Gate dam on the western edge of Pasadena, near the present location of the Jet Propulsion Laboratory in the foothills of the San Gabriel mountains. There they finally met with some success, in between accidents and misfires. By January 1938, they ran an engine fueled with a gaseous oxygen/liquid methyl alcohol mixture for 44 seconds at 75 psi. Malina and the talented Tsien also made

theoretical studies of the thermodynamic characteristics of solid propellant rocket motors and tested their theories by building and firing small rockets of their own design. The things took off, but like the Chinese and others had found out, they would only burn for a few seconds. Eventually the rocketeers garnered considerable publicity, not all of it warranted. Newspapers on both coasts printed sensational headlines predicting three–stage rockets flying to altitudes of five million feet (947 miles) at velocities of 11,000 miles per second (a rocket's escape velocity from the Earth's surface is about 7 miles per second, or just over 25,000 miles per hour).

JATO

"Will you please take a seat, Mr. Malina, and I will buzz Mr. Fleet for you," said the pretty, blue–eyed blonde secretary. Frank Malina wearily plumped himself down in the capacious chair and took his carefully prepared report out of the satchel. He and Hsue–Shen Tsien had stayed up past midnight at "The Boss's" home in Pasadena, and were dumbfounded when between 1:00 and 2:30 in the morning Dr. von Kármán found time to review their treatise. Bolstered by a couple of scotches, the Boss caught a couple of errors and recommended a few changes here or there, but all in all expressed his satisfaction with the report. It encapsulated the results of their thermodynamic analysis of solid–propelled rockets, and tabulated the most recent testing results in the Arroyo Seco facility. Frank Malina was encouraged that Reuben Fleet, President of Consolidated Aircraft Company of San Diego, had approached Dr. von Kármán and GALCIT with questions about the rocket program, hence the need for the report.

"Good morning, good morning!" Mr. Fleet greeted him with a firm handshake as Malina was ushered in. After exchanging a few pleasantries, Fleet got to the point. "So, tell me about these rockets of yours. I was just reading an incredible account in the *Los Angeles Times*…and, is it true your rockets can rise to altitudes of hundreds of thousands of feet?" "No, sir," responded Malina, his face getting a little red. "But I have this report here that states the facts and our analysis quite clearly." Malina handed him the report.

"Oh, and I see that Dr. von Kármán has his signature on the front page," said Mr. Fleet as he eyeballed the cover page. "Yes, sir, he is my thesis advisor, and …I am working on a dissertation on the performance of rockets, both solid and liquid fueled." Fleet took a few moments to review the report. Skipping over the details, he astutely homed in on the reported test results. "So, based on this test, you have calculated that these things can produce two pounds of thrust, for…what, 30 seconds?" "Yes, sir." "Call me Reuben," responded Mr. Fleet. "I have heard a bit about the kind of rockets that use liquid propellants, the sort that Dr. Robert Goddard from Clark University has tested," he said, looking up expectantly at Malina. Taking a cue, the young Ph.D. student responded, "he has moved to Roswell, New Mexico, to continue his research in private." Fleet leaned back in his executive chair. "Dangerous things, aren't they? But I don't know much about the solid propellant engines, how those work?" he asked.

"You mean, how are they *supposed* to work," Malina began. Fleet laughed. "Well, we call them rocket *motors*, that is the solid propelled rockets, as opposed to rocket *engines*, a term reserved for rockets that use an oxidizer and a fuel as propellants. And by the way, we approached Dr. Goddard at his out–of–the–way place in New Mexico, and he didn't seem interested in the least in cooperating or sharing technology with GALCIT. Too bad. Dr. von Kármán and I think that you can use the quick–burning kind of motors we have been able to make so far to generate hot, high–pressure gas for doing work in the actuation of various power or mechanical devices. What you might call variations of solid propellant rockets have been used for a long time as gun propellants for driving projectiles out of gun barrels and as explosives for mining, excavation, or munitions. To answer your question, a solid motor burns a black powder, starting from the inside surface of a propellant cavity. In other words, for the non–explosive motors we are trying to make, you put a mandrel inside the casing of the rocket, then fill the space between the mandrel and the casing with black powder, basically gun powder. If we could get a slow, controlled burn out of the powder, it would create hot gas for propulsion out of a rocket nozzle just like a liquid rocket."

"How does it stop, when it burns out all the propellant?" asked Mr. Fleet. "Yes. Of course you also have to protect the outside case that encloses the whole thing. Otherwise, the powder would burn right through it, create a big hole, and bang! The thing basically explodes, like a grenade, because this all happens in milliseconds.

We are not trying to make bombs, we are trying to make motors that generate controlled thrust for propulsion. If we could get the black powder to slow down, the thrust would be proportional to the inside surface burning area." Fleet thought for a moment. "Then a solid propellant rocket motor is much like a liquid–fueled rocket engine, it has a pressure chamber, a converging/ diverging de Laval nozzle for producing thrust, an outside motor case, and I suppose an igniter to set it off.? "Yes, Malina responded. "Wouldn't the thrust increase, though, as the motor depletes its propellant, gets lighter, while the burning surface inside the cavity gets bigger as the thing burns out the propellant, creating greater thrust?" Fleet intelligently asked. "Yes. We are still in the early stages of our research," said Malina. "In the future, I think that we will be able to control the thrust of a motor over its burning time by changing the internal configuration of the cavity. A solid motor can either burn radially, starting from the inside surface of a propellant cavity, or from the end, like a cigarette. Maybe we can combine both types of burning in a particular design. Our analysis also shows that a motor's thrust is directly proportional to the amount of burning surface inside it, and the burning surface, of course, naturally increases over time as propellant is consumed, exhausted, and the internal cavity grows in size. We have only been able to test with a powder mixture of nitroglycerin and nitrocellulose…" Fleet interjected, "I have heard that nitroglycerin can be unstable and nasty, it can explode on you if you're not careful. But dynamite, of course, is fairly stable and safe to handle if it is packaged correctly." Malina went on, "we have only begun to scratch the surface, as to what different solid propellant grain designs we can use, what the effects of different powder mixtures, different ingredients are. Maybe we can use ammonium perchlorate. I don't know. We need to do more tests."

"Have you found a lining of some sort that you can put inside the casing so that the propellant doesn't burn through, creating an asymmetric thrust and the thing gets out of control in flight?" Fleet asked, glancing at his watch. *This guy's pretty smart,* Malina thought. "Nope," Malina said. "First we have to get a controlled burn, instead of an explosion, out of the solid propellant. We need to get past the misfires and explosions, too."

"What are the advantages of a solid–propelled motor versus a liquid rocket engine?" Mr. Fleet queried. His secretary came in, reminding him of his next appointment. Malina continued, "we think that motors have one big advantage—they basically have no moving parts. In comparison to liquid rockets, solid rockets are relatively simple, the case also constitutes most of the structure, and they should require little servicing. You don't have the mess of two kinds of liquid propellants. The best oxidizer—oxygen—has to be kept at cryogenic temperatures, otherwise it boils off and turns to gas; you have to precisely mix the fuel and oxidizer through orifices in a metal injector plate which must meter fuel and oxidizer at the proper rates and in the right proportions. The injector also serves the structural purpose of containing the combustion gases at terribly high temperatures and pressures inside the combustion chamber. The engine nozzle erodes as it expels the hot gases. For solid rockets, you just need an igniter, a mixed, hollowed–out propellant grain, the exhaust nozzle, and an outside metal casing." Fleet asked, "suppose you *could* get a controlled burn. Would they be safer than liquid rockets?" "Really, no, I don't think so. We are still researching how you can keep solid propellants like nitroglycerin stable in storage. We have the thrust problem, it cannot be randomly varied in flight, or throttled, like we think that liquid rockets can." Fleet's forehead wrinkled as this sunk in. "Well, but are there any other advantages?" he asked. "Yes. Solid rockets should cost less than liquid rockets. A lot less." Reuben Fleet's face lit up, thinking about the potential business applications.

After Malina departed, Reuben Fleet consulted with his chief engineer, I. M. Laddon. Laddon had directed the development of the XP3Y–1, the predecessor to the famous wartime PBY Catalina flying boat. Consolidated Aircraft had recently won a Navy contract for 60 P3Y–1s, worth $22 million, the largest single military aircraft order since World War I. Both Fleet and Laddon agreed that solid fueled rocket motors, when the technology became sufficiently mature, could be fastened to both sides of a heavy seaplane, assisting in its takeoff from the water. It would be simpler than trying to use liquid rockets. Consolidated had the resources to perfect and produce larger solid rocket motors. If only Malina and his young rocketeering friends could demonstrate controlled burns, and solve some of the other problems.

Solid Rocket Motor Basics

Solid rocket motors (SRMs) are the simplest of all rocket designs with very few moving parts. With few structural components they are efficient in that the vast majority of their weight is actually usable propellant. They consist of an igniter, motor case (usually steel) filled with a mixture of solid compounds which burn at a rapid rate, exhaust nozzle, and mounting provisions which in some designs also serve as thrust skirts. The motor case is lined with a rubber-like organic material coating which assures good bonding of the propellant grain and acts as a thermal insulator to protect the case.

With solid propellants you don't have separate fuel and oxidizer to combust in a rocket chamber. The solid propellant is a mixture of combined fuel, oxidizer, and binder all in one, enabling operation in the vacuum of space. The binder is necessary to hold the whole package together. Ordinarily, in processing solid propellants the fuel and oxidizer components are separately prepared for mixing, the oxidizer being a powder and the fuel a fluid of varying consistency. They are then blended together under carefully controlled conditions and poured into the prepared rocket case as a viscous semisolid. This is called *casting*. A key advance was Jack Parson's idea of combining asphalt (as binder and fuel) with potassium perchlorate (as an oxidizer) to make the first castable composite solid propellant. For smaller rocket motors solid propellants are usually processed and formed by extrusion methods. In any case, the propellants are then caused to set in curing chambers under controlled temperature and pressure. They end up with a rubber–like consistency, like a pencil eraser.

SRM Operation

An electrical signal is sent to the igniter which creates hot gases which ignite the main propellant grain. Thrust is developed as the high thermal energy of the combustion gases is converted to kinetic energy in the exhaust. When ignited, a solid propellant burns from the center out towards the sides of the casing, expelling hot gases from the nozzle to produce thrust. The shape of the center channel determines the rate and pattern of the burn, thus providing a means to control thrust. SRMs can be ignited at a moment's notice and don't require tanking of liquids prior to operation—making them attractive for military applications in austere field conditions. Solid motors can be stored a long time – 5 to 20 years. On the downside, their efficiency (specific impulse) is generally lower than liquid systems, and they cannot be readily throttled. Unlike liquid–propellant rocket engines, once ignited solid rocket motors will burn until either all the propellant is exhausted or a thrust termination blow-out diaphragm stops the burn by dropping the pressure inside the case.

Basic components of a solid propellant rocket motor.

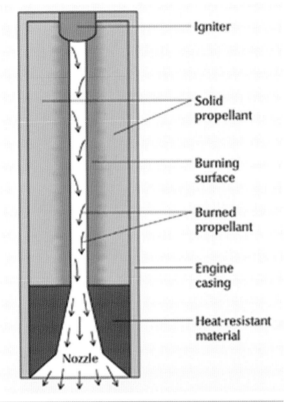

Simplified solid rocket motor schematic.

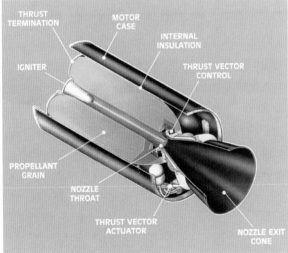

aluminum significantly increased the specific impulse of a castable composite propellant. JPL engineer Charles Bartley improved on Parson's asphalt by replacing it with a Thiokol polysulfide polymer. In 1948 Aerojet discovered that replacing potassium perchlorate with ammonium perchlorate helped reduce smoke while increasing the rocket's specific impulse.

For a while, von Kármán's Suicide Club languished. Amo Smith left Caltech for an engineering position at Douglas Aircraft. Weld Arnold dropped out of school and moved to New York. Hsue–Shen Tsien became engrossed in his dissertation. Forman and Parsons ran out of money and took jobs with Halifax Powder in the Mojave Desert. But good fortune would soon return, in the form of another Arnold, General Henry H. (Hap) Arnold, an Army Air Corps brigadier general who was commander of the First Wing, General Headquarters Air Force at nearby March Field, California. When he reported to March Field in March 1935, Gen. Arnold decided to visit his old friend Robert Millikan, whom he had met when Millikan led the Army Signal Corps's Science and Research Division during World War I. Ever alert to the possibilities of broadening the contacts of his school, Millikan introduced him to von Kármán and the GALCIT operation.

In contrast to Theodore von Kármán —a slender, soft–spoken, scholarly Hungarian — General Arnold was big framed and hearty. He had no technical education and traced his lineage to pre–Revolutionary America. Arnold rose through the ranks on the strength of dogged hard work and service as one of the nation's first military aviators. Yet both men understood the importance of science to military success, and they shared a certain "vision and judgment" about the future of technology. In 1935 to early 1936, Kármán tutored the general on aircraft stability, control, engine performance, and airship aerodynamics. When Kármán and Arnold met again in May 1938, Kármán showed the general two important sites: the rocket apparatus spread out beside the Guggenheim building, and the test range in the Arroyo Seco canyon. They also reviewed some of the promising theoretical findings of Malina, Amo Smith, and Tsien. General Arnold became intrigued with the possibilities rocket–assisted takeoff for heavy bomber aircraft. Arnold's influence over rocket development would be crucial in the years to come, as he rose to the highest ranks in the military: first as four–star General of the U.S. Army Air Forces in 1943, then five–star General of the U.S. Army in 1944, and finally five–star General of the

Solid propellants offer the advantage of minimum maintenance and instant readiness. Among the companies that jumped on the solid rocket bandwagon after GALCIT's pioneering research were, of course, Aerojet which von Kármán was instrumental in founding, the Hercules Powder Company, Thiokol Chemical Company, United Technology Corporation, and Atlantic Research Corporation (ARC)-- among others. In the United States, thousands of engineers and chemists improved the various binders, fuels, oxidizers, and additives that comprise today's solid propellants. To single out a few, at ARC two young engineers—Keith Rumbel and Charles Henderson—found that adding large amounts of

Air Force in 1949 (the U.S. Air Force was founded in September, 1947). As an aside, there have been only two five–star General of the *Armies* persons in U.S. history—George Washington was posthumously awarded this rank in 1976 as part of the American bicentennial celebrations, and John J. Pershing was awarded the rank after WWI. The slightly lesser five–star rank of General of the Army has been held by Arnold and four other WWII Generals—George C. Marshall, Douglas MacArthur, Dwight D. Eisenhower, and Omar Bradley. The equivalent five–star Fleet Admiral of the Navy has been held by four WW II Admirals—William D.Leahy, Ernest King, Chester Nimitz, and William T. Halsey (and Raymond A. Spruance almost made it). But Arnold was the only person in American history to hold *two* five–star rank General Officer positions.

Five–star General Henry H. Arnold

In August 1938 the Caltech rocket team's prospects further improved when Malina flew to San Diego and held a long conference with Consolidated Aircraft management, including president Reuben Fleet and his staff. They agreed that Malina write a feasibility report on the use of rockets for assisting heavy or overloaded aircraft takeoffs, and takeoffs from short runways. Getting heavily loaded seaplanes into the air was a particular problem. Within 10 days, on August 24, 1938, Malina completed a treatise entitled, "The Rocket Motor and Its Applications as an Auxiliary to the Power Plants of Conventional Aircraft." It concluded that solid rockets could indeed be adapted to boost aircraft to higher operating altitude and greater velocity. From then on, such booster rockets would be known as *Jet Assisted TakeOff* (JATO) units. JATO is actually somewhat of a misnomer, since it is technically more correct to use the term RATO, for *Rocket Assisted TakeOff*. But von Kármán and the GALCIT rocketeers decided it was too chancy to attach the "rocket" label, to avoid a bad connotation since the public still felt that rockets were unproven. The acronym JATO has stuck.

During the last months of 1939, Kármán and Malina decided to let Parsons try solid propellants for JATO. But the go–ahead was followed by months of failure. The team still couldn't get the black powder to ignite for more than a few seconds before it built up an excess of pressure in the combustion zone inside the rocket casing, causing dangerous explosions. Long, controlled burning was absolutely necessary for the safe use of solid JATO rockets on aircraft. Top authorities on sky rockets and explosives were called in to examine the problem. All reported the same conclusion to Dr. Kármán: a powder rocket, because of inherent instability, could not burn for more than two or three seconds, a fact known for centuries ever since the Chinese started experimenting with rockets.

One evening in the summer of 1940, Professor Kármán asked Frank Malina to drop by South Marengo for a discussion of the dilemma. The two chatted for a while, Kármán puffing on his cigar and enjoying a drink. The booze must have had a beneficial effect on him, as he suddenly had a Eureka moment. In the Guggenheim lab the next morning, he presented Malina with four differential equations describing the operations of a restructured burning motor. Malina took up the challenge, and found to his amazement that the pages held the answer to solid propellant rocketry. In theory, so long as the ratio of the throat area of the exhaust nozzle over the total internal burning area of the fuel remained constant, the powder would burn more slowly and evenly. Over the following months, Parsons returned to the Arroyo Seco hundreds of times and used this formula as the basis for countless experiments. He finally achieved reliable burning. The amateur chemist devised a paper–lined combustion cylinder in which he packed one–inch layers of black powder propellant blended to his own specifications. He used a small black powder charge to ignite the rocket, and the burning proceeded evenly down the length of the tube, creating the desired controlled explosive effect. The brilliant combination of Kármán's theory and Parson's chemistry resulted in the first proven controlled–explosion solid rocket motor.

and safer to handle than the mixture of nitroglycerin and nitrocellulose "double–base propellant."

Precise procedures for producing, mixing, and "cooking" and "curing" the solid propellant were critical to its success, and these factors continue to be crucial in making the more sophisticated solid propellants of

World's first successful JATO launch, August 16, 1941. The Ercoupe climbs steeply while a Porterfield airplane, which took off at the same time, is just getting up to speed on its takeoff roll.

The JATO rocketeers, March Field (near Los Angeles), California. Left to right: Fred P. Miller, John W. Parsons, Edward S. Forman, Frank J. Malina, Capt. Homer Boushey, and two Army Air Corps mechanics stand in front of the Ercoupe.

Thus encouraged, the Air Corps chose an Ercoupe, the lightest plane they could find, to conduct initial tests with JATO units strapped to racks on each side for takeoff. Jack Parsons prepared several dozen canisters, each developing 28 pounds of thrust over 12 seconds, for the initial flight tests at March Field. But a big problem ensued when Parson's propellant proved unstable in storage. The propellant tended to expand and crack, causing misfires and explosions. Kármán's rocketeers would have to pack the units with propellant just before flight, race to March Field from Pasadena, and load them immediately into the Ercoupe's JATO racks. Again, Parsons and his colleagues tinkered with a number of combinations and came up with a new type of propellant, consisting of a mixture of potassium perchlorate as the oxidizer and roofing tar as the fuel. They called this combination a "composite propellant," and its great advantage was that it could be cast as a single material. Properly mixed and cast, it was also stable in storage

today. First, Parsons heated the roofing tar to the liquid point in a mixer, then he added the granular perchlorate, and cast the mixture in the rocket chamber where it was allowed to cool to a hard solid. To get a good seal between the charge and the inside walls of the metal cylinder, Parsons developed a technique of pouring a little of the molten tar into the chamber to form a thin liner, Then the powder itself was poured in and allowed to cure. Similar procedures using large mixing bowls are followed in modern solid propellant production. Solid and liquid propellant rockets operate on the same principle. The key is to combust propellants in a chamber, generate exothermic chemical reactions which create heat and hot gas, and then convert as much of this energy as possible into kinetic energy. It is in the nozzle that this heat energy is converted into kinetic energy, and consequently nozzle design and performance is a major factor in overall rocket performance. Yes, innovations have been made in better binder production, mixing of ammonium perchlorate with aluminum powder and iron powder, chamber liners that don't debond, nozzles, and so on. But Parson's basic concept of casting propellant into large charges has made possible such outstanding rockets as the Polaris, the Minuteman, and the Space Shuttle reusable solid rocket boosters, giving the United States its excellence in large solid rocket motors.

Following the trail blazed by the GALCIT rocket pioneers, today's rocket motors depend on the combustion characteristics of the propellant, its burning rate, burning surface, and grain geometry. These design characteristics can become quite complex, and the applied science is known as *internal ballistics*. Grain geometry design is a science in itself. Success in rocket motor design and development depends significantly on knowledge of burning rate behavior of the selected propellant under all motor operating conditions and design limit conditions. Grain geometry design specialists can vary the burning rate by changing the propellant composition or characteristics. For example, they can add a burning rate catalyst, increase the percentage of catalyst, decrease the oxidizer particle size, increase the oxidizer percentage, increase the heat of combustion of the binder and/or the plasticizer, and even imbed wires or metal staples into the propellant. Motor designers can also vary the burning surface in a design and customize the thrust profile a

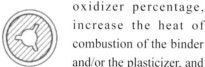
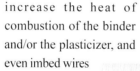

Solid propellant grain configurations

motor produces by incorporating simple geometric shapes into the internal cavity formed by the solid propellant itself. These shapes include, for example, a star pattern, rods, tubes, wedges, or slots.

On August 16, 1941, Air Corps Lieutenant Homer Boushey climbed into the cockpit of an Ercoupe at March Field, revved up the motor, and let off the brakes. As the plane rolled down the runway and gathered momentum, he snapped on an ignition switch, adding the thrust of six JATO canisters (three on each side) to the Ercoupe's own engine. The plane

shot off the ground at an impossible angle as if released from a slingshot, a large plume of white smoke trailing behind. The extra burst of power cut in half the normal distance required for the aircraft to take flight, yet did not materially affect flying performance. With bigger JATOs and more of them, even heavy bombers could operate from much shorter runways than ever before. The JATO had arrived.

As World War II wound up, contracts soon followed from the Army Air Corps and the U.S. Navy for more JATO rockets. During the last weeks of 1941, under General Arnold's suggestion, von Kármán and his rocketeers decided that a business independent from GALCIT and academia was necessary for mass production of the JATOs and other types of rockets. Caltech's charter committed it to nonprofit education and basic research, not private enterprise. Hap Arnold said that the Air Corps wanted a sharp distinction between scholarly and commercial activity. He and other airmen felt that an academic institution could not possibly go into mass production and compete on the open market for business, particularly with WWII looming. During the winter of 1941/1942, Dr. von Kármán and four principals from Caltech—Martin Summerfield (Kármán's former doctoral student), Frank Malina, Ed Forman, and Jack Parsons—were joined by Andrew G. Haley to organize a

Founding members of the Aerojet Engineering Corporation, standing in front of the Havoc Bomber: left to right (except for Col. Paul H. Dane—fourth from left, Director of the Army Air Forces Jet Laboratory at Wright Field, Dayton, Ohio): Theodore C. Coleman (Director of Aerojet), John W. Parsons (Vice-President), Edward S. Forman (Vice-President, Production), Andrew G. Haley (President), von Kármán (founder and Chairman of the Board), Frank J. Malina (Secretary-Treasurer), Martin Summerfield (Vice-President, Engineering), and T. Edward Beehan (General Manager).

company. Haley, an attorney from Washington, D.C, was the only one with business experience. Von Kármán wanted to call the enterprise "Superpower," but this sounded too much like Superman, so the group decided on the *Aerojet Engineering Corporation*. Aerojet was incorporated on March 19, 1942 (in Wilmington, Delaware for tax reasons) and was destined to grow in years to come into an aerospace giant in the field of rocketry.

In addition to building solid–fuel JATO units, the fledgling enterprise also perfected the technology for production of small liquid rocket engines that could be carried on planes, based upon Caltech research. As is always the case, perfecting sophisticated rocket technology is not easy. For some applications, newer liquid propellants have some advantage over the older solid propellants. They give combustion products of lower molecular weight and therefore provide higher nozzle exhaust velocities and greater thrust for the amount of propellant used. That is why liquid rockets are more often used in upper stages and on payloads in orbit, to minimize weight which becomes more and more critical the higher up you go in Earth's atmosphere. Besides solids, John Parsons had also been experimenting with liquid fuel oxidizers that were easier to handle than liquid oxygen. On the basis of his results, von Kármán and Malina decided to try red fuming nitric acid. They wanted a practical liquid rocket small enough to be carried on a plane and yet big enough to provide sufficient thrust. Looking at a picture of a Goddard rocket, von Kármán realized that its hideous trail of black smoke meant that the combustion was incomplete and therefore the rocket was inefficient. Surely Aerojet could do better.

But problems arose with the nitric acid oxidizer when mixed with gasoline or kerosene as the fuel. A rocket engine employing the mixture proved to be unstable, a phenomenon that came to be known as "combustion instability" Rocket firings would be plagued by violent chugging due to pulsating pressure in the combustion chamber. Sometimes the engine would blow up, at other times the flame would go out. What was wrong? *Unbeknownst to von Kármán and his colleagues, the Germans at Peenemünde were experiencing the same difficulty of combustion instability with their early V–2 rockets.*

For many months Von Kármán and Martin Summerfield researched the liquid–fuel combustion process. Eventually, Summerfield came up with the hypothesis that the instability in the engine depended upon the short delay between propellant injection and the start of combustion. When this time is short, burning is stable. When the time is long, as in the case of nitric acid and gasoline, the detonation wavefront moves too slowly and burning is unstable. Frank Malina also investigated the phenomenon, and decided that they should replace the gasoline they had been trying with aniline, even though it is toxic. The results were spectacular. Nitric acid and aniline ignited spontaneously on contact, and the combustion flame was absolutely steady. They had found the so–called "hypergolic combination" of fuel and oxidizer which ignite almost immediately on contact, so that no auxiliary igniter is needed. *Again, they found out after the war that the Germans had independently discovered the hypergolic principle at almost the same time and used it in a number of devices.* In fact, the more modern NASA Space Shuttle Main Engine (SSME) used hypergolic propellants—liquid oxygen and hydrogen.

Once the feasibility of solid propellant JATOs had been proven, the more volatile liquid fuel JATOs were tested at Muroc Dry Lake, California. In the spring of 1942, Aerojet conducted the first flight tests of JATOs using the nitric acid (oxidizer) and aniline (fuel) combination. They loaded two 1000–pound thrust liquid JATOs on the A–20B Douglas Havoc Bomber. Major Paul H. Dane revved up the engines and taxied down the runway. His assistant (Beverly Forman, a cousin of Ed Forman) actuated a switch, the rockets ignited, and the bomber took off with a roar, almost straight up. As smoke billowed forth, the A–20 continued to rise, like a bird scooped upward by a sudden draft. It was a wonderful sight, the first aircraft takeoff in America with a permanently installed rocket power plant, and the beginning of practical rocketry in the United States.

Havoc Attack Bomber piloted by Army Air Forces now—Colonel Paul Dane. The bomber (A–20B) achieved a sharp angle of climb in liquid fuel JATO flights on January 7 and 8, 1943.

THE AEROSPACE PIONEERS

Collectively, the progress of civilian aviation, military aviation, and aeronautical research set the stage for the aeronautical revolution that began in the 1930s. The design characteristics of the 1920s—fabric covered biplanes with radial engines—gave way to truly sophisticated airplanes of the 1930s with streamlined shapes, metal construction, retractable landing gear, and high performance. The national economy may have sagged during the Great Depression of the 1930s, but the aviation industry reached new levels of excellence as the 1930s witnessed an exciting period of growth in aviation. The aircraft pioneers of that era paved the way for modern reusable rocket vehicles, such as the Space Shuttle. Their names should be familiar to almost every educated person, and Boeing, Lockheed–Martin, and Northrop–Grumman are well–known aerospace companies today.

Donald Wills Douglas, the so–called "dour Scot," is famous for founding and growing Douglas aircraft without a government subsidy. Today, entrepreneurial space companies like the Space Exploration Technologies Corp. ("SpaceX") are following Douglas's pathway in the much more difficult arena of space. The DC–1 (Douglas Commercial–1) aircraft, designed by Art Raymond, was one of the first major commercial passenger carriers to emerge. A modified and larger version quickly followed, the DC–2. This plane won the then famous 10,000–mile London–to–Melbourne race, making the trip in 71 hours and 28 minutes. In 1935 Douglas produced his third model, the DC–3, which was a fattened DC–2 with berths and was first known as the Douglas Sleeper Transport. It was to become America's most famous airplane, and established Donald Douglas as a major airplane maker. In fact, in the mid–1990s the Air Force honored the DC aircraft series by naming its single–stage research rocket the DC–XA.

In 1934 James H. "Dutch" Kindleberger established North American Aviation (NAA), with the help of General Motors. It is said that Kindelberger built more planes in his 46-year career in aviation than any other man in history. In addition, the foundation he laid in North American produced the Apollo spacecraft and rocket engines that carried man to the Moon. His leadership took North American Aviation from the postwar era on to jet aircraft, nuclear energy, rocket engines, missiles, the triple-sonic Valkyrie B-70 bomber, and the X-15 research aircraft. NAA later metamorphosed into North American Rockwell, a business unit of Rockwell International Corporation.

Dutch Kindelberger

William Edward Boeing was born in 1881 in Detroit, Michigan to a wealthy German mining engineer named Wilhelm Böing who had made a fortune developing large low–grade taconite iron ore deposits and who had a sideline as a timber merchant. In 1900 William Boeing enrolled in Yale University, but he left Yale in 1903 to go into the lumber side of the business. Already a man of means, Boeing bought extensive timberlands around Grays Harbor on the Pacific side of the Olympic Peninsula in the state of Washington. He also bought into lumber operations. Even today, the smell of freshly cut wood and lumber mill operations permeates the area around Aberdeen and Hoquiam, two municipalities in the area. Boeing experimented with boat design, and became president of the Greenwood Logging Company.

William Boeing in Seattle, 1910. Mr. Boeing was a very private person.

Donald Douglas with one of his creations

It was in 1909 that Boeing traveled to the Alaska–Yukon–Pacific Exposition where he saw a manned flying machine for the first time. He was fascinated. In 1916 he went into business with George Conrad Westervelt as "B & W" and founded the Pacific Aero Products Co. When America entered the First World War in April 1917, Boeing changed the name of the company to Boeing Airplane Company, and obtained orders from the U.S. Navy for 50 planes. After the war, Boeing began to concentrate on commercial aircraft, secured contracts to supply airmail service, and built a successful airmail operation. In 1934, the federal government accused William Boeing of monopolistic practices. Under provisions of the Sherman Anti–trust Act (1890) and the Air Mail Act, the government forced him to break up his company into three separate entities: Boeing Airplane Company, United Aircraft Corporation, and United Air Lines. All three became successful. It was only after World War II that Boeing Company began to really grow.

Lockheed Corporation has an interesting history. The Alco Hydro–Aeroplane Company was established in 1912 by the brothers Allan and Malcolm Loughead. This company was renamed the Loughead Aircraft Manufacturing Company and located in Santa Barbara, California. But it didn't last very long.

The Loughead brothers – Allan and Malcolm—in cockpit.

In 1926, following the failure of the Loughead Company, Allan Loughead formed the Lockheed Aircraft Company (the spelling was changed to match his last name's phonetic pronunciation) in Hollywood, California. In 1929, Lockheed sold out to the Detroit Aircraft Corporation. The Great Depression ruined the aircraft market, and Detroit Aircraft went bankrupt. A group of investors headed by brothers Robert and Courtland Gross bought the company out of receivership in 1932. The

syndicate bought the company for a mere $40,000. Ironically, Allan Loughead himself had planned to bid for his own company, but had raised "only" $50,000, which he felt was too small a sum for a serious bid. In 1934, Robert E. Gross was named chairman of the new company, the Lockheed Corporation, which was headquartered at the airport in Burbank, California. The company remained there for many years before moving to Calabasas, California. Lockheed produced P–38 Lightning fighters during World War II and Constellation airliners after the war.

Allan Haines Loughead

Lockheed–Martin's second line of heritage, the Martin Marietta Corporation, began in 1912 when American aviation pioneer Glenn L. Martin organized a company to manufacture and sell airplanes. Four years later, the Wright Company acquired the enterprise to form Wright–Martin Aircraft Corporation. As we saw in the Orville chapter, the Wright Company had been reformed in 1915 after Orville Wright sold his sagging business (founded in 1909) to Wall Street investors. In 1917, with the help of several American industrialists, Martin incorporated a new Glenn L. Martin Company, which supplied its MB–1 bomber to the U.S. military. In 1928 Glenn Martin sold that company's manufacturing facilities and bought a 90 percent interest in pioneer automobile designer Louis Chevrolet's small airplane–engine company, incorporating as Martin Company. During World War II the company produced numerous small bomber aircraft, such as the twin–engine B–26 Marauder, and several rather unsuccessful commercial transports. In the 1950s Martin Company began work on the Pershing ballistic missile and the Titan intercontinental ballistic missile, later developing the latter into a space launch vehicle. In 1960 Martin's last airplane rolled off the production line, and the company devoted itself to missiles and space launchers.

In 1961 Martin Company diversified through its merger with the American–Marietta Company (incorporated 1930) to form Martin Marietta Corporation. Martin Marietta space activities included the construction of the two Viking landers, which touched down on Mars in 1976, and the Magellan spacecraft, which mapped the surface of Venus in the early 1990s, and the design and production of the space shuttle's external fuel tank. In the early 1990s Martin Marietta made two large–scale additions to its space–related assets. In 1993 it acquired General Electric's aerospace business, a move that was followed a year later by the purchase of the space systems division of General Dynamics, producer of the Atlas launcher and Centaur upper–stage vehicle. In 1995 Lockheed and Martin-Marietta merged to form today's Lockheed Martin Corp.

the wings. This load made the wings thick, which is all right for flight at 300 to 400 miles per hour, but by the late 1940s higher speeds of 500 to 700 miles per hour for crewed aircraft began to interest the Air Force. Indeed, the Bell X–1 had already pierced the sound barrier a number of times in continuing test flights.

On June 5, 1948, Captain Glenn W. Edwards of the U.S. Air Force decided to push the YB–49, successor to the XB–35 Flying Wing (X is experimental, Y is service, and B stands for production), as far as it would go in a dive. This pushed the plane beyond its placarded speed of 500 mph. The speed became too high for the structure and the plane began to shake violently and lose stability. The plane crashed and Edwards was killed. Capt. Edwards was honored posthumously by having the present Edwards Air Force Base named after him. But the following year the Air Force withdrew its support of the Flying Wing project (in favor of the more available B–36s instead), and it had to be abandoned. Although Jack Northrop's "dream machine"

Glenn Martin

Jack Northrop

One of the most interesting aviation pioneers was Jack Northrop, founder of the Northrop Corporation in Los Angeles. Northrop had never gone to college and had never formally studied aerodynamics, but he had an intuitive grasp of aviation and he produced some great and beautiful aircraft designs. He was employed for a time as a designer for the Douglas Aircraft Company and for Lockheed. In the late 1930s he established his own business as an aircraft manufacturer. One of his best designs was the "Flying Wing" which had no empennage (vertical tail). Northrop insisted that the crew, fuel, the entire plant including the engines, and everything else be installed in

was never realized, Northrop went on to become a great aerospace company, renowned for its expertise with one–of–a–kind difficult structures, such as used on the B–2 bomber to make it practically invisible to radar.

Leroy Randle Grumman and Leon A. Swirbul co–founded the Grumman Aeronautical Engineering Company in 1929. As risk–tolerant entrepreneurs, Grumman mortgaged his house and Swirbul convinced his mother to borrow money from her employers. While the employees in the plant felt comfortable calling the outgoing Swirbul "Jake," no one ever called Grumman anything but "Mr.

The futuristic–looking, jet–propelled XB–49 Flying Wing over Muroc Air Force Base, California in 1948. "The Flying Wing was doomed to failure because it appeared at the wrong time in aviation history," von Kármán remarked.

B–2 Spirit Bomber in flight over central California, 50 years later in 1998.

Grumman" out of deference to his reserved manner and respect for his skill as an engineer and designer. To family and close friends, he was invariably known as "Roy." Grumman won contracts from the U.S. Navy after Mr. Grumman invented an innovative retractable landing gear

Leroy Grumman

and a wing folding system that revolutionized aircraft storage and handling on aircraft carriers.

One of Grumman's strengths was its employee–friendly company culture and "hands–on" management style where the bosses could talk comfortably with both executives and factory floor workers. While continuing the company tradition of aircraft production for naval aviation, in a move towards diversification, Mr. Grumman entered the commercial civil aviation market and pushed for a change in priorities toward space projects, that culminated in the design and production of the Apollo program's Lunar Rover (aka Lunar Excursion Module). In 1969 the company was rebranded as Grumman Aerospace Corporation. In 1994 Northrop Aircraft merged with Grumman Aerospace to form today's Northrop Grumman Corporation.

In recent decades, large aerospace and defense contractors worldwide have merged to become humongous, company–gobbling global conglomerates. Driving factors include intensified competition in the international marketplace, economies of scale, and subsidies of domestic companies by their governments. The golden age of aerospace company startups ended in the 1930s in the United States. Today it is impossible for any startup company to replicate Boeing's multibillion dollar investments in commercial airliner production, for example. The commercial airliner and government–driven aerospace and defense sectors are dominated by a few giant companies: Lockheed Martin, Boeing, Northrop Grumman, General Dynamics, and Raytheon in the United States, BAE Systems in the U.K., and the Airbus Group in Europe. BAE Systems was formed in 1999 by the merger of

British Aerospace (BAe) and Marconi Electronic Systems (MES). In 1999 Deutsche Aerospace AG (DASA) merged with Aérospatiale–Matra and Construcciones Aeronáuticas SA (CASA) of Spain to create Airbus's predecessor, the European Aeronautic Defense and Space (EADS) Company. In 2014 EADS was reorganized as Airbus Group NV, and in 2015 it was renamed Airbus Group SE (Societas Europaea), as a corporate entity in the European Union. The size of these companies in terms of gross revenue is roughly the same, so who can claim the top spot may change from year to year. For airplanes, Boeing and Airbus have emerged as the sole two top competitors.

DC–3

One day in mid–1935, von Kármán was working with Douglas aircraft engineers on the difficult problem of air turbulence, the main source of drag on aircraft in flight. Theodore had just climbed into the ten–foot wind tunnel at Caltech with a wad of putty. In his own words, "Imagining myself being the airplane, I tried to feel where I might be pressed by an element of air. To show up the flow of air on the model, we used wool tufts and fine threads attached with tape to different parts of the model. We tried to observe where the tufts and threads ran smooth, and where they flapped violently, indicating the air was separating from the surface and wasting its energy in turbulent motion. In this way we found, as we suspected, that the air had rough going when it went past the corner between the wing and the fuselage, As soon as the corner was puttied in, the air smoothed out. It was really wonderful to see how well the fairing worked."

The world–famous Douglas Commercial (DC)– 3 aircraft. Note the "Kármán Fillet," the streamlined surface over the juncture between the upper wing and fuselage.

Models of the Northrop Alpha, a low–winged monoplane which preceded the DC–1, had started to sway and shake violently when subjected to winds of 200 mph in the Caltech wind tunnel. The experimenters eventually isolated the source of these vibrations as the turbulence created between the craft's wing and its fuselage. A sharp corner where the two came together caused the air to decelerate as it swept past, forming turbulent eddies. As these eddies broke from the trailing edge of the wing, they hit the tail, its vertical stabilizer, and the elevators, causing them to vibrate. Von Kármán studied the problem and came up with a solution based on his experience with the Kármán Vortex Street and eddy formation. Using putty, he applied a small fairing to the junction of the fuselage and the wing. Properly applied, this device caused a striking change in the Alpha's behavior. The air eddies smoothed out, and the tail no longer shook.

Based upon wind tunnel data and von Kármán's putty, the Caltech group designed a fillet for the Douglas DC wing, which was similar to Northrop Alpha wing. They thought it should perform the same air–smoothing function as the strip of putty on the model. It worked. As a result, the DC series of airplanes never suffered from the vibrations which plagued many other airplanes of similar design. This was one more remarkable example of how a wind tunnel finding could be applied in a practical way to save an aircraft.

Von Kármán and his Caltech group made more suggestions, which made the DC–3 a workhorse of aviation. They suggested the Douglas engineers incorporate the new engine cowling designed by NACA engineers, which would enclose the engines and serve to force air in a controlled manner around the cylinders. They also suggested that the nacelle which holds the engine and propeller be moved forward in the wing. Von Kármán also found that thin sheets of metal could withstand the buckling and aero loads of an aircraft in flight if one reinforced the metal by running stiffeners along it. Aircraft metal surfaces could carry far higher loads, in spite of the metal itself becoming stressed. Longerons and spars are now a staple of every aircraft (and the retired Space Shuttle) structural design.

PRIVATE, CORPORAL, SERGEANT

> *Starting in the late 1930s and through the 1940s, a number of contemporaneous factors laid the foundation for larger and more capable liquid and solid fueled rockets, which in turn paved the way for the Intermediate Range Ballistic Missiles (IRBMs) and InterContinental Ballistic Missiles (ICBMs) of the Cold War. The two superpowers— the United States and its allies, and the Soviet Union and the Warsaw Pact nations— confronted each other with nuclear weapons from the 1950s through the 1980s. Rockets supported two legs of the "strategic triad" on each side: the land based IRBMs/ICBMs and submarine–launched ballistic missiles (SLBMs). Nuclear weapon–equipped strategic bombers and fighters constituted the third leg of each side's triad.*

With Germany vanquished after World War II, both Cold War combatants frantically dismembered the monumental German rocket program and appropriated both its hardware and personnel, like two dogs quarreling over the same bowl of food. Although the post–war influx of German rocket technology helped kick start both the Soviet and U.S.–led rocket programs, each side had already laid a foundation with their own nascent propulsion and rocket research. On the U.S. side, the Corporal Army Missile Program exemplifies its early indigenous rocket program. From its small beginnings, the Corporal morphed into more sophisticated second and third generation systems courtesy of the great boost given to rocket science in America by captured German technology after the war.

Serious missile development in the United States began on July 1 1939 when the Army Air Corps initiated the GALCIT Project No. 1 under the leadership of Dr. von Kármán. The project ended up being extended by periodic renewals right through the war, until June 30, 1946. Parson's solid propellant mixture was named GALCIT 61–C, the only successful restricted–burning propellant then in service use. The GALCIT team continued development of the red fuming nitric acid (RFNA)/aniline liquid propellant rocket unit (years later an inhibitor was added to the RFNA so it became known as IRFNA). Eventually they designed and tested the first high–performance liquid rocket engine to operate at thermal equilibrium for 30 minutes. They also designed and tested the largest thrust solid propellant motor as of 1946, with 20,000–pound thrust.

Von Kármán's innovation of regenerative cooling for rocket engines deserves special mention. The liquid–fueled RFNA/aniline JATOs that were so successful in lifting the A–20 bomber were still more like a can of explosive with a 25–second duration. The Caltech rocketeers (aka the "Suicide Club"), much like the German rocketeers that were hobbled by the Nazis during WWII, aspired to something more beneficent than missiles as munitions. Wouldn't it be wonderful if they could build a manned rocket that could take us into high altitudes and eventually into space? To enable this vision, the Suicide Club needed higher pressures, hotter mixture ratios, and firing durations of at least several minutes and longer. Obviously, you can't strap a radiator to a rocket to cool the engine and keep it from melting. Or can you?

Rocket Engine Cooling

Von Kármán and his cohorts knew that the only coolant available for long space flights was the liquid propellant itself. Thinking of the rocket engine as a form of internal combustion engine, the rocket team learned that in both gasoline and diesel engines the heat that entered the walls of the combustion chamber would be about 25 percent of the total heat of combustion. This percentage approximately held for all makes of engine, for all sizes, and for all types of fuel. Radiator technology worked to cool car engines. It was reasonable to believe that the same percentage would apply to liquid–propellant rocket engines. However, von Kármán calculated that to remove this percentage of heat from the walls of a rocket combustion chamber would be impossible. The liquid circulating coolant would have to rise from its normal inlet temperature to perhaps 1000 degrees F or more, too high for any known liquid. They had reached an impasse.

Martin Summerfield studied the issue and came up with an analysis in which he showed that the amount of heat transferred to the walls in a rocket chamber and nozzle was nowhere near 25 percent, it was more like one percent. It turns out that in rocket combustion more of the heat is dissipated longitudinally down the nozzle, and a thin film–like boundary layer builds up on the inside of the metal surfaces of rocket combustion chambers and nozzles. This film enables a tremendous temperature gradient across its surface, somewhat like a heat shield, so that——thank God—it absorbs a lot of the heat generated by the combustion process. Although dissipating even the remaining one percent of a rocket engine's total heat energy from the chamber walls is very challenging, it is doable. Von Kármán's team proved so in tests with a monopropellant (nitromethane) type engine under the GALCIT contract. The technique of cooling a rocket engine by circulating the liquid propellant around the chamber walls, and around the injector before injection into the chamber, has come to be known as "regenerative cooling." It turns out that liquid hydrogen works very well as a liquid coolant because of its high heat transfer coefficient. Not only that, but a preheated coolant that has started to vaporize actually adds energy when it is expelled into the rocket exhaust. It also helps build up pressure in the combustion chamber.

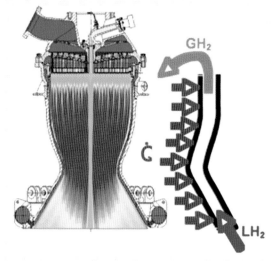

Cooling circuit of the Vulcain regen–cooled rocket engine. Vulcain is a family of European cryogenic first stage rocket engines for the Ariane 5. Notice how supercold liquid hydrogen (LH_2) turns into hot gaseous hydrogen (GH_2) as it circulates from the bottom of the nozzle to the top. The thin film of hot gas that forms on the inner chamber walls (red arrows) is too small to show even in this exploded view.

In 1943 the von Kármán rocketeers, led by Frank Malina, had begun calling themselves the Jet Propulsion Laboratory (JPL). In January 1944, Maj. Gen. Gladion M. Barnes of the Army Ordnance Department requested that Caltech undertake a research and development program on long–range, jet–propelled vehicles. This request led to the Ordnance/California Institute of Technology (ORDCIT) Project, the first of its kind in the United States. To counter the threat from German V–1 and V–2 missiles in World War II ORDCIT was intended to progressively increase the size and range of various missiles, commencing with the Private A and progressing through the Corporal and Sergeant.

On November 1, 1944, in the interest of more efficient administration, the GALCIT Research Project was reorganized and designated (following Malina's lead) as the Jet Propulsion Laboratory, Guggenheim Aeronautical Laboratory, Cal Inst. of Tech (JPL/GALCIT). As with

JPL founding fathers. Left to right: Dr. William H. Pickering (former JPL Director), Dr. Theodore von Kármán (JPL co–founder), and Dr. Frank J. Malina (co–founder, and first director of JPL). As Director of JPL from 1954 until his retirement in 1976, Dr. Pickering was closely involved with management of the Army's Corporal and Sergeant programs.

JATO versus RATO, JPL would have better been named RPL, but there was still a reluctance to use the "rocket" as rockets had a dubious "far out" connotation (imagine that!), so the JPL moniker stuck. Years later, with rockets gaining acceptance, the Air Force did institute a Rocket Propulsion Laboratory (RPL) at Edwards AFB, California. At any rate, JPL soon became a center for long–range missile development and research for the Army, and years later was destined to become NASA's center for interplanetary and space exploration. The JPL/GALCIT Project became known as JPL–1.

Private

The first missile to emerge from the JPL–1 program was a modest eight–foot weapon, the liquid-fueled Private A. It first flew in December 1944 and had a range of 11 miles.

Corporal

Under wartime pressures, the JPL–1 team soon began work on a larger and longer–range liquid–fuel missile called (naturally) the Corporal. Multiple contracts quickly followed and were named JPL–2, JPL–3, and JPL–4. JPL–4's primary purpose was to execute the development of a long–range guided missile. However, the contract encompassed such additional projects as fundamental research on propellants, matters involved in rockets and ramjet units, on remote control equipment, high–speed aerodynamics problems, materials, and provisions for the engineering, design, and fabrication of prototype missiles suitable for firing tests. Facilities

for the research called for under JPL–4 were covered by a separate contract, known as JPL–5.

But the war in Germany ended in May, 1945. As democracies are wont to do in peacetime, with the termination of hostilities the ORDCIT program was revised and the planned *Sergeant* missile was eliminated (only to be resurrected a few years later). The Corporal became a longer–term technology research and development effort to evaluate the basic principles of ballistic guided missile construction, flight, and guidance.

Frank Malina suggested an interim project, building a high–altitude sounding rocket, the original dream of the Caltech group in 1936. Out of this suggestion came a smaller, slimmer sister of the large Corporal, which was named quite logically the Women's Army Corps (WAC) Corporal. The official Army moniker for WAC was "Without Attitude Control." The WAC Corporal, 16

The 45–foot (14m) Corporal (top) was almost six times longer than its 8–foot (2.4m) predecessor, the Private.

WAC Corporal launch at White Sands Missile Test Range, New Mexico.

143

The U.S. Army, which after the war used captured V–2s for experimental flights into the high atmosphere, tried a more effective way. It replaced the payload with another rocket, in this case a "WAC Corporal," which was launched from the top of the orbit. Now the burned–out V–2, weighing 3 tons, could be dropped, and using the smaller rocket the payload reached a much higher altitude. Such was the "Bumper" rocket which in February 1949 reached an altitude of 393 km.

On February 24, 1949 a Bumper WAC Corporal "Round 5 " was launched from the nose of a reconstructed German V–2 at the newly built White Sands Proving Ground in New Mexico. Round 5 was the first U.S. missile to penetrate outer space. It attained a speed of 5,150 miles per hour and an altitude of about 244 miles, the greatest velocity and highest altitude ever reached by a man–made object up to that time. It also set precedents as the first missile to measure temperatures at extreme altitudes; the first to carry telemetry which transmitted to ground stations technical information concerning conditions encountered during flight; and the first to demonstrate the feasibility of the separation of two–stage rockets at very high altitudes. This was the first time radio equipment had ever been operated at such extreme altitudes.

One of the most important weapons of World War II was the V–2 long range rocket, developed by an elite group of German rocket scientists in Peenemünde, a secluded village on the Baltic Sea coast of Germany. Dr. von Kármán and his associates questioned the Germans in spring 1945. This V–2 is being placed in position for test firing at White Sands Missile Range (WSMR) in New Mexico in May 1946.

feet long and weighing 665 pounds, reached an altitude of 235,000 feet on October 11, 1945, with 25 pounds of payload. On this date the Army activated the 1st Guided Missile Battalion at Fort Bliss, Texas. Another Corporal attained a range of 63.5 miles and an altitude of 129,000 feet on May 22, 1947.

Suppose one wanted to use a V–2 rocket to send a small payload—say 10 kilograms—as high as possible. The usual payload of the V–2 rocket was one ton (1000 kg), and with that a height of about 100 km was possible. Reducing the payload to 10 kg would increase that height somewhat, but not by much, since the empty rocket, weighing about 3 tons, would also have to be raised to the full height.

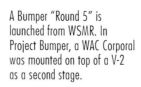

A Bumper "Round 5" is launched from WSMR. In Project Bumper, a WAC Corporal was mounted on top of a V-2 as a second stage.

First test firing of a Bumper "Round 8" at Cape Canaveral, Florida on July 24, 1950.

In 1950 the Army decided to convert the liquid–propellant Corporal into a tactical nuclear–armed ballistic

MGM–29 Sergeant Missile on guard at Camp Colbern, U.S. Forces South Korea, mid–1960s. It was capable of carrying a nuclear or conventional warhead.

controlled on Sergeant by drag-brakes, which counteracted the solid motor's thrust to reduce range from the maximum achievable. Sergeant deployment began in 1963. By 1964 it had replaced all of the Corporals in the field. The Sergeant's motor produced 200 kN (45,000 lb) of thrust for 34 seconds. Surplus rockets and the Sergeant's rocket motor (known commercially as Castor) became the basis for many sounding rockets.

U.S. Navy Aerobee

Not to be outdone, the U.S. Navy was also experimenting with rockets. The Navy adapted a few early test rockets to become "sounding" rockets for upper atmosphere research. The most versatile of these was the Aerobee, a larger version of the Army's WAC Corporal. It was first produced by Aerojet General for the Applied Physics Laboratory of Johns Hopkins University. Aerojet began work in 1946 and test fired the first complete Aerobee from the Naval Ordnance Test Facility at the White Sands Missile Range (WSMR) in New Mexico on November 24, 1947. It reached an altitude of 35 miles. The first instrument carrying Aerobee was the A–5 which

missile for use in the event of Cold War hostilities in Eastern Europe. As a quick–reaction tactical combat weapon, the Corporal was a logistical nightmare and not very effective. It burned highly toxic and difficult to handle red fuming nitric acid (RFNA) and hydrazine. Elaborate and time–consuming preparation before launch made its tactical responsiveness questionable; in fact, the Soviets would have been able to knock out the Army's Corporal battalions before they could even launch their first missile! The Corporal's accuracy was also abominable: until 1955 its in–flight accuracy to destroy targets was less than 50 percent, with only modest improvements thereafter. In fact, by 1959, the Corporal missile was used as a model for a child's toy!

MGM–29 Sergeant missile in flight.

Sergeant

In light of these shortcomings, the Army developed a solid–propellant replacement for the Corporal while it was still deployed. In January 1955 the Army selected JPL as development contractor and Sperry Utah as production contractor for the Sergeant (the next rank after Corporal). Sergeant's inertial guidance and solid propellant motor meant that logistics were vastly reduced. The missile could be launched within 90 minutes after reaching the launch location, compared to nine hours for Corporal. Range was

was launched on March 5, 1948 from WSMR. Carrying instruments for cosmic radiation research it reached an altitude of 73 miles.

Aerobees were also used to conduct one of the earliest U.S. physiological experiments on the road to manned space flight. On May 22, 1952, at Holloman Air Force Base, New Mexico, two Philippine monkeys, Patricia and Mike, were enclosed in an Aerobee nose section. Patricia was placed in a sitting position and Mike in a prone position to measure the effects of rapid acceleration on them. Fired to an altitude of 36 miles and a speed of 2,000 mph, these two monkeys were the first primates to reach so high an altitude. Along with the monkeys, two white mice, Mildred and Albert, also rode in the Aerobee nose. They were inside a slowly rotating drum in which they could "float" during the period of weightlessness. The section containing the animals was recovered safely from the upper atmosphere by parachute. Patricia died of natural causes about two years later and Mike in 1967, both at the National Zoological Park in Washington, D.C.

scientists for Gen. Hap Arnold. Von Kármán was amazed to discover his mentor and rival Ludwig Prandtl still presiding over aeronautical science at Göttingen. The Allies had spared the institute in order to capture it intact. Von Kármán questioned Prandtl not only about experiments in such fields as nuclear propulsion but about his collaboration with the Nazis in Berlin. Under pressure, the old man denied knowledge of the death camps, or of slave labor works such as at Nordhausen. Then, to his amazement, Prandtl asked who in the United States would fund his further research! For his part Dr. von Kármán was furious that a person of Prandtl's brilliance could be so naïve as to retain loyalty to the Nazi regime and ask for funding from a former enemy.

To complete this story, in 1946 a team of some 120 German rocket scientists and engineers led by Dr. Wernher von Braun was sent to Fort Bliss, Texas where the Army's 1st Guided Missile Battalion was based. In 1953, after eight years of development, a service version of the Corporal became the first tactical long-range rocket weapon to be made available to the U.S. Armed Forces. As Dr. von Kármán and his colleagues

Aerobee rocket launch, early 1950s. These relatively inexpensive rockets were used for 31 upper atmosphere experiments specially suited to their capability. Later improvements included the Aerobee-Hi with an enlarged fuel tank and the Aerobee—150, which had a liquid—propellant main stage atop a solid—propellant booster and was about twice the size of a WAC Corporal. It could deliver a 150—pound payload to an altitude of 170 miles. Aerobees were used for three flights in 1957 in conjunction with the International Geophysical Year.

The Kármán interrogation mission between bitter negotiations in post—war Germany, 1945. From left to right: Dr. Hugh L. Dryden, Ludwig Prandtl, "Major General" Theodore von Kármán, and the brilliant Dr. H.S. Tsien (also in improvised uniform). The Soviets invoked a similar practice of disguising their scientists in military uniforms to facilitate interrogations of captured German scientists.

At the end of World War II Dr. von Kármán donned a Major General's uniform and interviewed German

continued to improve on earlier rocket models, the chief

of the Army Ordnance Department, General Barnes (an outspoken advocate of ballistic missiles) asked von Kármán how far in rank his rocketeers intended to go. Von Kármán replied, "certainly not over Colonel," and then added with a smile, "that is the highest rank that works." The General laughed.

SCIENTIFIC ADVISORY BOARD

The prescient vision of General of the Army Hap Arnold manifested itself on a beautiful, cloudless day in September 1944. He had called von Kármán, who was convalescing from a serious carcinoma and intestinal operation, to meet him at La Guardia airport in New York City while he was changing planes. When Dr. Kármán arrived at the air terminal, a military aide escorted him to a waiting vehicle and they drove to the far end of the runway, stopping next to a parked official U.S. Air Forces staff car. The aide quietly disappeared. General Arnold motioned his chauffeur to leave, and welcomed Kármán in. Still the barrel–chested, open–faced presence, the General looked tired. Kármán was weak after major surgery and a long period of recuperation. They had to speak loudly to hear over the whine of engines and the winds buffeting the car. Arnold spoke first.

"We have won this war, and I am no longer interested in it. I do not think we should spend time debating whether we obtained the victory by sheer power or by some qualitative superiority. Only one thing should concern us. What will be the future of air power and aerial warfare in five years, or ten, or sixty–five? What is the bearing of new inventions, such as jet propulsion, atomic energy, rockets, radar, and the other electronic devices?"

Von Kármán listened with fascination. This was September 1944. The war was not over; in fact the Germans were to launch the Battle of the Bulge in December. Yet Arnold was already casting his sights far beyond the war, and realizing, as he always had, that the technical genius of men like von Kármán which could help find answers for him was not cooped up in military or civilian bureaucracy but was to be found in universities and in the people at large. "What do you wish me to do, General?" asked von Kármán.

"I want you to come to the Pentagon and gather a group of scientists who will work out a blueprint for air research for the next twenty, thirty, perhaps fifty years." Von Kármán was flabbergasted. What a challenge! Taken by surprise, he could only raise a feeble objection, "I do not like to work in the Pentagon." After more discussion, he softened his position. "I will do this only under one condition, that nobody gives me orders—and that I do not have to give orders to anybody."

Arnold reassured him, saying, "as far as taking orders goes, I will be your only boss. Nobody will be between us. As far as giving orders is concerned, you won't have to do that. Just tell me what you want, and I will see that it happens." Without really thinking the matter through, Kármán said yes. It afforded him a new, more direct opportunity to serve the country, and it also appealed to his adventuresome nature. They parted company. The following week von Kármán was discharged from the hospital, and in December he went to Washington to draw up a list of the people he thought would be suitable for this project.

Things moved fast as the astute von Kármán applied his decades of experience in dealing with uniformed officers and military institutions to win over the generals and decision makers in the Pentagon. Kármán quickly learned his way around the maze of corridors in the mammoth building, conferring with military and civilian leaders of principal importance to the scientific panel. With five–star General Arnold's backing, the Pentagon grapevine buzzed with stories of the new project. Von Kármán recruited an elite panel to what became known as the Scientific Advisory Group (SAG). In October 1944 von Kármán and three trusted colleagues——Hugh L. Dryden, Frank Wattendorf, and Dr. Vladimir Zworkin of the Massachusetts Institute of Technology—laid out a structure and drafted names for Arnold's SAG. Von Kármán interviewed in person, over the phone, and through the mail dozens of potential appointees to the SAG board. Fortunately, most of the first–class technical people who worked in wartime research had not yet secured peacetime positions and were available for the venture. Initially some three dozen leading scientists and engineers joined the SAG, including Dryden, Wattendorf, Zworkin, Dr. Hsue–Shen Tsien from Caltech, George S. Schaefer (a distinguished aeronautical engineer with Boeing Aircraft Company), some first–class electronics engineers (Drs. G.E. Valley, Ivan Getting, E.M. Purcell, and Lee Dubridge), and Dr. Norman Ramsey of Harvard, an expert in the nuclear field. Essentially, a robust SAG needed to encompass experts in all the scientific disciplines of importance to the future of air power in the United States. Arnold and Kármán also realized that a higher–ranked Military Director was needed for coordination and logistics support, and for securing secretaries, airplane transportation, materials, SAG meeting sites, and so on. They decided on Colonel Frederick E. (Fritz) Glantzberg, a veteran of the Ploesti oil field air raids in Romania. Col. Glantzberg would be assisted by a small cadre of some mid–grade military staff officers.

On November 10, 1944 General Arnold officially established Kármán's board of scientific advisors in the Pentagon. By early 1945 the SAG was starting to develop material in a number of specific areas, including high–speed aerodynamics, power, and communications. Many meetings were held. General Arnold addressed one of the early meetings with his vision of the future.

> *"I see a manless Air Force. I see no excuse for men in fighter planes to shoot down bombers. When you lose a bomber, it is a loss of 7,000 to 40,000 man–hours, but this crazy thing [the German V–2] they shoot over there takes only 1,000 man–hours. For twenty years the Air Force was built around pilots, pilots, and more pilots…The next Air Force is going to be built around scientists—around mechanically minded fellows."*

Then Arnold proceeded to tell the group what he wanted. "Search into every science to squeeze out basic developments that could make the United States invincible in the air," he said. "We want to fly over enemy territory and look through the leaves of trees and see whether they're moving their equipment," he went on. "Present radar won't permit this. So it's up to you scientists to tell us what developments have to be carried out in electromagnetic theory, wave propagation, and so forth, to produce a radar that will really penetrate the fog and reveal the men." *General Arnold's prescient words would ring true many decades later, when the Scientific Advisory Board (SAB, as the SAG came to be called) would grapple with sensing enemy "tanks under trees" and developing penetrating bombs to destroy deeply buried reinforced concrete bunkers and enemy command centers.* "Look twenty years into the future and forget the past; regard the equipment now available only as the basis for your boldest predictions." The General recited a long list of potential avenues of research gleaned from his talks with von Kármán and his long association with NACA: pilot–less aircraft, supersonic flight, bombs of far greater explosive power, air interceptors, aerial reconnaissance, air–to–ground communications, weather prediction, and atomic energy.

In July 1945, General Arnold asked Dr. von Kármán to write up a report on his investigations into German wartime scientific developments, and what the United States could do with its existing means of propulsion and detection, soon to be bolstered by technology and hardware captured from the Germans. The seminal Secret report called **Where We Stand** compared the two countries' state of technology in missiles and space. One key finding was that present technical competence could build ballistic missiles of 6000–mile range, enough to blitz any country of the world from bases in any other country.

In December 1945, following another reconnoitering trip to Europe, von Kármán and the SAG completed a preliminary draft of a seminal multi–volume study called **Toward New Horizons.** Completed in 1946, this report was a comprehensive look at the future of air power and the basic scientific potential which could change the future. *Toward New Horizons* was classified Secret and was not declassified until 1960. Together with a companion volume, **Science: The Key to Air Supremacy**, these products were the first exhaustive surveys of their kind in the history of the American military forces. In addition to describing World War II's technological character, these reports discussed the decisive contributions of organized science to the making of effective weapons. Never before had such large numbers of scientific workers united for the planned evaluation and utilization of scientific ideas for military purposes. And results were outstanding: the Allied side produced radar and atomic bombs; while the Germans countered with missiles and jet propulsion. *Toward New Horizons* showed not only that flight at supersonic speed was possible, but even gave the ranges over which supersonic aircraft could operate. Monographs explained the influence on the aircraft of the future of revolutionary power plants, such as the turbofan, turbojet, turboprop, pulsejet, and pure rocket. Rather than just looking at the performance of any one thing in its current form, the report tried to foresee and assess the potential for future advances.

The original SAG provided a long–range forecast of Research and Development needs for the Army Air Corps, and consisted of primarily those within the field of academia. In 1946 the Scientific Advisory Board (SAB) replaced the SAG, and held its first meeting in July. The Board included a Secretariat that would be considered part of the Air Staff in the Pentagon and would also report to both the Chief of Staff, Air Force (CSAF) and the Secretary of the Air Force (SECAF), whereas before they only reported to the Chief. The Board also expanded its membership to include those within the fields of academia, government, industry engineers and scientists. These individuals continue to serve today as a major force in determining U.S. Air Force research and development policy.

The U.S. Air Force took over the SAB after it was born as a separate military service in September, 1947. Today's SAB continues to provide a link between the Air Force and the nation's scientific community. It promotes the exchange of the latest scientific and technical information that may enhance the accomplishment of the Air Force mission. In addition, it may consider management challenges that affect Air Force use of scientific knowledge and technological advances. The Board's function is solely advisory, and provides independent findings and recommendations on science and technology for continued air and space dominance to the Air Force leadership—the SECAF, the Assistant SECAF (Acquisition), and the CSAF. They, or a designated representative(s) may act upon the Board's advice and recommendations. The SAB has up to 60 members who are distinguished in the science and technology communities, defense, and academia. The SAB is also authorized an undetermined number of non–voting consultants to provide technical expertise to the Board as required.

Fifty years after the *Toward New Horizons* study, the SAB produced in 1995 an unclassified 15–volume report, ***New World Vistas: Air and Space Power for the 21st Century***. This ground–breaking study documented in detail— in over 2000 pages of monographs— concepts and ideas which would produce a discontinuous or quantum enhancement of the effectiveness of the Air Force in the post–Cold War world. New World Vistas provided compelling reasons for pursuing these ideas, and established a path that stretched 50 years into the future. The definition of the path included suggestions for significant incorporation of commercial technologies and practices into Air Force operations, and suggestions for both change and reinforcement of the ways that the Air Force pursues science and technology goals. New World Vistas also published a classified volume and a volume of important ancillary information obtained during the conduct of the study. After the Cold War, no well–defined enemy exists. The new "enemy" could be anywhere, as evidenced by the vicious terrorist attacks on September 11, 2001 that destroyed the World Trade Center in New York City and part of the Pentagon in Washington, D.C., killing some 3,000 people. Military technology must now respond to diverse situations. Commercial technologies, developing at a rapid pace, often have significant military applications. Cost has become a major factor in the

Historic group photograph of the elite members of the Scientific Advisory Board (SAB), taken in the Pentagon at their first meeting on June 17, 1946. Dr. Theodore von Kármán is seated, center.

development of all systems. The Air Force must take advantage of new commercial technologies and must counter their use in adversary systems. It is essential that future systems be based on capabilities and cost, perhaps on an equal footing, rather than on solutions to specific problems.

New World Vistas was silent on two subjects which are still of importance today: *National Missile Defense* (at the time deemed embroiled in politics too complex to permit detailed concept definitions); and *Nuclear Weapon Technology* (nuclear technologies are developed outside of the Air Force, and today's nuclear forces are prohibited from pursuing new ideas of design or delivery).

NATO

In April 1949, the North Atlantic Treaty Organization (NATO) was organized as a defensive alliance against the expansion of the Soviet Union in central and Eastern Europe. In May 1955, the Soviets organized the Warsaw Treaty (formally the **Warsaw Treaty of Friendship, Cooperation, And Mutual Assistance** as a counterweight to NATO. The immediate reason given for the Warsaw Pact's formation was an agreement among the Western powers admitting West Germany to NATO. In reality the Warsaw Pact was only the first step in a more systematic plan to strengthen the Soviet hold over its satellites and as a lever to enhance the bargaining position of the Soviet Union in international diplomacy. The heart of NATO is expressed in its Article 5:

> *An armed attack against one or more of them in Europe or North America shall be considered an attack against them all; and consequently they agree that, if such an armed attack occurs, each of them, in exercise of the right of individual or collective self–defense recognized by Article 51 of the Charter of the United Nations, will assist the Party or Parties so attacked by taking forthwith, individually and in concert with the other Parties, such action as it deems necessary, including the use of armed force, to restore and maintain the security of the North Atlantic area.*

Fortunately, Article 5 was never invoked during the Cold War. NATO did invoke Article 5 for the first time in 2001, in response to the so–called "9/11" terrorist attacks of September 11 that were organized by the nefarious exiled Saudi Arabian millionaire Osama bin Laden.

After NATO was organized, von Kármán decided to organize a board similar to the Air Force's SAB, but now on a grander scale to support the NATO alliance nations. He declared that , "Progress in technology has become so swift that only a pool of nations could properly utilize scientific advances for mutual protection. With such an effort, it seemed to me, the international character of science could grow." While today we take for granted the idea of mobilizing science for defense, this was quite a revolutionary concept for many nations in 1949. Von Kármán invited the 12 NATO nations to a conference at the Pentagon in February 1951, where he and his associates proposed the *Advisory Group for Aeronautical Research and Development* (AGARD). Emboldened by Kármán's vision of international cooperation and scientific renewal, the first AGARD conferees adjourned their final session with a clear definition of AGARD's role: to "review advances in aeronautical science, exchange important information, and recommend how the scientific talents within NATO could best be employed in strengthening overall technical ability to solve mutual defense problems." AGARD would go on to produce a series of excellent scientific reports and assessments in support of this charter in the ensuing decades, and von Kármán would remain its chairman until his death.

AEDC

In 1941, with the advent of World War II, the U.S. Army constructed Camp Forrest (named for Civil War cavalryman General Nathan Bedford Forrest) in the vicinity of Tullahoma and Manchester, Tennessee. Between 1941–1946 it became one of the Army's largest training bases. In 1946, with the war over, the Army declared Camp Forrest and Tullahoma's neighboring Northern Field as surplus property. Buildings were sold at auction, torn down and carted away. Water and sewage systems and electrical systems were sold as salvage (today all that remains are roads, a few brick chimneys and concrete foundations). Soon after the close of the camp, the area was selected for the site of an Air Engineering Development Center for the U.S. Air Force (the Air Force did not become a separate military Service until September 1947).

In 1949, Congress authorized $100 million for the construction of the huge center. The Air Force selected a site for the new center at the Army's defunct Camp Forrest. The site was chosen for its abundance of land, water and electrical power from the Tennessee Valley electrical grid. It also helped politically that the region was depressed economically. Land was needed to buffer surrounding communities from potential test hazards and noise. Water was needed to cool rapidly flowing air and hot exhaust gases from jet engines. Abundant electricity was needed to power huge testing systems. Construction on the center started in 1950. On June 25, 1951, a year after General Arnold's death, President Harry S. Truman dedicated the Air Engineering Development Center in Arnold's honor, naming it the Arnold Engineering Development Center (AEDC).

The first jet engine test equipment installed at AEDC was acquired from the Bavarian Motor Works (BMW) in Munich, Germany. This was yet another spoil of war after World War II and the defeat of Nazi Germany. It took 58 railroad cars and two barges with another 450 tons by truck to move the equipment. After refurbishment, confiscated German equipment became the cornerstone for the AEDC's Engine Test Facility (ETF), which was completed in 1953. Two other large facilities were constructed and completed in 1959: the Propulsion Wind Tunnel (PWT) and the Gas Dynamics Facility (GDF). All three facilities remain active today as part of the multi–billion dollar AEDC complex, the largest of its kind in the world. For example, full–size jet engines can be tested at altitude conditions. Von Kármán's legacy lives on today—the GDF for testing aerospace designs at high speeds was dedicated to him on its completion.

PWT's huge wind tunnels have become hallmarks of the center and are perhaps the most heavily used facilities. PWT was used to investigate configurations for the Mercury space capsule, which sent Alan Shepard and John Glenn into space. AEDC was a key player in supporting Project Gemini, and it played a multi–faceted role in supporting the Apollo Program, which put 12 men on the moon. Apollo tests included aerodynamic assessments of the Apollo capsule and tests of Saturn 5 rocket upper stage engines. Since then there have been many new additions at AEDC including the J–4 Large Rocket Engine Test facility (1964), a 4–foot Transonic Tunnel (1968), ballistics ranges capable of studying projectiles and projectile impacts traveling at up to 20,000 mph. AEDC developed a laser-illuminated photography system which provides ultra–high–speed photographic exposures equivalent to 20 billionths of a second.

Aerial view of the sprawling AEDC complex, the most advanced and largest complex of flight simulation test facilities in the world with some 53 aerodynamic and propulsion wind tunnels, rocket and turbine engine test cells, space environmental chambers, arc heaters, ballistic ranges and other specialized units. Twenty–seven of the center's test units have capabilities unmatched elsewhere. The list of systems tested at AEDC reads like a "Who's Who" of aerospace—the U.S. Space Shuttle, F-35 Joint Strike Fighter, F–105 Thunderchief, C–141 Starlifter, C–5 Galaxy cargo plane, F–15 Eagle, F–16 Fighting

Falcon, B–1 Lancer bomber, B–2 Stealth bomber, A–10 Thunderbolt II, Pratt & Whitney F100 engine, General Electric J49 engine, MX missile, Sidewinder missile, Falcon guided missile, Navy Tomahawk Cruise Missile, Air Force Air–Launched Cruise Missile (ALCM), and the Global Positioning Satellite (GPS).

In 1956 Dr. von Kármán's efforts brought into being the International Council of the Aeronautical Sciences (ICAS) and, in 1960, the International Academy of Astronautics (IAA). One of the outstanding activities of the academy under his presidency was its sponsorship, in 1962 in Paris, of the *First International Symposium on the Basic Environmental Problems of Man in Space*, at which for the first time scientists from the United States and the Soviet Union, as well as other countries, exchanged information in this field. Between 1960 and 1963 he led NATO–sponsored studies on the interaction of science and technology.

A ladies' man to the end, von Kármán celebrates his 80th birthday with three attractive Hungarian friends in Paris, 1961.

President John F. Kennedy presenting the first National Medal of Science to Dr. von Kármán at the White House as admiring generals look on, February 1963. "I know of no one else who so completely represents all the areas involved in this medal—science, engineering, and education," JFK said. "I hope that my work has shown that the college professor is of use," von Kármán replied.

Theodore von Kármán was an optimist and believed in the future, despite the prevailing difficulties in the world. An indefatigable powerhouse, he continued to exert a powerful influence in world affairs until his death from heart failure in Aachen, Germany in the early evening of May 6, 1963, only days before his 82nd birthday. His legacy will never be forgotten. A crater on the Moon has carried his name since 1970.

СЕРГЕ́Й

"The road to space was paved with combat missiles"

Boris Chertok, guidance and control engineer

"The Earth is the cradle of humanity, but one cannot live in a cradle forever."

Konstantin Tsiolkovsky, father of modern rocketry

It should come as no surprise that the utility of missiles as weapons has served as the greatest impetus behind the development rocket science, past to present. World War II and the 44 years of the Cold War (1947–1991) served as the principal platform for advancing rocket technology, both liquid and solid–propelled.

DAWN OF THE SPACE AGE

October 5, 1957, 00:28 p.m. Local Time Soviet and now Russian practice has always been to record all space events in Moscow time (i.e., 10:28 pm October 4), despite the fact that the main launch facilities on the Russian steppes were two hours ahead. *We shall use the more accurate local time in this book.* Liquid oxygen and grade T–1 kerosene fueling operations on the modified (for civilian use) Soviet R–7 Semyorka ("the digit 7" in Russian) ICBM (InterContinental Ballistic Missile) vehicle were complete. The R–7 ICBM (military designation 'M1–1SP') was designed to carry a much heavier 3MT nuclear warhead (called 'Object D'), so its adaptation to civilian satellite use reduced its mass from 280 to 272.8 metric tons (tonnes) and its height from 34 m to 29.2 m. The genius behind the R–7 ICBM design: Sergei Pavlovich Korolyov (although also transliterated as Korolev, it is pronounced 'Korolyov' in Russian so

we shall take that stance here), the Soviets' head rocket engineer and designer. *Middle names are important in Russia. In conversation, colleagues or friends often refer to one another using first and middle names together.* Sergei Pavlovich was distinguished by a high forehead. His large head (containing an oversized brain, no doubt) was fastened to a short, bulldog–type neck. The civilianized R–7 had been vertically erected hours before on the pad at launch Site No. 1 in the then–secret Tyura–Tam (now Baikonur) Cosmodrome. The Cosmodrome is located at the center of an isolated 90km east–to–west by 85 km north–to–south elliptical restricted area in the desert steppes of Kazakhstan, near the Tyura–Tam (aka Tyuratam) railway station (about 200 km east of the Aral Sea, north of the Syr Darya river). It has traditionally been linked with the (also formerly secret) town of Dzhezkazgan. The facility, still in use today, derives its modern name from a wider area known as Baikonur.

U–2 spyplane photographs of the R–7 launch pad in Tyuratam (August 1957) and the surrounding area. The isolated desert of Kazakhstan was perfect at the time — the almost barren and enormous wilderness allowed radio signals to be transmitted from the ground station to rockets without any interference or interruption. Moreover the site was chosen for reasons of safety and secrecy. With new rocket technologies occasionally failing it was thought best to have the station well away from heavily populated areas.

Top row: R—7 undergoing final assembly and prelaunch processing inside the MIK 1 facility, and horizontal rollout to the launch pad at Site 1. Second row: Vehicle erected to the vertical position for launch, (with kerosene (orange) and Lox (white) propellant cars ready to commence fueling); and different views of R-7 preflight operations on the pad, and ignition.

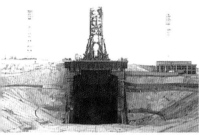

Top row: Russian practice was to bury the flame bucket (required to protect the vehicle from ignition shockwaves, high temperatures, and acoustic vibrations) underneath the pad. Bottom row: In the Russian zero—shock launch concept, there were no hold—down clamps. Once thrust built up, the rocket rose and the suspension arms rotated away on counterweights as shown in these photos.

Korolyov had led the decision to build the R–7 as a two–stage missile. His solution differed radically from the American one. In the United States missile design was given over to using rocket engines and their stages stacked on top of each other, a legacy from Hermann Oberth which he passed on to his pupil Wernher von Braun. This design fired the engines separately, and had the advantage of dropping off each stage as its propellants were depleted so that the engines on the next stage had less weight to carry. However, more and more stages is not necessarily better, there is a compromise. In the 1950s, this meant two stages. Some later designs, like the Saturn V and Scout missiles, used four stages. But staging is not like driving a car uphill and being able to throw off the luggage in the trunk, the back seat, and toss out many other weighty things as the car struggles uphill. The big disadvantage was that the amount of thrust available was only the amount of a single engine burning at a time (in going lighter, the US used only a single engine on the second stage). The end result was relatively little thrust, which limited payload size. With this design limitation the U.S. was technically challenged to develop lightweight materials and to miniaturize rocket components. US payloads had to be small.

Left: In a single engine with only one set of turbopumps, the RD-107 rocket engine combined four combustion chambers and nozzles, and two "verniers" in front (verniers are basically small rocket engines for steering).

Right: Similar RD-108 engine with four combustion chambers which powered the core (second stage) of the R-7 launch vehicle.

The Soviet R–7 vehicle consisted of a central, hammerhead–shaped core with four conical strap–on boosters which transmitted thrust to the core through a ball and socket joint. The four boosters together constituted the first stage, and the central core was the second stage. The boosters would drop off during flight, and the second (core) stage would continue burning. Each booster had a single RD–107 engine and the central core had a single RD–108 engine, for a total of five engines.

The Soviets adopted a rocket engine naming convention which continues in use to this day. In the RD–107, RD stands for ракетный двигатель, "rocket engine". The first digit "1" indicates LOX as oxidizer, a "2" means a nitrogen–based oxidizer compound like nitrogen tetroxide. The second and third digits are only numerical designators.

The Soviet's chief engine designer Valentin Petrovich Glushko, charged with developing liquid–propellant rocket engines and their serial production, had ingeniously developed the turbomachinery technology so that only a pair of high pressure LOX and kerosene pumps was required to feed four exit nozzles at once, as opposed to the standard American practice of only a single nozzle. So to the layman peering underneath the rocket it would appear that the Semyorka had 20 main engines in all. Each booster had two vernier rockets for steering and the central sustainer used four verniers, making for 12 vernier chambers in all. Verniers are small rocket engines, usually mounted on the rear and to either side of a space vehicle. When switched on they provide thrust for steering, much like a small auxiliary outboard motor engine could be used to steer a power boat around. The boosters and core stage all had to ignite simultaneously for launch. That meant that 20 rocket nozzles and the 12 verniers, 32 nozzles in all, would have to operate flawlessly to ensure a successful launch.

World's First Satellite

After technicians and the launch crew completed their tasks, the countdown began at 28 minutes past midnight.

8…7…6…5…4…3…2…1…IGNITION! Immediately a violent flash of light illuminated the night over the Kazakhstan steppe. Per standard procedure, the engines operated at less than full throttle on the launch pad to verify operation. Then suddenly the Tank Emptying and Synchronization System (SOBIS— *Sistema oporozhneniya bakov i sinkhronizatsii*)

RD–107 (4)

RD–108 (1)

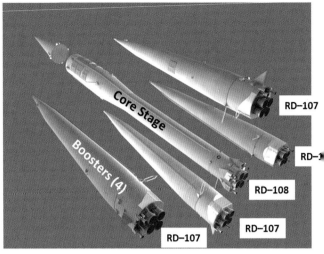

Core Stage

Boosters (4)

RD–107

RD–

RD–108

RD–107

RD–107

RD–107

RD–108

RD–107

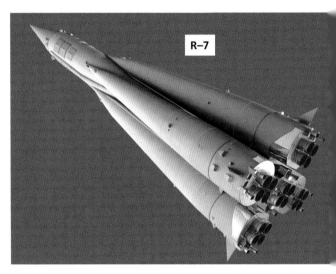

R–7

R–7 Vehicle Design Architecture. This time–honored clustered boosters and central core stage concept is still favored by the Russians today for Soyuz and Proton launch vehicles.

Korolyov's R-7 launches Sputnik 1 into the darkness from Baikonur, on Oct. 5, 1957. The four holding arms on the launch pad have swung away, releasing the Semyorka.

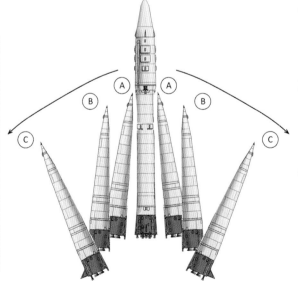

A. Lower connections are severed on the boosters. They begin to pivot around the upper connectors while still thrusting.

B. Upon reaching a certain rotation angle, upper connectors disengage, releasing boosters; oxidizer pressure valves open on boosters for retro thrust.

C. The now unpowered boosters fall away from the core stage, which continues a powered flight.

R–7 second stage separation sequence at altitude between its four strap-on boosters and the core stage:

tanks experienced a split–second delay. This delay caused an automatic system reset. Appraising the situation in an instant, technician Pavel Andropov stood poised to hit the destruct switch if the rocket malfunctioned in flight. In the last fractions of a second of the Block D time check, it completed thrust buildup. With all the engines at full throttle, the rocket lifted off with a roar. Glushko's four RD–107s and one RD–108 reliably put out 3904kN (877,654lbf) of thrust.

Launch information, good or bad, was being transmitted in real time from the launch bunker to Vladimir Bolkhovitinov, the commentator in the barracks at Site No. 2. He in turn relayed it to chief Soviet rocket designer Sergei Korolyov's office on the other end of the barracks via a high–frequency communications phone. Around 30 duty officers filled the reception area and Korolyov's office.

Sixteen seconds into flight, the ground controllers were dismayed when the *Semyorka's* Tank Emptying System (SOB—*Sistema oporozhneniya bakov*) failed, causing a slight increase in kerosene consumption. The SOB failure put everybody on edge. Would the world's first satellite launch attempt fail because the rocket engines ran out of fuel? Thrust times burn duration is a measure of the "oomph" that a rocket engine puts out to launch a payload into space. Not enough thrust or not enough burn duration, and the satellite would never make it into space. Nevertheless, as it rose, the *Semyorka's* flame gradually diminished and soon became indistinguishable against the background of the stars in the clear night sky. The four booster engines throttled back to 84% of full throttle before separation, and then they separated normally.

The core RD–108 engine had to keep burning at 100% full throttle, which was set by a flight integrator for 296.4 seconds after ignition, or 201 seconds after booster separation. How long would it last before running out of fuel? At 200 seconds the kerosene ran out. Freed from the load of the kerosene pump, the turbine began racing too fast. Just like with a race car, if a rocket's turbines turn too fast past a "redline", they could tear themselves apart and take down the entire space vehicle in the process. To

prevent this from happening, the RD–108 engine had a fail–safe emergency contact controlling the turbines' RPM. It sent a signal to shut down the engine, in this case about a second too early.

Everyone held their breath. There had been too many failures. Finally the signal came that the 83.6–kg satellite, called Sputnik–1 ("Satellite–1," or PS–1 for "Elementary Satellite–1") successfully separated from the conical payload shroud atop the second core stage. It had made it out of Earth's atmosphere. Shortly, Junior Engineer–Lieutenant V.G.Borisov relayed the information from the IP–2 'Tral' telemetry station (situated one km from the launch pad) that he heard Sputnik's celebrated "beep, beep, beep" radio signals. Reception lasted for two minutes while the PS–1 was above the horizon. Everyone—Korolyov, his guidance and control assistant Boris Chertok, and the others breathed a sigh of relief. But their worries were not over. Had it reached a stable orbit, or would it plunge back to Earth?

Sputnik 1 showing the outer sphere and pressurized inner sphere with battery case and transmitter.

A technician puts the finishing touches on Sputnik 1 shortly before the satellite's three-week orbit.

As the commentator exclaimed with glee that the signal was being received on the first orbital pass, someone down the hall was already shouting, "We did it! Everything is OK. It's beeping. The sphere is flying!" Observations during the first orbital passes showed that the satellite had been inserted into an orbit with an inclination of 65° 6′, an altitude of 228km at its perigee (the point of closest approach to earth), and an apogee (its maximum distance from the surface of the earth) of 947km. It took precisely 96 minutes 10.2 seconds to complete each orbital pass around the Earth.

Sputnik 1 was the first manmade object to circle the globe. The Sputnik 1 rocket core stage also reached Earth orbit and was visible from the ground at night as a first magnitude object following the satellite. Korolyov had intentionally requested reflective panels placed on the booster in order to make it so visible. The satellite itself, a small but highly polished sphere, was barely visible at sixth magnitude, and thus more difficult to follow optically without a telescope. Again and again the mass media mistook the core stage for the satellite, claiming that it was visible to the naked eye. Ahead of Sputnik 1 flew the third object – the 80cm–long conical payload fairing, a little bit bigger than the satellite.

There had been only one specification for Sputnik: its weight was not to exceed 100 kg. The designers quickly concluded that it would be advantageous to make it in the shape of a ball, 585 mm (23 in) in diameter. The spherical shape made it possible to utilize the internal space more efficiently while minimizing the body outside surface area. The satellite was designed rapidly and the parts were fabricated as the drawings were issued. A satellite 'twin' was mated with and separated from the missile body many times until the engineers were convinced that all the circuits operated reliably, the pneumatic locks activated, the nosecone fairing separated like it was supposed to, the antenna spike released from the stowed position, and a push–rod directed the satellite forward. The satellite had a one–watt transmitting unit that emitted alternate signals of 0.3–second duration on two frequencies: 20.005 and 40.002 MHz. When one signal was on, the other was in pause mode. The estimated continuous operating time was at least 14 days. Sputnik had two omnidirectional antennas, each with two whip–like parts 2.4 and 2.9 meters (7.9 and 9.5 ft) long. When assembled it looked like an exercise–ball sized polished aluminum sphere with four fishing poles sticking out the back end.

It was approaching midnight. An eerie quiet had descended on the Korolyov's office and the adjacent reception area. Everyone was tensely focused on the phone, waiting for Bolkhovitinov to give the word. They didn't have long to wait.

Beep…Beep…Beep…Beep

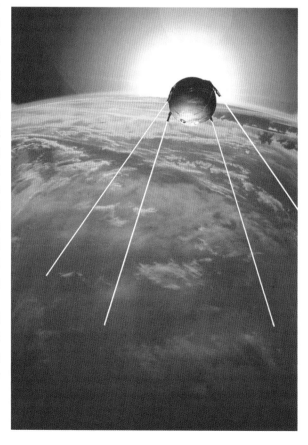

Artist's impression of Sputnik 1 in orbit.

At this time the world did not know that **Sergei Pavlovich Korolyov** was the genius behind the Soviet rocket program. His very name and identity were hidden behind a veil of secrecy—or, if you will, behind an Iron Curtain. Nevertheless his triumph echoed around the world. Surely he would have received a Nobel Prize had his identity been known. But his identity would remain shrouded in secrecy for decades, even within the Soviet Union.

On the first orbit, the Telegraph Agency of the Soviet Union (TASS) transmitted: "As result of great, intense work of scientific institutes and design bureaus the first artificial Earth satellite has been built." Amateur ham radio operators around the world easily detected Sputnik's beep–beep–beeps. When news of the satellite launch reached Washington, it was as if a bomb had exploded. It wasn't the scientific significance of the satellite's flight that shook the Pentagon specialists who had fought for a brink–of–war policy; it was the fact, now obvious to everyone, that the Soviet Union had produced a multi–stage ICBM against which air defense was powerless.

Sergei Korolyov in his prime.

Because of time zone differences, the world would celebrate October 4, 1957 forevermore as the beginning of the age of space, humankind's 'final frontier.' The Sputnik 1 satellite helped to identify the density of high atmospheric layers through measurement of its orbital change and provided data on radio–signal distribution in the ionosphere. Because the satellite's body was filled with pressurized nitrogen, it also provided the first opportunity for meteoroid detection, as a loss of internal pressure due to meteoroid penetration of the outer surface would have been evident in the temperature data sent back to Earth. The unanticipated announcement of Sputnik 1's success precipitated the Sputnik crisis in the United States and ignited the so–called "Space Race" within the Cold War. Sputnik 1's signals continued for 22 days until the transmitter batteries ran out. Sputnik 1 burned up on January 4, 1958 as it fell from orbit upon reentering Earth's atmosphere, after traveling about 60 million km (37 million miles) and spending 3 months in orbit. It was the first of a series of Soviet satellites collectively known as the Sputnik program.

BEATING THE AMERICANS

The Soviets were not done yet. After the successful orbiting of Sputnik 1 in 1957, it seemed that the Soviets had pulled ahead. Under Korolyov's design leadership, they aimed for higher and more-prestigious targets including the Moon and the planets Venus and Mars. As early as 1955 Korolyov had put forward a plan to build a multi-stage version of the R-7 rocket which would be capable of sending a payload to the Moon that would be capable of landing on the surface or going into orbit around it. For almost four post–war decades, 1945–1972, the Soviet Union would run the United States a close race to land on and explore the Moon. Within a month of Sputnik 1, the Soviets stunned the U.S. yet again with another space first: the first live animal to orbit the Earth, the dog Laika aboard Sputnik 2 on November 3. The Soviets would manage to beat the Americans into space, not just once, but many times. *However, these space firsts came at a heavy price as Avoiding Armageddon will explain:*

- 1957: First intercontinental ballistic missile, the R–7 Semyorka

- 1957: First satellite, Sputnik 1

- 1957: First animal to enter Earth orbit, the dog Laika on Sputnik 2

- 1959: First firing of a rocket in Earth orbit, first man–made object to escape Earth's orbit, Luna 1

- 1959: First data communications, or telemetry, to and from outer space, Luna 1

- 1959: First man–made object to pass near the Moon, first man–made object to orbit the sun, Luna 1

- 1959: First probe to impact the moon, Luna 2

- 1959: First images of the moon's far side, Luna 3

- 1960: First animals to safely return from Earth orbit, the dogs Belka and Strelka on Sputnik 5.

- 1960: First probe launched to Mars, Marsnik 1

- 1961: First probe launched to Venus, Venera 1

- 1961: First person in space (International definition) and in Earth orbit, Yuri Gagarin on Vostok 1

- 1961: First person to spend over a day in space Gherman Titov, Vostok 2 (also first person to sleep in space)

- 1962: First dual manned spaceflight and approach, Vostok 3 and Vostok 4

- 1963: First woman in space, Valentina Tereshkova, Vostok 6

- 1964: First multi–man crew (3), Voskhod 1

- 1965: First EVA, by Aleksei Leonov, Voskhod 2

- 1965: First probe to hit another planet (Venus), Venera 3

- 1966: First probe to make a soft landing on and transmit from the surface of the moon, Luna 9

- 1966: First probe in lunar orbit, Luna 10

- 1967: First unmanned rendezvous and docking, Cosmos 186/Cosmos 188 (it took until 2006 for the United States to duplicate this major space achievement)

- 1969: First docking between two manned craft in Earth orbit and exchange of crews, Soyuz 4 and Soyuz 5

- 1970: First samples automatically returned to Earth from another body, Luna 16

- 1970: First robotic space rover, Lunokhod 1

- 1970: First data received from the surface of another planet (Venus), Venera 7

- 1971: First space station, Salyut 1

- 1971: First probe to orbit another planet (Mars), first probe to reach surface of Mars, Mars 2

- 1975: First probe to orbit Venus, first photos from surface of Venus, Venera 9

- 1984: First woman to walk in space, Svetlana Savitskaya (Salyut 7 space station)

- 1986: First crew to visit two separate space stations (Mir and Salyut 7)

- 1986: First permanently manned space station, Mir, which orbited the Earth from 1986 until 2001

- 1987: First crew to spend over one year in space, Vladimir Titov and Musa Manarov on board of TM–4 – Mir

How and why did the Soviets appear to be ahead of the United States in the first years of the space race to land men on the moon in the 1960s? Well, in the first place, the United States has had a bad habit of not taking advantage of good old "Yankee ingenuity." Orville and Wilbur Wright had to take their invention overseas, where France and other European powers embraced manned heavier–than–air flight with a vengeance in the years before World War I. It was the Germans who literally leapt upon Robert Goddard's rocket technology in the 1930s, and far outpaced other countries in rocket science during World War II. Russia started a rocket technology program in 1933 while the United States officially did nothing. There are other examples. In 1993, the capricious U.S. Congress saw fit to cancel the Superconducting Super Collider (SSC), which would have been the world's largest and highest–energy particle accelerator complex by far. Now the lead in experimental particle physics research has passed to the Large Hadron Collider (LHC) in operation at CERN, where the groundbreaking discovery of the so–called Higgs fundamental particle in physics was verified in July 2012.

> *Although its importance is not publicly known, much less understood, the Higgs is a crucial building block in the "Theory of Everything," dubbed the "Standard Model" by the world's physicists. And that IS important. When the world understands how the universe works (and it will by the 22nd century, when the underpinnings of the Standard Model will be standard high school fare), you better watch out: anti–gravity cars, flying saucers (ours, not the aliens'), and eventually faster–than–light travel to the stars and their habitable planetary systems. By the way, in The Never–Ending Frontier you'll see why the very word "anti–gravity" is a misnomer. For now, please use "field effects propulsion" or a like term instead.*

The acronym CERN came from French for *Conseil Européen pour la Recherche Nucléaire* (European Council for Nuclear Research). CERN was a provisional council established by 11 European governments in 1952, with the express purpose of setting up world–class nuclear research facilities in Europe. After the provisional council was dissolved the new laboratory retained the CERN acronym, even though its name officially changed to the current *Organisation Européenne pour la Recherche Nucléaire* (European Organization for Nuclear Research) in 1954.

A second reason for its lead was that the USSR's (Union of Soviet Socialist Republics — Союз Советских Социалистических Республик, or СССР as it was then known) space program was actually *more competitive* than the American space program. Look at the facts. After Ike founded NASA in 1958, NASA was invested as the single coordinating structure for space exploration in the United States. While Soviet design bureaus had to jockey for position to obtain the best government contracts, NASA had no competition and could benefit from government largesse. During the Apollo program era that led to the first manned lunar landing in 1969, NASA was the beneficiary of practically unlimited funding to accomplish that goal. On the other hand the USSR's space program was split between several competing design groups led by Sergei Korolyov, Mikhail Yangel, Valentin Glushko and Vladimir Chelomei. And compete they did. Ironically, the United States—the world's leading space power—has faced the embarrassment of having to rely on its old foe, the Russian Federation, for access to the International Space Station after the Space Shuttle stand–down in 2011.

Thirdly, Soviet (now Russian Federation) launchers have followed a conservative engineering route to maintain high launch rates at modest costs. If a given design like the R–7 works, the Soviets do not change the basic concept, but rather evolve it step by incremental step. The Volkswagen serves as a good parallel. In 1937 the German government founded Volkswagen AG to mass–produce a low–priced "people's car." The Volkswagen (literally "people car") design configuration survived basically unchanged until 1974 when increasing competition from other compact foreign cars forced the company to develop newer, sportier car models, among them the Rabbit and its successor, the Golf.

The Soviets knew they had a good design with Korolyov's R–7. Why muck with it and start over? On January 1, 1958 Korolyov's design bureau began work on evolving the R–7 design. After Sputnik 1 it led into the Vostok (east), Voskhod (sunrise), Molniya (lightning), and

Soyuz (union) launch vehicles, and each of these had several variants. To create the Vostok the Soviets added a third stage (the Block E) to the R–7 first stage boosters and second stage core sustainer. Vostok launched the world's first lunar probes, the first man, and first woman into space. The Molniya vehicle was Vostok with a larger third stage (Block 1), and a new Block L fourth stage. The word "Molniya orbit" derived from this vehicle, denoting highly elliptical orbits it enabled. To create the Voskhod they just dropped the Molniya's Block L fourth stage. In 1966, about the time of Korolyov's death, the Soviets created the Voskhod's twin, the world famous Soyuz. Soyuz and its many variants remains the most–launched vehicle in the world today. But these are all knockoffs of the same basic Korolyov design. After 2,000 launches, the Russians have figured out how to get stuff into space. ***So far, every manned Russian or Soviet spaceflight has been launched by a Soyuz vehicle or one of its derivatives.***

In the fourth place, the Russians took a different, if less elegant, approach to the design and development of rockets than the Americans. Russian engineers and technicians did not bother trying to get the last ounce of perfection into the design, nor the last ounce of weight out of space hardware. Soviet engineers and scientists focused more on getting the job done, accomplishing the mission set out for them, without as many detours and embellishments as the Americans seemed wont to have. For example, the Russians would employ cruder manufacturing methods, not caring how "pretty" the welds looked as long as they worked.

Sputnik 2 provided an example of Russian design technique. In the rush to beat the Americans in the space race after Sputnik 1, Russian engineers and technicians rushed into design and production a new Sputnik 2 satellite, intended to carry a live dog, without any draft preliminary design! *They simply suspended all the rules that had been in effect for the development of missile technology.* The draftsmen and designers moved into the shops. Almost all the parts were manufactured using sketches. Assembly wasn't conducted so much according to blueprints and documents, but according to the designers' instructions and on–the–spot fitting. The total weight of Sputnik 2—508.3 kilos—was already a qualitative leap by itself. The Russians also had to make a last–minute decision to keep the *Semyorka's* core booster—basically a second stage— and the payload joined together in space. Unlike its predecessor, the second Soviet moon was not a special device, orbiting apart from its carrier. It was the last stage of the launching vehicle. The innovative Russians also reused

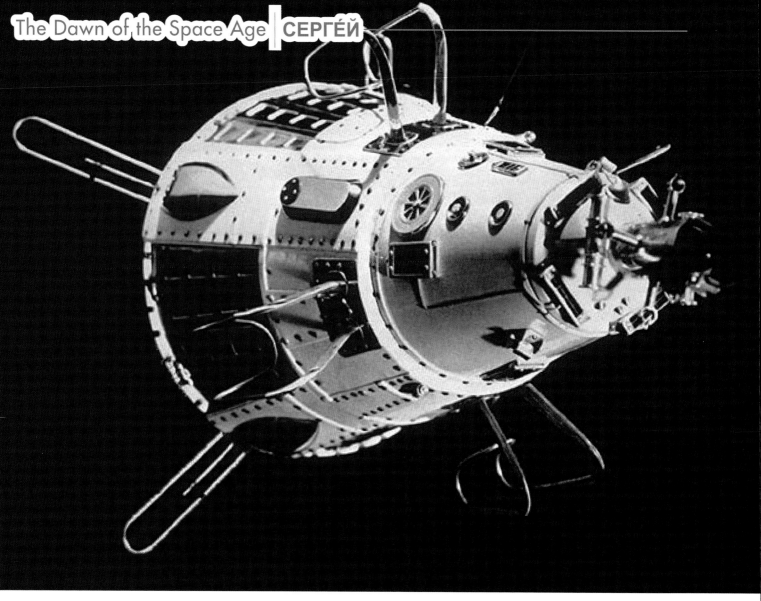

Sputnik 2 in the blackness of space

every bit of hardware they could. The cabin in which the dog would reside on orbit was derived from the nose sections of experimental ballistic missiles that had previously been used for carrying live dogs and other animals into the upper atmosphere in short and relatively low–speed flights which ended in a parachute landing. Telemetry and on–orbit parameters would reuse the *Tral* system already installed on the launch vehicle.

The Russians basically produced from scratch, without any preliminary experimental development, an experimental space laboratory making it possible to study a dog that everyone knew could not be returned to Earth. It would be a one–way trip for a poor female mutt named Laika (barker in Russian), a stray mongrel that had earlier been found wandering the streets of Moscow. Soviet engineers simply did not have time to create a reliable life support and thermal control system within such a short period of time under the urgency of beating the Americans into space and showing that Communism and a socialist economy was better than American democracy. The electric power

sources installed on the second stage's body to track the satellite were only sufficient for six days or so. Laika's cabin was equipped with a crude television camera, along with sensors to measure ambient pressure and temperature, as well as her blood pressure, breath frequency and heartbeat. These instruments allowed ground controllers to monitor how Laika functioned and died in space. Above the dog's cabin, the engineers mounted a spherical container that was developed for Sputnik 1. It held a radio transmitter and an instrument to register ultraviolet and x–ray radiation.

The 1120–pound Sputnik 2 was successfully launched on November 3, 1957, in commemoration of the 40th anniversary of their Great October Revolution—*less than a month after Sputnik 1.* This was a mind–boggling achievement! Today, it would honestly take NASA more time to debate and set up the first of three safety review boards for such a launch, never mind doing all the engineering and tests for carrying out a successful launch. But hold it; the ASPCA would probably have prevented the launch from ever taking place in the first place!

Sputnik 2 cutout model (without
antennas) on display at the
Polytechnical Museum in Russia.

Khrushchev sent the two mutts as gifts to an embarrassed JFK, the United States President.

Laika snuggles into
her place aboard
Sputnik 2

The pressurized cabin on Sputnik 2 allowed enough room for Laika to lie down or stand (but not turn around) and was padded. An air regeneration system provided oxygen; food and water were dispensed in a gelatinized form. She was fitted with a harness, a bag to collect waste, and electrodes to monitor vital signs. Early telemetry indicated Laika was agitated but eating her food, and that the temperature and humidity (of all things) were increasing relentlessly. By the fourth orbit none of the physiological sensors were returning data—she had died from overheating and possibly stress (but not suffocation as was rumored). *The animal rights organization PETA (People for the Ethical Treatment of Animals) would be pleased that had she lived Russian Mission Control planned to euthanize Laika with poisoned food, as she would have burned up in the atmosphere during reentry.* With all systems dead, the spacecraft continued circling the globe until April 14, 1958, when it reentered the atmosphere after 162 days and 2,570 orbits. Korolyov's R–7 rocket carried into orbit two more live dogs, Belka and Strelka, on August 19, 1960. Nikita

Another difference between the two space powers was in their handling of accidents. When the Soviets suffered space disasters and casualties, they were wont to investigate what went wrong, institute corrective measures, and move on. They had the advantage of being able to squelch public opinion and discussion in the state–controlled press, of muzzling dissidents by condemning them to the Gulag in Siberia. What's more, the Soviet citizens were used to this and resigned themselves to such lack of freedom after centuries of repression and totalitarian rule. In the United States, at least in the civilian space program which always seems to be under a microscope, space failures are more likely to lead to lengthy standdowns. The U.S. space program essentially shut down for 18 months following the Shuttle Challenger and Titan losses of the mid–1980s. The Soviets anticipated the deaths of some of their first space explorers as a necessary consequence of opening up the space frontier. As has been the case with nuclear reactor accidents, the Russian "pain threshold" seems quite a bit higher than in the United States. Perhaps this can be explained by the long–suffering history of Russian citizens. Most of Moscow (about ¾) was burned down by Napoleon in 1812. The country was devastated during both World Wars, suffering some 26 million casualties in WWII alone. More countless millions perished during Stalin's purges, massive famines, or in his dreaded "Gulag" prisoner death camps.

ROCKETEERING UNDER STALIN'S BOOT

Russia does not have the legacy of centuries of democratic rule that the United States has enjoyed since the American Revolutionary War. A succession of hard–boiled Communist party bosses ruled over the Soviets during the seven decades of the communist era (roughly 1920–1990). Before the communists, the Tsars ruled over the hapless Russians, and kings or queens before them. In the eyes of Western observers, it is simply a miracle that Russia was able to harness rocket technology and lay the groundwork for many of their space firsts during the reign of the brutal Soviet dictator, Josef Stalin

He was born as was Iosif Vissarionovich Dzhugashvili in December 1878, but around 1912 he adopted the name Stalin, deriving from Russian stal ("steel"). In 1913, Russia was notably backward in comparison with the world's leading industrial nations. The world's largest country held only fifth place in overall industrial output. By 1937, albeit through savage means, Stalin had increased the Soviet Union's total industrial output to the point where it was surpassed only by that of the United States. This industrial might enabled the Soviets to compete with and in some cases surpass the accomplishments of the U.S. in rocket science, maintain the threat of mutually assured destruction against its arch enemy throughout the Cold War, and almost run a dead heat against its archrival in both the nuclear arms race and the space race to land men on the Moon and gain priceless national prestige on the world stage.

> *But at what a price—very little of the benefits ever became available to the ordinary Soviet citizen in the form of consumer goods or amenities of life. A considerable proportion of the national wealth—a proportion wholly unparalleled in the history of any peacetime capitalist country—was appropriated by the state to cover vast military expenditures, an oppressive police and security apparatus numbering millions of people and informers, and further industrialization under a system that was much less efficient on the whole than capitalism and the free market.*

Prior to the revolution of 1917, Stalin played an active role in fighting the tsarist government. During the quarter of a century (1922–1953) preceding his death, Stalin probably exercised greater political power than any other figure in history. His position as Secretary General of the party's Central Committee during this time provided the power base for his unbridled dictatorship. Stalin industrialized the USSR, forcibly collectivized its agriculture, consolidated his position by intensive police terror, helped to defeat Germany in 1941–45, and extended Soviet controls to include a belt of eastern European states. Chief architect of Soviet totalitarianism and a skilled but phenomenally ruthless organizer, he destroyed the remnants of individual freedom and failed to promote individual prosperity, yet he created a mighty military–industrial complex and led the Soviet Union into the nuclear age.

Stalin shown on a 191? information card from the files of the Tsarist secret police in Saint Petersburg

Beginning in 1928, Stalin compelled some 25 million rustic households to amalgamate in collective or state farms within a few years. The reluctant peasants resisted desperately but Stalin had them attacked by troops and political police units. Uncooperative peasants, termed kulaks, were arrested en masse, shot, exiled, or absorbed into the rapidly expanding network of Stalinist concentration camps and worked to death under atrocious conditions. Stalin's crash industrialization, while successful in rapidly industrializing a backward country, also cost many lives. To cover his grandiose failures in these years, Stalin arraigned industrial managers in a series of show trials. Intimidated into confessing imaginary crimes, the falsely accused served as self–denounced scapegoats for catastrophes arising from the Secretary General's policies.

Sergei Pavlovitch Korolyov was born on January 12, 1907 in Zhytomyr, now recognized as part of Ukraine. His father, Pavel Yakovlevich Korolyov, was from Mogilev in Belarus; and his mother, Maria Nikolaevna Koroleva, was a daughter of a wealthy merchant in the Ukrainian city of Nezhin. Pavel had moved the family to Zhytomyr so he could teach the Russian language. Three years after Sergei's birth the couple separated due to financial difficulties. They divorced in 1915, and in 1916 his mother married Grigory

Mikhailovich Balanin, an electrical engineer educated in Germany. Although Pavel later wrote to Maria requesting a meeting with his son, Sergei was informed by his mother that his father had died. He never saw his father after the family break-up, and Pavel died in 1929 before his son learned the truth.

Korolyov grew up in Nezhin under the care of his maternal grandparents (Mykola and Maria Moskalenko) in an extended family household (his mother also had a sister and two brothers). As a child, he was stubborn, persistent, and argumentative; and grew up a lonely child with few friends. He began reading at an early age, and his abilities in mathematics and other subjects made him a favorite student of his teachers, but caused jealousy from his peers. The torment of classmates bullying and teasing him as a small child encouraged Korolyov's focus on academic work.

After getting a job with the regional railway, Grigory moved the family to Odessa in 1917, where they endured hardships with many other families through the tumultuous years following the Russian Revolution and continuing internecine struggles until the Bolsheviks assumed unchallenged power in 1920. Local schools were closed and young Korolev had to continue his studies at home. Grigory proved a good influence on his step-son, who suffered from a bout of typhus during the severe food shortages of 1919.

From a young age, Korolyov became enamored of the possibilities of spaceflight, and spent hours poring over the works of Tsiolkovsky, whose work was known in Russia. He no doubt dreamed of going to Mars and exploring the planets. In his late teens he designed and built gliders as a hobby. He studied at the Kiev Polytechnic Institute in 1924-1926, and continued his studies at the Bauman Moscow State Technical University from 1926-1929.

In 1930 Korolyov got his pilot's license, and he became interested in the use of liquid–fueled rocket engines to propel airplanes. On August 6, 1931 he wed Xenia Vincentini, a woman he had been courting since 1924 (he had proposed marriage to her back then, but she declined as she had her own career ambitions and wanted a higher education). That same year together with Friedrich Zander, a fellow space travel enthusiast, he participated in the creation of the Group for the Study of Reactive Motion, GIRD—*Gruppa issledovaniya reaktivnogo dvizheniya*, one of the earliest state-sponsored centers for rocket development in the USSR. In May 1932 Korolyov was appointed chief of the group.

During the following years, GIRD developed three different propulsion systems, each more successful than the last. In 1932, the military became interested in the efforts of the group, and began providing some funding. In 1933 they accomplished their first launch of a liquid-fueled rocket, which was called GIRD-X. With growing military interest in this new technology, the government issued a state decree to merge GIRD with the Gas Dynamics Laboratory (GDL) in Leningrad. *"State decrees" were used by Stalin and other Soviet leaders to promulgate a new policy or program.* The merger created the world's first Reactive Scientific Research Institute, RNII–1— Reaktivnyy nauchno–issledovatelskiy institut–1.

The word "reaktivnyy" (literally meaning "reactive") in Russian is commonly used to denote "jet propulsion" such as in the phrase "jet propulsion engine." However, strictly speaking, the word "reactive" encompasses not only air–breathing jet engines but also rocket engines which carry all of their own propellants. It was not uncommon in the 1930s and 1940s for the word "reactive" to denote both jet and rocket propulsion. In the particular case of the RNII, "reactive" was meant primarily to represent the development of what we now call rocket engines.

The year 1933 can be viewed as a starting point for organized Russian rocket science. The

Sergei Korolyov, aged 6 (1913).

initiator of the 1933 state decree was *not* Stalin, but Mikhail Nikolayevich Tukhachevskiy, the Deputy People's Commissar of Military and Naval Affairs (*Narkomvoyenmor*) and Deputy Chairman of the Revolutionary Military Council (*Revvoyensovet*), who was responsible for arming the Red Army. Russia at that time was still emerging from the shadow of a relatively backward, rural state predominantly populated by peasants. Russian cosmonautics (their equivalent to the U.S. astronautics) is indebted to Tukhachevskiy. He understood rocket building and that it had to be supported first and foremost by leading–edge technology and modern industry on a large scale; rather than by the enthusiasm of lone individuals like Goddard or Tsiolkovsky dreaming of interplanetary flights. A prominent military leader in the Soviet state system, Tukhachevskiy had the foresight to bring together under one roof Russian theoreticians and practical engineers: Yuriy Aleksandrovich Pobedonostsev, Mikhail Klavdiyevich Tikhonravov; rocketry enthusiasts (and future Academicians) Sergei Korolyov, Valentin Glushko, and Boris Rauschenbach; enthusiasts of solid–fuel rocket projectiles—the future *Katyushas* or "guards mortars" Ivan Terentyevich Kleymenov and Georgiy Erikhovich Langemak; and a number of enthusiastic proponents of space travel. Kleymenov, a military engineer, would head up the institute.

RNII conducted a wide–ranging research program. Korolyov was a hard worker and always spent long hours at his design office. He soon took a leadership role at RNII as combination deputy chief, chief engineer, and engineering project manager. He somehow combined charismatic leadership with a demanding, disciplinary style of management (hard–nosed bosses were the norm in those days). Korolyov led the development of various ballistic and cruise missiles with various types of engines, and a manned rocket–powered glider. He personally monitored all key stages of the programs and paid meticulous attention to detail.

Korolyov sitting in cockpit of glider "Koktebel."

On 10 April 1935, Korolyov's wife Xenia gave birth to their daughter, Natasha. In 1936 they were able to move out of his parents' home and into their own apartment. Both Korolyov and his wife had careers. The RNII team continued their development work on rocketry, with particular focus on the area of stability and control. They developed automated gyroscope stabilization systems that allowed stable flight along a programmed trajectory.

In late 1934, just when the worst excesses of Stalinism seemed to have spent themselves, Stalin launched a new campaign of terror against the very Communist Party members who had brought him into power. First he arranged the murder of his leading colleague and potential rival, Sergei Kirov. Stalin then used the show trials of leading Communists as a means for expanding a new reign of mass terror and bloodshed. He not only "liquidated" veteran semi–independent Bolsheviks but also many party bosses, military leaders, industrial managers, and high government officials who were already totally subservient to himself. Other victims included foreign Communists on Soviet territory and members of the very political state police organization, now called the NKVD (the *People's Commissariat for Internal Affairs*). All other sections of the Soviet elite—the arts, the academic world, the legal and diplomatic professions—also lost a high proportion of victims, as did the population at large, to a semi–haphazard, galloping persecution that fed on extorted denunciations and confessions.

The machinery of Stalin's state was so effective at controlling information and at brainwashing even the leading scientists that few understood the evil of the man. In fact, he was revered by most Soviet citizens as the Great Leader, a harsh but benevolent dictator who had masterminded the victory over Germany and would protect his people against a hostile world. If anything, Soviet scientists blamed the lower levels of government. The flimsiest of denunciations by somebody one hardly knew sufficed to send him to prison, and often to death. Knocks would come in the middle of the night, and one's friends would disappear, never to be heard from again.

In the 1930s, only the Soviet and German governments were supporting work on a broad spectrum of rocket–related subjects. From 1932–1935, Russia led the Germans in rocket science. Beginning in 1935, Germany caught up with and then surpassed Russia in the development of liquid–propellant rocket engines. The totalitarian Soviet state actually inhibited the development of large rocket technology for six years beginning in 1937, when Stalin executed the very initiator and patron of this field of research, Marshal Tukhachevskiy, and after him the leadership of RNII: Director Kleymenov and Chief Engineer Langemak. Grigoriy (aka Sergo) Ordzhonikidze, a Georgian Bolshevik and political leader, ended his life by committing suicide. Andrei Nikolayevich Tupolev, the principal figure in the early development of Soviet aviation and a leading aircraft designer, was led away to immediate imprisonment on the evening of October 21, 1937 by the NKVD. Tupolev and Leonid Lvovich Kerber were locked away with hundreds of other aviation specialists and ordered to carry on their aircraft design work. Tupolev's entire design team, and the Petlyakov and Myasischev design teams were imprisoned and forced to work in three NKVD–prison workshops (*sharashka*, in Russian шарашка, the "goofing-off place"). Sharashka were in fact secret research laboratories, where arrested and convicted scientists and engineers, some of them prominent, were anonymously put to work on projects assigned by the Communist party leadership. Tupolev had to work in the very buildings he had worked in prior to his arrest—later to become the Tupolev Design Bureau. There the men lived and worked, isolated from their families and allowed outside only in the "monkey cage"—a rooftop enclosure of steel bars. Once, when the aircraft of a *sharashka* design team flew over Red Square in a May Day parade, the jailed designers were permitted to view the fruits of their labor from the monkey cage. Sharashka colleagues would disappear in the night, or be summoned to NKVD headquarters for interrogation on design projects. Tupolev, Kerber, and most of their design team somehow survived. The group was moved several times during the war, the first time to avoid capture by advancing German forces. They even managed to design and fly both the Tupolev Tu–2 bomber and the Petlyakov Pe–2 dive bomber under these horrendous conditions.

Despite this political climate, by 1938 two of the finest rocket scientists in Russia (the Soviet Union actually took shape after WWII) had begun their rise to prominence: Valentin Glushko for liquid–propelled rocket engines, and Sergei Korolyov for the overall launch system. Distrust, hostility, and even hatred for the intelligentsia (including rocket scientists) was a common characteristic of communist leaders in those days. On March 23, 1938 Glushko became caught up in Stalin's Great Terror and was rounded up by the NKVD,

to be placed in the Butyrka prison. Granted, Glushko's personality was reputed to be bull–headed and he never lacked for an ego, but he didn't deserve to be locked away. He had become embittered by having to work in Korolyov's shadow, and was probably justified in feeling he did not get the credit he deserved. But for whatever reason, Glushko, Ivan Kleymenov, and Georgy Langemak denounced Korolyov and he was arrested by the NKVD on June 22, 1938 on false charges of sabotage and trying to slow down progress in RNII, when he was working night and day doing just the opposite!

Korolev later learned that he had been denounced by Glushko, and this may have been the cause of the lifelong animosity between the two men. In the paranoia of Stalin's iron rule, you didn't know whom to trust. There were rumors that Glushko and Korolev had been denounced by Andrei Kostikov, who became the head of RNII after its leadership was arrested (Kostikov himself was ousted a few years later over accusations of budget irregularities). Following torture in the Lubyanka prison to extract a confession (his jaw was broken by interrogators), Korolyov was tried and sentenced to 10 years in a labor camp. He was sent to one of the most feared parts of the Gulag network of labor camps, a gold mine in the desolate Kolyma region of eastern Siberia close to the arctic circle. Thousands of prisoners died each month in the freezing gold mine camps. On August 15, 1939, Glushko was also arrested and sentenced to eight years in the Gulag.

The infamous Gulag (an abbreviation of Glavnoye Upravleniye Ispravitelno–Trudovykh Lagerey, "Chief Administration of Corrective Labor Camps") was actually not a single prison, but consisted of many hundreds of prison camps, with the average camp holding 2,000 to 10,000 prisoners. Aleksandr Solzhenitsyn, author of The Gulag Archipelago, characterized the Kolyma Gulag as the "pole of cold and cruelty." During his brutal reign of terror, Joseph Stalin used the Gulag to dispose of his enemies, real or imagined (perhaps 2.5 million from the top strata of the military, government, and society in the Great Terror of 1937–1939 alone).

Unlike the sharashkas, most of the Gulag camps were "corrective labor colonies" in which prisoners felled timber, worked on general construction projects (such as the building of canals and railroads), or labored in mines. Most prisoners labored under the threat of starvation or execution if they refused. It is estimated that the combination of very long working hours, harsh climatic and other working conditions, inadequate food, and summary executions killed off at least 10 percent of the Gulag's total prisoner population each year. Over three million are estimated to have died in the Kolyma Gulag alone. Western scholarly estimates of the total number of deaths in the Gulag in the period from 1918 to 1956 range from 15 to 30 million! The exact numbers will never be known.

1941 he managed to be put in charge of a sharashka designing liquid–fueled rocket engines. Perhaps by now Korolyov had been hooked on the engineering of rockets after his work at RNII, because in 1942 he maneuvered to be moved to Glushko's sharashka, where he designed rocket-assisted take off boosters for aircraft.

Korolyov was kept in the sharashka and isolated from his family until 1944. He lived under constant fear of being

A map of the main Gulag camps which dotted the Soviet Union (left) from 1923 to 1961, based on data from the Мемориал ("Memorial") Human Rights Society in Moscow. There were many more smaller camps. Think about it.

Korolyov was worn down by 12-hour days of back-breaking work, a poor diet, the harsh climate and abuse at the hands of guards and genuine criminals in the Kolyma Gulag. After some months he was retried in late 1939 on "reduced charges" and his sentence reduced to eight years (1938–1944). He was transferred to Tupolev's sharashka, but by this time the poor man had lost most of his teeth to scurvy and was given potato juice to drink in an effort to nurse him back to health. *In 1929 Korolyov had earned his engineering diploma at Bauman Moscow State Technical University under Tutolev's tutelage by producing a practical aircraft design, and spent the first years of his career with Russia's best designers in the aviation industry.* Glushko was also liberated and put to work on various aircraft projects with other arrested engineers and scientists. In

executed for the military secrets he possessed, and was deeply affected by his time in the Gulag, becoming reserved and cautious. Under the pressure of fighting the Germans on the Eastern front and fearful of losing the war, Stalin released the best scientists and engineers—including Korolyov, Glushko, and Tupolev— on June 27, 1944 by special government decree (the charges against Korolyov were not dropped until 1957). Korolyov spent his first night at home telling the adult members of his family about the Gulag and his other ordeals. His daughter Natalya told Russian newspaper *Rossiyskaya Gazeta* that he finished the night by saying: "Never ask me about it again. I want to forget it all like a horrible dream." She added that her father had developed a loathing for gold, and would frequently say he hated it.

Joseph Stalin, 1943

secretary of the Central Committee, to begin yet another reign of terror in the Soviet artistic and intellectual world. Foreign achievements were derided and the primacy of Russians as inventors and pioneers in practically every field was asserted. Hopes for domestic relaxation of the Stalinist purges, widely aroused in the Soviet Union during the war, were thus sadly disappointed. It does credit to the courageous perseverance of Soviet rocket scientists and engineers that they were able to accomplish as much as they did under such dastardly circumstances over many years.

THE GREAT LEAP FORWARD

In spite of Stalin and his abominable mass repressions, the Russians did not idly stand by while the Americans were developing their MRBMs and IRBMs based on the preeminent German rocket technology of World War II.

What's an "MRBM"? It stands for Medium Range Ballistic Missile, with a range of 1000km or less (generally). The U.S. Sergeant missile was an MRBM. "IRBMs", or Intermediate Range Ballistic Missiles, are longer range, roughly between 1000 to 5500 km, but there is no hard and fast rule. As its name implies, a ballistic missile is propelled by a rocket engine for only the first part of its flight; for the rest of the flight the unpowered missile follows an arcing trajectory, small adjustments being made by its guidance mechanism. A guided missile is broadly speaking any military missile that is capable of being guided or directed to a target after having been launched. Besides MRBMs and IRBMs, there's lots of other kinds of missiles, both guided and unguided. We'll talk about them in a little while.

Stalin did not tolerate his enemies. He simply killed them. For example, Andrey Andreyevich Vlasov (1900–1946) was a Soviet military commander with anti–Stalinist sentiments, who after being captured with his army by the Germans in July 1942, formed the Committee for the Liberation of the Peoples of Russia and the Russian Liberation Army aimed at overthrowing Stalin's regime. The Germans allowed Vlasov's 50,000 troops to go into battle against the advancing Red Army near the end of the war. He was handed over to the Soviets in May 1945, tried, and summarily hanged in 1946.

After the World War II, Stalin became, if possible, even more suspicious and paranoid. In 1948 the defection of Titoist Yugoslavia from the Soviet camp struck a severe blow to world Communism as a Stalin–dominated monolith. To prevent other client states from following Tito's example, Stalin instigated local show trials, manipulated like those of the Great Purge of the 1930s in Russia, in which satellite Communist leaders confessed to Titoism, many being executed. He used his chief ideological hatchet man, Andrey Zhdanov, a

Like the Americans, the shrewd Soviets also raced to amass and exploit every scrap of captured German V–2 missile technology, documentation, and research facilities they could get their hands on after World War II. Ingenious German rocket advances, years ahead of both the Russians and the Americans, enabled a great leap forward in the development of rocket and missile systems for both countries. *The V–1 and V–2 defined the problems of propulsion and guidance that have continued ever since to shape cruise and ballistic missile development.*

For both the Soviets and the United States, the origins of their ballistic missiles lay in German rocket technology developments during World War II. Both countries raced to exploit captured German rocket parts, complete missiles, instrumentation, engineers, and rocket scientists—basically anything they could get their hands on. Although experiments were undertaken by both sides before WWII on crude prototypes of cruise and ballistic missiles, today's modern rockets and missiles are generally considered to have their true origins in the V–1 and V–2 missiles launched by Germany in 1944–45 during World War II. *The Wernher chapter following will discuss these and other German missiles in detail.*

While the Americans rushed their captured German rocket scientists and V–2 or other hardware to the continental U.S. (under Operation Paperclip) after the war, the Soviets were content to house their German rocket scientists, often their families as well, and invaluable hardware and design data in existing or refurbished East German facilities for the first few years after the war.

The Soviets first became officially aware of German long range rocketry research when British Prime Minister Winston Churchill wrote a personal Top Secret message to Stalin on July 13, 1944. Churchill informed him that the Germans were about to use their supersonic V–2 missile as a war weapon. Churchill was desperate to learn more about what the Germans were up to. He pleaded:

> *"There is reliable information to the effect that for a substantial period of time the Germans have been conducting missile tests from an experimental station in Debica, Poland. According to our information this projectile has an explosive charge weighing approximately 12,000 pounds, and the effectiveness of our countermeasures depends to a significant degree on how much we can find out about this weapon before it can be used against us. Debica is located on the route of your victoriously advancing troops and it is completely possible that you will seize this site in the next few weeks."*

> *"Although the Germans almost certainly will destroy or haul off as much of the equipment located at Debica as possible, you will probably be able to obtain a great deal of information when this area is in Russian hands. In particular, we hope to find out how the missile is launched because this will enable us to determine the missile launch sites..."*

The Germans had set up alternate missile test facilities near Debica after a massive armada of almost 600 four engined Allied bombers damaged their principal missile development and test facility on the northwest part of the island of Usedom, on the coast of the Baltic sea near the town of Peenemünde. On the night of August 17–18, 1943 the Allies rained down thousands of high–explosive and incendiary bombs on the top–secret German rocket facility, intending to kill as many of Peenemünde's expert technical and administrative personnel as possible. One wave of bombers followed another, carpet–bombing the Peenemünde production buildings, test–rig facilities, and laboratory buildings. A total of 1.5 million kilograms of high–explosive and incendiary bombs were dropped. The local air defense proved powerless against the huge bombing raid, but night fighter aircraft urgently called in from Berlin shot down 47 American B–17 Flying Fortresses. This one raid killed 735 Peenemünde residents, but only two key figures — chief engine designer Dr. Walter Theil and Erich Walter. Major damage was done to personnel housing and development works, but this damage was not as crippling to Germany's rocket program as systematic Allied raids against supporting assembly plants and hydrogen peroxide production facilities.

After hearing of the scale of the attack, Luftwaffe Deputy Commander Generaloberst Hans Jeschonnek, who was directly responsible for the air defense system of that area, committed suicide. However General Walter Dornberger and his protégé, the principal German rocket scientist Wernher von Braun, did not lose heart. They assured the chief (*SS Obergruppenführer*) of Himmler's security service, Ernst Kaltenbrunner, that the Peenemünde survivors would be able to overcome the aftermath of the catastrophe. Operations were slowed down but not halted. This experience proved that it is extremely difficult to stop experimental weapons development using unguided bombs dropped from the air alone. In the war against North Vietnam two decades later, the Americans again had difficulty destroying enemy troops hiding under trees or in caves with conventional bombing tactics. *However, today's weapons technology has progressed to the point that air weapons can be delivered with pinpoint accuracy. Now the biggest problem has become accurate and timely intelligence so that you don't hit the wrong target and kill innocent civilians.*

Unfortunately, intelligence is almost never perfectly accurate. In 1944 the Brits had only vague notions about the capabilities of German missiles. By August 1944 the Russian front lines on the eastern front had advanced to within 50 kilometers of Debica. In September the Red Army broke through to the site of the alleged Debica V–2 test range, which was in a marshy area. The Red Army also overran an A–4 missile research unit at Dembidze in Poland. The Soviets collaborated with British intelligence specialists who had detailed surveillance maps of the Debica area showing the coordinates of the V–2 launch site and numerous sites where test missiles had landed. The Russians found blown up aluminum tanks, pieces of exterior steel casing, and white shreds of prickly fiberglass. They didn't manage to get everything out of the swamp. Explosions of the propellant components had scattered missile parts all over the area. The Brits were very interested in the remains of radio equipment and control system instruments that had remained intact. They gathered several large cases of all sorts of parts to be sent immediately to Britain via Moscow, *where, of course, Soviet radio specialists took an opportunity to inspect the contents before they were transferred to the British mission*. The Russians also captured and sent what they could to Moscow. However, with the wartime demands on both Allies, it would take months to search for, dig out, pack, and transport the items to London or Moscow, respectively. Then there was the unpacking, cataloging, cleaning, reassembly with no assembly drawings, and checkout of those items that worked. Many, of course, did not work. Needless to say, it would take several trips back to Poland to uncover additional pieces from the marshes.

Today, we take for granted that missiles and rockets are going to reliably launch and perform their mission, not explode on the launch pad. But in 1944 rocketry was still a very new science. Just as with fearsome atomic weapons, engineers and scientists had to work by trial and error, with explosions in the air, failures of control surface actuators, gyroscopes that didn't work right, propellant and oxidant line valves that leaked, onboard electrical circuits that caught fire, and so on. Early rocket launches were often plagued by one failure after another. Some rockets failed to take off—immediately after ignition the circuit would reset itself. Some started to take off, only to immediately and disastrously fall on their tails, destroying the launch pad. Others successfully took off, but then exploded at an altitude of several kilometers due to fires in the tail section, or they crashed

due to control system failures, or they broke up in the air due to aerodynamic heating of the oxidizer tanks, for example. At the time, there were no textbooks or manuals on how to ensure the successful launch of rockets. V–2–era guidance systems were primitive and there were no multi–channel telemetry systems. Small wonder that the Russians and British found the Debica rocket test site littered with the bodies of failed rockets.

March 1945 A team of rocket specialists from the Soviet Scientific–Research Institute 1, (NII–1) was sent back to Poland. With assistance from members of the Polish resistance, more V–2 parts were identified and loaded aboard an Li–2 aircraft. During the war, the Russians had built Li–2s, basically a duplicate of the famous DC–3 (Douglas Commercial–3) under license from Douglas Aircraft Co. in the United States. This time, the LI–2 crashed near Kiev and only some of the parts made it to Moscow. However, it was sufficient for a team of ten specialists under General Victor Fedorovich Bolkhovitinov, a noted designer of Soviet bombers, to make the first reconstruction of the V–2 missile. Eventually, Russian rocket specialists were able to piece together enough missiles so that they were able to approximate their performance. Impressed by these successes, on April 19 the State Committee for Defense issued Decree 8206 ordering the formation of TsKB–1 – Central Design Bureau 1 – for the purpose of recovering liquid rocket technology from Germany, which was less than a month from defeat.

April, 1945. The Russians laid out all the missile parts in a large assembly hall in NII–1. Like most Soviet research facilities it was located in Moscow. One day several specialists were inspecting one of the reconstructed German missiles. Aleksei Mikhailovich Isaev, one of Russia's best aviation and rocket engineers, had crawled headfirst through the nozzle and into the combustion chamber. He was examining the details with a flashlight and couldn't believe his eyes. "This is something that can't exist!" he exclaimed. "A huge bipropellant injector, control vanes in the exhaust. An inertial guidance system…" Bolkhovitinov, head of the aircraft design bureau in which Isaev worked and a noteworthy designer of Soviet bombers himself, sat gloomily nearby. The Russians were dumbfounded at what they had found. They simply never imagined a liquid–propellant rocket engine of such proportions at that time! Only by collaboration between Moscow's various research institutes had the Soviets had been able to cobble together a reliable liquid–propellant rocket engine with a relatively puny thrust of 1.5 metric tons by 1943. Isaev dreamed of bringing the engine up to a thrust of 2–3 metric tons in a year or two. But now they had uncovered a giant V–2 engine with a thrust of (they would find out) 27 metric tons at liftoff, and almost 32 tonnes at altitude!

By late April the Soviet Army had entered Berlin and were battling their way to the center of the city where the German Chancellery was located. The western Allies were also encroaching Berlin from the other side. On the afternoon of April 30, Adolph Hitler and his new bride, Eva Braun, committed suicide—he by a gunshot to the temple, she by ingesting cyanide. Hitler's valet (Heinz Linge) and personal SS bodyguard staff were only able to partially cremate their bodies with petrol because of Soviet shelling on Hitler's bunker. They placed the remains in a shallow bomb crater, where they were found on May 2 by Ivan Churakov of the 79th Rifle Corps. They were repeatedly buried and exhumed by Russian SMERSH agents until they, their two dogs, and propaganda minister Joseph Goebbels and his family were secretly interred in an unmarked grave beneath a paved section of the front courtyard of SMERSH's Magdeburg facility. It wasn't until April 4, 1970 that their remains were secretly exhumed and thoroughly burned to ashes (the ashes scattered in the Elbe river) by the Soviets in East Germany.

> *SMERSH—Smert Shpionam (Death to Spies)—was the Soviet armed forces counter-intelligence agency whose primary task from 1943 to 1946 was to uncover spies and saboteurs in the military, screen all liberated Soviet POWs, and protect military factories.*

The Germans unconditionally surrendered to the Allies on May 8, 1945, a date which became known as V–E Day. Hitler's Third Reich lay in ruins. Fortunately for the Soviets and the West, at the instigation of Hitler the Germans had poured tremendous resources into perfecting the V–2 missile. They might have won the war if they had concentrated instead on their aircraft turbojet engine technology, which was also way ahead of anything the Allies could muster. As early as 1942, the Germans could have gone into series production with these engines. German officers testified after the war that they could have gone into large scale production of the twin–jet engine Me–262 Messerschmitt equipped with these turbojets, against

which the Allied propeller–driven aircraft were no match. Instead, only small amounts of Me–262s appeared at the front towards the end of the war.

Just after the war ended, on May 24, 1945 a team of Russian rocket scientists and engineers led by General Lev Mikhaylovich Gaydukov arrived in Berlin. They quickly discovered that the Germans had made tremendous advances in rocket science, far ahead of either the United States or the Soviet Union at the time. A week later Stalin decreed that all rocket institutes in the Soviet zone of Germany were to be taken over and put back into operation.

In mid–June a small group of Soviet scientists and engineers with a particular interest in rockets was moving through the federal state of Thüringen in central Germany (Thuringia is one of Germany's 16 Bundesländer, or federal states). They were simply amazed at the seeming lack of destruction as they cruised along through cities and towns. Then they reached a detour after they passed the town of Annaberg. "Halt!" shouted a group of German civilians behind an improvised barricade. The leader of the Soviet delegation, Colonel General Ivan Aleksandrovich Serov, stepped up to see what was the matter. The Germans started unintelligibly babbling and kept pointing up the road. Boris Chertok, one of the Soviet engineers who could speak halting German, strode up behind Serov. "What are they saying?" asked Serov. "Well, apparently the road is closed up ahead because the Americans have blown up a bridge and wiped a German village off the face of the earth." Finally, Chertok calmed the Germans down with a few packs of cigarettes, a scarce commodity in war–torn Germany. Then a couple of Germans accepted a generous helping of Russian vodka. More Germans joined in and things began to calm down. Between Chertok's passable German and the barely recognizable Russian spoken by one of the younger Germans, a schoolboy type, they found out what was the matter from one of the Germans standing nearby.

> *In April an American mechanized column had thrust deep into Thuringia, meeting virtually no resistance. They passed the town of Annaberg. Then upon entering the hamlet of Bad Berka, the advance guard was suddenly fired upon by automatic weapons and hunting rifles. A small detachment of Hitlerjugend (Hitler youth) had responded to Joseph Goebbels' desperate last–minute appeals and decided to become guerrillas. Without firing a shot, the powerful mechanized formation withdrew several*

kilometers. The young German combatants breathed a sigh of relief as they decided that their village had been saved from the occupying forces, but they were sorely mistaken. The timorous commander of the American unit reported that they were under fierce attack by a German battalion. He requested air support. With the Allies in complete command of the skies by this late in the war, a bomber formation was sent to his aid! An overkill of carpet bombing turned the ill–fated village and all of its inhabitants into formless heaps of smoking ruins. Only after this treatment from the air did the Americans continue their "victorious" advance.

Serov ordered the Soviets to make a small detour to get a look at the demolished "fortress." To their amazement they discovered an intense reconstruction project by the industrious Germans at the site of the former village. Now the Russians talked amongst themselves. If a Red Army unit had been there, these young upstarts would have been annihilated on the spot or taken prisoner. But the Americans didn't want to risk the life of a single one of their guys! Later that evening they shared stories among over vodka about the crazy Americans. Serov began by telling about how he had observed a military jeep full of obviously tipsy American soldiers, heavy pistols dangling from their broad belts, tearing around freshly occupied Germany at breakneck speed. It seems they were prowling about looking for new German 'girlfriends'.

EAST VERSUS WEST

The Soviet socialist state found it expedient to blur the distinction between their rocket scientists and key military leaders overseeing technical work in the design bureaus, particularly in the aftermath of World War II. Engineers holding responsible positions in technical organizations; for example, were sometimes authorized to don officers' rank insignia so as to shore up their authority in exploiting German missile technology. However, their new oversized uniforms and lack of battlefield decorations were often noticed by legitimate Red Army officers. These "trade–union officers" as they were known in the Soviet Army, represented various Soviet industries charged with the task of locating and removing to Russia machinery and equipment from freshly occupied East Germany. One of those cases involved Boris Yevseyevich Chertok who was a talented and pioneering guidance and control engineer, and a key member of Korolyov's team from 1946. He was granted a Major's uniform and shoulder boards. Another noteworthy Russian rocket scientist carried the rank and insignia of a Lieutenant Colonel, Aleksei Mikhaylovich Isaev, who was authorized to form his own design bureau, KB–Isaev in 1944 to engineer liquid–propellant rocket engines. Over the years KB–Isaev would reorganize, expand, and evolve into the large, highly skilled association of KB KhimMash, which today solves many complex problems in the creation of new technology—rocket and otherwise—for the Russian Federation.

Soviet Major
Boris Chertok

The Soviet military, like many armed forces around the world even today, did not have inflexible rules like the United States often does regarding time in rank—the "up or out" syndrome. Officers were, and still are, permitted to hold a rank as a Captain, let's say, for practically an indefinite period of time before being asked to retire. It was not uncommon for highly decorated post–war Soviet Generals to serve into their 80s as long as they were physically able—something quite unheard of in the United States *(there are rare exceptions; for example, RADM Grace Murray Hopper served in the Navy until age 80 when she retired in 1986 after 43 years of service).* Why not? The military benefits from their wisdom and vast experience gained over the years. The individual gets the satisfaction of being useful until he or she willingly wants to retire, instead of being forced out in the prime of working life—at age 45, for example.

July 1945. American forces had just pulled back from areas of East Germany they had occupied but were allocated to Stalin at the Yalta conference. The Russian territory now included the City of Nordhausen and a secret and infamous underground slave–labor V–2 production facility that was nearby. Although the Americans had stripped the facility of V–2 components and left only 'remnants' for the Russians, the remaining materials would prove more significant to Russian rocket science than the Americans thought.

The secret Nordhausen V–2 Plant

On their way to Nordhausen, "Major" Chertok and Engineer First Lieutenant Vasiliy Ivanovich Kharchev—the only real service officer in the small contingent of Russian rocket engineers—went to a café–cabaret that was still operating after the Americans had been ordered to leave. The café, which was located in a cozy bomb shelter, proved to be a noisy establishment that served beer and bootleg schnapps instead of appetizers and coffee. This night the café was filled with smoke, loud–mouthed American officers, and some black enlisted soldiers. A bosomy German brunette, dressed like a gypsy and well past her prime, was singing something unintelligible in a husky voice. Chertok and Kharchev made their way to the only free table in the smoky joint. Soon two more young Russian officers wandered in and sat down with them. Evidently, they were the first Soviet officers to go there. Suddenly an American officer jumped up and yelled a quick something in the direction of the bar. A fellow in white quickly flitted from behind the bar and deftly placed foaming beer mugs in front of the unsuspecting Russians. Then the singer came running over, gauged the Russians' ranks by their shoulder boards, and without asking permission planted a smacking kiss on Aleksei Isaev's cheek. "At last, the Russians have arrived!" she gleefully exclaimed in pretty good Russian. "What shall I sing for you?"

One of the American officers quickly barked something to her in the tone of an order. "He knows that I am Russian and wants me to interpret," she explained. "He welcomes the Russian officers to the land that they, the Americans, have liberated from our common enemy. Horrible crimes occurred here. He hopes that we will be friends. To victory and our comrades in arms!" All smiles, the American officer managed to add something from a bottle to his beer and the Russians' from a bottle that he was holding in his outstretched hand. Another of the American officers talked a lot, demanding the whole time that the singer interpret for the Russians. She did. This is what they learned.

Sergei Korolyov (left) with Aleksei Isaev (right)

"The Americans, who had advanced from the west as early as 12 April, that being three months before the Red Army, had reached the city of Nordhausen. It had been largely destroyed by Allied bombers. However, the British and Americans, with assistance from Soviet aviation, had been unable inflict any damage on Nordhausen's massive subterranean missile and jet engine factory, the so–called Mittelwerk. It had been shut down only 24 hours before their arrival. They were staggered by the sight of it. There were hundreds of missiles under the earth on special railroad flatcars; the factory and spur tracks were completely preserved. The German guards had fled, and the prisoners had not been fed for at least two days before the arrival of the American troops. Those who were capable of walking moved slowly. They approached the Americans to get food and did not hurry. It was as if they were doing everything in their sleep."

The singer interpreted further, "They told us that more than 120,000 prisoners had passed through the slave labor camp. First they built—gnawing away at the mountain. Those who survived and newly arrived prisoners worked in the subterranean factory. We found prisoners in the camp who had fortuitously survived. There were many corpses in the subterranean tunnels. Our soldiers were horrified when they saw all of this. We put many Germans to work cleaning up and bringing order. It will be easy for you to work there now. To our victory, to our friendship!"

The evening went on, with toast after toast. But the unsuspecting Russian officers had been duped. Some of their drinking companions that night were not American combat officers, but American spies in disguise who had been assigned to "wrap up" the seizure of German rocket engineers, look for remaining missile paraphernalia, and monitor the actions of Russians hunting for German secrets. The next day they were enlightened by SMERSH about American intelligence services conducting broad–ranging operations to seize German specialists (Russian–speak for scientists and rocket engineers). SMERSH had learned that the Americans seized over 100 fully assembled A–4 and V–2 missiles which had been piled up at the Mittelwerk factory. They hauled them off in special railroad cars, but they left behind sophisticated metal–cutting machine tools and enough assemblies for the Russians to put together 10 or maybe even 20 A–4 or V–2 missiles. Erstwhile Major Chertok, Lt. Kharchev, and Isaev (who had been awarded the honorary rank of lieutenant colonel) were told to coordinate their future actions on the secret recruitment of German specialists with the Soviet Military Administration in Germany (SVAG). This they did. The following day they left for what was left of the City of Nordhausen. And that is where things began to get interesting.

When the Russian officers arrived in Nordhausen on July 14, they found it heavily destroyed just like the bosomy German singer had said. The British and Americans, with assistance from Russian aviation, had sought to stop German missile production. Much to their angst, the Americans and subsequently the Russians had found the Mittelwerk subterranean missile factory practically untouched. The neighboring Dora concentration camp had also escaped the brutal Allied bombing attacks unscathed. The 77th Guard Division, part of the 8th Guard Army, had just taken over the city from the Americans and had already been billeted in the city and surrounding areas. The offices of the commandant and the town mayor, or Burgomaster (*Bürgermeister* in German) were already in operation. That evening,

Chertok, Kharchev, and Isaev were able to rendezvous with Isaev's team of rocket specialists who had arrived a day earlier and taken up residence in a remote and devastated villa close to Kohnstein Mountain, where the subterranean *Mittelwerk* factory was hidden.

In the morning of July 15, the Russian contingent of rocket engineers, scientists, and technicians found that a whole line of people wishing to offer their services had assembled outside their villa. They had been summoned by the Burgomaster and local authorities. An American soldier stepped out of the crowd, and amazingly, started speaking perfect Russian. "I am Anatoliy Shmargun, a Red Army officer and former German prisoner of war, liberated from the slave labor camp by the Americans. They gave me this uniform to replace my threadbare clothes." He then explained he was a first lieutenant in the Red Army and a political commissar who had been taken prisoner in 1944. After being moved to various places, he had been sent through Buchenwald to Dora. Lt. Shmargun did not look at all like all like the walking skeletons usually seen at death camps. Of course, he had no ID. Might he be a spy the Americans left behind? Isaev asked him the expected question: "Why did you survive?"

"Before the Americans arrived there was a lot of work," Shmargun began. "We were ordered to gather up and burn more than 200 corpses that were delivered from the factory to the camp. They needed us to be alive for that work. But we didn't manage to burn them all. About 100 bodies were still lying around when the Americans forced their way in. The Germans fled. The Americans fed us and gave us clothes. Some of the skeletons and I refused to leave with the Americans and decided to wait for our own troops."

"Now I can show you around the camp," he continued. "I know some Germans who worked at the factory and didn't leave. They have agreed to help in investigating everything that was going on there. I can be in contact with 'that side.' There were a lot of good guys among the American officers. There are also a lot of Russian girls in town. They were domestic workers and worked on farms. They know the language well, and can serve as interpreters until they are repatriated. I know places where the SS hid the most secret V–2 equipment that the Americans didn't find. We prisoners knew a lot." Shmargun certainly wasn't sounding like any spy, American or otherwise.

The horrors that Shmargun began to tell the Russians about, corroborated by live witnesses who had arrived from somewhere, were unbelievable. They began by inspecting the horrible Dora death camp. The liberated slave laborers showed them the area where the bodies had been placed before being fed into the crematorium and where the German guards had raked out the ashes. Now there were no traces of ashes anywhere. The rescued slaves explained that when the Americans had been there a commission had been at work documenting atrocities and war crimes. Now before everyone's eyes the Dora camp was being converted into a dormitory for displaced persons. It was apparent that the Americans had already put things in order there. All of the dead had been buried, and those who had survived had been treated and fed—the living dead were back on their feet. Special troop units were preparing the camp for Russians who were former prisoners or had been driven off into Germany. They would be sorted out and then repatriated. But Isaev and the other Russians just couldn't get the nightmare of the crematorium and piles of human ashes out of their minds.

A group of Germans, who had turned up as a result of the *Bürgermeister*'s efforts, was waiting for the Russians when they arrived in front of the entrance to *Mittelwerk*. A young, thin German with delicate facial features separated himself from the group. He boldly approached and introduced himself, "Engineer Rosenplänter from Peenemünde." He started rattling away in German, explaining that everyone had been evacuated from Peenemünde to Nordhausen, and they had settled not very far away in Bleicherode. At first von Braun and his protector, General Walter Dornberger, whom he knew personally, had lived there. They had left Bleicherode and moved farther west, he told them. Lt. Shmargun, who had picked up a considerable amount of German, did his best to translate for the Russians.

Before the arrival of the Russians, Rosenplänter told them, Americans had sent almost all of the rocket and missile specialists to the towns of Worbis and Witzenhausen. Rosenplänter and several dozen other specialists had refused to move, and the American officers, having checked with their lists, didn't make them go. Certain others had been taken despite their unwillingness. The agitated Rosenplänter was speaking too rapidly for Shmargun to translate, so someone drove back to the camp and brought a bilingual Russian girl, named Lyala, who could interpret more rapidly than

the Germans spoke. Now they were ready to explore the legendary subterranean chambers of the *Mittelwerk* missile factory. It took them almost two days.

Literally translated, *Mittelwerk* means "middle factory" or "factory located in the middle" (it was located in the middle of Germany). The construction of the factory began in 1942 under the codename *Mittelbau* (middle construction). This was before the successful launches of the V–2 (A–4) missile. Because of the local geography, the Germans didn't need to go extremely deep into the ground. The construction workers successfully used the natural terrain. The wooded hill that locals proudly called Kohnstein Mountain rose up almost 150 meters above the surrounding terrain four kilometers from Nordhausen. The limestone rock forming the interior of the mountain yielded easily to mining work. Four opened galleries had been cut in the mountain along the diameter of the base, each was a bit longer than three kilometers. Forty–four transverse drifts connected the four galleries. Each gallery was a separate assembly factory.

The two galleries on the left side of *Mittelwerk* were BMW–003 and JUMO–004 aircraft turbojet engine factories. These engines had already been made fit for series production in 1942. The Russians were amazed at the German jet engine technology—far ahead of anything they, the Brits, or the Americans had at the time. The Germans could have mass–produced twin–engine jet Me 262 Messerschmitts, which were equipped with *Mittelwerk* engines. Why did they only appear towards the end of the war, and then only in small numbers? In postwar memoirs, German generals noted that Hitler had been personally opposed to using these airplanes for a long time. *Thank goodness for Hitler's stupidity and stubbornness. We could have suffered millions more in casualties, if not lost the war altogether*, Isaev thought.

The third gallery at *Mittelwerk* was used for the production of the V–1 "winged bombs," which in modern terms would be called cruise missiles. The mass production of V–1s began in 1943. Only the fourth gallery was dedicated to the assembly and testing of A–4 missiles, which basically served as an early version of the V–2. Rolling stock bringing in materials could roll directly into each gallery from the surface. Rail cars loaded with the finished product exited at the other end. The gallery for A–4 missile assembly was more than 15 meters wide, and its height in individual bays reached 25 meters. In these bays they conducted the so–called vertical *Generaldurchhaltever–suchsprüfung*. Lyala translated this as "general vertical tests" and the Russians used it as the official term for this type of test for a long time (the literal meaning of the term *Generaldurchhalteversuchsprüfung* is "general endurance test"). Horizontal tests were usually carried out before vertical ones, although these did not have the prefix "general." The transverse drifts were where the assemblies and subassemblies were fabricated, integrated, inspected, and tested before they were mounted on the main assembly.

As they walked around in partial darkness, Rosenplänter explained that the lighting had been partially damaged by order of the Americans. Only "duty" lighting was on. Everyone had to be very careful while walking around the factory to not fall into some processing pit or hurt themselves on the remains of missile parts that hadn't been cleared away. A large number of missile components were scattered around in disarray, but it was easy to count dozens of "tails," side panels, middle sections, and propellant tanks. Then Shmargun directed their attention to an overhead traveling crane that traversed the entire width of the bay they were in. Casting a ghostly shadow, it was used for the vertical tests and for the subsequent loading of the missiles. Two beams were suspended from the crane over the width of the bay. They were lowered when necessary to a human being's height. Nooses were secured to the beams and placed around the necks of prisoners who were guilty or suspected of sabotage. The crane operator, who also played the role of executioner, pressed the raise button and immediately an execution of up to 60 people was carried out via mechanized hanging! Before their very eyes, the *polosatiki* (literally, "people in stripes," referring to their prison uniforms) as the prisoners were called, were given a lesson in obedience and fear in the macabre subterranean factory 70 meters below ground. Oh, but the Nazis were so brutally efficient.

A German who was introduced as an assembly engineer–tester said that the factory had worked at full power until May. During the "best" months, its productivity was as high as thirty–five missiles per day! At the factory the Americans had seized only fully assembled missiles, taking more than 100 which had been piled up at the factory. They had even set up electrical horizontal tests. The assembled missiles had been loaded into special railroad cars before the arrival of the Russians and hauled to the west—to their zone.

"But it's still possible to gather assemblies for ten and maybe even twenty missiles here," translated Lyala. The Germans said that the special equipment used purely for missile testing had been hauled away. But the ordinary machine tools and standard, general–purpose equipment in the shops had remained untouched. Even the most state–of–the–art metal–cutting machine tools had been passed over by the Americans in their haste to haul off the most valuable assemblies before they had to vacate the "eastern" German zone.

When the Russian team started asking technical questions, Rosenplänter offered to familiarize them with V–2 technology, but said that he did not know *Mittelwerk* that well and recommended another specialist from Peenemünde who had often visited *Mittelwerk* to perform monitoring tasks. Rosenplänter patently denied that they had anything to do with the atrocities that had taken place there. Like the Americans, the Russians were simply appalled at the German concentration camps and their use of slave labor to build German weapons of war, including the V–1s and V–2s. In fact Nordhausen would become known as the "City of Missiles and Death" because of the thousands of slave laborers who perished there assembling Hitler's weapons.

The Mittelwerk was part of a larger German wartime effort codenamed Mittelbau (middle construction). Mittelbau was a huge complex of factories, storage depots, facilities and prisoner camps, some underground, that were used from August 1943 until April 1945 to manufacture and test the V–2. Over 25,000 of the some 60,000 workers at Mittelbau were killed either by beatings, starvation, sickness, or by the brutal efforts of the SS to relocate them before the Americans arrived to rescue the survivors.

UNBELIEVABLE FINDS

June 1, 1945. A group of Russian rocket scientists and engineers had just reached the premier German rocket research center in Peenemünde. They were astounded at the variety of rocketry and test stands they found – not just the V–2, but *Rheintochter*, Rheinbote, *Wasserfall*, and *Taifun* missiles. Soviet security squads located 10 partially assembled V–2's in the surrounding area. Then word spread that the Soviet army had seized an entire abandoned V–2 factory in central Germany. It was clear that the German technological advances were so great that additional experts were needed from Moscow from the automotive and electrical industries. But even this did not prepare the Russians for what was to come next.

On June 21 Academician Aleksei Isaev and a group of colleagues from Moscow's Scientific–Research Institute (NII–1) were in Peenemünde scavenging through all kinds of trash trying to find any remnants of German missile documentation. They had not found anything for days. Towards the end of the day Boris Slepnev went behind a woodpile to relieve himself. He had just unzipped his pants when he saw it, half buried in the dirt. Quickly finishing his business, he bent over to retrieve what looked like a thin report with a diagonal red stripe across the front, and the frightening inscription *Streng Geheim* (Top Secret). "Holy crap!" he muttered to himself in Russian. "Hey guys, look what I found!" he shouted out.

Isaev and the others in the collective panel of Russian experts determined that this document was the design for a huge rocket–powered bomber. They couldn't believe their eyes. It was a very realistic, detailed design for a manned ICBM with wings, half a century ahead of its time. Isaev decided to immediately "classify" the report a second time, scribbling out the words "Top Secret, Special File" top and bottom on the cover in Russian. In the presence of witnesses, he slipped the report under the shirt of his most trusted collaborator, then instructed him to immediately return to Moscow in their B–25 "Boston" bomber that was parked nearby on Peenemünde airfield tarmac. He was to take the report directly to their group's "patron", General Victor Bolkhovitinov.

Once the report was safely in Moscow, Bolkhovitinov together with engineer Gollender, who had a good command of German, studied its sensational contents. The Soviets had stumbled across one of 100 printed copies of a Top Secret report entitled "*Über einen Raketenantrieb für Fernbomber*" (On a Rocket Engine for a Long–Range Bomber). The report had been issued in Germany in 1944. It was authored by the Austrian rocket engine researcher Eugen Sänger, who was already well known before the war, and Irene Bredt, a gas aerodynamics specialist. Dr. Sänger was known for his book *Raketen–flugtechnik* (The Technology of Rockets and Aviation), which he published in 1933. It had been translated and published in the Soviet Union. Back when he was a 25–year–old engineer, Sänger was captivated by the problems of rocket technology. He was one of the first serious researchers of gas dynamic and thermodynamic processes in rocket engines.

Bolkhovitinov and other NII–1 specialists were dumbfounded as they leafed through the top–secret report. Judging by the distribution list, it had been sent to the leaders of the *Wehrmacht* main command, the German ministry of aviation, to all institutes and organizations working in military aviation, and to all German specialists and leaders who were involved in rocket technology, including General Dornberger in the army department of armaments, who also served as chief of the Peenemünde center.

The problems of aerodynamics for an aircraft with a speed ten to twenty times greater than the speed of sound were new to Russian aerodynamics specialists. The German report went on to describe the launch, takeoff, and landing dynamics. In an apparent attempt to interest the military, the report included a highly detailed examination of bombing issues, considering the enormous speed of a bomb dropped from such an aircraft before it approached the target.

Subsequent analysis proved that this was not just any bomber. Half a century ahead of their time, the Germans had analyzed in great detail the technical capabilities for creating a manned winged rocket weighing many tons, comparable to the U.S. Space Shuttle or the Russian Energia–Buran system. The authors convincingly showed by constructing nomographs and graphics that with the proposed liquid–propellant rocket engine with a thrust of 100 metric tons (tonnes) it was possible to fly at altitudes of 50–300 kilometers at speeds of 20,000–30,000 kilometers/hour, with a flight range of 20,000–40,000 kilometers. The physical and chemical processes of high–pressure and high–temperature propellant combustion were studied in great detail, along with the energetic properties of propellants, including emulsions of light metals in hydrocarbons. The work proposed a closed, direct–flow, steam power plant both as a cooling system for the combustion chamber and as a means to activate the turbopump assembly. To launch the thing, two rocket engines would be used.

The total takeoff weight of the bomber was 100 tonnes, of which 10 tonnes was the weight of the bombs. The landing weight was assumed to be 10 tonnes. If the flight range were reduced, the weight of the bombs could be increased to 30 tonnes. They proposed that the subsequent work to implement the design of the rocket–propelled bomber be divided into twelve stages, in which the bulk of the time would be devoted to firing rig optimization of the engine, rig testing of the interaction of the engine and aircraft, launcher testing, and finally, all phases of flight tests.

The huge bomber would be catapulted off an inclined horizontal rail with a huge booster rocket that stayed on the ground. After only 11 seconds, it would break the speed of sound and reach 500 meters per second. Then a 100 tonne monster rocket would ignite and hurl the behemoth into space at 300 or 400 kilometers. The highest–thrust rocket bottle for the Russian BI bomber could only manage a pitiful 1.5 tonnes of thrust at the time.

"Here you see the trajectory projections for this beast," Isaev began. "This spaceship comes down at supersonic speed, but it doesn't break up in the atmosphere as you would expect it to. It just glances off it, like when you throw a flat stone across the Volga at an acute angle. It strikes, skips, and flies farther, at least several times or more." General Bolkhovitinov observed, "I remember how we used to compete at Serdolikovaya Bay in Koktebel. The one who got the most skips won." Isaev went on, "That's how these aircraft skip along the atmosphere, and they dive down only after they have flown across the ocean in order to slice their way into New York! What an impressive idea!"

Isaev was an engineer with an original way of thinking, who was captivated by extraordinary, new ideas regardless of who proposed them. He was simply captivated by this extraordinary find. "What have they invented here?! It's an airplane, but not our pitiful BI, with a 1.5 tonne bottle. This one had 100 tonnes of sheer fire! That damned engine hurls the airplane to a frightful altitude—300 or 400 kilometers! It comes down at supersonic speed, but doesn't break up in the atmosphere—it glances off it, like when you throw a flat stone across the water at an acute angle. It strikes, skips,

and flies farther! And it does this two or three times! Ricocheting!

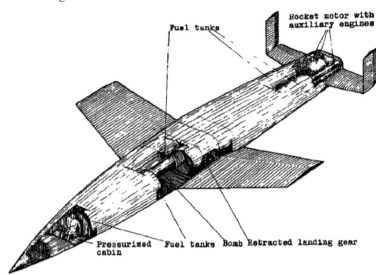

Perspective view of the incredible German "Rocket Bomber" with global reach

As other specialists studied the report and its contents, they ascertained that it was not for the design of the German MRBM A–9/A–10 missile, which was designed for a range of 800 kilometers. This report discussed the ranges required to strike New York! From today's standpoint, the layout of the vehicle described in the report—found in the woodpile in Peenemünde in May 1945—anticipated the structure of the American Space Shuttle and the Russian Energia–Buran system—by half a century!

It is interesting to note that Sanger had already shown by the early 1940s that launching space aircraft without auxiliary means was unacceptable. He proposed launching space aircraft using a catapult with a horizontal track that would enable the aircraft to reach a speed greater than the speed of sound. Commenting on the calculation and visual graphics of flight, Sänger and Bredt wrote:

"Takeoff is conducted using a powerful rocket complex fixed to the ground and operating for approximately 11 seconds. Having accelerated to a speed of 500 meters/second, the aircraft lifts off from the ground; with the engine at full power, it climbs to an altitude of 50–150 kilometers on a trajectory that is initially at an angle of 30 degrees to the horizon, and then becomes lower and lower. . . . The ascent lasts from 4–8 minutes. During this time, as a rule, the entire fuel supply is consumed. . . . At the end of the upward phase of the trajectory, the rocket engine shuts down; using its stored kinetic energy and potential energy, the aircraft continues its flight in its characteristic glide along a wave–like trajectory with attenuating

amplitude. . . .At a previously calculated moment, the bombs are dropped from the aircraft. Tracing a large arc, the aircraft returns to its airfield or to another landing pad and the bombs, which are flying in the original direction, come down on the target. . . .This tactic makes the attack completely independent of the time of day or the weather over the target, and deprives the enemy of any capability to counteract the attack. . . .The problem that we posed, which until now no one anywhere had solved, entails firing on and bombing targets located 1,000–20,000 kilometers away. A formation of 100 rocket–propelled bombers . . . within the course of several days would be capable of completely destroying areas approaching the size of world capitals, including their suburbs, located anywhere on the face of the earth."

Fast–forward 50 years. In modern–day Germany, they have recently designed an aerospace system that is called "Sänger" in honor of the pioneer of this idea. The largest German aircraft firms participated in the work on this program. They designed the spacecraft on the basis of forward–looking, but realizable technology. The modern Sänger is intended to transport various cargos into space while lowering costs and ensuring safety, reliability, and all–purpose use. It differs fundamentally from the 1940s design in that horizontal acceleration is not achieved by a catapult, but by a special booster aircraft that carries the actual spacecraft, which will be capable of inserting 10 metric tons of payload into near–Earth orbit—the same amount specified in Sänger's original design—at an altitude of up to 300 kilometers. Working in 1944, Eugen Sänger certainly could not have imagined the materials, engines, and navigation and control methods that German scientists with access to advanced space technologies are working on now.

In 1947, in conversations with Helmut Gröttrup, the Russians tried to determine Peenemünde's attitude toward Sänger's design during the war. The gist of his response was something like the following: First, the consensus was that work on Sänger's design might hinder the A–4 program and the other programs at Peenemünde that were purely rocket–oriented. Second, they believed that such a design would require at least four to five years of intense work before the first flight; and third, it was an aircraft—the design interested the *Luftwaffe*, but rocket technology was under the management of the infantry command. Even here institutional partiality was at work!

It is interesting to compare the different assessments of the development cycle for Sänger's aircraft. Peenemünde estimated up to five years, while G.N. Abramovich's subsequent assessment was up to ten years. German work on the "new Sänger" began in 1986 and they scheduled the first demonstration flight for 1999—a 13–year development cycle! And this was more than 50 years after Isaev's group extracted the top–secret report from the woodpile. Unfortunately current work on the project has been practically halted due to the European Space Agency's lack of funding.

August 1945 Another group of Russian officers was given intelligence by Smersh that a "forester's cabin" in a forest preserve not far from Nordhausen served as a hiding place for valuable Nazi artifacts that were hidden just before the arrival of American troops in the area. Did the Americans leave anything behind? At the risk of ending up in the American zone because this place was right on the border, the Russians pursued the scavenger hunt, armed with a map of the area. The Russians and Americans still hadn't laid a precise border over the hilly, forested terrain. Russian border guards and American soldiers were posted on the roads, but there were no checkpoints, barriers, or soldiers in the forest. After surrendering their entire supply of cigarettes to a German couple, they were led to a barn. The German proprietress shoved aside a pile of some sort of brooms, and pointed to a trapdoor in the wooden floor. They descended in the darkness to a dry cellar, where they discovered a treasure–trove of well–packed electrical equipment. In the light of the next day they discovered they had come upon a windfall of two sets of relay boxes and control panels intended for pre–launch tests and the launching of V–2 missiles. It was only later that the Russians found out that they had strayed across the border into American territory. Incidents such as this of both sides hunting for Nazi secret weapons and technology were being repeated all over Germany in the months after the war.

The Russians picked up another prized gem courtesy of Anatoliy Shmargun, the Russian POW who had served as their escort in the *Mittelwerk* subterranean factory in mid–July 1945. Lt. Shmargun led the Russian contingent to a distant wooden barracks hut, where in a dark corner, after throwing aside a pile of rags, he jubilantly revealed a large spherical object wrapped in blankets. They dragged it out, placed it on a nearby cot, and unwrapped the many layers of blankets. Boris Chertok, one of Russia's finest rocketeers, was stupefied—it was a gyro–

stabilized platform of the type that he had seen for the first time in Berlin at the *Kreiselgeräte* factory. At that time, "civilian" Colonel Viktor Ivanovich Kuznetsov, a gyroscope specialist who was also seeing the instrument the first time, had explained its layout to him.

"The German scientists seem to have mastered the difficult problem of attitude control of missiles in flight," Kuznetsov had told him. "We thought we were ahead by using accelerometers to measure the linear acceleration of our missiles after they are launched. But since our accelerometers are fixed to the missile and rotate with it, they are not aware of their own orientation in space. This can be thought of as the ability of a blindfolded passenger in a car to feel himself pressed back into his seat as the vehicle accelerates forward or pulled forward as it slows down. He feels himself pressed down into his seat as the auto accelerates up a hill, or he rises up out of his seat as the car passes over the crest of a hill and begins to descend. Based on this information alone, he knows how the vehicle is moving relative to itself, that is, whether it is going forward, backward, left, right, up (toward the car's ceiling), or down (toward the car's floor) measured relative to the car. But the passenger has no idea of the direction of his motion relative to the Earth, since he did not know what direction the car was facing relative to the Earth when he felt the accelerations."

"However," Kuznetsov continued, "by tracking both the current angular velocity of the system and the current linear acceleration of the system measured relative to the moving system, it is possible to determine the linear acceleration of the system in the so–called Earth–centered inertial reference frame, which is a 3–axis coordinate system [up, down, sideways] with its origin at the center of the Earth. Performing integration on the inertial accelerations—using the original velocity as the initial conditions—and applying the correct kinematic equations yields the inertial velocities of the system, and integration again—using the original position as the initial condition—yields the inertial position. This is like if the blindfolded passenger knew how the car was pointed and what its velocity was <u>before</u> he was blindfolded. Now he is able to keep track of both how the car has turned and how it has accelerated and decelerated since he began. In fact he can accurately know the current orientation, position, and velocity of the car at <u>any</u> time," Kuznetsov concluded.

"How has a gyro–stabilized platform, which still hasn't become a standard V–2 instrument, ended up in this death camp prisoners' hut?" Chertok incredulously asked Shmargun. Shmargun could not give a clear explanation. According to what he had heard from others, when the camp guards fled, some Germans who were neither guards nor *Mittelwerk* personnel brought a beautiful case to the barracks, covered it with rags, and quickly fled. By the time that the Americans arrived, the surviving prisoners had discovered the case and opened it; one of them said that it was very secret. They decided to put it away until the Russians arrived. They used the case to pack up various things that they had begun to acquire after liberation, and when they found out that Shmargun was staying to wait for the Russians, they revealed the secret to him and packed everything up in dirty blankets so that the Americans would be less suspicious.

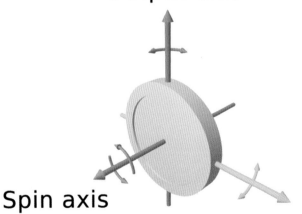

Output axis

Spin axis

Input axis

Diagram of a gyro wheel. Reaction arrows about the output axis (blue) correspond to forces applied about the input axis (green), and vice versa.

This operation went brilliantly. Now Isaev and his deputy Chertok were responsible for this priceless windfall. They wrapped it back up in the blankets, since no other container was available, transported it to the division headquarters, and asked them to store it there until it could be flown to Moscow. A gimbaled gyrostabilized platform would give Russian rocket science a tremendous boost. The German device would allow a vehicle's roll, pitch, and yaw angles to be measured directly at the bearings of the gimbals. *The gimbals are a set of three rings, each with a pair of bearings initially at right angles. With the gyroscopic action they let the platform keep the same orientation while the vehicle rotates around it. Two gyroscopes are used to cancel gyroscopic precession, the tendency of a gyroscope to twist at right angles to an input force. By mounting pair of gyroscopes (of the same rotational inertia and spinning at the same speed) at right angles the precessions are cancelled, and the platform will resist twisting.* For 1945, this was an ingenious solution to the problem of controlling a missile in flight.

THE PEOPLE

By the middle of the 20th century, the era of the lone space dreamers and rocket entrepreneurs like Jules Verne, Tsiolkovsky, Goddard, and Oberth was long since over. Sergei Korolyov, Valentin Glushko, and Werner von Braun in the United States were important figureheads in their country's respective space programs. But in the larger scheme of things, they were merely larger spokes in the wagon wheel of humankind's reach into space. At one time during NASA's massive Apollo moon project it was estimated that 300,000 government personnel and contractors were dedicated to the program. Envisioning, developing and designing, producing, operating and launching sophisticated and complex aerospace hardware into space takes thousands of knowledgeable people. First, the scientists and some farsighted entrepreneurial individuals to build the case for the expensive launching of something into space in the first place. Usually, space launches are justified because of their value in building on a previous knowledge base. The first exploratory missions to the planets in the 1960s built on previous knowledge and studies with earth–bound telescopes. Even Einstein said that he was able to achieve great things because he could "stand on the shoulders of giants." Second come the conceptual designers and engineers. They develop the "cartoons" of what the final spacecraft or new launch system might look like. Quite often the launch system is just an evolution of an existing one, like fitting a larger payload fairing to accommodate a science instrument or squeeze in multiple instruments, or adding an engine to the booster.

Then the real work begins, with hundreds of engineers and technicians actually producing the space system. At each step of the way, designs have to be checked out by testing and verification of performance and integrity. All the while there is interaction with the astronauts (or for scientific missions, the scientists) who must rely on the good work done by the design teams. Personnel must test components, subsystems, and flight systems on the ground, and prototypes have to be flown into space for tests before entrusting them with precious human cargo or irreplaceable science payloads like the Hubble Space Telescope or Chandra X–ray Telescope. For human flights astronauts must undergo many months of training, and so on. After flight ground operations personnel have to refurbish reusable components, like the Space Shuttle main engines or solid rocket boosters, in preparation for the next flight.

Every astronaut puts his or her life at risk with each launch. For scientific missions, one misstep can kill a whole project. Blunders like a misunderstanding between English and metric units led to the demise of the $328 million–dollar (in 1989 $$) Mars Climate Orbiter exploration mission to Mars. Finally, many millions more are spent in gathering scientific data, disseminating, and analyzing the results, a process that can take hundreds of scientists decades as with the Hubble and Chandra Space Telescopes.

Mid–July 1945 The Russians had assembled a crackerjack team of 12 German rocket experts in the hamlet of Bleicherode in the Russian zone, what would become East Germany. The eminent German rocket scientist Wernher von Braun himself had lived in Bleicherode for a couple of months after his escape from Peenemünde. His residence was the best villa in town, the luxurious Villa Franka on Lindenstrasse. The Russians quickly occupied the Villa Franka and converted it into an officer's club and rocket operations headquarters.

Led by the freshly minted 'Lt. Col.' Isaev and 'Maj.' Chertok the Russians decided to form a makeshift rocket research institute in Bleicherode. The Germans were delighted with this idea and announced that they could quickly put together specialists and an entire staff. They had to decide on a name for the institute. After a brief "Soviet–German" discussion, they came up with a name: 'Institut RABE' which the Germans said stood for *'Raketenbau und Entwicklung'* (Rocket Production and Development). *Only later did the Russians realize that 'Rabe' literally meant "raven" in German (ravens scavenge dead corpses) and they changed the name to the more innocuous "Zentralwerke" (Central Works).*

By working with the SVAG Directorate in Thuringia, the Russians soon struck pay dirt. Germans were offered a generously high pay scale, decent living accommodations, moral and material support, including letting them keep their families together. But they found the biggest payoff

in offering scientists and highly qualified engineers in specific specialties the opportunity to continue their work under the well–equipped auspices of SVAG. Prior direct involvement in rockets was not a prerequisite. This expanded the reach of the Russian net and quickly led to the discovery of prominent specialists who would make valuable contributions to Russian rocket science, even though they had not previously worked at Peenemünde.

Dr. Kurt Magnus soon joined the Russian team. He was a first–class theoretician and engineer in the field of gyroscopes and theoretical mechanics. Magnus quickly acquired an understanding of the state of struggling Russian gyroscopic platform technology and announced he would take on all gyroscopic problems. He summoned Dr. Hans Hoch, his colleague and friend at Hettingen University, who was a theoretician and brilliant experimenter in automatic controls. Magnus convinced Hoch to stay and work

Evening gathering of Institute RABE officers at the Villa Franka. Sitting (from left to right): L. A.Voskresenskiy, unidentified person, V. A. Bakulin, V. I. Kharchev, V. P. Mishin, Yu. A. Pobedonostsev. Standing (from second left to right): N. A. Pilyugin, A. G. Mrykin, S. G. Chizhikov, V. S. Budnik, S. P. Korolyov.

with the new Russian/German team at Bleicherode. Hoch was assigned to work on general stabilization theory and the "Mischgerät" laboratory. *A Mischgerät (mixing unit) was an amplifier–converter on the circuit connecting a missile's command gyroscopes with its control–surface actuators.* Unfortunately, Hoch's career would be cut short with his death in Moscow in 1950 from suppurative appendicitis. But Dr. Magnus would go on to become a renowned scientist and gyroscope expert, whose works were translated into many languages. His book *Kreisel*, published in 1971, is still a first–class textbook for those studying gyroscopic technology. Other prominent Germans who joined the team included Dr. Manfred Blasing, who was entrusted with the directorship of the control–surface actuators laboratory. Professor Waldemar Wolff took over the ballistic calculations associated with missile flight. Soon thereafter, an aerodynamics expert from Dresden, Dr. Werner Albring, joined Wolff. He took on the aerodynamic problems of missile flight in the atmosphere.

However, the Russian manhunt was still lacking in authentic Peenemünde missile specialists. Major Chertok, who had excellent connections, set up a maverick secret program to recruit the best German rocketeers over to the Soviet side. Chertok recruited his friend, Engineer 1st Lt. Vasiliy Kharchev, to head up this new program. Kharchev had recently graduated from the N. Ye. Zhukovskiy Air Force Academy in 1944, and possessed exceptional abilities and a penchant for coming up with new technical ideas. While the ideas weren't always feasible, they were very interesting and original. At Kharchev's suggestion, the program was code–named *"Operation Ost"* (East). Soon Operation Ost succeeded in enticing an authentic specialist on the combat firing of V–2 missiles, Fritz Viebach, to join the team. Then the Americans unexpectedly gave Operation Ost a boost.

"What the hell?" an incredulous Yuriy Novikov asked, more to himself than anyone. As the town commandant of Bleicherode in the Russian zone, he was supposed to keep order. His patrol had just stopped two Jeeps with American soldiers. A young, gagged German frau was sitting in the back of one of the Jeeps sandwiched in between two of the men. In the commotion she managed to pull the handkerchief from her mouth and started shrieking in German. Then two other women, who had been held down in each of the Jeeps by the

Americans also started to struggle loose, only adding to the ruckus. With weapons raised, the Russian patrol started to calm things down. Novikov understood from the young frau that the Americans had apparently burst into town and were trying to abduct German women. Novikov and his deputies had no choice but to arrest the lot and take them to the commandant's office. Once there the German women felt safer, but the Americans really started to raise Cain. Four American officers, each trying to out–yell the others above the din, were communicating through two interpreters that Novikov had brought in—a German interpreted from English to German and a Russian lieutenant from German to Russian, and vice–versa. Finally Novikov served the Americans some tea and offered them some Kazbek cigarettes. Then Chertok and his comrades strode in with some excellent German cognac, some snacks, and things began to calm down. The protesting Americans explained that the German women were simply the wives of German rocket engineers who were supposed to be sent to America with their spouses. At this point, Maj. Boris Chertok introduced himself as the Soviet representative for German missile specialists. "Why don't you come with us to the Villa Franka in Bleicherode and join us for refreshments," he said through the interpreters. The tired Americans responded with an "Okay," and the cortège set off for the Russians' villa.

Through their German connections, the Russians arranged for an excellent banquet and plenty of refreshments, including the alcoholic kind. Soon everyone was smiling. After numerous toasts, the Russians managed to get the Americans quite tipsy (Russians are renowned for their alcoholic tolerance). Excitedly slapping each other on the shoulders, both sides declared their friendship. The disarmed Americans volunteered that in September and October 1945 all of the German specialists that the Americans had named as war criminals would be sent from Witzenhausen, just outside the Russian zone, via France into the United States. Of course, everyone suspected that they would be subjected to interrogation. The problem was that several of their wives or mistresses had gotten left behind in the Russian zone, in particular in Bleicherode, and the German men had categorically refused to go without them. On behalf of the Army command authorities, the Americans requested that the Soviets help them return the German women to their spouses or lovers across the border.

The next day an agreement was reached. They composed a protocol in Russian and English and signed it. An American representative would present the Bleicherode town commandant with instructions as to which women he wanted transported from Bleicherode and the surrounding areas; this list would include the names of the German specialists associated with the women. The Russians would permit the transfers, but only under the condition that a Soviet officer be present to hear each woman voluntarily agree to leave—especially if she had children. 1st Lt. Vasiliy Kharchev was introduced to the American major who was taking part in the transfer. The operation to send off five women and three children took place peacefully, without any yelling. Kharchev explained to the alarmed women that they were free to act as they wished. He said the Soviet command was prepared to petition for their return to Bleicherode if they did not like it in the American zone. What could these women do? After all, they were not indigenous, but had been evacuated from Peenemünde, and their landlords were glad to see them go because it was considerably more profitable to house Russian officers.

The personable Kharchev quickly befriended the American officers guarding the Germans in the border towns of Worbis and Witzenhausen, and established rapport with the women. A Russian network of hastily recruited female "agents" sent a report that the wife of a German specialist wanted to meet with the Russians. The

Hellmut Gröttrup

meeting was set up near the border. Irmgard Gröttrup, a tall blonde in a light–colored sport suit, appeared with her 8–year–old son. "If we run into any trouble I'll explain that we were out walking and got lost," Chertok explained to his cohorts. Strictly speaking, Operation Ost had not been officially sanctioned by SVAG nor approved by his superiors in East Berlin.

"I want to make it clear that I will make the decision, and my husband will go along," the statuesque Mrs. Gröttrup peremptorily declared. *Is she a liberated female, or what?* Chertok thought. "You should know that I hate fascism, and I hate Hitler even worse. I have even been arrested on a number of occasions, and my husband as well. What will you promise us if we come over to your side?" To Chertok, it sounded as if she wanted to call the shots, and grab the initiative away from the Russians, who after all had just won the war. "That depends on who your husband is," he responded. "What position did he hold, and what expertise does he possess?" "My spouse was von Braun's deputy for missile radio–control and for electrical systems as a whole. He is prepared to come over to the Russians under the condition of complete freedom." "I need to know his name," Chertok responded. "And I need to receive the consent of a general in Berlin before we can give you an answer." "His name is Hellmut Gröttrup," she said. "First, we would like to meet with Herr Gröttrup," Chertok countered. "Well, you should hurry because we might be sent to America in a week or two," Frau Gröttrup proclaimed. Chertok conducted two negotiation sessions with Dr. Gröttrup. The good Doctor held out for a stipend of 5000 Marks per month (a princely sum at the time) plus a villa. Three days later, Chertok and Kharchev pulled off the transfer of the entire family: papa, mama, and the two Gröttrup children—without the consent of Berlin of course.

The Gröttrups settled in a separate villa and were offered a very high salary and food rations compared with those of the other Germans. But there was one condition: to keep order and to take part in the creative work of restoring missile technology, the Russians' own specialist Colonel Kuteynikov, who had a good knowledge of German, would live with them at the villa. The primary interaction with the Soviet management should go through him. However, the Institute RABE's German directorate and staff were not happy with this arrangement. They were unhappy that the institute that they had just created was being transferred to an associate of von Braun who would drive out everyone who disagreed with him. Gröttrup was clearly better informed than the others about the operations at Peenemünde. He had been close to von Braun, and he spoke very skeptically about the German contingent at RABE, except for Kurt Magnus and Hans Hoch. The others he simply did not know. To avoid stirring up passions, Isaev and Chertok agreed to create a special "Gröttrup Bureau" at the institute. Its first task was to compile a detailed report on the development of A–4 missiles and other projects at Peenemünde.

Helmut Gröttrup would prove out to be a prize catch for the Russians in the cooption of German rocket technology. But Frau Gröttrup was another matter. She grew even bolder than she had seemed during the first meeting. Soon thereafter she acquired two cows, "for the children and to improve the nutrition of the institute's Russian management." She managed to obtain orders for products in extremely short supply, which the stock clerk Semyon Chizhikov grudgingly had to pay for and deliver. Then an unexpected report from Colonel Kuteynikov sent the Russian team into a state of shock. Attached to the villa where the Gröttrup family had settled was a stable. The Frau was impatient to put it to use as it was originally intended. And so one night two rather decent horses appeared there. Colonel Kuteynikov, a man already advanced in years, evaluating the situation from all sides, reported that the Frau wanted to go horseback riding, not with her husband, but escorted by himself!

A CLOSE CALL

Recruiters continued combing the occupied zones for German rocket engineers. Not long after Herr Gröttrup's recruitment, Lt. Kharchev suggested that they go after Wernher von Braun himself. But upon asking permission for such an audacious operation, the local Russian military officials immediately distanced themselves, fearing a scandal with the allied forces. So this would have to be an officially unsanctioned, maverick operation. The local Russian authorities would look the other way of course. After all, von Braun, they allowed, would be a prize catch. Even if Lt. Kharchev and his miscreant cohorts were caught, what could happen to them? A slap on the wrist? Chertok, Nikolay Pilyugin, Leonid Voskresenskiy, and Kharchev met in secret and decided to take the risk. The daring–do Kharchev would drive up to the American border checkpoint and announce that he needed to cross into Witzenhausen to pay a courtesy visit to the American officers who had befriended the small contingent of Russian rocketeers in the Villa Franka where Wernher von Braun had stayed before. Then presumably Kharchev would go on a clandestine hunt for von Braun on the American side, hopefully before he was shipped out by the Americans.

Kharchev arrived all cheery at the border and offered to exchange wrist watches with the Americans. Then he parted with the only bottle of genuine Russian vodka in the entire Russian garrison. Surely this would endear him to the Americans. But Kharchev underestimated the Americans' hospitality in trying to pull his ruse off. "You are so kind. Let us take you into town as our guest," said the American master sergeant. "I want to introduce you to our commandant, Colonel Schraeder. A small contingent of Americans led the chagrined Russian army lieutenant like royalty to the commandant's personal quarters, a comfortable villa appropriated by the occupying American forces. Three soldiers escorted him up to the second floor. One of the guys knocked softly on one of the doors. "Uhh, who is it?" asked a nervous voice on the other side. "Sir, we have brought you Lt. Vasiliy Kharchev, a first lieutenant in the Russian army, who wanted to meet and greet his American friends," said one of his escorts. "Oh, come on in!" exclaimed a palpably relieved voice. Kharchev walked in with trepidation. The others quietly closed the door behind him, chuckling to themselves.

Kharchev couldn't believe his eyes, and had to almost avert his gaze. In the middle of the large bedroom was a wide king–size bed. On one side was Col. Rick Schraeder, in his hastily put–on skivvies. On the other half of the bed he saw an absolutely beautiful German blonde. She had hastily pulled up the bedsheet to cover her nakedness, but left one boob showing. Between the couple was a German shepherd. To all appearances they were having a late breakfast. There were bottles and all sorts of food on the bedside table. Then of all things, Col. Schraeder threw off the covers, shooed the dog away, and unequivocally told his astonished visitor to hop in! The German frau looked even more lovely in the buff. "For a Russian officer, my border neighbor, I begrudge nothing!" declared Schraeder with glee. "Aggh…uh…uh…" Lt. Kharchev stood speechless, his face turning redder by the second. Resisting the temptation with all his might, he stated in broken English that he was there 'on business.'

Finally, throwing on a robe, the good Colonel went out with Kharchev into the adjacent study. In good spirits, he invited his honored guest to share some of the best whiskey to be had in post–war Germany. Having come this far, the persevering Kharchev was not about to give up on his intelligence mission. He simply *had* to find out where von Braun was. *Who knows, maybe he wants to come to our side*, he thought to himself. To maintain camaraderie, Kharchev obligingly shared the bottle of whiskey with his host. Then they started on a bottle of fine cognac. Before he got in too big a fog, Kharchev good–naturedly argued his case. "We need to share the German specialists. They are war spoils," he pleaded. After a few minutes of banter back and forth he slyly turned the conversation to von Braun. To his chagrin the commandant just as good–naturedly explained that a special coterie of Americans was in charge of guarding him. "He is the most important war criminal we have," Col. Schraeder stated, "and is being very heavily guarded. We suspect that he was a Nazi as well." Kharchev knew he had to back down. A lowly lieutenant dare not butt heads with a full Colonel. So after the conversation was ended, Col. Schraeder dispatched his unexpected guest in good spirits. The Americans put Kharchev back into the same Jeep they brought him in with, and took him back as quick as the wind to the border checkpoint. Lt. Kharchev groggily hopped into his own vehicle. An hour later, he was red–facedly reporting his misadventure to Maj. Chertok, Pilyugin, and the others. At last, Pilyugin saw fit to reproach him. "What would have happened to us all if you had given into temptation? The commandant or somebody else there would have taken pictures of you in bed, and then presented the photos to 'the right people'! We could all end up in the clink!" Then Major Chertok added insult to injury. "You idiot! You let the king of rocketry himself, Wernher von Braun, slip through your fingers? You stupid…" "Wait a second," the hapless Kharchev protested. "Colonel Rick Schraeder said that he was under tight guard by the Americans, and I am just a lowly lieutenant in the Red Army…" "Enough!" exclaimed Pilyugin. All that poor Kharchev could do was to look down at the floor. After this reprimand, they put him to bed. Then Chertok and his other buddies burst into laughter at their friend's misfortune.

This story would eventually spread like wildfire among the Russian rocketeering community. Most of the listeners would guffaw and laugh. Korolyov in particular, his sides splitting with laughter, would declare, "What a fool Kharchev was in not accepting the American offer!" But in all seriousness, Korolyov did not lament that von Braun did not come over to the Russian side. Gröttrup was so highly regarded, and was being treated so well, what would happen if the much younger von Braun had turned up to compete against him? Upon learning of Lt. Kharchev's misfortune, Gröttrup was amused but eventually had to set the record straight. "Von Braun would try to foist his own ideas on us. Plus, it is simply out of the question that he would voluntarily come over to us. Yes, he is an excellent engineer, organizer of ideas, and a talented designer as well. But remember, like Colonel Schraeder said, he had also been a baron, a member of the Nazi Party, and even a *Sturmbannführer* (equivalent to the rank of major in the SS). He and his mentor, General Dr. Walter Dornberger, met several times with Hitler and received high state honors. Reich leaders Goebbels and Kaltenbrunner kept close tabs on von Braun's work. Unfortunately for him, ever since his youth von Braun has dreamed of space travels under the influence of Hermann Oberth's works, but he was forced by circumstance to apply his talent to purely military objectives."

The Russians would go on to find out that Wernher von Braun had been named technical director of Peenemünde–Ost (East) at the young age of 25, a rarity for the Germans. Von Braun was highly valued by the Third Reich, for his talent, self–initiative, and rare intuition, at least as regards the field of rocket science.

For his and Kharchev's failed attempt to kidnap Wernher von Braun, Russian rocketman Boris Chertok was fingered by his colleagues, only partly in jest, as the 'man that lost the moon race.' Even so, to this day, Russian historians blame von Braun for being a member of the Nazi Party and creating weapons of mass destruction (the A–4 and its progeny, the V–2). Historians have recorded Russia's enormous suffering at the hands of the Germans and the dreaded Third Reich. Over 23 million deaths. Going by the Soviet Union's post–war boundaries after the Yalta conference, this increased to an estimated 26.6 million casualties, easily an order of magnitude greater than those of the United States. The Russians believed Helmut and Frau Gröttrup when they said they had no idea of the horrors that took place on the forced slave laborers at the Mittelwerk. However, von Braun and the Peenemünde production personnel had been in Nordhausen more than once, and they of course must have seen everything. In fact, Helmut Gröttrup recalled that von Braun, during the hurried evacuation of Peenemünde, was clearly afraid of being captured by the Russian side. "We fired missiles against England," von Braun had said, "but it is the Russians who will extract revenge."

THE BUILDING YEARS

By August 1945 reconstruction of the V–2 had begun in earnest. The Zentralwerke's V–2 engine static test stands were put back in operation. On August 9, Chertok and Isaev led 150 German rocketeering specialists (out of a few hundred in East Germany) to Bleicherode. There a new group of 284 Russian scientists and engineers joined them. This stellar group included Korolyov, Glushko, Pilyugin, and two outstanding engineers: Vladimir Barmin and Vasiliy Mishin. Together these teams would form the nucleus of post–war Russian rocket development. In Prague a special train used to support V–2 flight testing was located, and Mishin was sent to fetch it. While drawings of most of the V–2 were located, not all could be found. Therefore Bleicherode set up a bureau to completely reconstruct the missile's technical drawings. It was

then to be put into production for the Red Army. By December 1945, 600 German and Soviet specialists were at work at Zentralwerke. In March 1946 the first two complete V–2's were ready for test. At the engine test stands German and Russian specialists tweaked the standard V–2 engine for better thrust and performance.

May 13, 1946. Boris Yevseyevich Chertok eyed the secret decree from the Kremlin with fascination. He was demonstrably a talented and pioneering guidance and control engineer. *So, they want to move our work deep in Russia, away from the prying eyes of Western observers, he thought to himself.* The secret decree, signed by Josef Stalin himself, created an array of new research institutes on Soviet soil with the objective of comprehensively exploiting German rocket technology. Stalin was astute enough to ensure that all the basic disciplines for developing rocket technology in the Soviet Union were covered. Dmitriy Federovich Ustinov, Head of the Ministry of Defense Industry, was put in charge of the entire effort. The new institutes were:

- NII–88, at Podlipki, General Gonor in Command, Sergei P. Korolyov Chief Designer, for rocket design

- OKB–46, at Khimki, Valentin P. Glushko Chief Designer, for duplication of the V–2 engine

- NII–885, Mikhail S. Ryazanskiy Chief Designer, Guidance Systems

- NII–885 at Monino, Nikolay A. Pilyugin Chief Designer, Control Systems

- NII–10, Viktor I. Kuznetsov Chief Designer, Gyroscopes

- GSKB, Vladimir P. Barmin Chief Designer, Launch equipment

- General Vasiliy I. Vosnyuk, Commander, Soviet Rocket Test Range (site to be determined)

Meanwhile, the unsuspecting German engineers and scientists at *Zentralwerke* pressed on to tease out the best possible performance from the captured or reconstructed V–2s. They conducted a series of 40 engine test firings between July and September 1946 using modified propellant mixer heads and varying mixture ratios. Through relatively simple measures they managed to increase the thrust of the basic V–2 engine from 25 to 30 metric tons. By the summer of 1946 5,000 German and Soviet workers labored on construction of the V–2 throughout East Germany (the newly partitioned Soviet zone). Buyers were sent to the Western zone and France to purchase certain necessary parts from the original manufacturers. The parts were smuggled to *Zentralwerke* without much trouble. By September 1946 the first ten 'N' series new–production V–2's were completed. Five were equipped for atmospheric measurements, and five for radio guidance tests.

Shortly thereafter, the Soviet leadership decided it was time to put their secret plan into action. Following months of preparation, the Soviet secret police struck on the night of Tuesday, October 22, 1946. A huge party was thrown for the 234 German specialists, with plenty of toasting in the Russian tradition in order to put the attendees under the weather. At 4 a.m. in the morning the hung–over specialists were awakened by Red Army soldiers banging on their front doors. They were asked to immediately 'volunteer' for five years work in the Soviet Union. On the following morning the specialists, their families, and their belongings were loaded on trains and shipped to Russia.

When the specialists first arrived in Russia, they were assigned to work closely with the Russian teams in Khimki, Monino, and Podlipki. Slowly they were all moved to Gorodomlya, an island used for research into animal diseases (and possibly biological warfare) in the 1930's. A first group of 234 German specialists was moved to Gorodomlya on May 22, 1947 and given the task of designing a 600 km range rocket, the G–1. Work had begun on this already in Germany but the initial challenge in Russia was that the technical documentation was somehow still 'in transit' from the *Zentralwerke*. The other obstacle was Russian manufacturing technology, which was equivalent to that of Germany at the beginning of the 1930's. By May 1948 the entire German team was in Gorodomlya, and no longer privy to what the Russians were actually doing with their designs, or what progress they were making. The first Soviet V–2 test stands were built in the Crimea, using equipment taken from Peenemünde. Later the focus of the effort moved to Zagorsk, nearer Glushko's primary facilities.

Following construction of the initial batch of 10 'N' series V–2 missiles in Germany, and the removal of available German rocket specialists to Soviet territory, Stalin spent some time before deciding what to do with them. It was not until July 26, 1947 that he issued a decree for testing of the captured or rebuilt V–2 missiles in the then new Soviet test range at Kapustin Yar. The Russians completed their rocket test stand in September 1947 with

The victorious Allies, principally the U.S., Great Britain, and Soviet Union scrambled to seize as much German rocket hardware as possible in the months after World War II. This photo shows a U.S. soldier guarding a captured Mittelwerk V–2 rocket motor in the German Dora production plant.

the objective of having the first launch by October, to coincide with the 30th Anniversary of the 1917 October Revolution. However the Soviet technicians could not get the rocket's igniter to work due to miswiring of the electric starter. In desperation, in mid–September 13 German technicians were loaded on a train and brought to the secret test site. They arrived on September 28, but the wiring problem was not identified and corrected until October 15. The first rocket successfully thundered aloft on October 18, although it immediately veered into the wrong direction. The Russians suspected German 'sabotage'. Nevertheless by November 13 all ten of the rockets had been launched, but only five were successful. A follow–on launch of 10 'T' series rockets, completed in Germany at Kleinbodungen, was just as dismal. Again, only five of them worked, with the others demonstrating a maximum range of 274 km and 86 km altitude.

The German team was back on Gorodomlya by December 9, once more in the dark about the program. The secretive Soviets never informed the Germans of the production of the Soviet copy of the V–2, which they named the 'R–1', or of its first flight test series beginning in September 1948. Without the Germans' knowledge, Glushko found that by using 96% alcohol fuel in place of the V–2's 75%, the basic engine would deliver 37 metric tons thrust. He conducted experimental tests of these tweaked V–2s and the all–Russian RD–100 engines in 1948–1949. Official state trials tests were conducted at Zagorsk in 1950. Glushko went on to build a seven metric ton variant of the R–1 rocket engine, the ED–140, in 1951.

April 14, 1948. The Soviets issued another secret resolution to put the R–1 into production. Aleksander Shcherbakov was charged with closing the 15–year manufacturing technology gap between Russia and Germany. To accomplish this the resources of 13 research institutes and 35 factories were tapped. Glushko was tasked with producing the Russian RD–100 copy of the V–2 engine. Prototypes had already begun factory tests at the end of 1947, with stand tests beginning in May 1948. During the test series the Russians gradually took over the testing. The German team was not even aware that tests continued beyond the end of 1948. R–1 test flight trials were accomplished swiftly – ten in 1948 and 20 in 1949. Two amusing anecdotes deserve mention. First, on the evening of November 1, 1948 a particularly vigilant Russian sentry guarding a test R–1 thought he heard something in the fog and fired a warning shot at the intruder. The next day, the launch team detected a strong scent of alcohol. Upon inspection they found that the missile's entire tail section was drenched with flammable alcohol from the fuel tank. The sentry had not fired his warning shot into the air, but straight into the filled alcohol propellant tank. Second, the Russians were famous for celebrating with "Blue Danube" after successful tests or launches. Blue Danube was rocket fuel (ethyl alcohol) tinted with manganese crystals!

During this time period the German team at Khimki was kept busy completing a build of ten KS–59 'Lilliputian' (Lilliput) subscale versions of a radical new flat–plate injector combustion chamber with a 60 atmosphere (6,080 kilopascal, or 882 psi) chamber pressure. The first engine was completed at the end of 1948. These were designed to test the design planned for Gröttrup's future G–4 missile. From the summer of 1949 to April 1950 100 tests were made of the engine. A wide range of propellants were used, including exotic fluorine compounds and suspended beryllium hydride fuels. The Lilliput endured them all.

The Cold and Korean wars influenced the Russians to hurry their missiles into production and deployment before they were even ready. After WWII Korea had been partitioned into two parts. Russian troops occupied Korea north of the 38th parallel of north latitude, an imaginary line cutting the country in about half. American troops occupied Korea south of the 38th parallel. Both North

and South Korea claimed the entire country, and their troops clashed near the border several times from 1948 to 1950. Nevertheless, the United States removed its last troops from Korea in 1949 and indicated early in 1950 that Korea lay outside the main U.S. defense line in Asia. The Communists believed the time was right for military action. On June 25, 1950 troops from Communist–ruled North Korea invaded South Korea without warning. The Communists almost succeeded in conquering the entire country, reaching the Pusan perimeter in September 1950. General Douglas Macarthur led the famous Inchon landing, a surprise move that turned the tide in the Allies' favor. UN forces mounted a counter–offensive, and advanced as far north as the Yalu River by October 1950. In November the Chinese entered the war, and by January 1951 retook all of North Korea and then some. Fearing an infectious spread of democracy throughout East Asia, the Soviets decided to accelerate their indigenous missile development.

However, at mid–20th century, it was not easy turning prototype missiles into reliable flight articles. Rocket technology was the equivalent of a toddler's first steps. A couple of steps forward, then one backward, then an accident and he falls. First, the materials. In the production of the A–4 missiles, the Germans used 86 brands and gauges of steel. In 1947, Soviet industry was capable of replacing only 32 grades with steel that had similar properties. The Germans used 59 brands of nonferrous metals, and the Russians could manage to find only 21 of them domestically. Most difficult of all were the nonmetals: rubber, gaskets, seals, insulation, plastic, and so on. The Russians needed 87 types of nonmetallic materials, but their factories and institutes were capable of providing only 48.

Then the failures. Many on the ground, and even more on launch and flight tests. The R–1's pyrotechnic igniter, located on a special device made of wooden strips in the combustion chamber, was supposed to burn a portion of the alcohol mixed with liquid oxygen vapor. After this, upon issuance of the 'preliminary' command, a significantly greater quantify of fuel was fed to the fire that had formed, and in seconds the preliminary stage was to form a steady, roaring jet. Then the 'main' command occurred. The main Lox and alcohol valves opened at full flow. The main–stage jet was generated with its characteristic roar, thrust increased, and the missile took off from the pad. However, on one test, instead after the ignition command the engine made a powerful "pop" sound much louder than a rifle shot. The shock caused the firing circuits to reset. The R–1 stubbornly refused to come off the launch pad. On the next test Nikolay Lakuzo manually retracted the armature of those relays that had popped loose during the pop. This forced the engine to ignite and go to 'mainstage' ignition, but like a bucking horse the missile rebelled. It suddenly pitched forward and went into horizontal flight, taking the heavy launch pad with it. All the observers jumped into open trenches that had been dug beforehand. Twenty meters later the pad dropped off and the missile went 10 kilometers horizontally, then it dove into the ground. Everything around the launch site was burned or swept away by the firestorm. Examining the mutilated launch pad, Glushko said sarcastically, "I didn't think that my engine could make launch pads fly too." The cause was identified as powerful pops from micro–explosions of fuel entering the combustion chamber after the 'ignition' command. Ground crews thought they eliminated the problem by fixing all defects in the ground–based cable network, and they verified correct operation of all electrical circuits on a dry missile. After several tries the missile still wouldn't take off even after the firing circuits seemed to work correctly. Almost all of 1950 was spent on the experimental development of a new shock–free liquid ignition system to replace the antiquated pyrotechnic one to battle the pops.

On another occasion, launch crews found that the lubricating oil in the bellows assembly of the main oxygen valve had stiffened in the cold, preventing its operation. They had to remove the main oxygen valves from all the missiles and send them to the factory in Khimki for degreasing. It wasn't all Glushko's fault— electricians, guidance specialists, and others had to share the blame too. Finally it took 21 instances of the engine failing to go into main stage ignition to get nine missile launches. And this was just the first of several series of launch tests.

Scenarios like the above were repeated numerous times, not just with the R–1 but also its successor the R–2. Despite these events the first production R–1s were accepted for service on November 25, 1950. In December Stalin rushed the first operational unit, the 92nd brigade (BON RVGK), to Kapustin Yar in December 1950. Each brigade was equipped with six launchers. In January 1951 the 23rd brigade deployed to Kamishin in Volgograd Oblast. Further deployments of this pathfinder unit were to Belokovorovich, Ukraine; Shyalyay, Lithuania;

Dzhambul, Kazakhstan, and Ordzhonikidze, the Far East, and the Primorsk area. The 77th and 90th brigades were formed at Lvov, Khmelnitskiy, and Zhitomir, Ukraine. In 1952 R–1 series production was moved to factory 586 in Dnepropetrovsk, and the first unit rolled off the assembly line in December.

All this sounds very good, but the R–1 was hardly an operational threat. In field service the rocket required 20 vehicles and four kinds of liquid propellants for the main engine, turbines, and starter (liquid oxygen, alcohol, hydrogen peroxide, and aqueous potassium permanganate). Six hours were required to prepare the rocket for launch, and its target accuracy was horrible: it could be as much as eight km off in range, and four km laterally. Soviet Red Army Generals also objected to letting their troops work with a rocket using alcohol for a propellant! The number of units fielded was small, reflecting the long delay in getting the missile into production. The field equipment was designed to also be used for R–2 missiles, which quickly replaced the R–1 in the field units. Even though the R–1 was almost immediately superseded by later designs, the effort laid the groundwork for the burgeoning Soviet rocket industry in the early years of the Cold War. Surplus R–1's were converted to use as a sounding rockets for military and scientific research missions: the R–1V, R–1D, and R–1Ye. The difficulties confirmed the Germans' 1946 assessment once again that Russian industrial technology was 15 years behind that of Germany. Despite the threatening supervision of the program by Stalin's secret police chief, Lavrentiy Beria, and the assistance of German rocket engineers, it would take <u>eight years</u> (1956) for the German technology to be fully absorbed and the missile to be reliable in service. This timing was almost identical to the American experience in exploiting German rocket technology after the war.

With all this help from the hardworking Germans, the Soviets managed to develop their first ballistic missiles for military purposes a few years ahead of the Americans. Wernher von Braun and some 115 German rocket scientists would unproductively languish at Fort Bliss, El Paso until 1950 when they were moved en masse to Huntsville, Alabama. This gave the Russians a head start in developing and flying their first missiles, as cumbersome and unreliable as they were, a few years ahead of the United States. In conjunction with the Germans they developed a series of missiles in rapid succession between 1946 and 1953, with the design, development, and testing times often overlapping each other. Korolyov designed the R–2 missile in 1947–1948 in competition with Gröttrup's G–1. The increased–thrust engine allowed the R–2 rocket to be stretched, with a lift–off mass 50% greater than the V–2. The Russians incorporated other improvements to double its range over the R–1 and V–2. In both the United States and Russia the development of ICBMs capable of reliably delivering nuclear weapons was still in its infancy, so they chose instead to arm the R–2 with a deadly Geran radiological warhead that dispersed a radioactive liquid at altitude, resulting in a lethal 'radioactive rain' falling in a wide area around the impact point. At least the R–2 design used undrinkable methyl alcohol in place of the ethyl alcohol used in the V–2 and R–1, eliminating the problem of the launch troops drinking up the rocket fuel!

The German and Russian teams competitively defended their respective G–1 and R–2 designs in late December 1948. The Soviet State Commission evaluating them found the G–1 as the superior design. Korolyov fought the decision for a long time, updating his R–2 design to include some of the G–1's features, such as the use of an integral fuel tank, and improved accuracy through a radio–controlled engine cut–off system. Finally the decision was 'reversed' and Korolyov's design was accepted for test. State trials flights were conducted from September 21, 1949 to July 1951. Barmin had already designed the mobile launch complex to accommodate either the R–1 or R–2 missiles. The R–2 system was accepted into the military on November 27, 1951. A 30 November 1951 decree authorized series production at factory 586 at Dnepropetrovsk. The first R–2 launch by a military unit was made with a prototype rocket in 1952. The first production rocket was rolled out there in June 1953, only six months after the first R–1 came off the assembly line at the same plant.

Official Soviet accounts and memoirs gloss over the German contribution to early Soviet rocketry. In the early post–WWII years, competitive rocket design competitions were held between indigenous Soviet teams (headed by Sergei Korolyov and German teams (headed by Helmut Gröttrup). In each case the German design was found superior by Soviet state commissions. In some cases, the Russians even used German drawings without modification, the German text being erased and Russian text substituted! German–led guidance teams under Gröttrup developed a radio–corrected guidance technique that the Soviets adopted for their first generation of ballistic missiles. Glushko used German–developed engines as the basis for those of the same missiles. The Germans had mastered the development of integral common–bulkhead propellant tanks, a feat it would take over a decade for the Soviets to duplicate because of their inferior manufacturing technology— lagging some 15 years behind the Germans. Germans also led the placement of the liquid oxygen tank forward of the fuel tank. It was *German* aerodynamic analysis that came up with the unique conical rocket configuration later adopted by Korolyov for the R–7 ICBM and the huge N–1 moon rocket.

Like its predecessor, the R–2 could hardly constitute an operational threat. In field use a crew of 11 was required for launch preparations. The radio–correction system required two truck–mounted stations. It took six hours to prepare the rocket for launch, including 15 minutes for the guidance system to be programmed. Once prepared, the rocket could be held in a ready–to–launch condition for only 24 hours before it had to be defueled and recycled. Like the R–1, the R–2 could be off in hitting its target by almost 7 kilometers! Nevertheless the Soviets deployed R–2s in rocket brigades equipped with six launchers (three divisions per brigade, each division with two batteries). The 54th and 56th brigades were formed for test launches at Kapustin Yar on 1 June 1952. Following receipt of production missiles in 1953, divisions were deployed to Zhitomir; Kolomiya; Medved, Novgorod Oblast; Kamyshin, Volgograd Oblast; Siauliai, Lithuania; Dzhambul, Kazakhstan; and Ordzhonikidze, in the Far East.

Versions of the R–2 for suborbital manned flight were studied by Korolyov in 1956–1958, but it was decided instead to move directly to orbital flights of the Vostok. However, some equipment tested on the R–2 found its way onto canine flights of Sputnik and Vostok. Reveling on their Sputnik successes, the Russians signed an agreement on December 5, 1957 to license production of the R–2 to China. A large team of Russian rocket engineers and technicians went to Beijing to set up the production line. They would transfer rocket technology know–how for the next four years. The R–2 provided the technological base for subsequent Chinese rocket programs.

Korolyov had been authorized to develop a longer–range IRBM back in April 1947, at the same time as the V–2–derived R–1 and R–2 designs. However, the R–3 specifications called for a monumental leap in technology: deliver a 3000kg atomic bomb to any point in Europe from Soviet territory, a range of 3000km. Again, Korolyov found himself in direct competition with a German design to the same specifications: Gröttrup's G–4. In selecting a final R–3 design, Korolyov examined and discarded several alternatives. These included use of 3 G–1s in a parallel–staging 'bundle' (or 'packet' as the Russians would later term it); winged missiles along the lines of the wartime A9/A10 or Eugen Sänger's Antipodal Bomber (the G–3); or sequential stages, as used on the earlier G–1 and G–2 designs. Korolyov even considered using balloon tanks of high quality steel (as would be adopted in the United States for the Atlas ICBM). But the Russians did not have such alloys available, and while the Germans had such technologies they kept the idea to themselves, 'in reserve'. The R–3 involved so much new technology that it was deemed necessary to build an R–3A intermediate experimental rocket, based on the R–2. This would be flown to test new construction methods, guidance systems, and higher energy kerosene instead of alcohol fuel. Glushko proposed to evolve the V–2 approach by using 19 of the Gröttrup team's newly designed ED–140 seven–tonne thrust chambers as 'preburners' to feed a main mixing chamber, producing a high–performance engine—the Glushko RD–110—of over 100 metric tons thrust. However during development problems of stability in the mixing chamber could not be solved, and this approach was abandoned (Glushko's designs for huge, high-thrust engines like the American F-1 on the Saturn V would be plagued by combustion chamber instability problems in future years, forcing the Soviets towards smaller, lower-thrust engines). Instead he scaled up the ED–140, as the Germans had planned. The first attempt was a leap to 65 tonnes thrust for the RD–105/RD–106 for the R–7. But this also proved unachievable, and Glushko settled for four chambers of 25 kilotons thrust each in the RD–107/RD–108 engines. Glushko adopted automatic thrust regulation in flight, a feature first introduced by Gröttrup's German team for the G–1 engine. Although the configuration was new and the performance spectacular, the per–chamber thrust was still the same as the V–2.

Once again, the Soviet State Commission ruled in favor of Gröttrup's G–4 configuration in place of Korolyov's R–3. Isaev for one found extensive problems with Glushko's engine design. The huge increase in thrust, performance, and use of new propellants seemed a leap too far. Somehow the ingenious Germans out–engineered Korolyov. Gröttrup's team finished their 20–volume design in June 1949, a mere three months after go–ahead! Their cylindrical single–stage design aerodynamic design was superior and stable in all flight regimes. They switched the Lox tank to the forward position in comparison with the V–2, eliminating instrument and engine pre–cooling problems. Also, Gröttrup's G–4 design involved fewer technical advances in engine design but greater improvements in payload mass fraction, a very key rocket performance parameter. **A missile's payload mass fraction is the ratio of the payload it can deliver to a target (or to orbit) to its total mass sitting fully fueled on the launch pad.** The R–3's design resulted in a payload mass fraction of only 4.2% to the G–4's 5.1%. A related parameter is the missile's **mass ratio: the ratio of its weight at burnout (including payload) to its total weight on the pad before liftoff.**

Propellant Mass Fraction and Residuals

A missile's propellant mass fraction (not to be confused with payload mass fraction) is the ratio of propellant consumed during flight to the missile's total weight on the pad before liftoff. The propellant consumed is also the total weight (called the gross weight) before liftoff minus its weight empty at the end of engine burnout. Of course, much like a car, a rocket never totally empties its fuel tanks. The amount of propellant left in the tanks (both fuel and oxidizer) after burnout is called the residual propellant. It can run from 3% down to 0.3%, depending on the expulsion efficiency of the tanks. It is like your car. Say you have a 20–gallon (76 liter) gas tank and you go until the car runs out of gas. The car could have between a pint to 2½ quarts (0.23-2.3 liters) of unusable fuel in the fuel lines or sitting at the bottom of the fuel tank. A related rocket parameter is called ullage. Unlike automobiles, liquid–fueled rockets must have an extra volume of gas above the propellant in sealed tanks. This space is necessary to allow for thermal expansion of the propellant liquids, for the accumulation of gases that were originally dissolved in the propellant, or for gaseous products of slow reactions within the propellant during storage. Ullage runs several times higher than the residual propellant by volume, usually between 3% to 10% of the tank volume.

The major reason the German design was better was that the missile (without propellant and payload) only weighed 2,760kg, whereas Korolyov's design had to lug along almost twice that amount, a whopping 5,480 kg. So even though the R–3 out–performed the G–4 as far as *propellant* mass fraction was concerned (88.2% to 90.8%), its structure was so heavy it lost out on the much more important measure of *payload weight* it could land on a target.

In the end neither the G–4 nor the R–3 missiles were put into production or deployed. Both became obsolete, superseded by better and newer designs. However the design concepts of the G–4 led directly to Korolyov's R–7 ICBM (essentially a cluster of G–4s or R–3As) and in the 1960s the N–1 superbooster Moon rocket. Notable unique features of these rockets borrowed from the G–4 work included conical rocket body forms, forward positioning of the oxygen tanks, elimination of drag–producing aerodynamic surfaces, and 'sharp point' warheads or payload shrouds.

The R–5 missile originated as "Theme N–1" of the original R–3 project requirement for a 3000 kg / 3000 km range IRBM. The original R–3A subscale technology demonstrator was authorized in 1949, before being cancelled in turn in October 1951. As a replacement, the Ministry of Defense Industry (MOP) issued a decree on October 21, 1951 for work to start on the R–5 as Theme N–2. Due to the extensive work already done under Theme N–1, the 6 volume R–5 draft project was completed by Mikhail Yangel and delivered on November 30, 1951. In 1952 MOP issued an official decree formally authorizing development of the R–5 missile with a 1000 km range. The R–5 IRBM original missile was planned to be equipped with a Pilyugin gyroscopic guidance system and to be built in several versions.

As alluded to above, the Soviets did not develop their missile and rocket programs by starting with the R–1 and progressing in strict numerical order to the modern R–110 liquid–propellant unguided surface–to–air missile (SAM). Rather, many designs with different R–numbers were often developed in parallel, with some being discarded, others being improved, receiving a new R–designation, and so on. As with the United States and its aircraft designations; SR– (strategic reconnaissance), F– (fighter), B– (bomber), C– (cargo), P– (pursuit), the Russians ascribed the simple word for rocket (ракета) to the R– designation. The word реагирующий (meaning "reactive" in Russian) was used to denote both jet and rocket propulsion in the 1930s and 1940s. They still use this word today to denote "jet propulsion" as in the phrase "jet propulsion engine." In at least two cases the Russians also ascribed the word for reconnaissance (разведывательный) to aircraft (the R–3 and R–5 aircraft). Early Soviet rocket designations also clouded the distinction between German and derivative Russian designs, for example they skipped over the R–4 and derived the R–3 design from the German G–4. These are summarized in the table and figure below.

> *Surface–to–Air Missiles (SAMs) are designed to knock out airborne targets. Ballistic missiles (sometimes called surface–to–surface (SS) missiles) are intended for ground targets. They can be divided into four classes. (1) Short range, artillery–type; (2) medium–range ballistic missiles (MRBMs); (3) intermediate–range ballistic missiles (IRBMs); and (4) the longest range—intercontinental ballistic missiles (ICBMs). There was, and still is, an ill–defined kind of blurry distinction between MRBMs and IRBMs. Roughly speaking, MRBMs have ranges between 500–1000km (310–620 mi) and IRBMs range from 1000–5500km (620–3420 mi), but these ranges can overlap. ICBMs have traditionally formed the mainstay of deliverable nuclear megatonnage (MT), with ranges exceeding 5500km (3,420 mi). There are at least four other kinds of missiles: (5) Submarine–Launched Ballistic Missiles (SLBMs) launched underwater from submerged submarines; (6a) Surface Launched Cruise Missiles (SLCM) generally fired from a sub or destroyer; (6b) Air Launched Cruise Missiles (ALCM) fired from aircraft; (7) Air–to–Air Missiles (AAM); and finally (8) Air–to–Ground missiles (AGM).*

Soviet R–1 to R–5 MRBMs and their German counterparts

Type	German Designation	Russian Equivalent	Propellants
1 stage MRBM	V–2	R–1	Lox/75% Ethyl Alcohol (drinkable)
1 stage MRBM	G–1	R–2	Lox/96% Methyl Alcohol (undrinkable)
1 stage IRBM	G–2	None	Lox/75% Ethyl Alcohol (drinkable) cancelled
1 stage IRBM	G–4	R–3/R–3A	Lox/Kerosene (cancelled)
1 stage IRBM	None	R–5/R–5M	Lox/96% Methyl Alcohol (undrinkable)

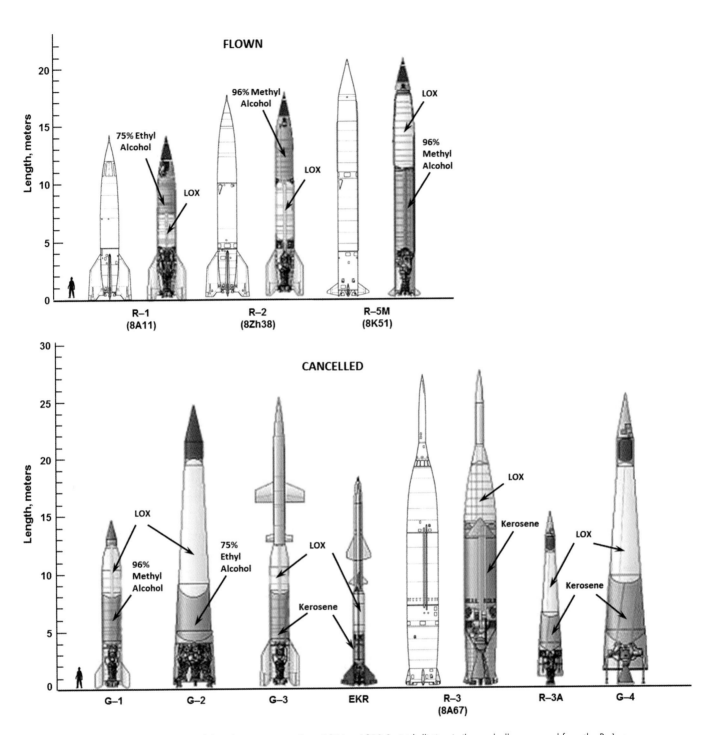

Early Soviet missiles (R—1 to R—5) and their German cousins. From 1946 to 1953 Soviet ballistic missiles gradually progressed from the R—1, a copy of the German V—2, to the R—2 and the R—5M, the first Soviet MRBM capable of a nuclear warhead. Each featured a single engine. The R—3 missile was an ambitious long—range project that was abandoned in 1951 although it allowed Soviet engineers to adopt and abandon certain key technological paths.

In the early days of military rocketry, from World War II until the early 1950s, most missiles were single–stage MRBMs. As technology advanced, single or two–stage IRBMs were developed (mid–1950s); then multi–stage ICBMs (late 1950s and 1960s). SLBMs came into widespread use following the first ICBMs. ALCMs and SLCMs came later (with the exception of the V–1 'buzz bomb' which could be considered a cruise missile since it was powered by a pulsejet). At first ballistic missiles were armed with either high–explosive or radiological warheads. Radiological warheads were used on the R–2 and R–5 prior to the availability of nuclear warheads. These so–called lateral warheads were designed for dispersal of radiological liquids over a wide area in the impact zone. Similar weapons were discussed at Los Alamos during World War II when it seemed that the atomic bomb would not be ready in time or not work. In this case radioactive waste could be dispersed over a wide area, effectively providing the fallout effects of a nuclear bomb without the blast and fire effects. With the availability of nuclear technology, most ballistic missiles became armed with nuclear warheads. Today's offensive strategic missiles are almost all of ICBM range, whereas all but the most modern SLBMs have historically been of intermediate range. Thankfully, such monstrous weapons have served such a strong deterrent role that they have never been used.

R–Series Missiles in the Soviet Union

1950 - R-1 the first ballistic missile
1952 - R-2 with detachable warhead
1955 - R-11 for ground forces
1956 - R-5M with a nuclear charge
1958 - R-11M mobile missile
1959 - R-11FM for submarines
1960 - R-7 the first intercontinental ballistic missile
1960 - R-7A
1965 - R-9A
1968 - RT-2 solid-propellant rocket
1972 - RT-2P

Just about the time that the Americans began to make good use of their German rocket scientists, the Russians did just the opposite and began to repatriate their Germans, starting on April 3, 1951. In contrast to the American approach, where the German expatriates were offered leadership positions in rocket development and eventually afforded U.S. citizenship status, the Russians increasingly treated them as second–class citizens. By

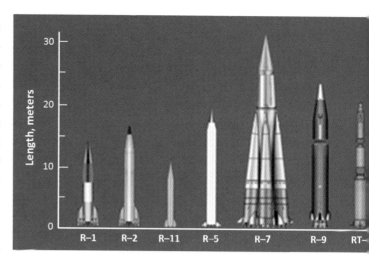

October 1951 the Germans were completely isolated on Gorodomlya Island and their productive work basically stopped. They were sent back to East Germany in several groups. The first batch left in December 1951, and the last group in November 1953. Much like the Captain of a sinking ship, Helmut Gröttrup and his family were among the last to leave the island, on November 22, 1953. Eventually Gröttrup made it to West Germany and was debriefed by the CIA in 1957, but he provided some deliberately false information and downplayed the importance of the German work in order to avoid Russian retribution. The full story did not come out until the end of the century when the Cold War was over.

German rocket technology gave the Soviet Union a tremendous boost in reaching nuclear parity with the United States after World War II. Pushed by Stalin, the Russians worked at a feverish pace on three major R&D "themes" (Russian–speak for technology development projects) in the 1948–1953 timeframe. Various Soviet organizations enabled these themes in: (1) designing a single–stage missile (the R–3) with a range of 3,000 kilometers; (2) developing a missile using storable *(meaning they can be kept in tanks at room temperatures)* propellants, and (3) conducting exploratory work on an intercontinental ballistic missile. These study projects were known as N–1, N–2, and N–3, respectively. On termination of these R&D themes, two new R&D studies were performed in 1953–55. The T–1 project focused on future ICBMs, and T–2 was dedicated to developing cruise missiles, respectively. All five of these studies were carried out simultaneously and in coordination by several leading missile development organizations, including NII–88, NII–4, and other institutes of the Soviet Academy of Sciences. This research led directly to the eventual creation of the first Soviet ICBM, the famous R–7 in 1957.

December 5, 1953 "Why must you insist on not using hydrogen as a fuel?" angrily voiced Korolyov. He was growing more impatient with Glushko by the day.

Glushko countered back. "Look, comrade, I have made tremendous progress with these complex and obsolete German engines, and it has been hard enough to deal with the plumbing nightmare of the V–2!" he bellowed. Valentin Glushko was generally soft–spoken, and it was known to almost everyone that he never raised his voice. In fact, he usually spoke in almost a whisper. Korolyov was astonished.

"It has taken us years, years! to first improve upon, and then develop our way past the heavy, complex, and undercooled German engines. It is very easy for you to say, 'use hydrogen because it has better performance,' but you are not the one spending nights and weekends trying to understand the combustion instability it engenders in the chamber. Sometimes it performs flawlessly, and then for no apparent reason the combustion process develops violent pressure oscillations that we can't control. You seem to think that propellant injectors function like those of the carburetor in your car. They do not. The injector performs many other functions related to the combustion and cooling processes, and is much more important to the function of a rocket engine than a carburetor is for an automobile engine. There is only one optimum condition for each of the following parameters: orifice size, angle of impingement, angle of resultant momentum, distance of the impingement locus from the injector face, number of injector orifices per unit of injector face surface, flow per unit of injection orifice, and distribution of orifices over the injector face. Worst of all, these parameters must be determined experimentally, Sergei Pavlovich! There is no magic formula! On top of that the pressure drop across the injector—another parameter we cannot predict—has a strong influence on combustion instability. Any pressure disturbance across the injector face, and poof! Combustion instability tears the engine apart. Sheesh! Rocket science is just such a difficult discipline … " His voice trailed off, and the expression on his face was so sad.

It still rankled Glushko that several years before— August 1946—while still in Germany, his old rival Korolyov was appointed chief of a department in the newly created NII–88 in Podlipki, northeast of Moscow. This organization was made responsible for the development and industrial production of missile technology based on German hardware.

"Well, alright then, if you must use liquid oxygen as the oxidizer and kerosene as the fuel, why can't you find a substitute for the hydrogen peroxide?" Korolyov asked. "It is very unstable. If you look at it cross–eyed, it breaks down and can create an explosion."

In addition to the main propellants, the Germans were the first to use refined hydrogen peroxide (H_2O_2) as an effective way to start rocket engine turbopumps. Not the 3% solution you buy in the store, but a concentrated 90% solution. It is an extremely strong oxidizing agent that can attack skin and eyes. Drying of spilled H_2O_2 on clothing can even cause fire. When injected with a catalyst like potassium permanganate ($KMnO_4$), H_2O_2 violently breaks down into water and oxygen. The hot steam–like gas vapor formed during this process in the closed chamber of a gas generator is directed at a turbine, which then drives the high pressure pumps feeding the main propellants—liquid oxygen and kerosene—into the engine combustion chamber. But this alone is not enough to start the engine. Since liquid oxygen and kerosene do not ignite by themselves, you need an igniter. During the start sequence, an electrical signal is sent to an initiator which releases the energy of a small amount of sensitive powdered pyrotechnic housed within the initiator, commonly called the squib or the primer charge. Next, the booster charge is ignited by heat released from the squib, and finally the main ignition charge propellants are ignited which sets the engine off. This whole train of events happens in milliseconds, faster than the human eye can blink.

Glushko took a seat across from Korolyov's desk. "Why must you be so imperious, Sergei" he asked. "You of all people should know that peroxide is the most efficient substance known to produce the high pressure steam it takes to drive the fuel and oxidizer pumps. Our engine tests show that the higher we can get the pressure of the mixing propellants in the chamber, the better the performance. Or do you not wish us to produce the best engine for the Motherland that we can?"

While both rocket engineers glowered at each other, Aleksei Isaev knocked on the door to Korolyov's office. Korolyov motioned for him to come in, welcoming a break in the impasse. "Is something amiss, comrades?" he asked. Silence. "Well, then I have great, great news!" he announced. "Valentin Petrovich, I can provide you with thin–walled copper combustion chambers, backed by steel supports to prevent them from bursting under pressure. This will save you a hell of a lot of weight, and provide better heat conductivity along the chamber walls in the bargain." Glushko's face warmed. At least someone brought in good news today.

"Also, we have been spending 18–hour days struggling with the chugging problem…" "Chugging problem?" Korolyov broke in. This was the first time he had heard of 'chugging' in connection with rockets. "You mean as in the sense of a steam locomotive chuffing along the track? You can't be serious, Aleksei Mikhailovich," he chuckled. "No, no, not at all like a train chuffing along the track. It is more like a kid jumping up and down on a pogo stick, but quite a bit more complicated and the pogo stick weighs many tonnes. Also, the frequency is ten times faster than any kid could jump on a pogo stick." Glushko felt impelled to explain. He loved showcasing his superior knowledge to Korolyov in the nitty gritty details of rocket engines.

"After years of research and testing, we still don't fully understand combustion instability. Low frequency instability, we have seen this to vary between 10 and 400 cycles per second. It seems to set in when combustion pressure oscillations in the combustion chamber couple with the propellant feed system, and then start to amplify and feed on each other within seconds. So, pretend that the entire missile, loaded with propellants is like a huge pogo stick. The kid jumping up and down, this is the combination of combustion and propellant feed lines instability. You could pretend that the coiled spring inside the pogo stick is bending or flexing of pipes, joints, bellows, or the long propellant tanks themselves. In fact, I think that the coupling could involve the whole vehicle and the propellants in the tanks as well, as they build up longitudinal oscillations. The pogo phenomenon, uncontrolled, could destroy the entire missile."

"Then what progress have you made in controlling the combustion instability in the chamber, which seems to be the root of the problem?" Korolyov asked him. His dark eyes, which had sort of a merry sparkle, looked at Glushko with a mixture of curiosity and attentiveness. "We are still doing experiments with injectors of many different types. The injection hole pattern, impingement pattern, hole distribution, and pressure drop all have a strong influence on combustion stability. We have abandoned the old V–2 showerhead type injector which employed nonimpinging streams of propellant emerging normal to the face of the injector. It relies on turbulence and diffusion to achieve mixing, which doesn't work as well as injecting impinging streams of propellants into the chamber. Remember, the injector acts like a

carburetor in a car. The injector has to introduce and meter the propellant flow to the combustion chamber, and atomize and mix the propellants in such a manner that a correctly proportioned, homogeneous fuel–oxidizer mixture will result, one that can be readily vaporized and burned."

At this point, Isaev realized it was time to showcase another of his great innovations in front of the Chief Designer. "Sergei, it may surprise you to learn that after numerous hot–fire ground tests, I can offer you a hard–won solution. My design group has designed a flat injector plate with mixing–swirling injectors. So you don't need separate fuel and oxidizer lines feeding into each sprayer orifice in the injector plate. I can show you a model of this new type injector here." He held up a prototype injector that was split in half to show how the two feed lines met in a Y–type intersection, with small swirling vanes machined inside the lower straight part of the Y. "It works!" Isaev proudly beamed. Glushko had suspected that Isaev was up to something, and now he was amazed. "What you have done, Aleksei Mikhailovich, is effectively cut the number of lines feeding propellants into each sprayer in half." The old V–2 engines required separate lines for oxidizer and fuel feeding into each sprayer orifice in the injector plate. The

Valentin Glushko a preeminent Soviet designer of rocket engines for missiles and launch vehicles, complemented Korolyov's skills in space vehicle design by providing him with the means to propel his vehicles. Although they had their differences, the two men respected each other.

plethora of lines resembled a forest behind the injector plate. Isaev's department enormously simplified this plumbing problem by essentially cutting the number of lines in half. Now the injectors would do their own mixing. Since lox and kerosene are not hypergolic (i.e., self–igniting on contact), rocket engineers could feed both propellants through the same line into the injector spray heads. "At last, I can dispense with the obsolescent German V–2 engines altogether!" Glushko exclaimed.

However, as much as they wanted to, neither Korolyov nor Glushko could shake off the German heritage of Russian rocket science just yet. In fact, Korolyov soon found himself caving in and adopting the superior German aerodynamic designs with fervor. The German G–4 concept had died with the departure of the Germans for East Germany. Korolyov did not push for R–3 production beyond the prototype stage, but rather used it as a technology springboard for the R–5, another missile he designed that was of German technology vintage. Like the R–3, it was never deployed.

SHROUDED IN SECRECY

During the Cold War many of Russia's towns and cities (at least 55), including some of its largest, were 'closed cities'. Anyone with a foreign passport was forbidden to enter, and many were out of bounds even to Russian citizens. These closed cities provided the technical foundation for Soviet military technology including chemical, biological and nuclear weapons research and manufacturing, enrichment of plutonium, space research, and military intelligence work. This meant that large numbers of highly qualified scientists and researchers were concentrated in these geographical areas, developing new technologies but isolated from the international research community. The workforce, all carefully screened and highly cleared individuals, were divided into the three classes: At the top of the heap were the elite nuclear theoreticians and physicists who conceived new concepts for weapons and blazed the trail for advances in nuclear weapons technology, like how to make the weapons smaller and more potent. The creators of the Soviet atomic bomb included the later

famous nuclear physicists Yuliy Borisovich Khariton, Yakov Zeldovich, Samvel Grigorevich Kocherants, and Andrey Dmitriyevich Sakharov. However in the mid–1950s even their very names were tightly guarded state secrets. Sakharov would later become known as the "father" of the Soviet hydrogen bomb. He became an outspoken advocate for human rights and reform in the Soviet Union before his death in 1989. Next in line were the weapon designers and engineers, who needless to say had to work closely with the elite group. The third group was the biggest by far: the technicians who actually got their hands dirty with fabricating, assembling, and testing weapon subsystems and the entire weapon itself.

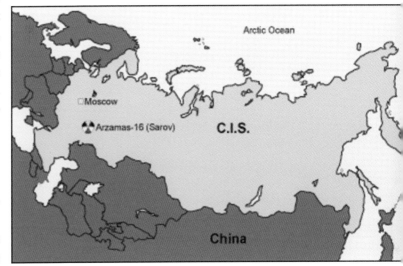

Sarov (population 92,100) is about 231 miles (372km) southeast of Moscow. The holy city originally took its name from the nearby Sarova monastery next to the Sarovka River. Until 1995 it was known as Kremlyov, while from 1946 to 1991 it was called Arzamas-16. Today the town is still off limits to foreigners as it is a major Russian center for nuclear research.

Such "secret cities" were known only by a postal code, identified with a name and a number. Originally, the number following the city was the distance in kilometers the facility was located from the city. In practice, the numbers were in some instances arbitrarily assigned, and changed from time to time, to obscure the actual location of the installation. Ten "non-existent" cities were dedicated to nuclear weapons development and production. One of these, the All–Russian Scientific and Research Institute of Experimental Physics (VNIIEF) was collocated with the small city of Sarov. VNIIEF was initially known as Arzamas–60, a postal code designation to show that it was 60 km south–southwest from the city of Arzamas. But the "60" was considered too sensitive, and the number was changed to "16." In fact, in 1947 the entire city of Sarov (Arzamas–16) disappeared from all official Russian maps and statistical documents! It was only with Glasnost and the fall of the Soviet Union in 1991 that all of the major heretofore secret cities were opened for collaboration in civil research. The slow process of breaking down the barriers of secrecy had begun.

The Soviet nuclear establishment at first underestimated the problem of creating high precision machine tools (recall that Russian manufacturing technology lagged the West), automation, and the complicated electrical instrumentation for nuclear warheads. A search for suitable facilities included the selection of the Ministry of Aviation Industry Factory No. 25. *By the way, the United States adopted a similarly simple numbering system, for example the Lockheed Martin Joint Strike Fighter manufacturing plant in Fort Worth, Texas was originally designated "Air Force Plant No. 4."* Yu. B. Khariton for one considered it unacceptable to develop critical nuclear weapons technology in a ministry other than his own. He maneuvered to have Factory No. 25 transferred from the Aviation Ministry to the Ministry of Medium Machine Building. The factory was converted into Branch No. 1, and was collocated with the KB–11 design bureau in the then top–secret city of Arzamas–16 which did not exist on any map. KB–11 had been formed in 1946 as the lead design bureau for developing the first Soviet atomic and thermonuclear weapons. In 1954 General Nikolay Leonidovich Dukhov was appointed the director and chief designer of Branch No. 1. He was already distinguished as a three–time Hero of Socialist Labor and former chief designer of the IS series of super–heavy tanks during WWII (the IS–1, –2, and –3 were named after Iosef (Joseph) Stalin).

After arriving at his new assignment General Dukhov knew he had his work cut out for him. Working on the relatively new field of state–of–the–art nuclear technology was a welcome change from the drudgery of engineering the many designs of tanks that would have to defend the Soviet homeland from incursions or even a full–scale attack by the NATO Allies through the so–called "Fulda Gap" in Germany. *Ironically, the West also had nightmares about hordes of Russian tanks going the other way and pouring through the Fulda Gap.* But unlike America, large portions of the Russian homeland had been invaded and occupied by foreign powers in the not too distant past: Napoleon, whose unruly army sacked and burned Moscow, and Hitler who besieged and starved out Leningrad, and came within a hair's breadth of sacking Moscow itself. Dukhov was very hands–on and a quick learner. He mastered what was for him a completely new field of technology extremely quickly, but not without the help of nuclear weapons physicists like Kocherants. *What better way to learn*, he thought, *than to ask and watch the experts themselves.* You certainly couldn't find this stuff in any textbook! On a cold, snowy day in February 1954 Dukhov button–holed Kocherants after his first staff meeting. He offered his colleague a Kazbek cigarette, and both men lit up. "Samvel Grigorevich, I need you to give me the 'Big Picture' on a typical atomic bomb design."

"Well, O.K.," Kocherants replied. "To make a warhead, the designers and technicians form a sphere from a noncritical mass of plutonium and then—" Dukhov raised his hand and interrupted him. "Whoa! Wait a minute. Excuse my ignorance but I thought that the atomic fission process requires uranium. Why plutonium?" Kocherants tried to hide his surprise. *Sheesh!* he thought to himself. *Where has this guy been?* After a moment he continued. "There is no uranium in the atomic bomb. It is the plutonium that forms a critical mass and explodes. This is a question of nuclear reaction physics. As you know, nuclear *reactors* use uranium. We produce plutonium from the uranium, but it is a complex and labor intensive process. To make a warhead, we first form a sphere from a noncritical mass of plutonium. Then we surround the sphere with a solid spherical mass of TNT or a mixture of TNT and another conventional explosive. The technicians have to finish the outside surface with great precision and install a large number of fuses that have to be actuated synchronously within a time window of microseconds. The explosion of the powerful conventional explosive is directed so as to form a converging spherical blast wave that squeezes the globular mass of fissionable plutonium and converts it into a supercritical state. However, this still does not guarantee a chain reaction. In order to start a full–fledged chain reaction, we need one more detonator. The neutron detonator sprays an intense beam of neutrons in all directions at the precise moment – measured in nanoseconds – inside the collapsing sphere."

At this point Lieutenant General Academician Yevgeniy Arkadiyevich Negin passed by and decided to join the conversation. Kocherants asked him, "Yevgeniy, you know more about the neutron detonator. Why don't you describe it for General Dukhov?" Negin was only too happy to oblige and show off his knowledge. "Integral to the neutron triggering of the atomic bomb is a "neutron gun," a complex electrical device in and of itself. We trigger all the bomb's detonators simultaneously with high voltage, up to 20,000 volts. Most of the nuclear warhead's mass is not taken up by the weight of the active plutonium, but by the heavy steel hull which is necessary to confine the omni–directional neutron beam. We coated the inside of the steel hull with a specially made reflective beryllium layer to deflect the high energy neutrons." Negin and Kocherants carried on with design details until Dukhov's eyes began to glaze over.

A few months later General Dukhov was stretching his tired feet over a chair next to his desk. After working a 12–hour day in mastering the intricacies of nuclear weapons development he felt comatose. Just as he was about to drift off to sleep, Andrey Sakharov, somewhat apologetically, knocked on his open door. "Yes?" he managed through closed eyelids. "Nik, I must talk to you in private!" Andrey half–whispered. He quietly closed the door behind him. "Our inside contact in NII–1011 has learned that they are competing for the Article 5G49 contract."

For identification purposes, typically all Soviet armament systems were given a secret code in the form of a number/letter/number alpha–numeric designation. For secrecy's sake, all atomic or hydrogen bombs were called "articles". Besides the simple "article" concept, there was also the more complex *"article in its entirety" since the "article" equipped with nuclear explosive could only be actuated by the second part, a case stuffed with all sorts of automatic control devices and circuits. Article 5G49 was the "article in its entirety" for a new megaton class of H–bombs.*

Dukhov suddenly snapped wide awake. "You mean they have already done it?" he asked. "Yes. As we feared, the Central Committee Presidium has set up a competitor organization in Chelyabinsk–70," Sakharov replied. "From now on, they will be competing against us for every single contract to develop new generations of articles," he added. Without delay, the general called Yuliy Khariton, the scientific director of Arzamas–16, to tell him the bad news.

The next morning Khariton convened a meeting behind closed doors of key nuclear experts in the KB–11 design bureau. Khariton opened the meeting. "Igor, first, I want to hear the latest news on fatalities or injuries at NII–1011." Igor Vasilyevich Kurchatov was the trusted leader of the first Soviet atomic bomb project. He was sure that Stalin would have had him shot had the 1949 test failed. Now he was Khariton's inside man at Chelyabinsk–70, the closed facility where NII–1011 resided. "The records I was able to scan said that 16 personnel have been exposed to life–threatening doses of radiation," Kurchatov responded. "Of these, so far five have died. I don't know how many others will succumb." The process for obtaining plutonium was (and still is) complex and labor intensive. The plutonium

Photo sequence of the first Soviet nuclear test, code named "First Lightning". Houses, a bridge, a simulated metro, armored vehicles, 50 aircraft and 1500 animals were positioned in the test grounds. The huge 17,000 sq/mile test site was located 95 miles west of the city of Semipalatinsk on the steppes of present day Kazakhstan. The device would later be weaponized as RDS-1, the Soviet Union's first nuclear weapon. The code designation RDS was actually arbitrary, one popular interpretation was "Reaktivnyi Dvigatel Stalina" (Stalin's Rocket Engine), another was "Russia Does It Alone". *The whole focus of the Soviet program at this point was to set off a Soviet atomic blast at the earliest possible time whatever the cost.* At Project leader Lavrentiy Beria's insistence , the RDS–1 was an exact copy of a United States design called the "Fat Man", mainly thanks to extensive espionage.

was produced from uranium in nuclear reactors. Lack of knowledge of the dangers posed by radiation exposure in the 1940s/early 1950s resulted in quite a few fatalities from radiation exposure. Even many of the leading specialists who first developed the chemical process for extracting plutonium were subject to lethal doses in those early days. "Very well, then. Any comments?" Khariton asked the group. Nuclear scientist Kirill Ivanovich Shchelkin volunteered, "I think that the authorities will chalk up the radiation overdoses to lack of experience, and they probably won't penalize the Chelyabinsk facility as much as we would like." The Soviet government had set up a competitor organization, NII–1011 in Chelyabinsk–70, which competed through the 1960s with KB–11 in Arzamas–16 for contracts to develop new generations of thermonuclear weapons.

It did not take long for the Soviets to end the U.S. monopoly on atomic bombs. On August 29, 1949 they detonated *First Lightning*, a plutonium implosion atomic bomb at the Semipalatinsk Test Site. On the heels of the U.S., just four years later on August 12, 1953 the first Soviet hydrogen bomb was successfully tested (dubbed *Joe–4* by the Americans). By uncanny coincidence, the United States was developing its own city–busting hydrogen weapons almost exactly in parallel with the Soviets (on November 1, 1952 the Americans exploded a hydrogen–type device code–named Mike). Small wonder. Soviets had never really fallen far behind the Americans in the nuclear arms race, courtesy of spies like Klaus Fuchs, Harry Gold, David Greenglass, and the infamous husband/wife pair Julius and Ethel Rosenberg, among others (on June 19, 1953 both Rosenbergs were executed in the electric chair). In fact, by the 1950s both countries were running practically neck and neck in an escalating nuclear arms race.

FROM DICTATORSHIP TO TOTALITARIANISM

In some measure the rapid industrialization of Russia and its expansion after WWII to encompass the Soviet Union (aka Union of Soviet Socialist Republics – USSR) under one monolith can be credited to Joseph Stalin. During the latter part of Stalin's brutal dictatorship he presided over an empire that occupied nearly one–sixth of the Earth's land surface, with a land area seven times as large as India and two and one–half that of the United States. The USSR extended more than 6,800 miles (10,900 km) from east to west, covering 11 of the world's 24 time zones, and enjoyed the world's longest coastline and the longest frontiers.

1. Armenia
2. Azerbaijan
3. Belarus
4. Georgia
5. Kazakstan
6. Kyrgystan
7. Moldova
8. Russia
9. Tajikistan
10. Turkmenistan
11. Ukraine
12. Usbekistan

13. Estonia
14. Latvia
15. Lithuania
Baltic States

The USSR consisted of a tightly knit and controlled federation of 15 Soviet Socialist Republics (S.S.R.'s)– Armenia, Azerbaijan, Belorussia (now Belarus), Estonia, Georgia, Kazakstan, Kirgiziya (now Kyrgyzstan), Latvia, Lithuania, Moldavia (now Moldova), Russia, Tajikistan, Turkmenistan, Ukraine, and Uzbekistan. In addition, the Karelo-Finnish SSR was in existence until July 1956. And then there was Poland, home of the Warsaw Pact, where the first V-2 missile parts had been discovered towards the end of WWII. Ukraine represented a big chunk of USSR missile and rocket infrastructure in itself. The all–powerful Soviet state harnessed the social, technological, and economic resources of its SSRs, making it a peer superpower to the United States. The capital was Moscow, then and now the capital of Russia, the largest SSR. *Today's Russia by itself is still almost twice the area of the U.S.*

By 1952 Stalin's health was deteriorating. The long era of Stalin's iron rule (recall that Stalin derives from the word Russian word for steel, *stal*) was about to come to an end. His main problem was high blood pressure. As his grip on power weakened, the Politburo was abolished and was replaced by a larger Presidium of the Central Committee with 36 members. More stress was laid on "collective leadership" within this body.

Every year the Communist Party of the Soviet Union (CPSU) elected a 27 member Central Committee. In March 1919 the party decided this body was too large to determine policy. It was therefore replaced by a five man Politburo (increased to nine in 1925 and 10 in 1930). Its first members were Vladimir Lenin, Leon Trotsky, Joseph Stalin, Lev Kamenev and Nikolai Krestinsky. In 1952 the Politburo was abolished and replaced by a larger 36–member Presidium of the Central Committee, with 36 members. After the death of Joseph Stalin the Presidium was reduced to 10 members. In 1966 the Politburo received a new lease on life, and the Presidium was abolished.

At the end of February 1953 Stalin fell into a coma. After four days he briefly regained consciousness and the leading members of the communist party were called for. While they watched him struggling for his life, he suddenly opened his eyes and cast a glance over everyone in the room. It was a terrible glance. Then he raised his left arm. His nurse, who was feeding him with a spoon at the time, took the view that he was pointing at a picture showing a small girl feeding a lamb. But his daughter, Svetlana Alliluyeva, who was also at his bedside, took a different view. In her words

"He suddenly opened his eyes and cast a glance over everyone in the room. It was a terrible glance. Then something incomprehensible and awesome happened. He suddenly lifted his left hand as though he were pointing to something above and bringing down a curse upon all of us. The next moment after a final effort the spirit wrenched itself free of the flesh."

Stalin then stopped breathing and although attempts were made to revive him, his doctors eventually accepted he was dead of a cerebral brain hemorrhage. None too soon. While he was ill, Stalin had received a letter from a Dr. Lydia Timashuk claiming that a group of seven doctors, including his own physician, Dr. Vinogradov, were involved in a plot to murder him and some of his close political associates. He had the doctors named in the letter arrested and tortured, whereupon they confessed to being involved in a plot arranged by the American and British intelligence organizations. One of his last

dastardly acts was to order Lavrentiy Beria, the head of the NKVD, to instigate a new purge of the Communist Party. The members of Presidium began to panic as they saw the possibility that like previous candidates for Stalin's position as the head of the Soviet Union, they would be executed. Fortunately for everyone, Stalin died before these events could come to pass.

After Stalin's demise a power struggle ensued in the Soviet leadership. Beria became one of four deputy prime ministers, and maintained his hold over the former NKVD (now the MVD) security apparatus, which at that time combined both the secret political and regular police functions.

Like the multiple plot twists in Leo Tolstoy's labyrinthine novel, War and Peace, state–run organizations in Russia, the Soviet Union, and now the Russian Federation have historically undergone a very complicated maze of renaming, subordination, splitting up, or consolidation. Their names and abbreviations only make sense in Russian. Adding to the confusion is that these evolutions have been driven by the force of personality and political intrigue in the respective leadership du jour. Three of the principal organizational domains include (1) first and foremost state "security" apparatus for political control, suppression, and censorship all across the board; (2) design bureaus and scientific research institutes, including those involved with rocket science and technology; and (3) political leadership of the country, involving the Communist Party and the military as instruments of power. Unraveling these mazes and making sense of all the historical interplay between them, for example the design bureaus and scientific research institutes, would be a subject suitable for a Ph.D. dissertation.

In March 1946 all of the Russian People's Commissariats (NK) were redesignated as Ministries (M). Leaving out many details that would confuse even a historian, the NKVD (People's Commissariat for Internal Affairs) was renamed the MVD (Ministry of Internal Affairs) of the USSR. Think of the MVD as performing the "regular police" functions. Part of the former NKVD (the NKGB–the People's Commissariat for State Security) was spun off to form the separate MGB (Ministry for State Security). Think of the MGB as the performing the dreaded secret political and police functions. During the continuing struggle for power in post–Stalinist Russia, in March 1953 Beria tried to succeed Stalin as sole dictator by merging the MGB (secret police, run by Semyon Ignateyev) into the MVD (regular police, run by Beria) so that he could control both.

By July 1953, however, Beria had been defeated by an anti–Beria coalition who conspired against him. The coalition was led by a cabal of three men thirsting for power: Georgy M. Malenkov, Stalin's heir apparent; Vyacheslav M. Molotov (of Molotov cocktail fame); and Nikita Sergeyevich Khrushchev. Khrushchev maneuvered to gain an upper hand over his cohorts by his unparalleled control of the Communist Party machinery. In September 1953 he replaced Malenkov as First Secretary. Beria was arrested, deprived of his government and party posts, and publicly accused of being an "imperialist agent" and of conducting "criminal antiparty and antistate activities." Convicted of these charges at his trial in December 1953, Beria was immediately executed. His MVD was split in two. The reformed MVD retained its "regular" police and law enforcement powers, while the second, a new agency called the KGB (transliteration of "КГБ", Russian abbreviation of Комитет государственной безопасности —Committee for State Security) assumed fearsome internal and external security and intelligence functions, and was subordinate only to the Council of Ministers (somebody has to be in charge!). Who would emerge as sole victor of the remaining triumvirate—Malenkov, Molotov, and Khrushchev—would soon become known.

WORLD'S FIRST NUCLEAR DEATH MISSILE

It is no accident that the Soviets leapt ahead of the United States early in the Cold War Space Race. Their pursuit of excellence in rocket technology was relentless. Unfettered by the shackles of Stalin's dictatorship, the top Soviet leadership—the 10–member Presidium (aka the Council of Ministers) and the Central Committee of the Communist Party of the Soviet Union (CPSU)—decided to take a fateful step to ensure the primacy of Soviet nuclear power. They realized the urgency of getting an armada of nuclear–tipped missiles deployed before the United States. They communicated this urgency to the Soviet nuclear and rocket science establishments by issuing a Top Secret state decree on April 10, 1954 to proceed full steam ahead on development of thermonuclear weapons

and the means to deliver them against the United States and its Allies. As an interim step, the decree mandated the development of the R–5M IRBM as a means to deliver nuclear warhead strikes on Europe. The Soviet approach was to achieve noteworthy development milestones quickly through massive, concentrated effort, then move to the next milestone with minimal continued testing.

April 12 1954. An entourage led by Dmitriy Ustinov, Head of the Ministry of Defense Industry, hosted a joint meeting with Glushko, Korolyov and their associates. After being briefed by them on the status of Soviet rocket developments, Ustinov stood up. He had a boyish look about him, with a thick cowlick projecting forward from the top of his forehead. "Comrade Valentin Petrovich, the Ministry has decided not to press forward with your bulky and unwieldy concept. Instead, you and Korolyov are to proceed with the design, development, and testing of the R–5M. "But, Sergei and I have put in a lot of hours, days, weeks, and months evolving Gröttrup's G–4 design. We propose vast improvements. Instead of Herr Gröttrup's V–2–derived Lox/Alcohol engines, I have designed a cluster of four RD–110 engines burning higher–performance Lox/Kerosene around the G–4 core vehicle. At least on paper we can deliver a 3000kg atomic bomb to a range of 3000km, like was originally spec'd out for the R–3 in April 1947. This puts all of Europe at our disposal," protested Glushko. Ustinov countered, "The explosion of the RD5–6 400 kiloton warhead at Semiplatinsk on August 12, 1953 proved the design of a lightweight nuclear warhead. We don't need to deliver a 3000kg warhead. By the way, in our encrypted transmissions and official classified documents, it fizzled and gave a yield of 4 kT." A common Soviet disinformation practice was to give nuclear weapon yields incorrectly, by a factor of ten or in different units. The idea was that if you were a spy, you would be deceived, but if you were in the know you'd recognize the error and the reason for it. This deception worked more than once against American and British intelligence services. Now Korolyov spoke up. "Since September 1953 I have been working on a reentry vehicle for this warhead. The RV [reentry vehicle] uses a sharp–nosed 'fast point' in contrast to the German 'slow point' design—a blunt nose and conical sides. My heat transfer specialist thinks that he can resolve the thermal problems of the sharp point." Ustinov added, "A slow–falling warhead might be attacked by American fighters, or even intercepted by their ballistic missile defense. The secret part of our Five Year Plan calls for abandoning the Geran and Generator radioactive warheads in favor of much more powerful nuclear weapons." *The Geran was supposed to disperse a radioactive liquid at high altitude, but the Generator was even more deadly in that it used a number of small containers rather than a single chamber. Each of them was to burst above the Earth's atmosphere.*

Ustinov went on with his speech. His voice took on a commanding tone. "As of today, all R–5 and G5 work is to be suspended in accordance with the decree of 10 April [Russians always put the day before the month. Inviolable State "decrees" were the way the totalitarian Soviet government set state policy]. The R–3, R–4, and R–5 are to be put on a slow track to oblivion. The 8D52M engine [the code name for Glushko's RD–103M] will be used in the new R5–M single stage design to boost a one megaton warhead to 1200 kilometers, enough to strike at strategic targets in Europe. Korolyov has finally worked all the bugs out of its separable reentry warhead design. Korolyov, you and your teams will use the new southern facility at Dnepropetrovsk for development, design and full–scale production of the R–5M."

October 1954. A small cadre of Soviet rocket specialists led by Korolyov arrived at KB–11 to interact with the developers of nuclear warheads with the aim of developing the complete package of Soviet nuclear delivery systems. Unlike the United States, the Soviets almost exclusively relied on liquid–fueled missiles for their IRBMs, ICBMs , and SLBMs that would be launched from submarines. Korolyov's associate Igor Nikolayevich Sadovskiy and General Designer Dmitriy Ilyich Kozlov were named to lead the project for the nuclear–tipped version of the R–5.

Korolyov was no fool. He realized that it was paramount to improve the accuracy of Soviet IRBMs for them to pose a credible threat to the West. He supplemented Nikolay Pilyugin's gyroscopic guidance system with radio control of the pitch angle of the missile in flight. Later he and control specialist Boris Chertok installed a combined auto–stabilizing guidance system into the R–5M incorporating autonomous inertial control plus lateral radio correction. In–flight control of the missile was maintained with four aerodynamic fins located on the aft bay, and four jet vanes located on the perimeter of the single combustion chamber of the engine. The accuracy of the R–5M was "improved" to 1.5 km downrange and 1.25 km cross–range from the aim

point, for a CEP of 1.4 km, still horrendous by today's standards but quite an achievement for the Russians in 1956, the year the R–5M was deployed. This was a substantial improvement over the accuracy of the R–1 and R–2 missiles. *Circular error probable (CEP) is the radius of a circle into which a bomb will land at least 50% of the time.* Korolyov's R–5M design weighed almost 29 tonnes, over twice as much as the R–1, and could deliver a nuclear weapon over a range of 1200km, doubling the performance of the R–2. In the meantime the ultra–secretive Soviet nuclear establishment was working desperately to downsize their nuclear weapons, in particular H–bombs, from room–size to 1350kg, the maximum payload that the R–5M could carry.

The first phase of R–5M flight trials began on January 21, 1955 and continued into July. Of the 17 missiles launched, 15 missiles managed to reach the target. Two of the missiles deviated from the planned flight path by more than the seven degrees permitted and the range control officer had to shut the engine down using the APR —Avtomaticheskiy Podryv Rakety — system. The APR was a then–new automatic missile destruction system that the Soviets felt compelled to develop lest a nuclear missile stray from its path or otherwise malfunction. Of all systems the APR had to function flawlessly, and without scattering radioactive plutonium all over the place, let alone igniting a nuclear blast over their own territory. The second R–5M flight test phase in August–November 1955 consisted of 10 successful launches at ranges of 1083 to 1190 km.

Five R–5M IRBM missiles were selected for the final series of test launches beginning on January 11, 1956 from the Kapustin Yar launch facility. The payload sections of four were equipped with functioning mock–ups of a nuclear warhead. Actually these were not mock–ups since they were equipped with everything that was required for a nuclear explosion except for the products initiating the chain reaction. Ground personnel checked out the integration with the missile's systems, the preparation process, and the inflight operating reliability of all the automatic controls, including verifying the reliability of the APR through telemetry. The first four mock–ups launched normally. As luck would have it, the last launch, the fifth, was the worst of the lot. It succeeded only in demonstrating multiple failures in the missile's electrical and automatic controls systems. The launch team considered themselves lucky that the APR worked to destroy the wildly pirouetting missile after the stabilization controller failed.

Korolyov convened a failure review team meeting that night in his hotel room, which also served as his office. "Boris, didn't your team conduct the stand–alone tests on all the control systems?" he asked irritably. "Yes," Chertok answered. "Then what about the integrated horizontal tests on the entire missile?" "Well, sort of…" "Sort of?" Korolyov asked back. His voice grew louder. "You DID use the electrical equivalent of the payload section, right?" Chertok's expression looked like that of a fox caught in the henhouse by a farmer's bright searchlight. "Well, not quite, Sergei Pavlovich." Chertok sat down.

"Explain yourself. And it better be good. So help me, if you're responsible for delays in the missile's preparation and flight certification…it will be YOUR ass in the fire, after Nikolay Pavlov reports to State Commission Chairman Mitrofan Ivanovich Nedelin that the warhead was prepared for rollout, but that your team is causing a launch delay!" Korolyov was stretching the truth a bit but his face didn't show it. He knew it would be his neck on the line first if his team was responsible for launch delays. Also, he kept it to himself that the nuclear team's neutron gun had failed too. Chertok took a few moments to compose himself. "We did conduct thorough tests of the control systems, and all were passed. Except for an anomaly in the pitch axis of the stabilization control circuit. After the first test failure, Leonid Voskresenskiy conducted repeated tests of the stabilization controller over the roll, pitch and yaw axes. He could not duplicate the failure, so we called it an anomaly and certified the system ready for flight."

Voskresenskiy volunteered a technical solution. "I recommend replacing the amplifier–converter and repeating horizontal tests, but this will require three to four hours." "Great. Another delay." Korolyov shot back sarcastically. "Will someone please tell me why we lost temperature telemetry on the payload before the missile had to be destroyed?" Yevgeniy Shabarov, another member of Korolyov's failure review team spoke up. "I have been worried about our interfaces with the nuclear payload folks all along. Not the technical interfaces but the human. The nuclear prima donnas want to control everything. Nikolay Dukhov and his team of nuclear engineers, technicians, and physicists insist on maintaining tight control over their side of the interface.

Dukhov told me that 'all we want is an electrical signal from the missile that it was to be destroyed, and we will handle the rest.' The nuclear team assembles and prepares the warhead for integration with the missile in a special building with particularly high security. A tall fence surrounds the area around the building and it is guarded by special troop units of the KGB. I have never seen such tight security. Not only is the entire facility Top Secret, but there are special access rooms where none of our team is permitted. Aside from our on–the–job contact, we almost never run into the nuclear experts. They stay in a separate hotel constructed for them. Even when it comes to the motor transport service that the firing range services provide for us in full, they have their own."

Boris Pervyshin, a younger member of Korolyov's team, raised his hand to speak. "During the last launch, I wanted to double check the payload fairing's inner diameter and the spacing of its bolt holes to ensure a successful mating with the missile body. They told us that the nuclear "article in its entirety" would be tested at first without plutonium. So I didn't think it would be a problem. I started making my way toward the device when a hand grabbed my arm from behind. "Nyet." Two armed guards stopped me cold and told me to step back. 'But we have to integrate the "article in its entirety" with our missile,' I protested. They called General Dukhov in. He made me stand at attention and basically said, "Let me explain a few things to you. Our nuclear experts' work procedures call for a triple check of all assembly and testing operations. The head of assembly or testing has to hold the instructions and listen as the tester reads aloud the steps of an operation. For example, take the cover on your interstage adapter over here," he told me, pointing to an adapter ring on the forward end of the missile. "The tester says, 'Unscrew five bolts that secure the adapter ring cover.' The performer of the operation unscrews them. A third participant in the operation reports, 'Five such–and–such bolts have been unscrewed.' The controller, a military acceptance operative, has to report that he accepts the operation's execution. Then a notation to that effect is made in the appropriate document. Only after this can the entire team move on to the next operation. This is how we do it around here. Work goes slowly and scrupulously, with the mandatory reporting aloud about the execution of an operation and a notation to that effect in a special process logbook." Then one of the ever–present KGB agents added for good measure, "You are not to even come near the 'article in its entirety.'"

"Now let *me* explain a few things," Korolyov began. "During my first meetings with Khariton, Kocherants, and Gen. Dukhov, they announced that the conditions that would affect their 'article'—vibrations, loads, temperatures, and atmospheric pressure—needed to be cleared with them. They told me our team has to guarantee the proper conditions, not only in flight, but during all instances of ground preparation. General Dukhov told me that the "article" has problems with low temperature. It turns out that on days when the temperature drops below freezing it is necessary to install an insulating cover on the payload section and maintain a specific temperature. So you darn right they don't want the temperature sensor to fail in flight. We shall fix that."

"I will speak to Khariton and General Dukhov himself," he continued. "From now on the nuclear and rocket teams will have to work together on the payload section's interior layout, especially the various sensors and the telemetry system which determine the conditions that the warhead is exposed to in flight. Shabarov, I want you to maintain contact with the nuclear experts' facility and to observe them throughout the preparation of the entire payload section. I will speak with your namesake, warhead deputy chief designer Yevgeniy Negin to get you in. You are to report directly to me. I will get all of us moved to the new secret building for the firing range, which is only three kilometers from the R–5M launch site. This will be our "home" for the next few weeks. Boris [Chertok], you will be deputy technical director for missile preparation at the engineering facility. Leonid [Voskresenskiy], you will be my other deputy technical director, responsible at the launch site for preparatory operations and launch execution. Our missile team and General Dukhov's nuclear payload team have to ensure flawless integration between the missile and its subsystems on the one hand—not just structural, but also electrical systems, hydraulics, signals, thermal, and so on—and the entire payload. Not just the warhead. We will work hand–in–glove with Dukhov's team to ensure a successful launch each and every time. Understood?"

General Nikolay Dukhov, Yuliy Khariton, and Sergei Korolyov implemented the concept of "combat readiness" into the R–5M launch preparation process. Each team developed a process–oriented plan of actions for the missile, payload, and launch operations. Korolyov's process called for all sorts of tests before the nuclear experts attached the warhead at the launch site. They rehearsed all procedures not just during the day, but also at night using portable lights. Nikolay Pavlov was Korolyov's counterpart on the nuclear side. It was his job

From left to right: the R–1, R–2, R–5M in preflight with a launch gantry, and R–5M with launch gantry removed, leaving it sitting on a "launch table" for quick liftoff.

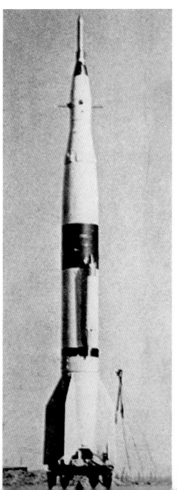

to supervise the preparation of the payload section and warhead. The work at KB–11 reached a fever pitch as the nuclear weapon designers struggled to keep up with the next R–5M launch which would be armed with a live weapon, the first nuclear test of Article 5G49.

Recovered R–5M IRBM reentry vehicle (the top stage with the nuclear warhead). Note the pointed heatsink nosecone design and the drag flaps on the rear. IRBMs do not have the range or altitude of ICBMs, consequently heat is less of a problem. However, the slow velocity of this warhead made it an attractive anti-air target.

February 2, 1956, *Kapustin Yar Area 4N.* Shortly before dawn the R–5M test missile, looking like Ichabod Crane's headless horseman without its payload section, was hauled to the launch pad in a special transport assembly along with the 8U25 portable launch stand, the same as the R–5. In those days the smaller IRBMs employed a portable steel launch table assembly, also called a 'firing table', which placed the missile and payload in a vertical position for launch. Pavlov's technicians placed the prepared payload section containing Article 5G49 'in its entirety' inside a special thermally insulated, specially guarded vehicle that delivered it to the launching pad. Korolyov's team performed the laying operations, poured electrolyte into the onboard batteries and installed them. Both teams integrated the payload horizontally with the missile body right at the launch site. Technicians checked out the "ground–to–missile" power switchover and the "abort launch" system in case it was needed. The transport assembly erected the missile and payload into a vertical position for launch. The launch team attached umbilicals, electrical cables, and checked all connections. Next came LOX and methyl alcohol fueling operations which took about two hours. Around 1:05 p.m. the standby–for–launch command was issued, then the final countdown sequence commenced. At 1:13 p.m. the R–5M thundered off the launch pad, the first nuclear–tipped IRBM in history.

Without breaking up this time, the R–5M delivered its 80 kT nuclear warhead a distance of 1200 km, from the Kapustin Yar launch facility (Area 4N) to a point near Priaralsk Karakum, 150 km northeast of the Aral Sea. The impact fuse went off and the surface nuclear explosion marked the beginning of the nuclear missile era, the 'hottest weapons' of history's 'coldest war.' No publicity followed this historic event. American technology at that time did not have the means to detect missile launches. Therefore, they recorded the nuclear explosion as a routine nuclear weapon ground test. It was heavily classified at the time that the prototype warhead was a fizzle when it exploded – planned yield was 70 kT, but actual yield was only 300 metric tons. The problem was traced to a failed heating element on the warhead. However, in some official documents the warhead's design and test yields were miscast as 300kT, to mislead the spies in accordance with Soviet practices of deception.

The R–5M was the end of the road in being the ultimate extrapolation of German V–2 technology in Russia. It was introduced into the Strategic Rocket Forces on June 2, 1956; about a year before the U.S. Air Force's 1,500–mile Thor IRBM would be declared operational. After it reached initial operational capability (IOC) the R–5M received the secret alphanumeric code name "Article 8K51." Soon Cold War planners would be able to choose from three warheads: a 40 to 80 kT nuclear fission warhead, a 300 kT boosted fission or fusion warhead, or a one–megaton (MT) blockbuster fusion warhead. The Soviets formed specially trained military engineering brigades due to the nuclear warhead. Launch preparations had to be made meticulously although the final launch procedure was automated. Initially it took 30 hours to prepare the rocket for launch, but after several years of service this was reduced to between 5 and 6 hours from the normal readiness condition. The rocket had to be launched quickly after filling its uninsulated liquid oxygen tank. A minimum operational R–5M field site required only a large presurveyed clearing with soil stabilization or possibly a poured or prefabricated concrete apron. The missile's allowable hold time in the most ready prelaunch condition (reaction time equal 15 min) was about one hour (whereas today's missiles can stay on hold almost indefinitely). The Soviets deployed a total of 48 R–5Ms between 1956 and 1957, primarily at sites close to the western borders of the Soviet Union where they could be aimed at the most important targets in Europe. In 1959 R–5Ms were put on alert for the first time, and the R–5M remained in service until 1967. No further deployment was carried out because the more effective single–stage R–12 IRBM subsequently replaced the R–5M missile.

The marriage of rocket science with nuclear weaponry heralded the Cold War doctrine of "mutually assured destruction," or MAD. Fortunately for the world, MAD would work for decades to avert a world–wide nuclear conflagration. And once again, military purposes were the main impetus behind the advancement of rocket science as we know it today.

RISE TO GREATNESS

Korolyov had reached a turning point. He had developed the first Soviet strategic missile with a nuclear delivery capability and (for the time) acceptable accuracy. Besides the R–5M, by this time Korolyov and the Soviet rocket establishment had begun working on the then–new short range R–11 MRBM and the R–11FM SLBM. They were also hard at work on R–5M improvements, the M–5RD and R–5R among others. However, the R–5M's reach was only 1200km. This was not enough. To feel secure, the Soviets had to threaten the American homeland itself with annihilation. On May 20, 1954 the Presidium and the Central Committee of the CPSU updated their April 10 decree with a resolution for the development of a new ICBM, to be given the lucky number seven (Semyorka) in the R–series, with a range of 8,000km (4972mi) and a 5500kg payload. The R–7 project was to take top priority. The secret decree also specified a maximum miss distance of plus or minus 10km (6.2mi). Korolyov was dumb–struck. He would have to use cryogenic and hard–to–handle liquid oxygen propellant because of its higher energy. And only the Soviet nuclear establishment knew that the old 3000kg nuclear fission warhead for the R–5M had been superseded by a new design. Tests during 1953 at Semipalatinsk had demonstrated the possibility of building an H–bomb of vastly greater power to defend against the Americans.

The total warhead mass would have to be increased to 5500kg, of which 3000kg would be the nuclear device itself. This is the weight of three compact cars!

However, Korolyov had a few things on his side in facing this daunting task. Much like the charismatic Wernher von Braun would do in America a decade later, he had the genius to marshal tremendous organizational efforts, pick the right people for the right jobs, foster inventiveness in himself and others, engage in collective brainstorming, and above all incentivize everyone to perform heroic work for the Soviet motherland. Korolyov was also greatly assisted by Valentin Glushko's superlative engine designs. Despite Stalin's excesses and working under a totalitarian regime in very primitive conditions relative to America, Soviet rocket scientists were every bit the equal of the Americans. In fact, somehow Korolyov and his associates were able to produce a draft R–7 missile design in two months, a record time even for the Soviets!

Sergei P. Korolyov at the Kapustin Yar firing range in 1953.

First, the Soviets had to find a suitable testing and launch range (firing range) for the R–7 and its successors.

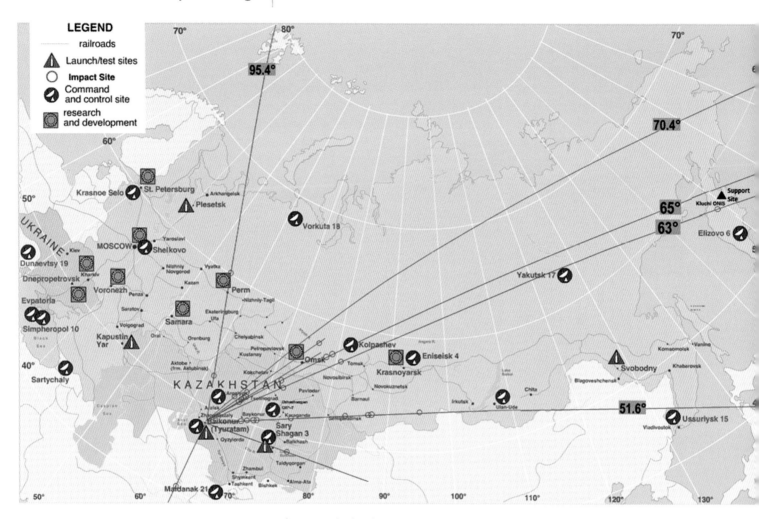

Map of Russia's rocket launch sites and operational research facilities.

A reconnaissance commission led by Vasiliy Ivanovich Voznyuk was set up for this purpose. Korolyov appointed Leonid Voskresenskiy to represent his interests. After some heated arguments, reconnaissance flights, poring over maps, and trips to four candidate sites; they adopted a remote location in the semi–desert region of Kazakhstan by the Tyuratam railroad station on the Moscow–Tashkent railway. The Tyuratam launch facility took its name from the nearby village of Tyuratam, in the Kzyl–Orda region on the bank of the Syr–Darya river. Construction of the Tyuratam launch center began near an abandoned open–pit copper mine and former Gulag prisoner labor camp on May 31, 1955. The new town of Zarya sprang up overnight about 35km south of the facility to provide housing, schools and support infrastructure for workers. The expense of constructing the launch facilities and the several hundred kilometers of new road and train lines made Tyuratam one of the most costly infrastructure projects the Soviets ever undertook. Like many Soviet military and space launch facilities its location and very name were state secrets. The Soviets soon changed its name to "Baikonur" to mislead the West by suggesting that it was near Baikonur, a small mining town about 320 km (199 mi) to the northeast of the launch center in the desert area near Dzhezkazgan.

However, the United States was not fooled. Besides having good ground intelligence, an American U–2 spy plane was able to find and photograph the Tyuratam missile test range on August 5, 1957, just two months before Sputnik. On January 28, 1958 the growing town of Zarya was renamed Leninsk in an effort to hide its true identity. Leninsk achieved city status in 1966 (the city was renamed Baikonur by President Boris Yeltsin in the 1990s). What came to be called the **Baikonur Cosmodrome** quickly grew to accommodate missile firing ranges and combat missile positions spread out over hundreds of kilometers on the vast Kazakhstan steppes. The Soviets added launch facilities for "civilian" spaceflight to compete against the United States in the race to space. Today Baikonur stretches like an ellipse, 90km east to west by 85km north to south with the Cosmodrome main buildings in the center. Missiles of different designs can be launched independently from dozens of launch sites, although some of them have fallen into disrepair.

Baikonur remains one of Russia's three major launch complexes today, the others being the Plesetsk Cosmodrome, used primarily to launch military spacecraft, and the new Vostochny Cosmodrome which at this writing is under construction in the Amur Oblast, in the Russian Far East. When completed in 2018, it is intended to reduce Russia's dependency on the Baikonur Cosmodrome in Kazakhstan. Plesetsk is commonly used for military satellites placed into high inclination and polar orbits since the range for falling debris is clear to the north which is largely uninhabited Arctic and polar terrain. It is situated in a region of taiga, or flat terrain with boreal pine forests in Arkhangelsk Oblast, about 800 km north of Moscow. Sary–Shagan is used for antimissile defense and laser weapons tests. Kapustin Yar, in Astrakhan Oblast between Volgograd and Astrakhan, is a development site for new technology and occasionally used for test rocket launches.

Rocket engines traditionally require the most development and testing time by far than any other system on a missile. Glushko decided that an ICBM using German ideas was impossible. He had developed the RD–110 with a cylindrical burner but it suffered combustion instabilities, a perennial problem for Glushko. He returned to his experiments from the 1930's with cylindrical burners – combining his ideas with German ones. He used slots for cooling the nozzle for example. In the early 1950s he built various 0.6 meter diameter experimental chambers, resulting in an RD–106 LOX/kerosene engine. However, the RD–106 never became operational. Glushko was unable to achieve stable combustion in such a large chamber, a huge scale–up from the ED–140 German design. Even worse, the 1954 R–7 payload specification required a 50% increase in thrust per stage. At the end of 1954, Glushko abandoned the RD–106 and came up with the ingenious innovation of using a <u>single</u> turbopump to drive <u>four</u> V–2 size combustion chambers and nozzles. A single gas generator used the decomposition of hydrogen peroxide (H_2O_2) into a hot gas to spin up a steam turbine. Peroxide–powered steam turbines were a German development for the V–2, adapted from the water pumps commonly used in fire engines. A common drive shaft drove pumps for kerosene, liquid oxygen, hydrogen peroxide (for steam generation), liquid nitrogen (for tank pressurization) and pumps for the vernier engines. Kerosene was used to cool the engine nozzles and then passed on to the mixing head. Glushko used igniters

to initiate the combustion process because kerosene and LOX are not hypergolic. The United States did not (and still doesn't) have a comparable four–chambered system.

Glushko's engine team had to resolve innumerable problems. Rocketry is a complicated science taken to the extreme. For example, during engine test bed development two zones of unstable operation were revealed: a "lower" zone (at 40–70% of nominal pressure in the combustion chamber) and an "upper" zone (where pressure value exceeded the nominal pressure by 5–7%). The "lower" zone was dangerous at the engine start. Glushko's team designed a special start oxidizer valve to accelerate the pressure increase and ensure stable engine start. At first they couldn't eliminate the dangerous "upper" unstable engine zone. Extensive testing revealed that increasing the kerosene fuel flow rate through the injectors formed a "screen" which moved up the lower limit of the upper unstable zone. Unfortunately, every silver lining has a cloud. Increase the fuel flow rate too much and you significantly decrease engine performance and burn time (rocket engines prefer to run "lean" with a high oxidizer to fuel weight ratio). They found the best way to solve this dilemma was to precisely adjust how the injectors were produced and the propellant mixing head assembled. The engine production team subjected each engine to rigorous propellant mixture ratio control testing, rejecting chambers disposed to unstable operation.

One day in October 1956 Glushko and Korolyov were engaged in one of their usual lively discussions. "Valentin, you simply must synchronize the thrust of all the engines during their buildup. I can't tolerate your damned engine pressure and thrust transients!" Korolyov pleaded. By now, he was becoming famous (albeit his name was unknown outside secret circles) as the patriarch of Soviet rocket science. "I categorically refuse!" said an exasperated Glushko. "Our synchronization system is designed to regulate the thrust in flight at a steady–state output [known as main stage engine operation]. Our engine specialists, the best in the world, are unable to control a rocket's thrust buildup transient, let alone all the engines at once. Sheesh!"

Korolyov had come prepared with a proper retort. "Varying the thrust from the strapon boosters could lead to huge destabilizing moment values. For good flight dynamics reasons, we just can't rigidly attach the strapon boosters longitudinally to the central sustainer. If any strapon booster's thrust varies just a bit from the others, it will come off the cluster at launch and create havoc. With explosive propellants, this portends an instant disaster and destruction of the entire launch pad! Is that what you want?"

It usually didn't take much to get the fiery Glushko hot under the collar. "Look, Sergei Pavlovich. You insist that we drop off the boosters and then fire the central sustainer in space. Fine. But we have absolutely no experience in firing a powerful liquid propellant engine in space. For one thing my tank people say that in the weightlessness of space the fuel tends to cluster around the edges of the tank, forming a void in the center. Even with positive thrust—which is impossible after the boosters drop off—I will not guarantee that my engine would fire reliably somewhere out there far away, under unknown conditions. Even for the first stage, no matter how accurate Pilyugin's control system is in commanding the booster engines to shutdown, we always get an uncontrolled residual fuel burnoff, causing an asymmetric after-effect burn. This is simply a fact of liquid rocket science."

Korolyov sat there thinking, his brown eyes glistening with intelligence. "O.K., then we fire all five engines simultaneously on the ground. And I will task Pilyugin's control group and Chertok's guidance engineers to synchronize the consumption of all the propellant components from all of the strapon boosters so that they run out of fuel at the same time." Boris Chertok's heart sank. He looked over at Nikolay Pilyugin, who took a hipshot. "To do this we have to have control over the total consumption and the ratio of kerosene–oxygen in each engine," he said. Glushko voiced one more objection. "If you decide to fire all the engines at once on the ground, the central second stage will have to operate for over 250 seconds. This is over twice as long as we figure the graphite–tungsten vanes in the nozzle exhaust can withstand the horrible heat. I have to have them in all the engines, and the aerodynamic fins on the strapons for control, otherwise your missile could fall out of the sky on Moscow!" *The original plan had called for three jet vanes and one aerodynamic fin on each strapon booster for control.*

The four Soviet rocketeers—Korolyov, Glushko, Chertok, and Pilyugin—sat silently for a few moments. "You are not the only ones with problems, you know," Korolyov whispered. "My designers and I have to come up with an interior structure capable of supporting the R–7 missile's massive weight. To overcome the tendency of the outer rocket tube to buckle due to the weight of the interior fuel tanks, we reinforced the outer tube and dispensed with the interior fuel tanks by making the single outer tube itself serve as fuel and oxidizer tanks, separated by hemispherical bulkheads." *American ICBMs would also incorporate this innovation.*

"It is not just the engine shutdown problem. Valentin, you *have* to coordinate the firing of all your rocket engines simultaneously," he said, glowering at Glushko. "Our early tests showed that treating the structure of each engine in the R–7 cluster as independent of the others results in uneven fuel depletion and loss of precious thrust. If there is too much thrust on one side, the whole missile can veer off–course, and may have to be destroyed. This is not a simple problem. It involves precisely timing the exact instant that each engine ignites and ramps up to full–stage thrust. My designers also had to internally connect the fuel tanks in order to ensure that all four of the sustainer rocket engines would burn all the available fuel evenly before shutoff."

Over the succeeding weeks the R–7 teams, with Korolyov riding herd on them, had to solve these and other problems at breakneck speed, one after the other. Under the direction of Konstantin Davydovich Bushuyev and Sergei Kryukov, Pavel Yermolayev's team designed the engine cluster's load bearing system so that the assemblage of strapon boosters transferred their thrust to the central sustainer through an arrangement of upper load–bearing contact points. *The locus of points that transfer applied forces or loads like this is called the "load path" in a structural design.* Another team designed an explosive bolt release system that simultaneously separated all four booster engine clusters from the central core without disturbing the missile's guidance. Flight tests confirmed that with this design solution they separated cleanly, without impacting or damaging the central missile body or its payload. Aleksey Sergeyevich Abramov developed the electronics for a system to regulate the propellant component consumption ratio and to synchronize the consumption between all the strapon boosters. Yuriy Portnov–Sokolov developed a new system called SOBIS, the *Tank Emptying and Synchronization System.* Konstantin Marx, famous for his inventiveness, took on the study, design, development and testing of sensors measuring the levels of liquid oxygen and kerosene in the tanks.

Vasiliy Mishin led a team including Boris Chertok, Mikhail Melnikov, Ivan Raykov, and Boris Sokolov to kill two birds (the control vanes and aero fins) with one stone. They bypassed the recalcitrant Glushko and got rid of the control vanes on all the engines in the bargain. They solved the aftereffect burn problem—which alone could cause a range error exceeding 50km for the ICBM. And they dispensed with all of the aerodynamic fins.

In November 1956 Mishin and his team, accompanied by Glushko, strode into Korolyov's sparsely appointed office. "We have come up with an ingenious solution to the difficult problem of guiding the missile in flight," Mishin proudly announced. *Missiles that primarily travel through the Earth's atmosphere (like SLCMs, ALCMs, SAMs, AAMs) can use fins and moving ailerons in the fins to let air rushing over these surfaces control flight. However, a rocket entering space obviously cannot use airfoils for control.* "Until now, the only way we knew of guiding the beast was to mount the main engines on pivots in order to change the direction of thrust. With the large engines of the future, this will become increasingly unwieldy. It is like having to control a speeding locomotive by turning the whole thing in the air. Why not mount a series of smaller rockets around the periphery of the missile instead? Then no matter how big Glushko's engine cluster becomes, by gimbaling the smaller vernier rockets, pointing them off–axis, I can steer the whole huge clustered engine missile." Glushko broke in. "And how do you propose to control your so–called vernier rockets?" he asked. "By using radio signals from the ground, of course," Mishin replied. "However, we realize that coordinating all of the verniers together so that we don't veer the rocket off–course, this will be a very difficult and challenging problem. Each of the 16 vernier rockets on the R–7 will have to be controlled by radio signals from ground control stations."

The tightly coordinated Mishin, Glushko, and Korolyov design teams rapidly implemented this revolutionary concept. The missile's entire ascent would be guided by a number of specifically arranged auxiliary (vernier) control engines which would use the same propellants as the main engines and would receive power from the turbopump assemblies. They added two small chambers on each engine of the strapon boosters and four more small vernier thrusters on the sustainer engine. Each engine would be radio–controlled from the ground. After shutdown of the second stage main engine, the control engines would continue in operation. A radio station would continue taking precise radar Doppler velocity measurements. When a specified velocity was reached, ground control would issue control engine shutdown commands and there would be virtually no aftereffect burn.

Another big problem was controlling the bird before blastoff. After thorough analysis, calculation, and much discussion Korolyov approved Vladimir Barmin's design of a huge framework to hold the missile down during a special automatic launch sequence. A massive ferroconcrete launch pad was built over a 45–meter deep flame pit. The rocket hung over an opening, supported by four retractable trusses attached at the missile's waist to load–bearing points at the top of each strapon booster. The boosters would fire first. Thrust buildup transients would be taken care of by electrically controlling the booster engines to an intermediate thrust level that was less than the weight of the entire cluster. The launch framework would hold the missile steady while the thrust variations between the engines worked themselves out. Only after their stable operation was electrically monitored was the central sustainer engine ignited (30 years later the Space Shuttle launched in similar fashion, with the three main engines igniting before the two solid rocket motors were set off). As the central engine gained thrust, the missile would begin to lift off and it would safely separate from the launch system as the three big hold–down arms fell away. In flight, the strapon booster engines would build up to full thrust nominal mode and the missile would pick up speed.

At the request of Ryazanskiy, Borisenko, and Guskov—the primary radio system developers—two new Radio–Control Ground Stations were to be symmetrically placed along both sides of the R–7's ground trace (the area directly below its flight path on the ground) from 150 to 250 kilometers downrange from Tyuratam. One would serve as the main base station and the other as the relay station. A third radio control ground station, required for accurate range control, was situated from 300 to 500 kilometers downrange.

Korolyov's design solution also increased the thrust his missiles needed to loft the Soviets' heavier nuclear bombs and civilian payloads like Sputnik. However, Korolyov had to solve the problem of carrying both stages into space, as Sputniks 1 and 2 had to do in 1957. The boosters had to burn for a precise time between 104 and 130 seconds, shutdown at the same instant, and the central (sustainer) engine would keep burning an additional 155 to 216 seconds depending on the mission and flight profile.

The Americans would call this a "stage and a half" design. The four boosters constituted the first stage because they shut off three minutes or so before the central engine in flight. The single central engine they would call a "sustainer" half–stage because it sustained flight to orbit. The Atlas missile, for example, would be a "stage and a half."

Building on his R–5 experience, Korolyov's design teams shaved off every kilo they could off the "dry" (without propellant) stages. The R–7 ICBM would require 32 precisely controlled combustion chambers (20 for the main engines and 12 for the vernier engines). Thirty–two chambers required systems to control the preparation of turbopump assemblies for startup, the opening of dozens of valves in a required sequence, and simultaneous engine ignition and subsequent operation in all modes. "The difficult we can do tomorrow. The impossible will take us a little while," Glushko often told Korolyov.

Fabrication of the R–7 components was essentially complete by late February 1957. A missile production line was set up. A stream of R–7 parts was sent by special train to Tyuratam. Receiving inspectors checked everything. Experienced technicians worked around the clock to assemble parts into subassemblies, subassemblies into assemblies, and integrated assemblies into missiles while others conducted form, fit and function tests. Korolyov's team bench tested the engine release mechanism. The first missiles came off the assembly line and were prepped for flight.

out the flame and the missile miraculously separated clean. After 98 seconds of perfect flight, the booster engine that leaked kerosene experienced premature cutoff. Another 5 or 10 seconds and it might have made it. However, with thrust askew the missile went into an uncontrolled, tumbling flight and crashed 400 km from the site. On June 10, another Semyorka failed to even get off the ground after three tries. Despite temperatures that reached 45C (113F) in the shade, the overworked Soviet rocket establishment prepared for a July 12 launch attempt in the dry heat. Back then indoor air conditioning was unheard of in the Soviet Union. This time a tumultuous dance of flame appeared under the entire missile. A split second later, flame engulfed the missile over the strap–on boosters from top to bottom. But in an instant the engines built up thrust and a stream of air pulled the whirling flame downward into the enormous concrete escarpment and flame bucket underneath the missile. Then about 35 seconds after a triumphant liftoff the missile unexpectedly began to spin about its longitudinal axis. A guidance and control system failure. All four strap–on boosters fell off the core. All five hot, smoking engine assemblies began somersaulting as

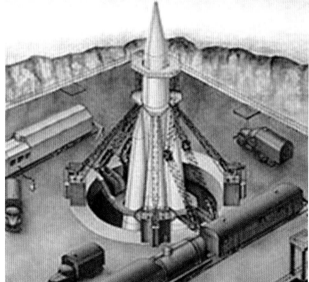

R–7 ICBM launch complex at Site 1 (foreground) and processing facilities at Site 2 (background). Above-ground launches made it vulnerable to what the Soviets feared would be a first strike by U.S. forces. Korolyov proposed at least partially burying the missile for protection as shown on right.

On May 15 the maiden launch of the first R–7 failed. A kerosene leak caused a fire that engulfed the entire lower part of the missile before liftoff. As the engines built up to main stage, a stream of air seemed to put

they and the dummy payload fell to the ground. Another failure. But the Russians pressed on, with summertime temperatures reaching 50C (122F) in the dry desert heat of the Kazakhstan steppes.

Korolyov had other troubles. His associate and competitor, Mikhail Kuzmich Yangel, had been authorized to conduct development and test flights of the R–12 ICBM simultaneously to the troubled R–7 so that one or the other should succeed. After all, it was a matter of national survival of their homeland from the ravenous West in the eyes of worried Soviets.

No wonder. Russia has been plagued by invasions and incursions for a thousand years. In the 10ᵗʰ–11ᵗʰ centuries constant incursions by nomadic Turkic tribes like the Kipchaks and Penchnegs caused a massive migration of Slavic populations to safer, heavily forested areas of the north. Mongol invasions in the 13ᵗʰ century resulted in the destruction of Kiev and the death of about half the population of the Rus. The invading Mongol elite, together with their conquered Turkic subjects (Cumans, Kipchaks, Bulgars) became known as Tatars. They formed the Golden Horde khanate, which pillaged modern–day southern and central expanses of Russia for over two centuries. Napoleon had sacked and caused the burning of Moscow 1812 (Russian Count Fyodor Vasilievich Rostopchin actually gave the sabotage order), and the memories of Hitler's recent invasion and siege of Leningrad (now Saint Petersburg) were only too fresh.

The R–12 was first tested on June 22 at the Kapustin Yar launch site, and the results looked promising. Korolyov argued that the R–12's hypergolic *storable propellants* (so–called because unlike supercold cryogenic propellants, they can be "stored" for longer periods at room temperature) were dangerous because they are toxic and ignite on contact, while oxygen and kerosene are inert (until ignited). For his part, Yangel maintained that his propellants could be stored in a missile indefinitely, while missiles using liquid oxygen could only be fueled immediately prior to launching because the extreme cold would destroy the fuel gaskets given time. A launch abort required unloading the cryogenic oxygen from the R–7 missile. At any rate, after all the R–7 failures Korolyov was worried that Yangel's team would be given a go–ahead to supplant the R–7 as Russia's first ICBM. A similar situation was about to play out in America, where the Army's Jupiter C launch vehicle would compete against the U.S. Navy's Vanguard for the honor of putting the first U.S. satellite into orbit.

Mikhail Kuzmich Yangel, a leading missile designer in the Soviet Union. Yangel's bureau was part of the General Machine-Building Ministry headed by Sergey Afanasyev. For his outstanding work, Mikhail Yangel was awarded the Lenin Prize in 1960 and USSR State Prize in 1967. He was also awarded four Orders of Lenin, Order of the October Revolution, and numerous medals. He died in Moscow in 1971.

Yangel's R–12 IRBM in process of fueling. It entered the Soviet Strategic Rocket Forces in March 1959 (though Western intelligence believed that an initial operational capability was reached in late 1958). It could deliver megaton–class nuclear warheads and constituted the bulk of the Soviet offensive missile threat to Western Europe.

AUGUST 21, 1957

Once again the weary launch crew prepared the R–7 for what they didn't know would be its first successful test flight. Its target was thousands of kilometers away in the Kamchatka peninsula. On start–up, fuel was allowed to flow by gravity, and the engines were ignited. The inflow of fuel turned the turbine blades, which began to pump hydrogen peroxide into the steam generators. Once steam was produced, the turbo pump powered up to 8300 rpm. The supporting trusses held the rocket down until full take–off thrust was developed, about 10 seconds after ignition. Miraculously, the launch and flight proceeded normally, and Pilyugin's radio guidance system behaved itself. During flight the R–7's radio telemetry systems measured readings from over 700 onboard sensors. Telemetry electronics were held in sealed containers within the propellant tanks. Rectangular pads on the outside of the sustainer and booster stages covered a collection of Tral telemetry antennas. Portnov–Sokolov's SOBIS system flawlessly synchronized the fuel consumption of the four strap–on boosters to keep the rocket's weight in balance. The missile carried a dummy telemetric warhead containing the Tral–G, RTS–5, transponders for a radar system called Binokl, and an interferometric angle measurement system called Irtysh. On the rocket, a Tral–V system reported data on the lateral boosters. Tral–T and the RTS–5 systems transmitted status data on the central sustainer during flight. The Tral system worked by pulse–time modulation (PTM), encoding analog parameters during the time interval between radio pulses. A total of 6000 measurements per second were sent on 48 channels, and channels were often multiplexed to send many sensor readings at once. The RTS–5 system, designed to send rapidly–changing data such as engine vibration, sent 50,000 measurements per second on 8 channels using pulse–duration modulation (PDM).

The new radio receiving stations along the ground track of the R–7, from Baikonur to the Kamchatka, received and recorded information in case there was a failure. The rocket also contained the radio guidance system, which worked with two RUP stations (radio control points) some 250km

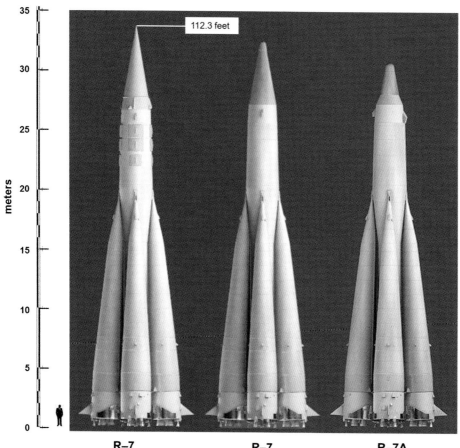

112.3 feet

R–7
Test Version

R–7
Operational Version

R–7A
Modified Version

**Booster Engine
Arrangement
(all versions)**

The R–7 ICBM (8K71) family. Note the gradual migration away from pointed nose cones in the design, as the Soviets gained knowledge similar to that of the U.S. in how to survive the tremendous heat of reentry.

R–7 ICBM test launch at Baikonur, August 1957. Note the characteristic "Korolyov Cross" shape made by the four R–7 boosters as they separated from the still–burning core stage at altitude.

away downrange. The speed and course of the rocket were measured by trilateration. Everything worked as planned. Commands were sent to stabilize the trajectory and precisely time engine cut–off. Sensors measured the thickness of the asbestos/phenolic–resin heat shield, external pressure at the tip, pressure at various points on the sides, internal temperature and pressure, acceleration and angular velocity. After 120 seconds the strap–on rockets fell off into the desert and 2.5 minutes later the second stage separated. The dummy nuclear warhead flew from Tyuratam to the Kamchatka peninsula 6275km (3900 mi) away. However, instead of reentering like it was supposed to, the warhead burned up 10 km above the target.

Korolyov had two worries. Well, surely he had more than two. In 1946 his marriage with Xenia Vincentini began to break up. Like von Kármán, Korolyov was as fond of women as they were of him. Vincentini was heavily occupied with her own career, and at about this time Korolyov had an affair with a younger woman named Nina Ivanovna Kotenkova who translated English periodicals for his design bureau. Vincentini, who still loved Korolyov but was angry over the infidelity, divorced him in 1948. Korolyov and Kotenkova got married in 1949, but he is known to have had affairs even after his marriage to Kotenkova. It's a wonder she stayed married to him.

Korolyov was also suffering from a kidney disorder, a condition brought on by his detention in the Gulag. Like many Soviet Chief Designers, he was a workaholic. He was warned by the doctors that if he continued to work as intensely as he had, he would not live long. However Korolev reasoned that once the Soviets lost their leadership in space, the capricious Khrushchev would likely cut off the funding for his programs. So he continued to work - now even more intensely than before.

Would the R–7 be benched in favor of Yangel's R–12? More importantly, would the United States get to orbit first? He had reason to worry. The American IGY (International Geophysical Year) committee had agreed to an IGY rocket conference beginning in October, and the Americans would be presenting a paper entitled, "Satellite Over the Planet" on October 6th. Korolev was convinced that they would key a satellite launch to an important event such as this. Staying up all night after the August 21st test, he discussed a launch schedule with his aides to beat the United States by putting up the world's first satellite on September 17th to commemorate the 100th birth anniversary of the father of rocketry (Goddard conveniently aside), Konstantin Tsiolkovsky.

"The Semyorka is reliable," Korolov pleaded before the Council of Ministers. "The Americans have conducted three successful suborbital flights. We think they are testing reentry nosecones." *And I wonder why the latest photo showed them as being so blunt, instead of streamlined with a sharp nose like the German V–2,* he thought to himself. *It just doesn't make sense.* Wiping his brow, he continued. "Come on, comrades. Almost a year ago, on September 20, 1956 they lifted a payload to an altitude of over 1000 kilometers and a range of over 5200 kilometers from Cape Canaveral. Then on May 15 this year they lofted another payload to over 500 kilometers and a range of over 1000. And now I've heard that they launched successfully a **third time** with a four stage configuration on August 8. This time it went almost 500 kilometers and over 2,000 kilometers downrange from the Cape. Dr. von Braun's rocket team is poised at this very moment to beef up their Jupiter–C launch vehicle and heft a large payload into orbit." *Von Braun's team modified the Jupiter-C into the Juno for civilian launches.*

The recalcitrant Council of Ministers remained as divided as before when they authorized Yangel's bureau to develop the R–12 IRBM as an insurance policy should Korolyov's R–7 fail. With only one card left to play, Korolyov challenged: "I propose that we put the question of national priority in launching the world's first artificial Earth satellite to the Presidium of the Central Committee of the Communist Party. Let them settle it." This put the ministers in a quandary. A successful U.S. launch after the ministers had forbidden a Soviet one would open each one up to political suicide. The crafty ministers passed the risk right back to Korolyov by approving his launch plan. Korolyov quickly amended it for the launch to be on the same day as the projected American launch: October 6th.

On September 7 the R–7 flew successfully a second time, bolstering Korolyov's confidence for a successful satellite launch the following month. But as the deadline neared, Korolyov remained on pins and needles, worried about an imminent U.S. launch. Finally, his patience and nerves exhausted, Korolyov informed Moscow of this intention to move up the launch to October 4 (Moscow Time), two days before the Americans were scheduled to present their paper at the IGY–sponsored rocket conference.

Korolyov needn't have worried. All of the American space race teams were mired in frustration:

- Delays in getting the U.S. Atlas ICBM tested resulted in an announcement on September 1 that Atlas's initial operational capability would be delayed to June 1959.

- The luckless Thor had experienced four unsuccessful flight tests and was facing the constant pressure of Jupiter program successes. The prime contractor, Douglas Aircraft, was facing disaster.

- Von Braun's team was forced to seeth quietly on the sidelines. Their proposal to use a Jupiter C launch vehicle to launch the first U.S. satellite had been rejected in 1955 by the Eisenhower administration in favor of the Navy's Project Vanguard, using a booster produced for civilian space launches (as described in the next chapter).

- The Navy's Vanguard team was beset by daunting obstacles: unproven technology, lack of funding, shoddy workmanship and poor quality control by the prime contractor (the Martin Company), and not least, public scrutiny of launch failures.

Korolyov's R-7 became the biggest leap in the world's rocketry since the German Aggregate A-4, precursor to the V–2. Ironically, developed to be the first Soviet ICBM, the R-7 grew obsolete as a weapon even before it started flying. Yet, as a launch-vehicle, it continued serving Russian space program more than half a century after it was originally conceived. In the 21st century, R-7-derived space boosters have remained the only vehicles delivering Russian manned spacecraft into orbit. The operation of the International Space Station has also depended on the R-7-based boosters, launching crew, supply ships and lifeboats for the outpost's crews.

THE PROBLEM OF REENTRY HEATING

Using missiles to deliver nuclear warheads required solving many complex problems. One critical problem was how to have the warhead survive the heat of reentry as its ballistic arc brought it back from space into the atmosphere. The atmospheric blanket around our planet ensures that all but the largest meteoroids are destroyed before they hit the ground. During the Cold War both America and its NATO Allies, and the Soviets and their Warsaw Pact Allies, knew that the key to delivering thousands of invincible nuclear bombs to the other side lay in solving the ICBM reentry heating problem. Basic calculations showed that the kinetic energy of a nuclear warhead returning from suborbit at 11 km/sec (7 miles/sec) was sufficient to completely vaporize the warhead. Despite these calculations, the military stakes were so high that simply assuming atmospheric reentry's impossibility was unacceptable. Both sides embarked on high–priority programs to develop reentry technology. Everyone knew that meteorites were able to successfully reach ground level. A way had to be found to engineer a solution. As early as March 1920 Robert Goddard had described the concept of the ablative heat shield in his "Report Concerning Further Developments" to the Smithsonian Institution:

> *"In the case of meteors, which enter the atmosphere with speeds as high as 30 miles per second, the interior of the meteors remains cold, and the erosion is due, to a large extent, to chipping or cracking of the suddenly heated surface. For this reason, if the outer surface of the apparatus were to consist of layers of a very infusible hard substance with layers of a poor heat conductor between, the surface would not be eroded to any considerable extent, especially as the velocity of the apparatus would not be nearly so great as that of the average meteor."*

The kinetic energy (one half its mass times its velocity squared) of a NASA Space Shuttle reentering from orbit was 2.12 gigajoules. This is a tremendous amount of heat, enough to bring over 6,000 tonnes of water to a boil! However, it is a common misconception that the outer shell of a reentry vehicle (RV) itself has to absorb or dissipate all of the heat during reentry. Fortunately, Mother Nature has been very kind.

Using the Shuttle example, most of this heat is transferred by convection into the air around the Shuttle, or is dissipated in the vehicle's wake. Only about 1/1000 of the heat goes into the Shuttle's thermal protection system itself, which shields the underlying structure from the extreme heating.

Another misconception is that an aerodynamically sleek–looking missile with a sharp–pointed nosecone, like the Germans used in their V–2, is best. It isn't. A traditional pointed nosecone transfers the entire energy of the nosetip shockwave to the RV surface. This transfers the energy as heat to the payload. Using mathematics in 1951, two scientists, H. Julian Allen and A. J. Eggers, Jr. of NACA made the counterintuitive discovery that a blunt shape (high drag) made the most effective heat shield. From simple engineering principles, Allen and Eggers showed that the heat load experienced by an RV was inversely proportional to the drag coefficient, i.e. the greater the drag, the less the heat load. By making the RV's nose blunt, air can't "get out of the way" quickly enough, and acts as an air cushion to push the shock wave and heated shock layer forward (away from the vehicle). Since most of the hot gases are no longer in direct contact with the vehicle, the heat energy would stay in the shocked gas and simply move around the vehicle to later dissipate into the atmosphere. This results in the RV gathering less energy–creating heat.

Korolyov had to make adjustments to his sharp–pointed R–7 and its heat shield, but he didn't have the benefit of the Americans' classified research. It would take until March 29, 1958 for the Soviets to solve the reentry heating problem. Nevertheless, the R–7 made a second successful test flight on September 7, 1957. The stage was set for the historic Sputnik flight on October 4. It was not a case of the Russians simply improving WWII German rocket designs. To get there, both Korolyov and Glushko had to discard and get past the old German rocket heritage, much like Dr. Wernher von Braun and his rocket teams would have to do in America.

Korolyov had reached the pinnacle of greatness. His design led the pathway for the Vostok, Soyuz, Molniya, and Progress flight vehicles. In many ways his concepts

still fly today. The same vehicles with modifications have served for decades and launch rates have been so high that most failures are by now assumed to be random and the vehicle's launch schedule continues almost uninterrupted. Today, Soyuz is marketed internationally by Starsem, a joint Russian/European consortium. Starsem harnesses the power of four leading space organizations: Airbus Defense & Space, Arianespace, the Russian Federal Space Agency (Roscosmos), and the Progress State Research and Production Space Center (TsSKB–Progress). Over ten versions of space rockets based on the R–7 rocket design have operated for over 60 years. Through 2016 about 1865 Soyuz rockets had been launched with an unparalleled success rate of 98% for production models. In fact its reliability has kept increasing over time so that today Soyuz and Progress flights are almost routine, like taking an airline flight. The indomitable Glushko also shares the limelight. Over 8870 engines (and counting) have been utilized during the Soyuz flights, and over 10,300 engines have been produced including the engines for test bed development. Glushko's liquid–propellant engine is the only one in the world possessing such statistics.

Meanwhile, American rocket engineers did not just sit idly by while the Soviet Union made tremendous strides in rockets, missiles, and the exploitation of space to demonstrate the superiority of the Communist social order. No way!

John Glenn entering his cramped capsule

WERNHER

"The rocket will free man from his remaining chains; the chains of gravity which still tie him to this planet. It will open to him the gates of heaven..."

"We can lick gravity, but sometimes the paperwork is overwhelming"

"Man is the best computer we can put aboard a spacecraft...and the only one that can be mass-produced with unskilled labor."

Wernher von Braun

FEBRUARY 20, 1962

9:47 a.m. Eastern Standard Time. Over 10 months had passed since Gagarin's historic flight the previous April 12. The United States needed a space hero badly, but *Friendship 7* had endured 11 frustrating delays in its countdown. Malfunctions, improvements, and bad weather made 40 year-old John Herschel Glenn's long wait to fly in space seem endless. An hour earlier he had once again eased himself and his bulky spacesuit into the cramped *Mercury 6* capsule, codenamed *Friendship 7*, perched above a Mercury-Atlas 6 rocket. As the countdown finally reached zero at 9:47 a.m., Glenn felt a jolt as the rocket ignited below him and thundered him into space for America's first manned orbital flight. Or so he hoped. It was a standing joke that the wives of astronauts assigned to the Mercury Program had taken out extra life insurance policies on their husbands.

Five minutes after liftoff the engines cut off, and the previously felt gravitational forces of nearly 8 g disappeared--Glenn was weightless. Over the Indian Ocean on his first orbit, Glenn became the first American to witness the sunset from space. "This moment of twilight is simply beautiful. The sky in space is very black, with a thin band of blue along the horizon," he radioed down to NASA's Mercury mission control center (MCC) in Cape Canaveral, Florida. NASA's MCC was not moved to the Manned Spacecraft Center (MSC, now the Lyndon B. Johnson Space Center (JSC) in Houston, Texas) until 1964.

Glenn in orbit

Two—stage Atlas D launches Glenn into orbit

On the nightside of Earth, nearing the Australian coastline, Glenn made his planned star, weather, and landmark observations. Within voice radio range of the Muchea, Australia tracking station in those days, there were no communications satellites in orbit, much less the GPS tracking systems we take for granted today. Glenn and fellow astronaut Gordon ("Gordo") Cooper engaged in a long space-to-Earth conversation. "I feel just fine, my stomach is settled down" Glenn told him. The two other American astronauts who had flown in space, Alan Shepard and Gus Grissom (we'll meet them later) had already warned Glenn about that strange feeling one gets in the stomach—akin to the "butterflies" some athletes get before a key competition, but worse—in weightless space. "I see a very bright light down there. Hmm, looks like the outline of a city," Glenn continued. "You're probably seeing the lights of Perth and Rockingham," Cooper answered. Glenn was getting very excited (well, who wouldn't!). "Hey Gordo, I can see stars as I look down towards the real horizon! I'm seeing a haze layer

about seven or eight degrees above the horizon on the nightside. The clouds are reflecting the moonlight. It looks so pretty up here, like a dream…That sure was a short day. That was about the shortest day I've ever run into!" he excitedly told Cooper.

As he orbited onward above the Pacific over Canton Island, Glenn experienced an even shorter 45-minute night. He prepared his periscope for viewing his first sunrise in orbit. As the day dawned over the island, he radioed Cooper again. "What the? …I'm seeing literally thousands of little specks, brilliant specks, floating around outside the capsule. Maybe the spacecraft's tumbling, or am I looking into a star field?" "Check it," Cooper replied. Glenn took a quick hard look out of the capsule window. No, a momentary illusion. But he had to report everything he saw or experienced, no matter how weird. There were still folks who thought that men would die floating around in space. Just like that.

The specks didn't go away. "They look like luminescent fireflies, streaming past the spacecraft from ahead." *Fireflies?? This is crazy! There aren't supposed to be any fireflies in space!* Glenn thought to himself. "They seem to flow leisurely past the window, but not to be originating from any part of the capsule." Down at Mercury MCC, some thought at first that Glenn had discovered life in space, others that he was hallucinating. As *Friendship 7* sped over the Pacific Ocean expanse (it spans about 13,000 miles (20,921km) and covers a third of the Earth's surface) into brighter sunlight, the "fireflies" disappeared. The best theory now is that they were small flakes of ice that became dislodged from the surface of the spacecraft by the bright sunshine. By the way, you don't want to be caught in the open sunshine in space without a spacesuit. Your body will get severely burned almost immediately, you'll roast at 250°F (121°C), and your blood will boil. But not to worry, you'll probably die from asphyxia first.

Near the end of the first orbit, Glenn reported that the thrusters on the automatic attitude control system were malfunctioning and, like any good pilot, he switched to manual control (for the duration of flight) to regain control over the spinning craft. Glenn and his fellow American astronauts had successfully argued for such a "brick and a stick" system in their spacecraft. They were right. Glenn had saved his life with his foresight. Mercury MCC decided to cut Glenn's planned six orbits back to three.

Then, trouble again. At Mercury MCC, engineer William Saunders intently eyed his telemetry control console. 'Segment 51', an instrument providing data on the spacecraft landing system, was presenting a strange reading. Was it a false reading? According to the signal, Friendship 7's heat shield and the compressed landing bag were no longer locked in position. If this was really the case, the all-important heat shield was being held on the capsule only by the straps of the retropack. So-called retro-rockets (retros) have to fire in the opposite direction of orbital motion to slow a reentering spacecraft down. On the Mercury spacecraft, the retros were built into a package called the *retropack*.

False reading or not, Mercury MCC couldn't take a chance. Almost immediately Flight Director Christopher Kraft ordered all tracking sites to monitor the instrumentation segment closely and, in their conversations with Glenn, to mention that the landing-bag deploy switch should be in the "off" position. Although Glenn was not immediately aware of his potential danger, he became suspicious when site after site consecutively asked him to make sure that the deploy switch was off. Engineers had figured that a spacecraft's heat shield had to withstand temperatures of 2800°F (1538°C) during descent from orbit. A loose heat shield threatened *Friendship 7* with incineration during its imminent reentry.

Meanwhile the operations team had to decide how to get Glenn and his capsule safely back through the atmosphere with a loose heat shield. Kraft and mission operations Chief Walt Williams weighed the information they had received. It would be safer to keep the retropack. Wally Schirra, another Mercury astronaut serving as the California communicator, passed the order to Glenn to retain the retropack until he was over the Texas tracking station.

Now came one of the most dramatic and critical moments in all of Project Mercury. In the Mercury MCC, at the tracking stations, and on the recovery ships ringing the globe, engineers, technicians, physicians, recovery personnel, and fellow astronauts stood nervously, stared at their consoles, and listened to the communications circuits. Glenn, reduced to using the "brick and a stick" he had fought for, while checking his course through the window engineers hadn't wanted to give him (damn those engineers!), knew that his angle of reentry was crucial. If he came in too steeply, he would die in a ball of fire, as the temperatures would be too high for even an intact heat shield (the 2800°F estimate was based on an optimal reentry angle). Too shallow an angle, and a slow death by asphyxiation awaited him after skipping off the Earth's atmosphere into a skewed orbit without a way to maneuver back for reentry. Glenn had to thread a needle of life and death—and he had to do it at almost 18,000 miles per hour (28,970kph). And unlike Yuri Gagarin's flight, the whole world was watching.

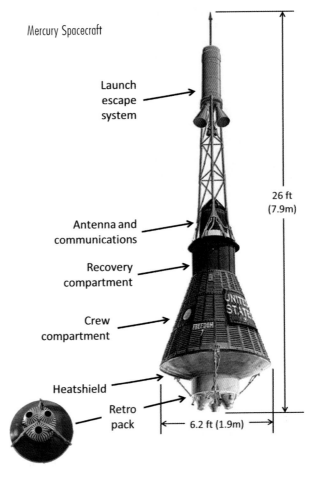

Mercury Spacecraft

Launch escape system

Antenna and communications

Recovery compartment

Crew compartment

Heatshield

Retro pack

26 ft (7.9m)

6.2 ft (1.9m)

blocked radio transmissions. Until he emerged from the fireball, there was no way to know whether he and the capsule had burned up.

Friendship 7 came now to the most fearful and fateful point of its voyage as the terrific frictional heat of reentry enveloped the capsule. Glenn experienced his worst emotional stress of the flight, his heartbeat a-thumping. "That's a real fireball outside," he radioed the Cape, with a trace of anxiety evident in his tone. No one heard him. "I heard noises that sounded like small things brushing against the capsule. Then a strap from the retropack swung around and fluttered over the window, and I saw smoke as the whole apparatus was consumed. I thought the retropack had jettisoned. Flaming chunks of retrorocket, heat shield, and spacecraft came off and flew by the window," he said later. With the heat shield and his ablation protection disintegrating, he feared he was doomed. But Glenn kept his cool and maintained the best control he could over the spacecraft.

God must have been looking after him. The heat shield stayed in place. At 28,000 feet (8534m) the drogue automatically shot out. Moments later, Glenn's voice reassuringly crackled through loudspeakers at Mercury MCC. With immense relief, he watched the main chute stream out, reef, and blossom at less than 17,000 feet (5182m). After a five hour flight, Friendship 7 splashed into the Atlantic about 800 miles (1287.5km) southeast of Cape Canaveral near Grand Turk Island. The nation could breathe easily again. Because retrofire calculations had not taken into account the spacecraft's weight loss in consumables, Glenn landed somewhat short--about 41 miles (66km) west and 19 miles (31km) north of the predicted area. The Noa, a destroyer code-named Steelhead, picked up Glenn only 21 minutes after splashdown.

Friendship 7 capsule containing Glenn is recovered from the Atlantic. The yellow coloring in the water is not a shark repellent. It's a dye to make the capsule more visible to the recovery crew.

Astronaut John Glenn (waving to crowd) is honored with a ticker tape parade on March 1, 1962 in New York City. Next to him are his wife Annie, and Vice President Lyndon Johnson.

Glenn and Friendship 7 slowed down during their long reentry glide over the continental United States toward a hoped-for splashdown in the Atlantic. As Friendship 7 slammed into the atmosphere, the reentry fireball built up a plasma sheath around the spacecraft. At its very hottest—the most crucial time--all communications were lost with the spacecraft. The Americans did not know yet that this was a normal part of reentry, as the hot plasma surrounding the capsule

AMERICA GETS ITS SPACE HERO

John Glenn became the third American in space and the first to orbit the Earth. He was hailed as a national hero. He didn't have to go to Washington, DC to visit U.S. President John Kennedy--Kennedy came to him. At Cape Canaveral, JFK was there to greet him, and presented him with the Space Congressional Medal of Honor. After receiving his medal, Glenn took time to show his wife Annie and his children his capsule, and how he had 'threaded the needle' with it as he reentered the Earth's atmosphere without burning up. He was given huge tickertape parades in various cities. Onlookers in New York City threw a record 3,500 tons of paper as Glenn passed by. "My flight was but one step," he modestly told a crowd there on March 1, 1962. He retired from the U.S. Marine Corps as a Colonel in 1965, and went on to serve six terms as a U.S. Senator from Ohio.

But there was another hero. An unsung one who had more to do with America's push for space supremacy than Glenn or any other astronaut, living or dead. His name, like Sergei Korolyov's, is not exactly a household word today.

A GENIUS

Wernher Magnus Maximilian Freiherr von Braun (pronounced "Vernher von Brown" in German) was born as the second of three sons into a tradition-rich aristocratic family in Wyrzysk, then a part of the German Empire, in March, 1912. Through his mother, Emmy von Quistorp, he could trace their line of ancestry back to 1285 and medieval European royalty. The "Freiherr" in his name (literally "Free Lord") was considered about equal to the title Baron. Young Wernher received a very careful education not only in the usual scholastic subjects, but also in art, music, classical literature, and religion. He learned to play the cello and the piano at an early age and wanted to become a composer. An accomplished amateur musician, he could play music by Beethoven and Bach from memory. To this end he took lessons from Paul Hindemith, the famous German composer.

Starting in 1925, von Braun attended a boarding school at Ettersburg castle near Weimar where at first he did not do well in physics and mathematics. But after his Lutheran confirmation, his mother gave him a telescope, and he discovered a passion for astronomy. When Wyrzysk was given to Poland in 1918, his family, like many other German families, moved to Germany.

They settled in Berlin, where the 12-year-old von Braun exhibited an interest in explosives and fireworks. He was also inspired by the speed records established by Fritz von Opel and the daredevil Max Valier. At the age of 13, von Braun exhibited an interest in explosives and fireworks. His father, Magnus von Braun, could not understand his son's consuming interest in so dangerous a hobby. He feared his son would become safecracker.

One day in 1925 the young teenager obtained six skyrockets, strapped them to a toy red wagon and set them off. Streaming flames and a long trail of smoke, the wagon roared five blocks into the center of the von Braun family's home town, where they finally exploded. As the smoke cleared, the toy wagon emerged as a charred wreck. Young von Braun emerged in the firm grasp of a policeman. The youngster was taken into custody by the local police until his father came to collect him. Despite being severely reprimanded by his father, the youngster's interest would not be denied. A few months later, another of von Braun's creations went haywire and with quite a racket crashed into a neighbor's house, barely missing a window.

The three von Braun brothers (left to right):

Wernher (age 12), Sigismund (13), and Magnus Jr. (5). They all inherited the title of baron at birth.

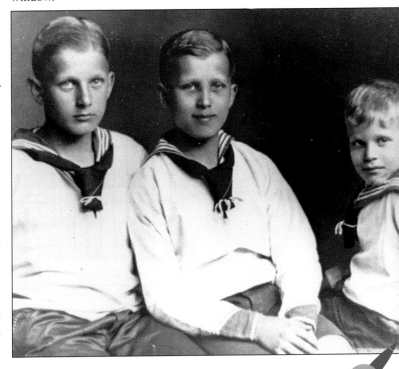

In 1928 his parents moved him to the Hermann-Lietz-Internat (also a residential school) on the East Frisian North Sea island of Spiekeroog, where he acquired a copy of the book *Die Rakete zu den Planetenräumen* (The Rocket into Interplanetary Space) by rocket pioneer Hermann Oberth. The idea of space travel had always fascinated von Braun, and from that point on he applied himself to physics and mathematics in order to pursue his interest in rocketry. He soon rose to the top of the class.

The thing about Wernher von Braun was not just that he would become a great rocket scientist. It was also his breadth of knowledge, his social skills, his organizational acumen, his everything. An avid reader since his boyhood, von Braun not only devoured the books written by Jules Verne and Hermann Oberth, but also those by Johann Wolfgang von Goethe, Friedrich von Schiller, and Immanuel Kant. It was particularly the latter who played a decisive role in the development of von Braun's religious thinking. Here he found the bridge between the visible world around us, which never ceased to impress him most profoundly, and some invisible world from which, as he was convinced, must come the master plan for this marvelous system of natural laws and orderly evolution.

Starting in 1930, 18-year-old von Braun attended the Technical University of Berlin. He joined the *Verein für Raumschiffahrt* (VfR), the "Society for Spaceflight") and assisted the VfR "gang" in liquid-fueled rocket engine tests at the VfR's leased *Raketenflugplatz* (Rocket Flight Field) near Berlin. He also studied at ETH Zurich. By age 22 the boy genius would earn his doctorate in physics. Two years later he was directing Germany's military rocket development program, the best in the world at the time.

June 22, 1932

Klaus Riedel, Rudolph Nebel, and Wernher von Braun finished preparing a Repulsor rocket in the middle of the night (the Repulsor and other VfR creations have been described in the Robert chapter). The whole rocket was 4 meters long, and the main body only 6 centimeters in diameter. It weighed only 12 kilos fully fueled, had an engine with a water-cooling jacket in the nose, and a recovery parachute and flare stuffed into a tail compartment with little bitty ineffective fins (by today's standards). Common practice at the time used a "nose drive" engine, with the combustion chamber and

de Laval nozzle in front, followed by the propellant tanks in tandem, then the tail and any guidance fins. Besides the backward configuration (engine in front) and having to carry its launch platform, a rocket with such a high slenderness ratio (400cm/6cm = 67) would be difficult to control and vulnerable to bending from a number of sources: asymmetric thrust loads, guidance maneuvers, even strong wind gusts. Basically a poor design in the eyes of modern rocket hobbyists, but this was the early 1930s and rocket science was still in its infancy.

They carefully slid the Repulsor into an aluminum launch rack and mounted the whole affair atop Nebel's open-topped car. To maintain maximum secrecy, the German Army Ordnance Command ordered them to appear at 4:00 a.m. outside the Kummersdorf Army Proving Grounds, 25km southwest of Berlin. They would be met there by an army contingent dressed in civilian clothes. Lt. Col. Karl Becker headed the army group, assisted by three Captains (Ritter von Horstig, Walter Robert Dornberger, and Erich Schneider) and Dr. Erich Schumann, a physicist who directed a small research branch in Section 1 at the Ordnance Command and held a professorship in the University of Berlin. In April 1930 Dornberger had been appointed to the Ballistics Council of the *Reicheswehr* (Germany's 100,000-man armed forces) Weapons Department as assistant "examiner" to secretly develop a military liquid-fuel rocket suitable for mass-production that would surpass the range of German Army artillery.

The groups drove in separate cars to the Kummersdorf-West test stand in the dark, on poor roads that may have damaged the fragile rocket. On arriving, Von Braun and the others carefully unpacked their creation. Captain Dornberger had warned the civilian rocket enthusiasts that he didn't think their design was mature enough for field testing. Just months earlier, an errant Repulsor had leaked gasoline onto a shack owned by the police, causing it to burn down. The whole thing was captured on tape by Max Valier's film crew, to the embarrassment of the VfR amateurs. But the German rocketeers had pressed on.

Finally, at 6:30 a.m. von Braun enunciated the countdown, while Riedel stood nearby, ready to close the electrical contact switch. "…Drei…Zwei…Eins.. Zündung!" von Braun shouted. Within a second, flames appeared at the nozzle, then the rocket started rising from its rack, but in slow motion. *Too slow!* to von Braun's horror. Like a wobbly old man getting up from his chair, the whole assembly swung lightly back and forth as it slowly gained altitude. At about 300 meters, the engine and its attached propellant tanks and tail turned over into an almost horizontal trajectory and headed for the low cloud deck covering the range. It pierced the clouds and reached about 600 meters in height before crashing headlong into the ground about 1300 meters away. No parachute. No flare. Just smoking wreckage on the ground.

The intrepid Nebel had promised nothing less than an 8km flight (even von Braun disagreed, arguing that the rocket only had enough fuel for 4km), then he backed down to 3.5km. The Repulsor rocket was to eject its parachute and a red flare at the peak of its trajectory, in which case the army had promised to pay Nebel 1,367 Reichmarks for expenses. Otherwise, nothing. After the test, von Braun took Capt. Dornberger aside.

"A test is successful if you learn something valuable from it," he argued. Dornberger responded, "I think you should meet with Lt. Col. Becker, at your convenience. We have something we want to tell you."

In the aftermath, von Braun did visit with Becker and the Army Ordnance Command. There he found that Ordnance had a distaste for Nebel and his antics. "We just don't trust Nebel," Captain Schneider told him. Lt. Col. Becker added, "We are greatly interested in rocketry, but there are a number of defects in the manner in which your organization is going about its development. For our purposes, there is far too much showmanship. You would do better to concentrate on scientific data than to fire toy rockets." He went on, "we don't like Nebel's publicity-seeking approach. For that matter, the whole idea of the VfR group, to raise funds from the public and charge admission for viewing rocket launches. The glare of being in the limelight is unbecoming, you can't hide your disasters and mishaps. But we mainly object to disclosure of critical, militarily useful advances to the open public." "What do you want from me?" von Braun replied. For a 20-year-old, the youngster had an unusual astuteness and commanding presence about him, the kind that would cause everyone in the room to take pause. He already knew the answer, but just waited for Lt. Col. Becker to say it. Instead, Capt. Walter Dornberger spoke up. "The *Rakentenflugplatz* group of amateurs has an unsystematic approach. What we need first is accurate measurements and data. How do you measure your propellant consumption, your combustion pressure, your thrust? But Wernher, what we want most, is for you to help us learn from the test failures. Because, like

you said, a test is a success if you learned why something important failed."

Von Braun found Becker, in contrast to Nebel's description, collegial, knowledgeable, and scientific. The two established an immediate personal connection. For the officers in the Ordnance Command, von Braun's class background and parentage counterbalanced his youthfulness. But it was his intellectual ability that really won them over. Dornberger for one realized von Braun's unequalled brilliance. It was essential to entice von Braun to take technical leadership of Dornberger's *Waffenamt Prüfwesen* (Weapons Proof) organization under the *Heeres Waffenamt* (Army Weapons Department). In particular, Dornberger was "…struck by the energy and shrewdness with which this tall, fair, young student with the broad massive chin went to work, and by his astonishing theoretical knowledge."

August 1932

Nebel refused to give up. After hearing von Braun voice his concerns, he contacted the German Army and proposed the use of liquid fuel rockets as war missiles. He arranged for Army representatives—the same ones as attended the June 22 test fiasco-- to observe a demonstration launch at Kummersdorf, the Army's solid fuel rocket test site. He wanted to reinstate his reputation with a successful launch. To ensure success, Nebel arranged for the Heylandt Company, which was working under an army contract, to provide a new engine that Walter Riedel (no relation to Klaus) and Arthur Rudolph had designed.

Once again, the other Riedel (Klaus), von Braun, and Nebel prepared the rocket for launch. This one was a bit more stout: 3.5 m long and 10 cm in diameter. It had a mass of 20 kg fully fueled before launch, an empty mass of 10 kg, and a thrust of 60 kgf. Once again, the trip to the primitive test stand at Kummersdorf West, the countdown to ignition, the launch. Unfortunately, the failure this time was even more miserable. The Repulsor only reached an altitude of 20 to 70 m before veering off horizontally into a forest. Von Braun and Riedel had calculated an exhaust velocity of 2000 m/s, but only 1700 m/s was demonstrated. To Nebel's view, the Army seemed unimpressed. However, in private they wanted to recruit the best German rocketeers away from amateur rocketry and into making weapons of war for Germany. They were impressed that von Braun and Riedel had been able to calculate and measure the rocket nozzle's exhaust velocity so accurately.

KUMMERSDORF

A youthful von Braun, mid–1930s

Von Braun had been won over. He understood that the amateurs, enthusiastic though they were, lacked the resources and a systematic approach to solve the difficult problems of rocket science. Little was known about aerodynamic, navigation, guidance and control principles. The bizarre-looking vehicles that resulted from the "nose drive" showed that the Raketenflugplatz gang had not mastered stability and control in flight. But that was less important than getting a vehicle off the ground without endangering the onlookers too much. The endless problems with propulsion—burnthroughs, leaks, explosions, valves and lines frozen by liquid oxygen— were much more pressing. *A coordinated, national effort is the only way*, von Braun thought. On October 1 he received a contract as a civilian employee to Kummersdorf, charged with the development of liquid rockets as a variation of long-range artillery. Col. Becker also provided him with a grant to continue his graduate education in physics at the University of Berlin. By November he was working directly with the Ordnance Command on a <u>real</u> rocket with 300kg$_f$ thrust.

Whereas at the Raketenflugplatz only the thrust and duration of engine burn could be measured, at Kummersdorf fuel consumption and flow rate, combustion temperature, and other parameters could be simultaneously measured to begin to put rocketry on a scientific footing. Young von Braun would work with Capt. Walter Dornberger. On December 21, 1932 he was just beginning his rocket research at Kummersdorf. While Dornberger watched from a safe distance, the intrepid von Braun held a flaming gasoline can at the end of a four meter-long pole. As he dangled it by the rocket engine, the fuel and oxidizer tanks exploded. Von Braun was knocked back. "Wernher! Are you all right?" Dornberger shouted. He ran as fast as he could towards the stricken rocket genius. "Oh mein Gott!" he exclaimed as he helped von Braun to his feet. "Your eyebrows. They are almost singed off!" Von Braun was much more concerned with the irreparable damage done to the test stand than to his eyebrows. They would grow back. The stand was a smoldering pile of ruins. Dornberger couldn't believe it when von Braun nonchalantly said, "Oh, well. Back to the drawing board. We must address the nasty problem of autoignition."

In three weeks, the diligent Germans rebuilt the test stand. The second try worked flawlessly for a few seconds, developing 140 kgf of thrust. But then the aluminum combustion chamber burned through, highlighting that the chamber cooling problems had not been resolved. A year of hard work followed, with engine burnthroughs, redesigns, successes, and failures. Early rocket pioneers had no instruction manual. The redesign process was tedious and largely empirical, involving endless variants.

In 1933, von Braun's military rocket development program, the first in the world, had to go beyond the solutions that the rocket boys at Raketenflugplatz had invented to the four problems that confronted them. The VfR group blazed the trail when they began using aluminum for the obvious purpose of saving weight and thus increasing the performance of their amateur rockets. But von Braun had to do better. The aluminum chambers were burning through and causing explosions. He needed help from German industry. Grabbing a phone book, he got in touch with welding experts, instrumentation firms, valve factories, and pyrotechnical laboratories. In April, he and his superiors contacted a firm that specialized in aluminum anodizing, which hardens the surface through the electrolytic formation of an oxidation layer. This advance proved crucial in increasing the durability of their engines. In turn, the firm led them to Zarges, a small manufacturer in Stuttgart who would become the primary contractor for engine and propellant tank construction for the next three years in Germany.

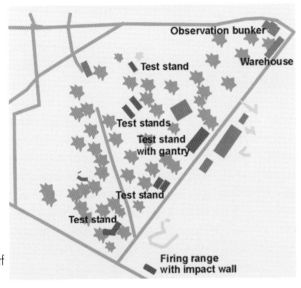

Kummersdorf
Layout

A second problem was how to fully automate the process of ignition that had plagued the rocket program from the beginning. Von Braun was not blessed with hypergolic propellants, like the hydrogen and oxygen on the Space Shuttle, that ignite spontaneously on contact. Von Braun came up with the next best thing, hypergolic igniters. His team innovated a mixture of hydrazine-hydrate (N_2H_4-H_2O) as fuel with 80% hydrogen peroxide for oxidizer. Foolhardy technicians had to insert two thin tubes supported by a wooden stick into the thrust chamber from below. Upon an ignition signal, a ground-mounted supply unit, including remotely operated valves, fed the two hypergolic components into the injection region, where they burned with a spontaneously igniting hot flame. Then the technicians ran for their lives. The problem was that if ignition was delayed even by milliseconds from the time the propellants went through the injector and entered the thrust chamber, an explosion usually resulted. By today's standards, their ignition method was clumsy, to say the least. The thin feed tubes frequently clogged. The engine had to eject a considerable amount of inert solid material (including the tubes and the stick) during the critical engine starting process where conditions are iffy at best. For repeated starts, the unwieldy ignition system would have to be

mounted to the vehicle, adding inert flight weight. But the Germans had to live with it until they could invent a better way.

A third, huge problem was propellant tank pressurization. In those days, the pressure of the burning gases in rocket combustion chambers was around 10 atmospheres (slightly over a megapascal, or 147 psi). As we saw in the Robert chapter, the tanks have to be pressurized way more than this to overcome the pressure drops through all the valves, propellant ducts, and the injector. That meant the fuel and oxidizer tanks had to withstand at least 15 atmospheres or so (1.5MPc, or 220 psi), which made them quite heavy (think how much a scuba tank weighs). To make matters worse, as rockets got larger, structural weight exceedances were magnified exponentially thanks to physics and the rocket equation. In other words, a huge rocket like the Saturn V could barely get off the ground in proportion to its size, whereas a small toy rocket just zooms off the test stand because of its smaller weight. It basically amounts to the thrust to weight (t/w) ratio. Let's say you want to launch a 1000kg rocket. If its thrust is only 900 kgf (kilograms-force) it will just sit there. It has to overcome the gravitational force of 1000 kgf holding it down, and by more than just a little. It all depends on the rocket's size and mission. The huge Saturn V moon rocket had a thrust-to-weight ratio of less than 1.2, the Space Shuttle's t/w was 1.5. If a small toy rocket had a t/w of 1.5, it would just wobble off the test stand and fall over.

The Germans tried to increase liquid oxygen evaporation (and hence the pressure) in the LOX tank with small burning cartridges. But putting hot gaseous oxygen in the alcohol tank led to explosions. They could use a bottle of compressed nitrogen or evaporated liquid nitrogen as pressurizing gas, but that meant separate tank, valves, and lines which incurred a weight penalty. Furthermore, as the fuel or LOX drained from the tank, the gas would expand and the pressure would drop, resulting in a drop in the rate and pressure of propellant delivered to the engine over time. That meant a slow drop in thrust because thrust is directly proportional to the chamber pressure which in turn is proportional to the pressure in the propellant tanks. The rocket pioneer Herman Oberth had already suggested a solution. They would have to develop complicated turbopumps for larger missiles to boost up the pressure. The modern Space Shuttle Main Engine and its derivative on NASA's Space Launch System use this method.

The fourth and biggest challenge was the design and construction of the rocket itself. The Germans would occasionally pick Robert Goddard's brain by contacting him directly with technical questions. Von Braun used Goddard's plans from various journals and incorporated them into building the next generation of rockets, the "Aggregat "series (Aggregate or Assembly), a nondescript term used for secrecy reasons.

THE GERMAN AGGREGATE ROCKETS

Wernher von Braun led his team for over 12 years (1932–1945) as they developed the *Aggregate* series of rocket designs as prototypes or test articles with the goal of developing the world's first single–stage ballistic missile. Von Braun's close associates knew that he had his eyes set on opening up the space frontier. But he realized early on that he would have to kow–tow to the German military if he was to progress beyond the *Rakentenflugplatz* group of amateurs to fulfill his dreams. He undoubtedly felt patriotism towards his country in engineering weapons of war for the German army, just as did American, British, or Soviet aerospace engineers.

One can speculate as to whether von Braun's conscience bothered him when the V–2s were killing civilians in London or Antwerp. Peenemünde was under daily threat of American bomber attacks by day, and British bombings by night (on August 17, 1943 Peenemünde was indeed massively bombed). During the Cold War military personnel on both sides were ready to kill not thousands, but millions with the proverbial push of a button (actually it was with Top Secret launch codes, fail–safe weapons, keys and switches, two or more person policies, etc.). It is well known that von Braun despised fascism and the bellicose Nazis. It is easy to castigate one's fellows in hindsight. If one were to do so, *all* of the Peenemünde engineers would have been guilty and the world's rocket science would have been set back a decade or more.

> *The venerable V–2 is the basis for most of the rocketry that exists in the world today. It represented a quantum leap in technology for its time.*

In any case, you could sum von Braun and his team's achievements with one word: *V–2.*

The enterprising German rocketeers embarked on a step–by–step development program to achieve the awesome (for their time) capabilities of the V–2. After years of development and many tests, the final prototype in the Aggregate series—the A–4—was renamed the V–2 that was introduced in the Theodore chapter. Its development followed the V–1 (for *Vergeltungswaffen–1*, "Vengeance Weapon–1", a name chosen by the German Propaganda Ministry because of the devastating Allied bomber attacks during the war).

Hitler delayed development of the V–2 in the years 1939–1942 by not giving it his support. He changed his mind in mid–1943 and ordered the highest priority be given to perfecting the missile as a war weapon. In 1944 Germany went all–out in producing about 6,152 V–2s in a little over a year (Feb 1944 – Mar 1945), at great cost to its armaments industry which—in hindsight—could have been producing more effective weapons. In only seven months, -September 1944 – March 1945–Germany managed to launch 3,170 V–2s against Allied targets. That sounds like a great number, but it only represents 51.5% of the missiles that were built and many were not successful in hitting their targets. In fact each V–2 successfully launched killed only three people on average. Its principal effect was psychological because, unlike the subsonic V–1 that you could hear coming because of the loud buzzing noise it made, like an angry bee, no one could hear a V–2 coming down because of its supersonic Mach 3.3 speed.

Roadmap and Overview

Mastering the complexities and intricacies of rocket science and engineering required endless research and learning what works and what doesn't through test after test after test, first at Kummersdorf and then at Peenemünde as described below. Nobody in the world had ever done this stuff before. Almost everything von Braun and his team did was a world first, and paved the ground for those brave enough to follow.

The next page shows a timeline and roadmap for the Aggregate rockets. The highlighted V–1 and V–2 were the only ones the Nazis used in World War II. Von Braun's team of Peenemünde rocketeers put most of their effort into the A–4/V–2, but small research teams worked on many more designs that were not authorized. In one way or another, the victorious Allies picked up on all their amazing technology advances.

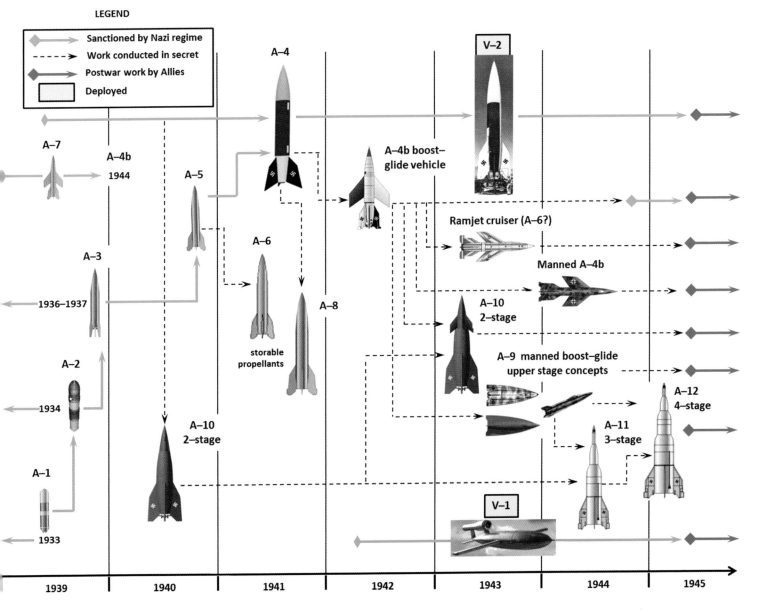

Although von Braun's team didn't have a direct hand in developing it, the V–1 weapon is also shown for completeness. The lines and arrows show how you have to keep building and building on previous achievements—or learn valuable lessons from failures—and keep chugging. Over 12 years Von Braun's rocketeers developed a plethora of Aggregates, some good, some bad, and many that never got past paper designs despite their best efforts. *Lots and lots of designs. When they had to "pack their bags" after losing the war to the Allies, they destroyed or burned many one–of–a–kind designs, reports, and documents to prevent their capture. That still left them with 14 tons of the most important ones they had to hide. But they couldn't take the many railroad box cars–fulls of partially finished missiles, spare parts, test rigs, and so on. The Allies got them, as we saw in the last Chapter.*

- The A–1, A–2, A–3, A–5. and A–4 can be grouped in that order as development versions of the production version V–2 rocket that was launched against the Allies in WWII. The A–5 was a subscale model of the full–size A–4.

- The A–6 was derived from the A–5 subscale model, but with alternate propellants (probably storables).

- The unmanned A–4b was an A–4/V–2 with wings. The A–7 was built as a subscale model of the A–4b. Adding wings increased the A–4's range by 87.5%, everything else being equal (payload, propulsion, gross weight). Only one prototype flew, but it lost a wing coming down so it never demonstrated the increased range.

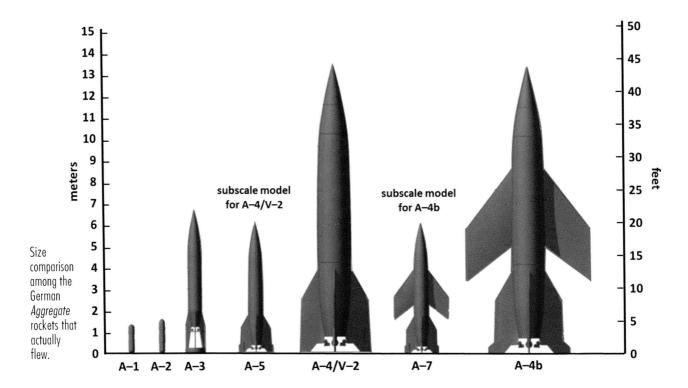

Size comparison among the German *Aggregate* rockets that actually flew.

A like number of Aggregate vehicle types were more ambitious designs that—fortunately for the allies—never made it off the drawing board. All were designed for vertical launch. They will be described later in this chapter, and can be enumerated as follows:

- The A–8 was a designed as a stretch version of the V–2 (about 9.6 tonnes heavier) using storable propellants (so called because they can be kept at ambient temperatures versus supercold cryogenic temperatures). Von Braun dubbed his creation the "Super V–2." Although it never flew during the war, some members of von Braun's rocket team got together in France after the war and evolved the design, leading to today's Ariane series of European civilian expendable launch vehicles. Arianespace launches Ariane rockets from Kourou in French Guiana, where the proximity to the equator gives a significant advantage for the launch.

- The A–9s were a family of derivatives of the A–4b with longer range and better performance. Manned versions featured a pilot in a pressurized cockpit and landed horizontally. Putting a man onboard made it 3.5 tonnes heavier and decreased the range. The pilot had to make an unpowered landing and use a drag chute like the Space Shuttle. Von Braun tinkered with the propulsion system in an effort to increase its range, for example, he designed a variant with a ring of 10 solid propellant rockets.

– The A–6 designation was apparently reapplied to a version of the A–9 with a ramjet propulsion system. It was vertically launched into space. After reentering the atmosphere and beginning the supersonic glide phase, the ramjet would be ignited so that it could operate for 15–20 minutes as a supersonic cruise missile. The pilot could return to base and land horizontally.

The rest of the Aggregate rockets were larger multistage launch vehicle designs that can best be thought of as a set of Russian *Matryoshka* dolls, a set of wooden dolls of decreasing size placed one inside the other. Here's how it works:

- The A–10/A–9 was a two–stage intercontinental boost–glide missile. The A–10 was the boost stage and used either six A–9 combustion chambers or one large combustion chamber feeding into a single nozzle. It carried a modified A–9 as nested second stage, which also served as its payload.

- The A–11/A–10/A–9 was a three–stage ICBM or a satellite launcher. The A–11 was the boost stage and used six A–10 engines. It carried an A–10 nested within its propellant tank. The second stage was the A–10 boost–glide missile, which in turn carried an A–9 inside as its payload. To achieve orbit, the A–9 would have to be modified. A payload of only a few hundred kg could be orbited at 300 km altitude.

- The A–12/A–11/A–10/A–9 was the largest Matryoshka doll. As a four–stage vehicle consisting of the A–12, A–11, A–10, and A–9 stages, it could place up to 10 tonnes into a 300 km low Earth orbit.

Another point of confusion is that scholars frequently use the reverse nomenclature A–9/A–10, A–9/A10/A–11, A–9/A–10/A–11/A–12 to denote multistage Aggregate models, which begins with the smallest (top) stage. Although perhaps more accurate in depicting von Braun's Matryoshka doll nested vehicle paradigm, to avoid confusion we will adopt the simplified notation A–10, A–11, A–12; where A–## is simply the largest outer "doll". The A–11 and A–12 were intended as orbital launch vehicles to take the Third Reich to the planets. *To reiterate, German rocket engineers were way ahead of everyone else in the world at this time in rocket science, design, and engineering.*

Aggregate Rocket Family*

NAME	USE
A–1	Rocket technology development prototype. First in series of rockets leading to V-2. Exploded at Kummersdorf in 1933 during a test run. Considered aerodynamically unstable (a stabilizing flywheel was mounted forward) and no launch attempts were made.
A–2	First flight test rocket in the series that led to the V-2. Two were built, dubbed Max and Moritz. Both were successfully flown in December 1934.
A–3	First large rocket attempted by Wernher von Braun's rocket team. It was equipped with an ambitious guidance package consisting of three gyroscopes and two integrating accelerometers. The rocket was intended as a subscale prototype for the propulsion and control system technology planned for the much larger A–4. All four of the launches in 1937 were failures, and a total redesign, the A–5, was developed.
A–5	Subscale test model of A–4 (V-2). Replaced the A–3 in this role after its unsuccessful test series. The A–5 used the same powerplant as the A–3, but had the aerodynamic form of the A–4 and a new control system. It had many successful flights between 1938 and 1942.
A–4/ V–2	Full scale development began in 1939 and used the preceding Aggregate rockets for testing. In the years 1939–1942 development was hindered because Hitler didn't think the technology was mature enough to be militarily effective. The A–4 was successfully launched in October 1942, despite Hitler's lack of support for the program. After a briefing in mid–July 1943 he changed his mind and ordered that all resources be devoted to the perfection and production of the A–4 as the V–2, a weapon of war.
A–7	Subscale test model of the A–4b rocket, tested 1940, resurrected 1944.
A–4b	Unmanned, winged boost-glide version of the V-2 missile. Research and design began in 1939 and wind tunnel tests at Peenemünde followed in 1940 using the A–7 subscale model. Development slowed because of other demands on von Braun's team. Work slowed further in mid–1943 when Hitler ordered that all efforts were to be focused on the V–2 as a weapon-in-being. Nevertheless von Braun managed to continue some development and flight tests by disguising boost–glide missile work under the cover name A–4b (i.e. a modification of the A–4, and therefore a production-related project). Two test flights: one failure, one partial failure in January 1945.
AGGREGATES DESIGNED BUT NEVER FLOWN	
A–8	Designed by von Braun in 1941 as 'Super V–2' stretched version of the V–2 ballistic missile that used storable propellants. Never reached the hardware stage, but design continued after the war in France which gave rise to the Ariane family of rockets which are in common use today.
A–6	Version of the A–5 subscale test model of the A–4 (V-2) using alternate propellants (1941). Later applied to a manned boost–glide cruise missile like the A–9, but ramjet–powered (1943). It could return to base under ramjet power after performing its mission.
A–9	Family of manned boost–glide cruise missiles patterned after the A–4b. Variants were designed with different propulsion systems. A modified A–9 served as the smallest "Matryoshka doll" of the A10/A–11/A–12 series.
A–10	Intercontinental boost-glide missile. The A–10 was the world's first practical design for a transatlantic ballistic missile. Design of the two stage missile began in 1940 and first flight would have been in 1946. Work stopped in mid-1943 because of Hitler's dictum forbidding non–V–2 development; then resumed in late 1944 under the code name Project Amerika. No significant hardware development was possible after the last A4–b crashed.
A–11	Winged orbital launch vehicle. The A–11 was planned at Peenemünde to form the basis for launching the first earth satellite— or as an ICBM.
A–12	Would have been the world's first true orbital launch vehicle, as sketched out at Peenemünde. It would have been a four-stage vehicle and calculations suggest it could have placed 10 tonnes into low Earth orbit.

* The logical order shown did not follow a numerical sequence. Some models like the A–7 were developed first, then put on hold for various reasons. Von Braun had to use the A–4b nomenclature to disguise his work.

By June 1933 von Braun's Kummersdorf team of rocketeers had completed the drawings for the first missile in the *Aggregate* series, the A-1. Von Braun designed the A-1 around an engine developed by Arthur Rudolph at the Heylandt Company. A derivative of the VfR group's Repulsor rocket, it burned LOX/alcohol and developed 3 kilonewtons (kN) of thrust. The size of a toy hobbyist's rocket, the 1.4m long A-1 blew up on its first flight attempt, taking the test stand with it. Another A-1 was destroyed on the ground by a mechanical failure of the LOX tank. The Germans had persistent difficulties with the fuel and oxidizer valves. They had to invent better, more precise propellant valves that wouldn't freeze up, overheat, or otherwise get stuck. If the valve operation timing wasn't perfect the rocket's ignition would be delayed, resulting in a so-called "hard start." Delays by even milliseconds past a certain threshold had the same effect as a faulty igniter—a big explosion as a hot ball of gases burst open the combustion chamber.

Even if the engine worked flawlessly, early rockets had a bad habit of careening wildly out of control. A rocket is no good if it deviates from a specified trajectory and goes wherever, like an inflated toy balloon when you let it go. Artillery shells or bullets fired from rifles or guns are spun longitudinally to make them stable in flight--that is why rifle and gun barrels are rifled. Today's satellites are often spin-stabilized. But unlike a solid rocket, a liquid fueled one cannot be spun on its axis. Centrifugal forces would disturb the propellants in the feed lines and tanks. Like water in a spinning bucket, propellants would rise up on the wall of the tanks, making it more difficult to feed them to the thrust chamber. As a crude interim solution, Dornberger proposed that only part of the vehicle be spun up. Thus inside the nose of the A-1 was a heavy rotating wheel, that used the principle of the gyroscope to stabilize the missile by brute force. Before launch, the A-1's gyroscope would be spun up to 9,000 rpm by an electric motor on the ground, then left to run solely on its momentum during the rocket's brief flight.

Gyroscopes

A gyroscope's axis, like that of a spinning top, will tend to remain fixed in space. If perturbed by an external force, it will move or "precess" at a right angle to the force exerted. The behavior of a gyroscope can be most easily appreciated by consideration of the front wheel of a bicycle. If the wheel is leaned away from the vertical so that the top of the wheel moves to the left, the forward rim of the wheel also turns to the left. In other words, rotation on one axis of the turning wheel produces rotation of the third axis. The gyro's resistance to precession is directly dependent on its angular momentum, a product of its mass and rate of rotation.

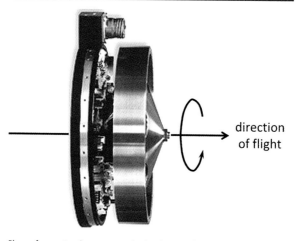

direction of flight

Photo of a reaction/momentum wheel such as used on the A–1. When spun up to high speed, its gyroscopic effect resists changes to the orientation of the spin axis. It stabilized the longitudinal orientation of the A–1 missile in flight—in other words, kept it going in the same direction.

The A-1 was destined to never fly. Unsure of its stability and other problems, the Germans gave up on it and moved on to the A-2 in 1934. This time von Braun's group separated the tanks and placed the gyro rotor between them. Moving the heavy gyro to the middle had the advantage of moving the vehicle's center of gravity backwards, thus shortening the moment arm of any deviations of thrust away from the rocket's longitudinal axis. This increased the rocket's stability in the critical early part of its flight, when aerodynamic forces were weakest because of the vehicle's low velocity. However, stability actually decreased in the latter part of the flight because the rocket's center of gravity moved closer to its center of aerodynamic pressure (in the extreme, this is what causes an airplane to "stall" in flight). Separating the tanks in the A-2 design stopped another problem: oxygen leakage into the fuel tank caused by vibration-induced cracking of the oxygen tank. As any astronaut will tell you, rockets shake and vibrate in flight, sometimes violently. .

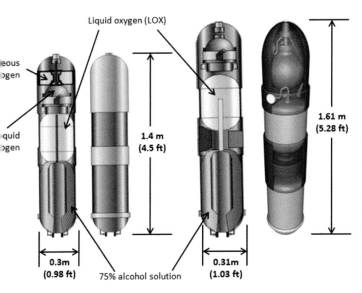

Liquid oxygen (LOX)

eous
ogen

quid
ogen

1.4 m
(4.5 ft)

1.61 m
(5.28 ft)

0.3m
(0.98 ft)

75% alcohol solution

0.31m
(1.03 ft)

From left to right: cutaway views and concept of the Aggregat–1 (A–1) and Aggregat–2 (A–2).

JULY 27, 1934

Herr Professor Doktor Erich Schumann (grandson of the classical music composer Robert Schumann) was frowning. "Did you hear about Dr. Kurt Wahmke and his two technicians at Kummersdorf? On July 16 they were killed and another technician injured when they were testing a premixed 90% hydrogen peroxide (H_2O_2)/ethyl alcohol propellant. The fuel tank exploded." "I heard and saw the explosion. I ran over to help, but it was too late," a teary-eyed von Braun replied. "When we just lose hardware in a test or mishap, that is the name of the game. But when humans lose their lives…" his voice trailed off.

Dr. Schumann drew everyone's attention back. "Wernher, you have impressed the doctoral committee with your creative doctorate thesis, and defended it well in June. Your seminal research has opened the door to Germany's exploitation of the new field of rocketry, which promises to make long range artillery obsolete. While American is sleeping, Germany is poised to become the world leader in this field. But why do you insist on not joining the National Socialist German Workers Party, when that is the future of Germany?" he entreated. Von Braun just sat there, eyeing his new Diploma. After the usual headings of the newly established Physics Department II of the University of

Berlin, it read, "Wernher Magnus Maximilian von Braun has distinguished himself and is awarded this Doctor of Philosophy in Physics with Aerospace Engineering Option," followed by the usual signature blocks for the academic leaders, provosts, and president of the University. Now Major Walter Robert Dornberger had been looking on in silence. "I, too, am distressed by your decision," Dornberger finally said.

Promising to resurrect a revitalized Germany from the humiliation of the Treaty of Versailles after World War I, Adolph Hitler's National Socialist German Workers Party (NSGWP, *Nationalsozialistische Deutsche Arbeiterpartei* or Nazi Party) allowed rocketry to become a national agenda (although it took Hitler himself several years to become convinced rockets were real weapons). Even in abridged form, von Braun's publicly available dissertation, "Regarding Combustion Experiments," was impressive, exceeding the University of Berlin's exacting standards. But Major Dornberger had been invited to a closed door meeting to view von Braun's real thesis. His subject was so secret that even the title was classified: "*Konstruktive, theoretische und experimentelle Beitraege zu dem Problem der Fluessigkeitsrakete*" (Construction, Theoretical, and Experimental Solution to the Problem of the Liquid Propellant Rocket), which he finished on April 16, 1934.

"Please, let me think about it," von Braun finally entreated. *I dare not tell Dr. Schumann and Maj. Dornberger what I'm actually thinking*, he thought.

Throughout his younger years, von Braun had not shown signs of religious devotion, or even an interest in things related to the Lutheran church or to biblical teachings. In fact, he was known to his friends as a *fröhlicher heide* (merry heathen). But the excesses of the Nazis were too much. The way they swept into power through murder and intrigue. Hitler's diabolical preoccupation with exterminating every last vestige of Jews and other "non-Aryan" German citizens, when it was the very Jews who had gotten German technology this far.

With the rise of the Nazi Party, the physics community in Germany suffered a terrible blow. Fully 25 per cent of academic physicists in Germany, almost all Jewish, found themselves forced from their positions shortly after Hitler came to power. By 1934, one of every five institute directorships in Germany was vacant. Not only was the number of physicists who left Germany large, but the quality was truly astounding. Fascism flushed away the cream of

European physics: Albert Einstein, Hans Bethe, Edward Teller, Leo Szilard, Eugene Wigner, John von Neumann, Michael Polanyi, Theodore von Kármán, George de Hevesy, Felix Bloch, James Franck, Lothar Nordheim, Enrico Fermi, Niels Bohr and Eugene Rabinowitch. Along with some sympathetic non-Jewish scientists such as Erwin Schrödinger and Martin Stobb, these men were to become the driving force behind atomic research in Britain and the U.S.

Then the most recent atrocity, von Braun lamented. The *Röhm-Putsch* (Night of the Long Knives) where between June 30 and July 2, 1934 the Nazi regime carried out a series of political executions numbering in the hundreds. Most of the killings were carried out by SS, the elite Nazi corps, and the *Geheime Staatspolizei* (Gestapo), the regime's dreaded secret police. Over a thousand Germans were arrested on trumped-up charges. Hitler targeted members of the *Sturmabteilung* (SA), the paramilitary Brownshirts, and their leader, Ernst Röhm. Publicly Röhm remained loyal to Hitler and was his Chief of Staff. Privately, he was highly critical of Hitler's policy towards the *Reicheswehr* and maintained absolute control over the SA. Hitler saw the independence of the SA and the penchant of its members for street violence as a direct threat to his regime.

From 1919 until their renaming to *Wehrmacht* ("Defense Power") in 1935 after the Nazis came to power, the German armed forces had been known as the *Reicheswehr* ("National Defense"). The Wehrmacht consisted of the Heer (Army), the Kriegsmarine (Navy) and the Luftwaffe (Air Force). The *Schutzstaffel* ("Protective Squadron") or SS, started as a small paramilitary section of Heinrich Himmler's *Allgemeine SS* ("General SS"). Under Himmler it was renamed the *Waffen-SS* ("Weapons SS") and grew from three regiments to a force of over 38 divisions (almost a million strong) which served alongside the regular army, but was never formally part of the Wehrmacht. It was Adolf Hitler's will that the Waffen-SS never be integrated into the Army. They were to remain the armed wing of the Party and were to become an elite police force once the war was over. For this reason, although operational control of the Waffen-SS units on the front line was given to the Army's High Command, in all other respects they remained under the auspices of Heinrich Himmler's SS organization, and behind the lines these units were an instrument of political policy enforcement, including the implementation of the Nazis' "Final Solution" to the Jewish problem.

On July 13, 1934, Hitler justified the purge in a nationally-broadcast speech to the Reichstag (the German parliament)

"In this hour I was responsible for the fate of the German people, and thereby I became the supreme judge of the German people. I gave the order to shoot the ringleaders in this treason, and I further gave the order to cauterize down to the raw flesh the ulcers of this poisoning of the wells in our domestic life. Let the nation know that its existence—which depends on its internal order and security—cannot be threatened with impunity by anyone! And let it be known for all time to come that if anyone raises his hand to strike the State, then certain death is his lot."

Concerned with presenting the massacre as legally sanctioned, Hitler had the cabinet approve a measure on July 3 that declared, "The measures taken on June 30, July 1 and 2 to suppress treasonous assaults are legal as acts of self-defense by the State." *Bullshit*, von Braun thought. *The sight of the Nazi swastikas all over the place makes me sick.*

By the end of July the A-2 design was still not finished. Difficulties in getting proper atomization and evaporation of the propellant droplets drove von Braun and his co-workers toward longer and longer combustion chambers to give the propellant mixture more time to burn completely. In fact, incomplete burning was one of the main causes of suboptimum engine performance. Von Braun's pioneering work helped lay the foundation for today's combustion science, which affects not only rockets, but anything that uses propellants to fly in the atmosphere: airbreathing missiles, supersonic transports like the Concorde and Tu-144, and future hypersonic-speed vehicles cruising in the atmosphere at Mach numbers above 5. Each of these vehicles requires both fuel and oxidizer. Airbreathers only have to carry their fuel along because they get oxygen from the surrounding atmosphere. Furthermore they benefit from the fact that most engines require several times more oxygen by weight than fuel.

Rocket Combustion

Airbreathing vehicles, missiles, solid and liquid propellant rockets get their best performance by getting the propellants to burn completely. Given enough time for the chemical reactions to fully occur, the products of combustion (what comes out of the nozzle) will be in chemical equilibrium at an equilibrium temperature that engineers call the adiabatic flame temperature.

Getting complete combustion requires just the right proportion of oxidizer to fuel by weight—this is the so-called stoichiometric oxidizer to fuel ratio. It is like having just the right mix of ingredients for a recipe. The stoichiometric mixture ratio varies. For example, the stoichiometric ratio for gasoline engines is about 3.4. But air only has 23.2% oxygen by weight, so a gasoline engine has to take in 3.4/0.232 = 14.7 pounds of air per pound of gasoline for optimum efficiency (as an aside, the ratio is slightly lower for modern gasoline engines due to additives). Any mixture less than an average of 14.7 to 1 is considered to be a rich mixture, any more than 14.7 to 1 is a lean mixture (from the point of view of the fuel). Diesel engines have better fuel economy because they typically run 40% lean, so their mixture ratio is about 24.5. Gasoline engines can't do this, because at lean mixtures gasoline produces much hotter combustion gases than does a stoichiometric mixture, so much so that pistons can melt as a result.

Von Braun's Aggregat series of rockets used a mixture ratio of about 1.3 by weight (1.3 kilos of LOX were required for every kilo of alcohol fuel). Strictly speaking the ratio was 1.7 accounting for the fact that the fuel was only 75% ethyl alcohol mixed with 25% water.

Von Braun and his team also had to grapple with complex aerothermodynamics. Aerodynamics is the study of vehicles in the sensible atmosphere, for example, the DC-3 which Douglas aircraft was developing at the time. Von Braun had to add the thermal part because of aeroheating, where vehicles are going so fast through the atmosphere that their surfaces get red hot. For example, the supersonic SR-71 (Strategic Reconnaissance-71) spy plane's cockpit windows got so hot that crewmembers could use them to heat up their sandwiches! A more extreme example is the Space Shuttle, where the nose and leading edge surfaces got white-hot on reentry and it blazed like a blowtorch. Early on, von Braun collaborated with Dr. Rudolph Hermann at Aachen University to test a pointed, slender body with fins, for drag and stability up to the design limit Mach number.

Avoiding Armageddon will explain how it was not until the 1950s that a design solution for aeroheating on reentry at hypersonic speeds was discovered, and for years it was a closely guarded secret. After the seminal report by Harvey Allen and Alfred Eggers, blunt nosecone shapes came into wide use in the United States for heavy reentry vehicles.

PEENEMÜNDE

In December 1934 the 107kg (300kgf thrust) A-2 reached over 2.2km in two flight tests on the island of Borkum in the North Sea. By the mid-1930s it had become clear that the heavily populated area around Berlin where the Kummersdorf facility was located was not suitable for the launch of larger and more powerful rockets. Kummersdorf was too small and congested for advanced engine and flight testing. The staff at Kummersdorf already numbered around 80, and more engineers and laborers were needed. A new site was sought on the coast from which rockets could be safely launched, tracked and/or recovered during testing, and which was also secure and secret. Wernher von Braun and his colleagues now had ample funding and could build modern design offices, manufacturing facilities and huge test stands. These were needed for the final development of and testing of such precursors to the V-2 as the A-3 and the A-5, which had been started at Kummersdorf and could now be completed at Peenemünde.

As commander, Walter Dornberger became a masterful salesman, administrator, and political infighter for the rocket program. Wernher became Technical Director of all engineering activities in Peenemünde and was now in a position to hire highly skilled personnel for the operation of all manufacturing, test and launch facilities.

The working environment at Peenemünde was contrary to normal rules of military life. The Peenemünde engineers had very little direct contact with the regular Army officers, even socially. Although enlisted men were often supervising officers, it worked. Most people at Peenemünde thought the working environment was "nothing special", but under von Braun it really was special. They often held "parties with rocket fuel" using the ethyl alcohol delivered as propellant for the rocket. In common with later rocketry projects, the engineers were under tremendous strain and worked very hard but remained enthusiastic.

Von Braun's management style stressed the responsibility of individuals, not committees. He was able to talk to everybody at every level. But often his practice of going around intervening management levels and giving orders to someone else's subordinates caused friction.

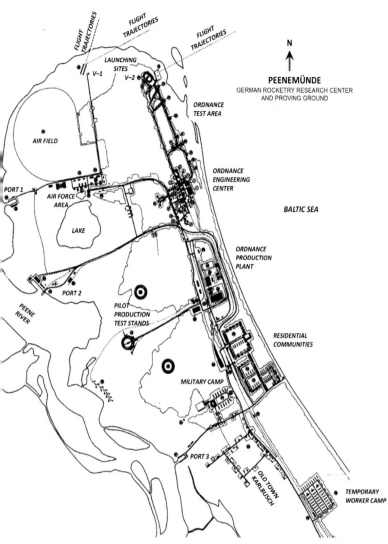

Peenemünde rocketry research center and proving ground as it looked during WWII. Until after the war, the world would not realize how far ahead of everyone else Germany would get in rocket science.

Dornberger and von Braun's rocket team moved to *Werk Ost*. Dr. Walter Theil, Deputy Director under von Braun and five staff stayed behind at Kummersdorf to work on V-2 engine development until the summer of 1940, when the test stands at Peenemünde were finally completed.

Wernher von Braun in German army uniform, early 1940s

Meanwhile, like the Americans <u>after</u> WWII in the wake of the Cold War, the Germans <u>before</u> WWII did not waste any time and pressed their rocket development at breakneck speed. Hitler's Third Reich had a world to conquer and transform into a pure, if not super Aryan race.

The Holy Roman Empire of the Middle Ages could be called "First Reich." The German Empire of 1871-1918 was equivalent to "Second Reich," but until 1943 the Germans called it Deutsches Reich (German Reich). In 1943 the official name became Großdeutsches Reich (Greater German Reich). In colloquial English usage this last Reich, where Reich stands for "sovereign state," became simply Drittes Reich (Third Reich). Today, it is doubtful there will ever be a "Fourth Reich" because of its connotation with Nazi Germany.

The Luftwaffe, too, had decided to invest funds in rocket research and soon a joint operation with the German Army resulted in the selection of a secluded area near the small fishing village of Peenemünde on the Baltic coast. It was perfect—400km of ocean to the east for use as a missile shooting range, and room along the flight path on the coast for tracking radars. In August 1936 the Germans broke ground on "*Heeresversuchsanstalt Peenemünde* (HVP)" (Army Research Center Peenemünde). The huge facility covered the northern peninsula of Usedom (aka Island of Usedom because it is separated by rivers from land) and took almost four years to complete. The Luftwaffe would occupy HVP-West (aka *Werk West*), and the Army would possess HVP-East (*Werk Ost*) and HVP-South (*Werk Süd*). In 1937-1938, most of now Colonel

THE V–1 FLYING 'BUZZ BOMB'

Over in the *Werk Süd* part of Peenemünde, the Luftwaffe was also conducting rocket development, of a sort. About the time rocketry was getting started in Germany (1932), the German engineer Paul Schmidt developed and patented a pulsejet engine. This curious engine operated by creating 500 fuel-air explosions per minute (about 8 Hz). Its characteristic buzzing sound gave rise to the colloquial names "buzz bomb" or "doodlebug" (a common name for a wide variety of insects). It could only operate at a rather low altitude range: 200m to 2km. The Luftwaffe developed the world's first cruise missile, the *Fieseler Fi-103,* around this pulsejet engine. The Argus Motoren company began production of Schmidt's pulse jet engine in 1942. The Fi-103's guidance system was rather crude. When a small propeller in the nose had turned a preset number of times, corresponding to the desired range to a target, the counter pushed the missile's rudder hard over, resulting in a dive to the ground.

As its design rapidly matured, the German Propaganda Ministry renamed the Fieseler Fi-103 as *Vergeltungswaffen–1*, or "Vengeance Weapon–1" (the A–4 became the V–2 as the next formidable weapon). Most V-1s were fired from catapult ramp launch sites along the Pas-de-Calais coast of northern France and subsequently from other sites in German–occupied Western Europe.

The V–1 was about 8.32m (27.3ft) long, including the tailpipe of its jet engine, and had a wingspan of 5.37m (17.5ft) and was 1.4m high. It could be built for a tenth the cost of the awe–inspiring (for its time) V–2 missile. Germany is said to have produced almost 30,000 V–1s during the war. It carried an 850 kg (1,870 lb) explosive warhead at 576 km/hr (360 mph) with an with an average range of 240km (150 mi). The British knew when a subsonic V–1 was coming because of the loud buzzing noise made by its pulsejets.

9,521 V–1s were hurled against London in a span of five months, from June to October 1944. Only about 2,419 reached London, killing 6,184 people and injuring 17, 981. From July 1944 to January 1945, the Luftwaffe launched approximately 1,176 V–1s from modified Heinkel He-111 H-22s flying over the North Sea. In October, 1944, under pressure from invading allied

100m–long starting ramp for V–1 "Flying Bombs" (left), rear view of V–1 showing launch ramp section

V–1 "Dootlebug" cutaway view and in flight

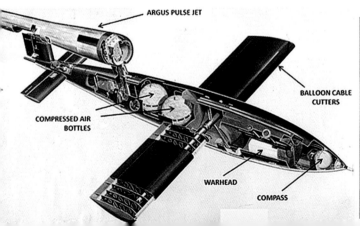

ARGUS PULSE JET

BALLOON CABLE CUTTERS

COMPRESSED AIR BOTTLES

WARHEAD

COMPASS

239

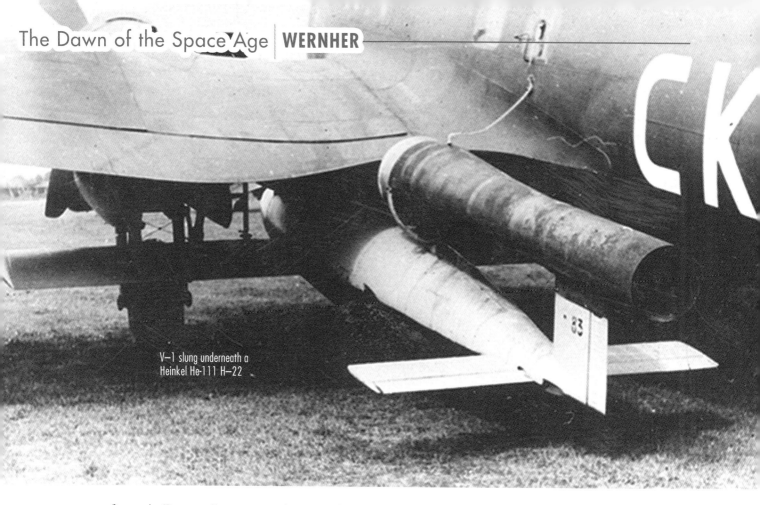

V–1 slung underneath a
Heinkel He-111 H–22

forces in France, Germans on the ground switched targets and fired another 2,448 V–1s against the port of Antwerp in Holland and other targets in Belgium. The attacks stopped when the last site was overrun on March 29, 1945. In total, the V–1 attacks killed about 9,000 people (about the same as the V–2) and injured some 22,892, mostly civilians. *As shown later, accounting for their lower cost, V–1s were about 2.4 times as efficient as the V–2 in terms of Deutsche Marks spent per person killed in war.*

However, the V–1 rockets had several problems, because they:

1. Were often prone to explode prematurely, occasionally resulting in the loss of the aircraft from which they were dropped. The Luftwaffe lost 77 aircraft as a result of the launch of these sorties:
2. Flew at a relatively slow 160 m/s (576kph) and low altitude below 2km, making them vulnerable to British flak batteries and interceptors.
3. Provided the enemy with a forewarning of attack by its characteristic engine "buzzing" noise and the cutoff of that noise when it went into its terminal dive.
4. The V–1 usually had to be catapult-launched from fixed "ski ramps." In contrast, the V-2 was mobile, more accurate, could not be intercepted, and gave the enemy no warning of an attack in its supersonic ballistic course to the target.

The British deployed barrage balloons against the V–1 missiles, but the Germans ingeniously equipped the leading edges of the V-1's wings with balloon cable cutters as shown on the previous page, and fewer than 300 V-1s are known to have been destroyed by hitting the cables.

Barrage balloon over London

A–4 STEPS TO DEVELOPMENT

A–3

In the meantime, Von Braun's team hurriedly moved on to the A-3, a much more ambitious design than the A–1 or A–2; the first large rocket attempted anywhere in the world. It was some 67cm (2.2 ft) in diameter, 7m (23 ft) long, weighed 740kg (1631 lb), and generated 14.7 kN of thrust. Several times larger than the A-2 in every respect, it was equipped with a bigger guidance package consisting of three gyroscopes and two integrating accelerometers. A notable A-3 innovation was a liquid nitrogen tank (inside the liquid oxygen tanks) which was heated by small electrical elements to provide gaseous nitrogen to pressure-feed the propellants. A small film camera recorded instrument readings in the nose. The A-3's expected longer range required temporarily moving the launch site from Kummersdorf to the island of *Greifwalder Oie* because they couldn't wait for the marvelous Peenemünde facilities to be completed. Unfortunately, all four test launches attempted in December 1937 were failures.

> *Early rocketry was an inexact science, with progress registered through trial and error. However, the German team learned more from their failures than their successes. Von Braun would later quip that, "Our main objective for a long time was to make it more dangerous to be in the target area than to be with the launch crew."*

The A-3 was intended as a subscale prototype for the propulsion and control system technology planned for the much larger A-4. But it developed stability and aerodynamic control problems and had an unsuccessful test series. The Germans replaced it with a better subscale test model, the A–5. It was slightly smaller than the A–3 [5.8 m (19.1 ft) long] and retained the A–3's propulsion system, but incorporated the aerodynamic form of the A–4 and a better control system.

A–5

Development of the aerodynamics and control systems for the A–4 took hundreds of tests of the A–5 - in wind tunnels, air-drops, and powered flights. This was also a grueling trial and error process. How could a missile with a range of 320 km be guided accurately to its

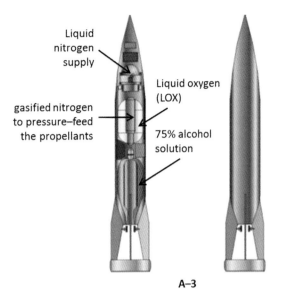

Liquid nitrogen supply

gasified nitrogen to pressure–feed the propellants

Liquid oxygen (LOX)

75% alcohol solution

A–3

A–5

Concept and cutaway view of the Aggregat 3 (left) and A–5 testbed development prototypes. 25 versions of the A–5 flew from 1938 to 1942 from Peenemünde, some several times. The test flights were essential in refining the aerodynamics and other technology for the A–4.

target? Ernst Stienhoff's guidance team had to pioneer a control system better than the A–3's that integrated bits and pieces of known technology into a heretofore untried application. The gyros had to do a better job of orienting the missile along a predetermined path in a vertical plane pointed at the target. The accumulating accelerometers had to precisely determine when the missile had reached the correct velocity and then cut off the engine. It was thought that this would provide sufficient accuracy, although operations would indicate otherwise.

The turbopumps to feed the propellants to the engines proved relatively easy - to von Braun's surprise, high-volume low-weight pumps were already well developed for fire engines in Germany. He simply adapted the pumps from the high pressure fire engine water pumps which were already in use in Germany at the time.

A–4

March 23, 1939. "Heil Hitler!" announced the contingent of army soldiers escorting the Führer. All the others, including von Braun, snapped to attention and returned the stiff-armed Nazi salute. Adolph Hitler wanted to become more familiar with liquid rocket engine technology to assess its potential as a war weapon. The army decided to stage demonstrations and exhibits at the old Kummersdorf proving grounds near Berlin because Hitler's busy schedule made a trip out to Peenemünde inconvenient. "Sir, I am Colonel Walter

Dornberger, your commander here," said Dornberger. Dornberger politely introduced key members of his team, starting with the fair-haired von Braun, then Walter Theil, deputy director under von Braun, Arthur Rudolph, Ernst Steinhoff, Rudolph Hermann, and key others. *My guys are such a tight-knit, cohesive group,* Dornberger thought to himself, smiling. The astute Hitler homed in on von Braun. "Well, I see you are now one of us. I see from your dossier that you joined the Party in…in November 1937. Very good. Very good. Now I think you should become a bonafide officer in the SS." When Hitler "suggested" something, von Braun and everyone else in the room knew it was practically an order. "Jawohl, mein Führer!" he replied. Thankfully, Hitler had glossed over von Braun's brief membership in Himmler's Waffen-SS horseback riding school back in 1933-1934.

Dornberger led the contingent to a conference room where a color-coded cutaway model of the precursor A-3 rocket was set up. He let von Braun take the floor. Von Braun and Dr. Theil after him explained its systems in meticulous detail. Then they began describing the role of the A-5 design as a subscale test model of the A-4, their ultimate masterpiece. "These A–3 ground–truth test data," Dr. Theil lectured, "back our decision to use its rocket propulsion engine on the A–5." Stienhoff stood up to explain the challenging guidance and control aspects. "We worked many months on resolving the A–3's stability and control problems." He pulled out some drawings and put them on the table. "We installed a new Siemens guidance and control package. This configuration design drawing shows the position of the gyroscopes and accelerometers, which integrate the three–vector state data of the missile during flight..." He droned on and on, putting Hitler to sleep.

When von Braun stood up to animatedly describe the A–5's new aerodynamics, Hitler perked up a bit. "These two posters show how the A–5 has the same outer mold line design and four fins as the A–4, which will be the production model. Professor Goddard's work provides little theory and no practical experience in supersonic aerodynamics. We are paving new ground. The A–5 has to be controlled when rising vertically at near zero speed, where aerodynamic surfaces are ineffective. Then it has to remain controllable and stable at subsonic, transonic, and supersonic speeds up to Mach 4. Development will take many months."

Not comprehending the technical details, Hitler looked bored and asked no questions. Later, at lunch

he asked, "I want to know about the Aggregate series development schedule." Dornberger explained, "Mein Führer, we will have the first A-4 flying by early 1942. With the provision of adequate resources by the *Wehrmacht* (Germany's military machine after 1936) we can begin large-scale production of the A-4 ballistic missile in early 1943 at the latest. With it we can bomb New York and other American cities and objects. The subscale A-7 can serve as a pathfinder for a winged missile to convert rocket speed and altitude into aerodynamic lift and range. In several years, we could develop multistage Aggregate vehicles that could even strike America from the Fatherland, and…" "Whoa, how far along are you with these fantastic creations?" Hitler interrupted, chuckling. *I can't believe these guys!* he mused to himself.

"Well, we are encountering problems," Dornberger confessed. "As we briefed you, we learned a lot from the, uh, from the four launch failures of the A-3. All the flight tests confirmed the shortcomings of the control system. It became evident that the step from the A-3 to the A-4 was too large. We decided that a number of unsolved problems could be tackled to better advantage with a smaller rocket, so we have built an intermediate rocket model, the A-5, to test fly first. It has the aerodynamic form of the A-4, utilizes improved gyroscopes activating exhaust jet vanes for controls, and parachutes for recovery. We have conducted numerous ground tests and last summer we successfully launched four A-5s. One reached an altitude of 12km." *Bah*! thought Hitler. "What is the range?" "The A-5's is about 20km." "No, I mean the A-4." "Our design estimate is about 350km, more if we can improve the design."

"And you can have it operational by when?" Hitler asked. "Mid-1942," Dornberger responded. "What is the impact on schedule if you substitute synthetic *Eisenbled* (an alloy) for the light metal alloys in the rocket frame?" "None if the Wehrmacht can make it available in quantity." Hitler then went on a rambling discussion of the deceased Max Valier (we met him in the Robert chapter) and his exploits, dismissing him as a dreamer. Dornberger argued, "but, mein Führer, think about the early days of the Zeppelin, when Lilienthal (whom we met in the Orville chapter) made the first primitive experiments. The state of rocket development is just as immature today—you have to give it a chance." "Airships are dangerous and filled with explosive gas. Look at the Hindenburg disaster," Hitler countered. *On*

May 6, 1937 the German rigid airship Hindenburg, filled with hydrogen gas, caught fire and was destroyed while attempting to dock with its mooring mast in Manchester, New Jersey. Of 97 people on board, 35 died in addition to one fatality on the ground.

The Führer witnessed a few ground test firing demonstrations that afternoon. Finally he departed with handshakes and few words. He may have been impressed, but did not show it. Afterwards, the rocket team was dismayed. Dornberger believed Hitler was enthralled with artillery and tanks (the "Blitzkrieg"), didn't understand the possibilities, and didn't believe the time had come yet for the development of rockets as weapons. Hitler was wrong.

In September 1939 the Luftwaffe dropped an A-5 model from 7,000m in order to test its stability after breaking through the sound barrier. The Germans performed many other flight tests, including pitchover maneuvers and gyroscopic control, and stability tests. Developing fins that would not break off at supersonic speeds was a real challenge. Several A-5 rockets were successfully fired from the Island Oie near the Peenemünde test facility and recovered. They attained altitudes of 12km and ranges of 18km. These flight results provided now a solid base on which the final version of the A-4 could be designed and developed.

In the wartime pressures of building rockets for the army, von Braun had forgotten all about Hitler's "suggestion" that he join the *Schutzstaffel* (SS) Nazi corps. Finally one day in the spring of 1940 Heinz Müller, an *SS-Standartenführer* (Colonel) came to von Braun's office at Peenemünde. After polite introductions, SS-Col. Müller got to the point. "Reichsführer-SS Heinrich Himmler sent me to urge you to join the Schutzstaffel."

Reichsführer-SS was the highest rank of the SS, the equivalent of a General Field Marshall in the German Army. There was never more than one Reichesführer-SS simultaneously in the entire SS, with Himmler holding the position as his personal title and rank from 1934 to 1945.

"I have the orders right here, bestowing on you the rank of *Untersturmführer* (Second Lieutenant)," Müller finished, handing the paperwork to a reluctant von Braun. For his part, von Braun wished that Müller would just go away. But he just stood there, an expectant look on his face. Finally, von Braun told him, "thank you, Standartenführer Müller. I will tend to this later when I have more time, right now I have a pressing launch and test schedule to meet, and we are falling behind." Unperturbed by von Braun's cool reception, SS-Col. Müller said, "Don't forget, Dr. von Braun, to pick up your new uniform and insignia at Werk West." *Peenemünde West was the Luftwaffe Test Site.*

After he left, von Braun felt there was little he could do. The matter was settled. Or was it? He picked up the phone and called Col. Dornberger. "Walter, I do not like the imposition that Himmler is putting on me. It seems that I have no choice. He sent his henchman, Standartenführer Müller, to tell me not only had the orders been signed by Reichsführer-SS Himmler himself, but that I could 'go pick up my uniform and Untersturmführer insignia at the military clothing store in Werk West!'" There was a pause on the phone. "Listen, Wernher. This is a losing battle for you. But first let me ask you, what do you really want to do with rockets?" asked Col. Dornberger. Thinking a moment, von Braun responded, "I want to build rockets for spaceflight. Since a boy, I have been fascinated by this technology, which gives us power to break free of Earth's gravity. With rockets we can reach the moon, Mars, and eventually the other planets. That is my dream for Germany." "Well, let me bring you back to earth, my friend," Dornberger counseled him. "To get your dream, you have to play the game. The game is that we build weapons for the *Heer*."

Germany's Wehrmach consisted of the Heer (Army), the Kriegsmarine (Navy) and the Luftwaffe (Air Force).

Dornberger continued, now in a fatherly tone, "To build weapons for the Heer, you have to continue our mutual work. But to do that, you have no alternative but to join the SS. I must order you not to even mention space travel in Hitler's presence." Wernher's father, Magnus, advised him the same. On paper, von Braun joined the SS and began as an honorary Untersturmführer. After consulting with the other rocketeers at Peenemünde, von Braun decided that his SS membership would protect their work. By virtue of his outstanding technical leadership, he was honorarily promoted three times by Himmler, the last time in June 1943 to SS-Sturmbannführer (Wehrmacht Major).

Many German academics, scientists, and technicians were members of the Nazi Party, often because party membership brought benefits such as research grants and promotions. The Party often bestowed honorary rank as a reward.

Over the next few years Von Braun sought cooperation with universities, especially for research and recruitment. Unlike the Top Secret Manhattan Project to develop the atomic bomb in the U.S. where military authorities tried to prevent project scientists from even talking to each other, the rocket program at Peenemünde was thrown open to the entire scientific community in Germany to tackle the problems of early spaceflight. One of the students recruited in 1943 was Georg von Tiesenhausen, a freshly minted engineering graduate from the University of Hamburg. "The main professors, the lead investigators, became our laboratory directors. We worked closely with the universities all over the country. We gave them the list of problems, and they had to solve them," he later recalled. Dr von Tiesenhausen would design the Lunar Rover, which carried the Apollo astronauts over the lunar surface on some of the later missions of the Apollo program. Another standout recruit was Ernst Stuhlinger, who had earned a Ph.D. in physics from the University of Tuebingen in 1936. He developed guidance and propulsion systems with von Braun. Stuhlinger was also instrumental in the development of ion engines for long-endurance space flight and a wide variety of scientific experiments. His textbook, Ion Propulsion for Space Flight (1964) is a classic.

Von Braun preferred direct private contacts to the more rigid structures of the German bureaucracy. At Peenemünde he established a flexible management system that could respond to external constraints. He envisioned major projects on a vertical axis; and technical support laboratories superimposed on a horizontal axis. Every project manager had direct access to all laboratory facilities. Technical departments were not dependent on the fortunes of any given project, yet had the flexibility to adapt to changing demands. The research team assembled at Peenemünde included men of exceptional talent. Many of them had advanced degrees and practical experience in industry before joining von Braun. Few had worked in rocketry, but expertise in fields like physics, chemistry, mechanical engineering, and electrical engineering suited them to work on various aspects of rocket development. Von Braun remarked, "It's not about the infrastructure, test stands, and facilities; it's about the guys that know how to run them."

V–2 DEVELOPMENT

Going from the prototype A–4 to the production model V–2 was a huge step, involving 65,000 changes to the initial design drawings, which themselves numbered in the thousands.

This remains a fact of modern engineering. Most of the manpower and cost goes into making a working prototype model (whether an airplane, car, ship, or train) ready for production.

The first A-4 flight test model was completed on February 25, 1942. It slipped out of its "corset" after being fully tanked at Test Stand VII, fell two meters, smashed three fins, and came to rest on the rim of the engine nozzle. After being repaired, the rocket failed during the first test firing on March 18 and was junked. The next A-4 was damaged during its fourth ground firing test on May 20. After repairs, it was launched on June 13. A roll-rate gyro malfunctioned immediately after launch. After the rolling missile became supersonic, the propellant feed system failed, telemetry ended at 54 seconds (24.1km downrange), and the missile fell cartwheeling into the Baltic Sea less than a kilometer away and exploded. On August 16 the next test missile penetrated the sound barrier for the first time, but then veered from course and its nosecone broke off after 45 seconds of flight. But like Goddard, the persevering Germans kept at it. After all, they had shown that a missile of this size could be made to work. Hundreds of test firings from 1938 to 1942 brought improvements in stability, propulsion, gas stream rudders used for steering, the wireless communication system. and instruments to plot flight paths.

By now the facility at Peenemünde employed 1,960 scientists and technicians, and some 3,852 other workers. Von Braun's team had suffered through many mishaps, failures, and explosions. Hundreds of test firings from 1938 to 1942 brought improvements in stability, propulsion, gas stream rudders used for steering, the wireless guidance communication system, and instruments to plot flight paths. The entire rocket team conducted extensive component testing of combustion chambers, pumps, valves, guidance and control elements.

Dr. Wernher von Braun at Peenemünde. In January 1943 he became engaged to Dorothee Brill, a physical education teacher in Berlin. However, the engagement was broken off due to his mother's opposition. Later in 1943, while preparing V–2 launch sites in Northeastern France, von Braun had an affair in Paris with a Frenchwoman.

October 3, 1942, the Real Dawn of the Space Age?

Von Braun's overworked rocket team had worked long hours and gotten little sleep. Today would serve as a dress rehearsal for V-2 operational launch procedures. The A–4 was transportable despite its large size (14m or 46 ft high). A major advance was its use of turbopumps for pressurizing propellants, which permitted light-weight propellant tanks. Pruf Stand (Test Stand) VII, the largest at Peenemünde, was capable of static firing or launching rocket engines of up to 200 tonnes thrust, but today the single-stage A–4 would generate only 24.9 tonnes at liftoff. The day was sunny, with few clouds and a cool breeze from the Baltic Sea. "Well, at least we are launching from lucky seven," said technician Fritz Schenck. Seven has always been considered lucky, recall the R-7 ICBM in Russia, the Friendship 7 spacecraft and Mercury 7 astronauts in the U.S.

A parade of launch support and logistics vehicles arrived at Test Stand VII. Von Braun, Colonel Walter Dornberger, and propulsion engineer Konrad Dannenberg were waiting for them. Schenck screwed down the legs

of the *Abschussplattform* (firing table) to lift the weight of the launch platform and missile. The towing dolly was removed, then he carefully adjusted a dial-sight to level the platform. A *Meillerwagen* transporter-erector vehicle was winched back to the firing table. Its arm held the A-4 securely in place in the horizontal position with huge clamps. The arm also functioned as a mobile servicing tower, with built-in plumbing for permanent delivery of alcohol and LOX when fueling. Rolf Cuyler and Dieter Ostheim swung out hydraulic extending supports and screwed them down at the rear of the Meillerwagen. Jürgen Ostheim started a small Volkswagen engine to operate two hydraulic rams, which Norbert Diefendorf controlled to raise the arm of the *Meillerwagen* and the A-4 into the vertical position. Fritz Baldwin connected electrical cables which ran from the Steyr power supply vehicle to the *Feuerleitpanzer* launch control vehicle and the A-4's electrical cable mast.

A–4 on Meillerwagen

Meanwhile a survey crew was busy measuring to make sure the rocket was true to the vertical. Technicians removed protective engine jet covers from the venture in the combustion chamber. Others carefully bolted the fragile carbon graphite exhaust rudders in place. The missile stood proudly, awaiting its launch command. Manfred Keil had even affixed a Nazi flag to the tip of the nosecone. It looked for all the world like a Buck Rogers rocket: pointed nose, proportional body, and four tail fins. The A-4's physical configuration was a mock warhead in the nose; below it was the instrument compartment, and below that the 75% ethyl alcohol/25% water fuel tank. Next came the LOX tank and, between the tail fins, the compartment for the rocket motor and auxiliary equipment. While LOX entered the rocket motor directly from above, the alcohol/water fuel mixture was piped into a cooling jacket surrounding the engine at the bottom. Thus, the motor was cooled by its own fuel, which also entered at the top after having done its duty as a coolant. The auxiliary equipment consisted of two centrifugal pumps driven by a steam turbine, and the equipment to produce the steam for the turbine. There were also two auxiliary fuel tanks: one for high-strength hydrogen peroxide (H_2O_2), the other for potassium permanganate ($KMnO_4$). When these two substances were brought together in the steam generator, the hydrogen peroxide was converted into the steam which drove the turbine. In German vernacular, the LOX was "A-Stoff," fuel was "B-Stoff," H_2O_2 "T-Stoff," and $KMnO_4$ "Z-Stoff" (*Kraftstoff* is fuel in German). The rocket was painted in black-and-white chessboard pattern to aid in determining if it was spinning around its longitudinal axis for stability.

A Fueling Troop arrived at the scene. A *Hanomag* towed an alcohol bowser, followed by an *Opel Blitz* alcohol tanker towing a trailer pump, another Hanomag towing the LOX trailer, and finally an Opel Blitz H_2O_2 tanker. Alcohol fueling took 10 minutes. When it was almost competed, the LOX tanker was towed to the other side of the A-4.

LOX was always pumped into the rocket no more than an hour before firing to keep the internal valves from freezing. During LOX fueling, H_2O_2 was manually pumped into a pre-measured container mounted to the Meillerwagen, which in turn emptied into the A-4's H_2O_2 tank by gravity. Claus Getman climbed up to the A-4's midsection joint and adjusted the tension created by eight tonnes of propellants, while another removed the $KMnO_4$ catalyst from its heater and emptied it into the missile manually. The $KMnO_4$ was kept heated to quicken its spontaneous reaction with H_2O_2 when powering the A-4's 580 HP steam turbine. Fueling completed, the LOX tanker pulled away and the support vehicles retreated to a safe distance.

The igniter was prepared for insertion into the combustion chamber. Schenck readjusted the A-4's orientation with the dial site on the firing table. Then the arm of the Meillerwagen came down. Now the members of the firing troop took cover in slip-trenches. The Feuerleitpanzer firing control vehicle was positioned about 130m away from the rocket, down in a protective trench. Technicians pressurized the A-4's onboard air and nitrogen batteries to 200 bar (2900 psi). The gyroscopes were powered up with 28-volt, 60-amp DC. After so many failures, von Braun decided to serve as launch control officer himself in hopes that Lady Luck would ensure a successful launch. He entered the Feuerleitpanzer. Inside, he asked Dieter Strauss,

A-4 test rocket undergoing fueling and prelaunch preparations, and launch from a V-2 launch pad in Peenemunde, 1943.

Top row: Sequence of video stills of the world's first space launch on October 3, 1942. Although it didn't reach orbit, the missile did exit the sensible atmosphere. Lower left: Video sequence of post–war tests at White Sands Proving Ground, New Mexico. Lower right: Peenemünde test launch on June 21, 1943. The missile exploded after 75 sec. *World firsts are often ill–documented. The video of Neil Armstrong's first steps on the Moon in 1969 would also be grainy and poorly captured. Ditto for the first video of the Moon's far side in 1959, you could barely tell it had all those craters.*

manning the steering table, "Steuerung klar?" Eying his instrumentation, Strauss responded, "Steuerung klar!" Everything got very quiet and tense. This was it. Von Braun called out, "X1" (t-minus one minute). He stepped onto a small perch in the Feuerleitpanzer to oversee the launch site. The seconds ticked by. "Schlüssel auf Schießen!" von Braun ordered. "Ist auf Schießen, Klarlampe leuchtet!" responded Fritz Dering, manning the propulsion controls.

Pressurized air at 32 bar (464 psi) pressed the H_2O_2 and $KMnO_4$ into the turbine where they spontaneously ignited to produce high-pressure steam. Within milliseconds the turbine started rotating at 3800 revs per minute to power the A-4's fuel and oxidizer turbopumps on a common shaft. The fuel pump squished 58 kg/sec of alcohol/water fuel at 23 bar (334 psi) through 1224 precisely engineered injection orifices; while the LOX pump forced 72 kg/sec of LOX at 17.5 bar (254 psi) through 2160 injection orifices into the chamber. The igniter performed its task of igniting the propellants, and the chamber built up 15 bar (218 psi) at 2500 °C, producing 1.5 to 2.5-tonnes of thrust, not enough

to lift the A-4. Von Braun, Dering, and Strauss monitored everything to make sure a precisely established sequence of commands operated properly. After three seconds, von Braun barked his last order, barely heard over the roar of the engine, "Hauptstufe!" After that, Dering pushed a button. The propellant pumps and steam turbine ramped up to 5000 rpm and began to scream as the engine reached mainstage. They could feel the earth shaking and vibrating under the pressure of 25 tonnes of thrust. The cables disconnected electromagnetically. The air batteries and nitrogen batteries kicked in to provide electrical power during flight. The rocket rose majestically straight up with its almost 2:1 thrust/weight ratio. Dering jumped up to the launch table and turned the spanner of the high pressure bottles down. The A-4 kept rising and rising, through some thin clouds— its exhaust flame white-hot and intense. Some soldiers started slowly walking to the launch site, which ironically looked very empty.

To Dannenberg, observing through binoculars, the trajectory looked a bit too steep. The missile was supposed to turn slowly towards the target. Werner Dahm, one of von Braun's leading aerodynamicists, and Walter Theil had figured the A-4 could reach up to 206km (the altitude of an orbiting satellite) if launched vertically. But this launch was supposed to test the A-4's guidance system and downrange capability. Each of the four tail fins had an air vane which acted like an aileron to control the missile in flight and a heat resistant graphite vane in the nozzle exhaust jet for fine trajectory control during powered flight. The vanes relied on radio beams transmitted from ground stations along the missile's flight path for command and control, while a simple analog computer adjusted the azimuth. Two free gyroscopes provided lateral (vertical and horizontal) stabilization. A simple Doppler tracking system measured its velocity. At 58 seconds into flight a radio signal commanded engine shut down. By now Dannenberg had lost the A-4 in the clouds, but a small onboard pitching motor did its duty, precisely moving the eight vanes to pitch the missile along an approximate parabolic flight path.

Everyone waited for observers near the target area to report where the missile landed. After some tense minutes, they got the call. The flight was a success! It had reached 88km altitude and landed 190km downrange. Everybody at the test stand whooped and hollered. For the first time in history, a manmade object had penetrated into outer space! That very afternoon, Dornberger gathered the key Peenemünde personnel. **"We have invaded space with our rocket and for the first time - mark this well - have used space as a bridge between two points on the earth,"** he told them. **"We have proved rocket propulsion practicable for space travel. This third day of October, 1942, is the first of a new era of transportation, that of space travel."** When Professor Oberth got the news, he echoed Dornberger's words. "We have known for a long time that rockets work, if they are built right. We must not forget over this success that our goal is not the launching of rockets; it is the travel through space, and the exploration of the unknown out there!" It was time for celebration. That evening the Peenemünde group made good use of their stock of ethyl alcohol rocket fuel, partying well into the night.

While the above trial operations and launch procedures seem cumbersome, during the war well-trained German troops were able to erect within a time period of two hours a battery of three V–2 missiles, including fueling them and rotating them on their launch table into the proper flight direction for firing.

V–2 PRODUCTION

Hitler had blocked development and full–scale production of the V–2 in the years from 1939–1942 by not giving it priority in the defense ministry amidst other pressing armaments programs, so the necessary engineers and production rocket engines could not be obtained. He was enamored of tanks, artillery, and the Luftwaffe; and didn't think the time was right for rockets. Again, he was wrong.

To further frustrate the Peenemünde crowd, the Army put Gerd Degenkolb in charge of organizing V–2 production. A sworn enemy of Dornberger, he set up a Nazi-supported bureaucracy in parallel to that of Dornberger, requiring the approval of the Army weapons bureau on any decisions. He even sent four engineers to spy at Peenemünde, asking them to provide him recommendations on the reorganization of the place. In order to productionize the A–4 design, Degenkolb began authorizing many detailed changes. Degenkolb was not an engineer. He didn't understand that going from A-4 prototype test vehicle to V–2 production required much development and testing as we have discussed many times. Every change had to be proven in test first, and only incremental steps could be taken.

Degenkolb's predecessor, Detmar Stahlknecht, had planned to produce 300 V–2 missiles per month by January 1944, and 600 per month by July 1944. Degenkolb unrealistically decreed that 300 per month be achieved by October 1943, and 900 per month by December 1943. It is a wonder that the Peenemünde engineers were able to accomplish what they did.

July 7, 1943 Major General Walter Dornberger, von Braun, and Ernst Stienhoff, an expert in missile guidance and control at Peenemünde, had finally been allowed to present their case for increased funding to the Führer himself. *Well, at least I've earned that second star,* Dornberger mused to himself. His wife had helped pen the star on him at a ceremony last May. *That and my doctorate, life isn't so bad after all.* Dornberger was smart. He had completed an engineering course with distinction at the Berlin Technical Institute, earned a Master's degree in mechanical engineering from the *Technische Hochschule Charlottenburg* in Berlin, and was awarded an honorary doctorate for his accomplishments in assembling the team of highly talented engineers under von Braun's direction and provided them with the funding and staff organization necessary to do great things.

Hitler had not deigned to review the V–2 program—let alone actively support it—for over four years since his visit to Kummersdorf in March 1939. The trio, with Steinhoff at the controls, had boarded an He-111 airplane—the kind that the *Luftwaffe* would use a year later to launch a barrage of almost 1,200 V–1 missiles against London—and headed for the Führer's bunker in East Prussia. They didn't know what to expect. Would Hitler throw them out of his office?

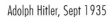

Adolph Hitler, Sept 1935

victory would give the Red Army the strategic initiative for the rest of the war. The Axis would suffer 500,000 dead, wounded, or captured soldiers, including 170,000 Germans killed. Hitler would had a lot of blood on his hands, second only to Stalin.

"Here is the A–4 in production in the vast assembly hall at Peenemünde," Dr. von Braun began. "Now the *Meillerwagen* towing it to the huge launch complex... the men loading the B-Stoff and A-Stoff...prelaunch

A war-weary Adolph Hitler, July 1943

Things didn't start out well. Their appointment was for 1130, but then it was delayed until 1700. Maybe the Führer would dispense with tossing them out of the office, and not even bother meeting with them in the first place. Finally, one of Hitler's attendants ushered them into his almighty presence. Dornberger was shocked at the terribly haggard and changed appearance of the Führer.

Small wonder. By mid–1943 the war was not going well for Germany, and in particular for Hitler's military judgment. Why didn't he trust his own Army High Command? He should have known better than to meddle in their military and tactical planning because he didn't trust them. In February 1943, Hitler's repeated refusal to allow their withdrawal at the Battle of Stalingrad led to the total destruction of the 6th Army. Over 200,000 Axis soldiers were killed and 235,000 were taken prisoner The Allies were poised and ready to invade Sicily. Hitler had just received word that the Battle of Kursk had begun 450 kilometers (280 mi) south of Moscow. In what would become the biggest tank battle of World War II, a decisive Soviet

preparations...the countdown and finally launch." Von Braun stopped while Steinhoff brought in a model and plans they had brought with them for the hardened production/launch bunker that was being built on the English Channel. Von Braun patiently went through all the aspects while Hitler paid closer attention than before.

"I love it!" Hitler exclaimed. The Peenemünders were taken aback. "I want to build not one, but three such facilities." "But mein Führer," Dornberger argued, "mobile launchers would be militarily less vulnerable and less costly." Hitler was unconvinced. "The seven–meter thick bunker walls," he declared, "would draw every allied bomber like flies to honey. Every bomb they drop there will be one that does not fall on Germany. Walter, can you increase the payload to ten tonnes?" "Why?," von Braun asked. "Because I want to obliterate London with a nuclear warhead," Hitler responded. "In the meantime, while we rebuild the heavy water facilities in Germany, is a 2,000 per month production rate possible?" *The German nuclear energy project needed heavy water (deuterium oxide) to produce nuclear weapons. In Operation Gunnerside, on February 28, 1943 a team of Norwegian* commandos trained by the British Special Operations Executive succeeded in destroying the Norsk Hydro heavy water production plant in Vermork, Norway. Operation Gunnerside was the most successful act of sabotage in all of World War II.

"It will take four to five years to develop a missile with greater payload," Dornberger replied. "Back in 1936 when we set the specifications for the missile no one dreamed of the short-term availability of nuclear energy. It will take years to develop nuclear weapons. As to production, it is limited by Germany's industrial capacity to produce the alcohol fuel." *V-2 production consumed a third of Germany's fuel alcohol production and major portions of other critical technologies. To distill the fuel alcohol for one V–2 launch required 30 tonnes of potatoes at a time when food was becoming scarce.*

Hitler was visibly upset that the V-2 would not turn out to be a war-deciding weapon and expressed his displeasure. "Mein Führer," Dornberger pleaded, "the Vengeance–2 will be a great psychological weapon. The British, the Americans, everybody will know it is unstoppable, something against which there is no defense."

On July 8 Hitler held a private conference with General Dornberger. He said, "I have only had to excuse myself to two men in my life - and one of them was (Field Marshall Walther) von Brauchtisch, who always championed the importance of your work. I did not listen to him when he told me again and again how important your research was. The second man is yourself. I never believed that your work would be successful. If we had this weapon in 1939, Britain would have conceded, and there would have been no war."

Hitler ordered that the V-1 and V-2 missile programs be given the highest priority in the German defense ministry. Immediately needed staff and materiel began flowing into both programs. Karl Otto Saur, Albert Speer's head of the Central Office of the Ministry of Armaments and Munitions, immediately ordered Hitler's insane production goal of 2,000 V–2s per month, despite the fact that there was no prospect of producing enough alcohol fuel or training enough launch crews to actually fire the missiles at such a rate. Hitler threatened any industry leader who did not commit to meeting this crazy production goal with immediate replacement. *German alcohol production would limit the maximum number that could ever be produced to 900 per month.*

Hitler went even further, ordering all work at Peenemünde on von Braun's imaginative new designs stopped. All efforts were to be focused on perfection and development of the V–2 as a weapon-in-being. Von Braun had to stop all efforts at accomplishing his dream: manned space travel.

By early August 1943, Dr. Walter Theil was on the edge of a nervous breakdown. Continuous changes to his engine design were affecting other parts of the rocket, resulting in drawing changes simultaneously with efforts to mass-produce detailed parts. The engine itself was considered too complicated to fabricate in production, requiring thousands of hand-assembled tubes to introduce fuel and LOX into the chamber. The V–2's new Mischduese injector plate engine promised better performance, but had combustion instability problems. Theil and his exasperated team declared that V–2 development could not be completed before the war's end. They recommended stopping production plans, and with Dr. Eberhard Rees as their spokesman threatened to lead an engineering "revolt" (good for them). Theil resigned with notification that he intended to lecture thermodynamics at a local college. Von Braun argued that Theil and his engineers stay the course. He demanded that production continue.

Ironically, just days later, on August 17, 1943 the Peenemünde facility came under direct attack by the Royal Air Force (as discussed in the Сергéй chapter) and Dr. Theil was killed. He deserved much credit for the development of the V–2 engine. His death was a big loss—had he lived, the Peenemünde team could have resolved the Mischduese propellant injector problems. Von Braun had to put the complex 18 injector 'basket-head' design into production instead.

Martin Schilling replaced Dr. Theil, who had been working on a better propellant injector design for the V-2 engine. The engine group was desperate to replace the V-2's thousands of cumbersome tubes with a simpler injection system: rows of simple bored holes on a flat injector plate at the head of the thrust chamber. Professor Georg Beck at the Technische Hochschule in Dresden developed a ring-pattern injector that worked well in subscale engines. But the design proved unstable in the 25.4 tonne thrust full size engine. They relearned the lesson that subscale demonstrations do not guarantee success in the flight version, especially in rocketry where unexpected problems can crop up. Schilling decided to stick with the kludged A-4 prototype engine design that used 18 separate 1½ tonne chambers.

Labor for V–2 production became another pressing problem in 1943. Using slave labor, Germany began building underground production factories for both missiles and aircraft to hide them from Allied carpet bombings (so-called because they covered an entire area with explosives, like a carpet). The most important V-2 production sites were the central plants, called Mittelwerk, in the southern Harz Mountains near Nordhausen, where an abandoned gypsum mine provided an underground cavern large enough to house extensive facilities in secrecy. Slave labor from Dora carved out an underground factory in the abandoned mine, which extended a mile into the hillside. Foreign workers under the supervision of skilled German technicians assumed an increasing burden. At Mittelwerk 90% of the 10,000 laborers were non-Germans.

Hitler was desperate for a "wonder weapon" to turn the tide in Germany's favor and became concerned about V–2 development after July 1943. This piqued the interest of Reichsführer SS Heinrich Himmler, who conspired to take control of the rocket program and research activities at Peenemünde as a means to expand his power base. In February 1944 Himmler asked von Braun to leave the Peenemünde rocket team and come to work for him and the SS. Of course, von Braun refused. Dornberger also resisted Himmler's advances.

March 12, 1944

"Papa, the V-2 is a weapon of war! It is meant to kill people," von Braun decried. One of von Braun's most trusted lieutenants, Walter Riedel headed up the Technical Design Office at Peenemünde and was nicknamed "Papa" only because he was 10 years older than von Braun. As the war was not going well, the Peenemünde team was having a motivational get-together at guidance engineer Helmut Hoelzer's home. "The warhead is 738kg of high explosive Amatol. It is capable of leveling a city block! We should be developing the A-4 as a space vehicle rather than a weapon of war." "Keep your voice down," Riedel whispered. "You know we simply have no choice. As scientists none of us embrace Hitler's crazy ideas. Hitler does not like scientists in the first place because we are an independent sort and fail to rally around his flag. What's more, he has absolutely no interest in rocket technology for space flight." Helmut Gröttrup had been

looking in silence, making sure there were no snoops around. Now he joined in. "Pssst! Wernher, it will be so easy to adapt the A-4 to civilian use. Maybe we can take advantage of Hitler's preoccupation with the horrible battle losses going on. It is evident to everyone that we are losing this war. Here is a conceptual adaptation of the payload to a satellite." He pulled a sheet of paper from inside his shirt front. It contained detailed specifications for a hypothetical satellite payload. After a quick perusal, von Braun knew that Gröttrup's analysis was dead on. "Schon gut!" von Braun excitedly exclaimed. "Shhh!" counseled Riedel again. "Hannes and I will study the manufacturability of this concept tonight." With that, von Braun and the clandestine clique quietly dispersed.

No one had noticed Aloïsia Putzkammer, a young female dentist who overhead their conversation. The dirty rat, she noted their names and reported them to Himmler. Early in the morning of March 15 the Gestapo pounced without warning on von Braun and his unsuspecting cohorts. They were arrested on trumped-up charges of treason for not supporting the war effort. The Gestapo imprisoned von Braun in Stettin (present-day Szczecin in Poland) and accused him of sabotage by declaring that his main interest was in the developing the A-4/V-2 for space travel—not as a weapon. Also, since von Braun was a pilot, who regularly piloted his government-provided airplane, it was suggested that he was planning to escape with A-4/V-2 secrets to the Allies. Von Braun stayed in jail for two weeks until Munitions and Armaments Minister Albert Speer personally intervened with Hitler to release him as essential to V-2 production.

As we saw in the Серёй chapter, atrocities were perpetrated at V-2 production facilities at Nordhausen and the nearby concentration camp at Dora where some 20,000 died as a result of execution, starvation, and disease. 350 slave laborers were hanged, including 200 for sabotage. Many others were shot. These facts stimulated controversy that would plague the rocket pioneers who left Germany after the war.

Similar ambiguities clouded the issue of responsibility for the slave labor at Nordhausen. The manufacturing facilities were far from Peenemünde, under the supervision of Himmler's SS. Alwin Sawatzki controlled the Mittelwerk and he derived his authority from SS-General Hans Kammler, who called the shots. Himmler and Kammler dictated production schedules and allocated V-2s for deployment and for testing. Neither Dornberger nor von Braun had direct authority over Mittelwerk, but both men visited the plant several times and observed conditions. In August 1944, von Braun himself picked labor slaves from the Buchenwald concentration camp to work on V-2 production. He later reported, "They were in pitiful shape ... It was hellish. My spontaneous reaction was to talk to one of the SS guards, only to be told with unmistakable harshness that I should mind my own business, or find myself in the same striped fatigues! I realized that any attempt of reasoning on humane grounds would be utterly futile."

In April 1944 Arthur Rudolph, chief engineer of the Peenemünde factory, learned of the availability of concentration camp prisoners (mostly Russians, Poles, and French) and enthusiastically endorsed their use, helping win approval for the initial transfer of some 1,400. Perhaps in the beginning Rudolph was actually sympathetic to the pitiful concentration camp victims, hoping to rescue them from a certain death sentence in the camps to a productive life in the V-2 production factories. The first prisoners began working in June. Over time Rudolph had to have known about the brutal treatment and conditions under which the slave laborers worked on the V-2 production lines. How complicit he was in setting the insane V-2 production schedules by which slave laborers were worked to death is unknown. Perhaps like von Braun he was intimidated by the Nazis into compliance.

Von Braun was interested in producing space vehicles to enable manned exploration of the planets, not as weapons of war. However, unlike Rudolph, Dornberger and von Braun could influence V-2 production only indirectly, by lobbying for greater resources. It is tempting to criticize in hindsight or to pass blame after the fact, but one must realize that times were different back then. There is a saying: "During the war, everyone was a Nazi. After the war, nobody was a Nazi." Uncooperative rocket scientists were always under the threat of imprisonment by the Nazis. The arrest of von Braun and his men was intended as a warning to key members of the Peenemünde team that nobody was immune from the force of SS control. Rudolph and others at Mittelwerk were frequently reminded that they too could join the forced labor teams if they did not fully cooperate with the SS authorities. So, little wonder that they complied.

V–2 Detailed Design Schematic

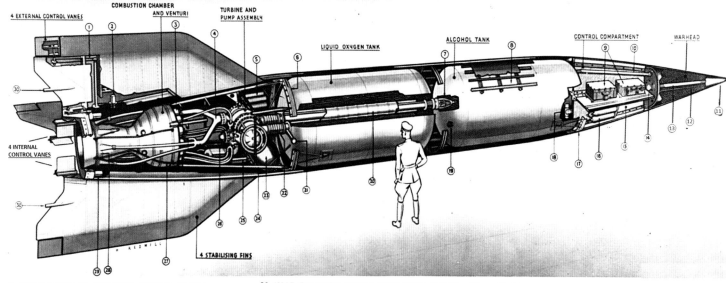

1 CHAIN DRIVE TO EXTERNAL CONTROL VALVE
2 ELECTRIC MOTOR
3 BURNER CUPS
4 ALCOHOL SUPPLY FROM PUMP
5 AIR BOTTLES
6 REAR JOINT RING AND STRONG POINT FOR TRANSPORT
7 SERVO-OPERATED ALCOHOL OUTLET VALVE
8 ROCKET SHELL
9 RADIO EQUIPMENT
10 PIPE LEADING FROM ALCOHOL TANK TO WARHEAD

11 NOSE PROBABLY FITTED WITH NOSE SWITCH, OR OTHER DEVICE FOR OPERATING WARHEAD FUZE
12 CONDUIT CARRYING WIRES TO NOSE OF WARHEAD
13 CENTRAL EXPLODER TUBE
14 ELECTRIC FUZE FOR WARHEAD
15 PLYWOOD FRAME
16 NITROGEN BOTTLES
17 FRONT JOINT RING AND STRONG POINT FOR TRANSPORT
18 PITCH AND AZIMUTH GYROS
19 ALOCHOL FILLING POINT
20 DOUBLE WALLED ALCOHOL DELIVERY PIPE TO PUMP

21 OXYGEN FILLING POINT
22 CONCERTINA CONNECTIONS
23 HYDROGEN PEROXIDE TANK
24 TUBULAR FRAME HOLDING TURBINE AND PUMP ASSEMBLY
25 PERMANGANATE TANK (GAS GENERATOR UNIT BEHIND THIS TANK)
26 OXYGEN DISTRIBUTOR FROM PUMP
27 ALCOHOL PIPES FOR SUBSIDIARY COOLING
28 ALCOHOL INLET TO DOUBLE WALL
29 ELECTRO-HYDRAULIC SERVO MOTORS
30 AERIAL LEADS

V–2 Performance

The production V–2 was indisputably the forerunner of modern space rockets and long–range missiles. It was 14m (46ft) long, 1.65m (5.4 ft) in body diameter, increasing to 3.55m (11.6ft) including the tail fins. At launch it weighed 12.7 tonnes (metric tons, or 28,000 lb) and burned a mixture of 75% alcohol/25% water (for fuel) and liquid oxygen oxidizer. It could carry a 1,000 kg (2,200 lb) payload.

In production, the V–2 generated 27 tonnes (59,600 lbs) of thrust at the start, which increased to about 31.8 tonnes (70,100 lbs) at altitude. At engine cutoff (around one minute), the V–2 was going at 4,828 kph (3,000 mph, or Mach 4.05). From this point onward the V–2 was unpowered and at the mercy of gravitational and drag forces, like a cannon ball launched over the horizon. On its ballistic trajectory the V–2 kept rising to the vacuum of space (88 km, or 55 mi). As it arced over towards its target, there was no air to slow it down, so the pull of gravity increased its velocity to 5,400 kph (3,355 mph), or Mach 4.5. For its time, this was a stupendous velocity of about a mile/sec; no other missile could even come close. In diving into its target, atmospheric drag would gradually slow it down to about 3,960 kph (2,461 mph), or Mach 3.3. Its explosive warhead (Amatol Fp60/40) weighed 738 kg (1627 lb). enough to level a city block. The V–2's operational range was 320km (200 mi), about that of a modern Scud–B.

Replica of the V-2 at the Peenemünde Museum in Germany

EFFECTIVENESS OF THE V–1 AND V–2 AS WEAPONS OF WAR

V–1

Unlike the V-2, the V-1 was a cost-effective weapon for the Germans as it forced the Allies to spend heavily on defensive measures and divert bombers from other targets. More than 25% of the Allies' Combined Bomber Offensive's strategic bombs in July and August 1944 were used against V-weapon sites, often ineffectively. Gauging its merit in terms of Deutsche Marks spent per person killed, the V–1 easily won out over the V–2. They both killed about the same number of people. 13,145 V–1s and 3,170 V–2s were launched to do so. But since the V–1 caused only a tenth as much to produce, the number launched was equivalent to only 1,316 V–2s. In other words, it cost 2.4 times as much for the V-2 to kill a person as the V–1, admittedly a gruesome statistic.

V–1 Missile Statistics

Number Built*	30,000
Ground launches	
London	9,521
Holland (Antwerp) and Belgium	2,448
Air Launches (London)	1,176
Total Launches*	13,145
People killed*	9,000
People injured*	22,892

* Approximate, exact numbers are in dispute.

V–2

On Sept. 8, 1944 the first operational V-2 exploded in a suburb of Paris, killing six people and injuring 36. The second and third struck London a few hours later. This began the infamous "London Blitz" when Hitler thought he could terrorize Britain into submission. The last barrage of 217 V–2s were fired from The Hague in the Netherlands on March 31, 1945. In sum, the German Wehrmacht launched about 3,170 V–2s against Allied targets in World War II, resulting in the death of an estimated 9,000 civilian and military personnel, while 12,000 forced laborers and concentration camp prisoners were killed producing the weapon. *The V-2 is the only weapon system to have caused more deaths by its production than its deployment. It also still holds the world record for the missile weapon causing the most deaths, mayhem, and destruction.*

A section of London east of St. Paul's Cathedral after the raid of 29 December 1940

V–2 Missile Statistics

Used in tests	632
Launched in combat*	3,170
Built but unfired*	2,350
Total built*	6,152
Civilian and military casualties*	9,000
Forced laborers and prisoners killed in production*	12,000

* Approximate, exact numbers are in dispute.

But how effective really was it? V–2 launch positions proved to be the Achilles heel of the German's missile strategy. All of the actual launches were from launch "tables" at fixed positions (a launch table is a rudimentary launch pad without a launch tower). The missiles were transported there and set up using the A–4's *Meillerwagen* transporter–erector. The British and Americans conducted reconnaissance and then organized air strikes against the rocket launchers, which were difficult to camouflage. The Germans had plans to make some of the V–2s mobile as a counter–measure. The missile was to have been launched from a simple rack mounted on a railroad flatcar. The mobile launcher consisted of alcohol and liquid oxygen tanks, launch equipment, and the equipment needed to perform prelaunch checks. However, the Germans did not quite succeed in bringing their mobile launchers to the point of combat–readiness before the war ended.

V–2 launch sites (some omitted for clarity)

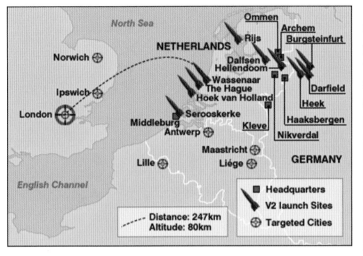

Another defensive tactic devised by the Brits was to use radars to detect V–2 launches and apply updated data to grid coordinates to plot and determine the missile's trajectory. Then at a precisely–timed moment they would mount a massive antiaircraft artillery barrage to destroy the incoming V–2. If all else failed, the Brits had cadres of human observers and telephones positioned on the coasts. They even tried sound wave amplification to detect incoming aircraft or V–1s using a system of amphitheaters positioned near the coasts.

Nazi agents in Britain were the only source of information to the Germans as to where the missiles actually hit. But most of these agents had been turned by British intelligence and were sending back false reports as to the impact points of the rockets. These false reports indicated that the missiles were going long and impacting beyond London. As a result of corrections due to this false information, the German average impact point kept moving farther and farther east as the campaign went on. As a result the average V–2 impact point ended up on the eastern edge of Greater London's Air Defense Zone.

Another problem was that the V-2 lacked a proximity fuse. It just had an impact fuse, which reduced its effectiveness. It would bury itself in the target area before or just as the warhead detonated, a ground burst. A properly calibrated proximity fuse would have enabled air bursts, which like nuclear airbursts would spread the effects of the explosion more evenly over a larger area. Furthermore, the V–2's guidance systems were too primitive to hit specific targets, and each one cost the Germans about the same as a four–engined bomber, which were more accurate (though only in a relative sense), had longer ranges, carried many more warheads, and were reusable.

In fact, each V–2 successfully launched killed only six people on average. Its principal effect was psychological. Unlike bombers or the V–1 flying bomb with its characteristic buzzing sound, the V–2 traveled so fast that it exploded on the ground with no warning and no possibility of defense.

> *It is interesting that the most damage occurred not to the Allies, but to Germany. Historians agree that, despite the scope of the V–2 mobilization effort by the fanatical Hitler, in the end the inaccurate missile did not change the course of the war and proved to be an enormous waste of resources. Each V-2 cost as much to produce as a high-performance fighter airplane. German forces on the fighting fronts were in desperate need of airplanes, and the V–2 rockets were doing no significant military damage. The 6,117 V–2's that were built represented a huge and debilitating financial and manpower drain on Nazi Germany, on a ranking with the U.S. Manhattan Project to build an atomic bomb.*

At war's end hundreds of V–2s and trainloads of spare parts were captured by the Allies and put to good use in research that led to the development of their missile and space exploration programs. Personnel and technology from the V-2 program formed the starting point for post-war rocketry development in America, Russia, and France.

NAZI DREAM WEAPONS

The V–2—the world's first guided missile— was just the beginning. Before Peenemünde had to be abandoned under the threat of advancing Russian troops in 1945, its scientists and engineers conceived some fantastic concepts for their time. These were not just paper sketches or cartoons drawn on napkins. Under the duress of wartime emergencies, including allied bombings of their research facilities, they somehow managed to develop detailed engineering drawings, specifications, scale models, and mockups for weapons ranging from boost–glide cruise missiles to a 5,000km range ICBM carrying a one tonne warhead. They developed and flew Surface–to–Air Missiles (SAMs), launched missiles from submerged submarines, investigated the use of nuclear energy for future spaceflights, and performed preliminary design work for sounding rockets. They also designed—in detail—a stratospheric rocket that could travel from Europe to America in 40 minutes—comparable to the "Rocket Bomber" described in the Серğй chapter. They also developed satellite launch vehicles capable of putting 10 tonnes (22,000 lbs) into low Earth orbit! Wow.

Unmanned A–4b Boost–Glide Missile

Back in 1940, the Luftwaffe had proposed adding wings to von Braun's A–4 design to make a cruise missile out of it, with longer range because it could glide to its target after the engine's power was cut off. Luftwaffe engineers assisted von Braun's team to develop a 5.91 meter–long (19.4 ft) subscale model, the A–7. Wind tunnel tests were performed at Peenemünde to gather data on candidate designs.

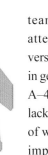

A–7 concept

However, von Braun and his rocket team were unable to devote much attention to developing the winged version. Instead, they became immersed in getting the Aggregate test vehicles and A–4 rocket developed (despite Hitler's lack of support in the beginning) because of wartime urgency demands the Nazis imposed upon them. Tests using A–7 models were soon dropped. Nevertheless, von Braun and other innovators on the team managed to perform some work on such imaginative designs.

Things got worse after the Peenemünde leadership's apparently successful meeting with Hitler on July 7, 1943. Shortly afterwards, Hitler ordered all work not focused on the V–2 stopped. For 15 months, until things eased up in October 1944, the intrepid von Braun and his colleagues handled this dilemma by hiding their advanced development work under the cover name A–4b. When Reichsführer Himmler's henchmen came snooping around, they could claim they were performing valid wartime production–related work by modifying the approved A–4. Discussion of spaceflight at Peenemünde was only done privately. Such discussions were risky and could only take place within a small, selected group of people.

But then in October 1944, with the Allies overrunning launch sites in Belgium and Holland, SS-*Obergruppenführer* Hans Kammler ordered work on the A-4b resumed in order to have a long-range version of the weapon that could reach England and Allied continental rear areas from Germany. Now unhindered, within weeks von Braun's crew reviewed the 1940 Peenemünde wind tunnel results and built a prototype winged version of the missile. Tests optimized the wings at 52–degree sweep and 13.5 meters^2 wing area. It had the same gross weight, LOX/Alcohol propulsion system and thrust, and payload (1,000 kg or 2,200 lbs, of which 738 kg was the explosive warhead) as the V-2; but its empty weight was 1,350 kg greater because of the wings. Von Braun traded off the empty weight increase against the increased range afforded by the wings, everything else being equal. The winged A–4b won out with a range of roughly 600 km (373 mi), an 87.5% range increase over the V–2's 320 km.

During its first launch on December 27, 1944, the steering failed at 30 m and the missile crashed a short distance from the pad. Several more test articles were on hand. The Germans were in a hurry with the Allies

Winged A–4b boost–glide missile on *Meillerwagen* transporter-erector vehicle, and on test stand at Peenemünde before successful launch, 24 Jan 1945. It had the same thrust as the A–4 and V–2, 27 tonnes at liftoff, increasing to 31.8 tonnes (70,140 lbs) at altitude.

closing in, but testing could not resume immediately due to a shortage of alcohol fuel.

A month later, January 24, 1945, a second launch attempt from Peenemünde was successful. The A–4b reached 80 km (50 mi) altitude and 1200 m/s (2700 mph), or Mach 3.62. *By the way, this set world speed and altitude records for winged flight.* It then flew stably in supersonic flight. The automatic guidance system was designed to keep the missile on course in both supersonic and subsonic flight regimes. However a wing broke off shortly after the beginning of the glide and the A–4b failed to reach its designed range of some 640km. This concluded work on the A–4b, the increasingly chaotic situation in Germany preventing further flight tests of anything von Braun could dream up.

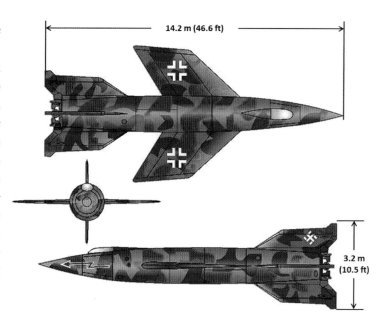

Manned A–9, baseline version. Three–view artist's concept.

Manned Boost–Glide Missiles

Von Braun's highly talented engineers designed derivatives of the winged A–4b that were manned, others that served as the top stage of multi–stage ballistic missiles, and one version that used a futuristic ramjet for cruise propulsion. The idea was the same: get more range out of the V–2 by putting wings on it. The mission profile was to launch it vertically under its own power as high as possible into space, cut the power off, and glide for long distances to a target, dropping a weapon, and landing horizontally at a conventional airfield using landing gear and a drag parachute. Unlike the A–4b, they were designed with a man onboard to make them reusable. None of them ever flew, but the ambitious designs were decades ahead of their time and presaged vehicles such as the SR–71 and U–2 reconnaissance planes.

A–9

The A–9s were another family of derivatives of the A–4b with longer range and better performance. First they tried taking the A–4b and simply putting a man in it. But manned capability is not simple. It greatly complicated the design, requiring a life support and pressurization system, an ejection seat and parachute, canopy, instrumentation, landing gear, and so on. The structure would have to be beefed up to make the manned cruise missile safer to fly, a lesson learned from the A–4b's loss of a wing in test flight.

A first version of the A–9 was designed using the same propulsion system as the A–4b: 75% alcohol/25% water mix for fuel, LOX oxidizer, 31.8 tonnes (70,140 lbs) thrust at altitude, and 1 tonne payload including the pilot and his pressurization/life support system. Adding landing gear and putting a man onboard increased the A–4b's gross weight by almost 3.5 tonnes (7,000 lbs). Its warhead delivery capability would be cut in half (from 1,000kg to 500kg) and the range reduced to about 550 km from the A–4b's 620 km. Was it worth it?

The Peenemünde engineers weren't sure. So they tried beefing up the propulsion to up the range. They designed an A–9 boosted with a ring of 10 solid propellant rockets which would achieve Mach 6 cruise at 20 km. This fixed the range problem: it increased to 950 km (590 mi) for the same payload.

Another A–9 design would have been capable of delivering express cargo 600km from the launch point within 17 minutes.

A–6

Another version would use a ramjet. It was apparently designated the A–6 to distinguish it from the others. The rocket engine would accelerate the vehicle to supersonic speed and an apogee of 95 km. After re-entering the atmosphere and beginning the supersonic glide phase, the ramjet would be ignited. Burning synthetic petrol or coal slurry, it was designed to cruise for 15–20 minutes at 2,900 km/hr (1800 mi/hr), or Mach 2.4. This would have allowed the aircraft to return to base, unlike the A–4b which would be restricted to a

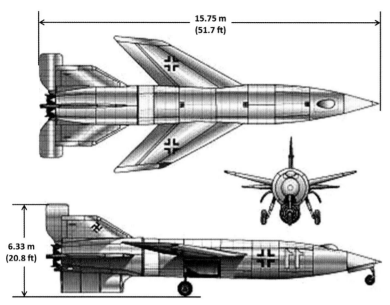

15.75 m
(51.7 ft)

6.33 m
(20.8 ft)

A–6 Ramjet–powered boost–glide cruise missile, three–view artist's concept.

SURFACE–TO–AIR MISSILES

In cooperation with the Nazi Ministry of Aviation [Reichsluftfahrtministerium, (RLM)], Peenemünde engineers also designed a series of SAMs, by far the most advanced in the world at the time. If they had reached operational status, the Allies would have been powerless to stop them. The principal ones were the *Wasserfall* guided SAM, the solid motor powered *Taifun*, and the Schmetterling Hs–117.

Wasserfall W–10

Of all of the German SAM systems, the Wasserfall had by far the greatest potential, and had it entered operational service in early 1944 could have inflicted heavy losses on Allied heavy bomber fleets. Fortunately, the Nazi leadership were obsessed with bombarding Britain with strategically useless A-4/V-2 ballistic missiles, and put their funding into that program, launching some 3,170 weapons for negligible military effect. Had the prodigious effort invested into the A-4/V-2 program been put into the Wasserfall, the course of the war could have been very different. The Luftwaffe had ambitious plans for the Wasserfall, envisaging 200

one–way mission. One pilot was provided with a pressurized cockpit. The reusable A–6 would be launched vertically but land on a conventional airfield using landing gear and a drag parachute. It was proposed to the German Air Ministry as an invulnerable photo reconnaissance aircraft, anticipating the SR-71 aircraft by 20 years! The Ministry rejected it, probably because it was simply too far out for its time.

Clockwise from lower left: Wasserfall W–10 SAM launch from Peenemünde on 23 September 1944, Wasserfall cutaway view, Wasserfall concept of operations, and (lower middle) U.S. Army Hermes missile launch. Note the similarity between Wasserfall and Hermes, which was developed almost a decade later. The Army added a second pair of fins to the middle of the fuselage to provide extra maneuvering capability.

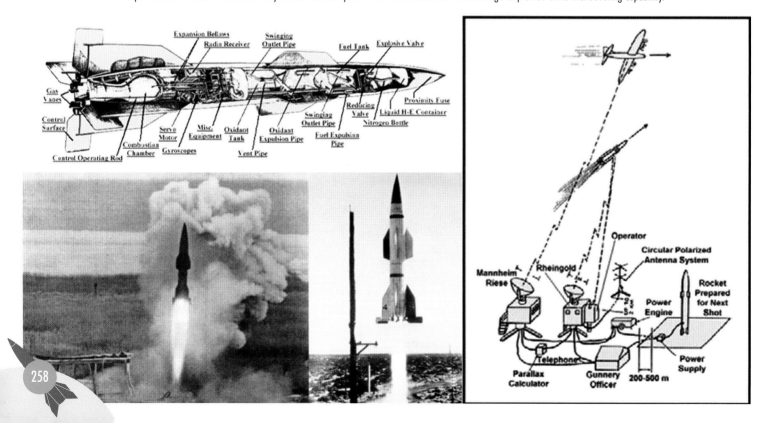

batteries installed in three SAM belts across Germany, requiring monthly production of 5,000 Wasserfall reload rounds. The never-implemented plan would have seen the first operational battery in November 1945, with 20 sites operational by March 1946. The Wasserfall proved most effective in providing a baseline for postwar US and Soviet SAM designs.

Schmetterling Hs-117

Of all the experimental German antiaircraft SAMs of World War II, the TV guided Schmetterling (Butterfly) came closest to deployment in batteries to knock down Allied aircraft. An operator using a telescopic sight and joystick guided the missile by radio control. It originated in 1941

Wasserfall SAM launch sequence.

Taifun

In 1942 Karl Heinrich Scheufeln and a small team at Peenemünde developed a design for the Taifun SAM. Scheufeln wanted to introduce an extremely simple system in case von Braun's more complex projects "did not work out." He wanted an inexpensive rocket that could be fired in salvoes at American B–17 bomber formations. Powered by pressure–fed storable propellants—nitric acid and a synthetic fuel—the Taifun would reach an altitude of 25 km and a range of 12 km. Unguided, it would be stabilized by four fins at the base. A cordite charge would pressurize the hypergolic propellants. The Electromechanische Werke in Karlshagen built a few production examples. The Taifun was to be launched in salvos from 46 launching rails at once against large Allied bomber formations. Von Braun's team engineered it to reach the unbelievable speed of Mach 3! The original plans called for 2 million to have been produced by January 1945, but the engine experienced lengthy delays and none were ever deployed operationally (thank goodness).

when Henschel & Son's talented missile designer, Herbert A. Wagner, proposed several antiaircraft projects, including the Hs-117 SAM and the Hs-293 anti–ship guided missile. RLM approved the Hs-293 but rejected the Hs-117 design

Two Taifun SAMs on display at the RAF Museum Cosford, Shropshire, England.

because they saw no need for more anti-aircraft weaponry. However, by 1943 the large-scale bombing of Germany caused the RLM to change its mind, and Henschel was given a contract to develop and manufacture it. Although smaller than the Wasserfall, the specification was nearly the same, calling for intercept of aircraft flying up to 760 kph.

fighters without aborting the attack. Nevertheless the Hs–293 was able to sink or damage at least 32 Allied ships in 1943–1944, killing thousands of sailors.

Schmetterling Hs–117 SAM on display in a hangar in Chantilly, Virginia (left). Video stills of launch during WWII in Germany (right).

Flight tests began in May 1944. Altogether there were 59 launches, of which 29 were considered successful. Mass production was ordered in December 1944. Flight testing was considered completed in January 1945 and a prototype for mass production was built. Production was to have begun of 3,000 missiles per month, with deployment beginning in March, 1945 --an unrealistic timetable typical of Germany's desperate programs late in the war. SS-*Obergruppenführer* Hans Kammler cancelled the project on February 6, 1945.

AIR–LAUNCHED MISSILES

The Henschel Hs-117H was an air-launched variant of the Hs-117 and the world's first air–to–air missile (AAM). The Hs-117H could be launched from a range of up to 6.2 miles, and could reach targets 5 km (16,400 ft) above the parent aircraft. Work on the Hs-117H continued into 1945 but it was never used operationally.

The Henschel Hs–293 was designed by Herbert Wagner in 1940 as an anti-ship guided missile. Over 1,000 were built, from 1942 onwards. The Hs–293 was basically a radio-controlled glide bomb with a solid rocket motor slung underneath it. The motor provided only a short burst of speed, making range dependent on the height of launch. One drawback of the Hs–293 was that after the missile was launched the bomber had to fly in a straight and level path at a set altitude and speed parallel to the target so as to be able to maintain a slant line of sight for radio guidance of the missile, and it could thus not maneuver to evade attacking

Henschel Hs–293 anti–ship guided missile on display at the Deutsches Technikmuseum in Berlin. It was a radio–controlled glide bomb with a solid rocket motor slung underneath it.

SLBMS

In 1944, the Germans developed plans to attack targets in the United States directly with V–2s. *Project Laffarenz* conceived of employing Germany's latest type XXI snorkel–equipped U–boats to tow three V–2s, each in special displacement containers, across the Atlantic within striking distance of the American coast. The containers, equipped with special ballast cells, would be trimmed to neutral buoyancy and towed to the launch location by the subs. The cells would be flooded to elevate the container into a vertical or angular position and held there by special stabilizers. The U–boat would then pump fuel into the V–2 and fire the missile. By late 1944, at least one of the containers reportedly was completed at the Baltic port of Elbing, but it was never tested with a live firing.

MULTISTAGE AGGREGATE VEHICLES

Germany's most ambitious plan to surpass the V–2 were the multistage Aggregate launch vehicles. There was never any doubt that manned space travel was Von Braun's life goal. The technology needed for manned flight presented many technical challenges. He realized early on that only multistaged liquid propelled rockets could achieve his dream. Rockets certainly needed lighter propellant tanks, but there was a practical technical limit to this, and in any case, there still had to be a payload. An Advanced Projects Group under Dr. Ludwig Roth led the development of multistage ICBMs and other futuristic concepts under Dr. von Braun's direction. Roth's Advanced Projects Group began studying multistage vehicles in 1940. They adopted the nested vehicle "Matryoshka doll" paradigm introduced in the Сергей chapter. The smallest doll would be a modified A–9 vehicle, with successively larger carriers.

Alternate Propellants Study

Von Braun knew that liquid oxygen/liquid hydrogen was the ultimate propellant combination, but also that learning how to handle liquid hydrogen would be a long-term affair. In 1940/1941, a one-year research study showed that other propellant combinations could produce no more than a 20% improvement in specific impulse compared to the existing single-stage V-2 technology. Therefore a multistage rocket was the only way to achieve orbital spaceflight.

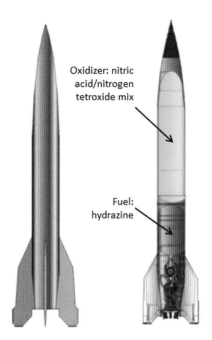

Oxidizer: nitric acid/nitrogen tetroxide mix

Fuel: hydrazine

Trade study V–2 derivative concept using alternate propellants. A–8 storable propellants version shown.

A–10

Using catapults and wings an A–9 might nearly achieve 1000 km range, but the only solution for transatlantic missions was the two-stage A–10—the *world's first practical design for a transatlantic ballistic missile*. Design of the two stage missile began in 1940 and its first flight was planned for 1946. The first stage was a 200-tonne thrust monster rocket with six A-4 combustion chambers feeding into a single expansion nozzle. It would have a total mass of 87 tonnes, of which 62 tonnes would be propellant. *At this time, the Americans and Russians were having a hard time getting*

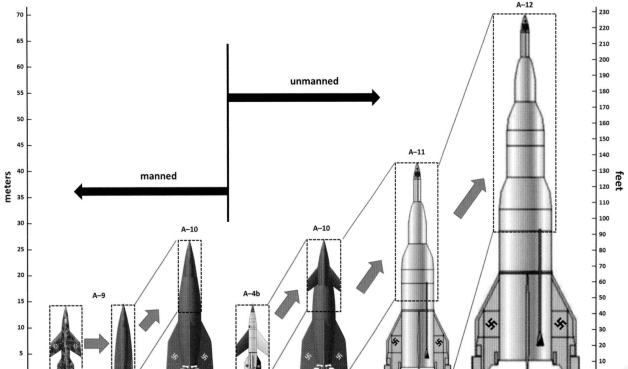

Progressive design approach for multistage nested Aggregate vehicles.

one-tonne thrust rockets to work on a test stand, let alone fly.

The second stage would be a modified A–9 vehicle with a warhead. It would serve as the payload and the smallest Matryoshka doll for the A–10. The piloted A-9 second stage had two fuselage strakes instead of wings, because wind tunnel tests showed that these provided better supersonic lift and solved the problem of transonic shifting (backward, which tends toward instability) of the aerodynamic center (ac) of lift, as we touched on in the Theodore chapter. A secondary benefit was better packaging of the A–9 into the forward interstage of the A–10.

Modified A–9 side and top view. A–9 in flight to a target

Several design alternatives were considered:

1. The A–10 first stage booster would burn for 50 to 60 seconds, taking the A–9 to 1200 meters/sec (Mach 3.6). The booster would have separated and been recovered with the aid of special parachutes. The A–9 would have continued under its own power, reaching an apogee of 55 km, followed by a long hypersonic glide in the atmosphere. The second stage would be equipped with air brakes for deceleration over the target, followed by a parachute for recovery in the water. The A–9 would reach a hypersonic velocity of Mach 8.45 (2800 m/s), with a range of 4,100 km and a total flight time of 35 minutes.

2. The A–10 booster would separate at about a 110–mile altitude and be recovered. The second stage A–9 would burn its engine to an altitude of about 215 miles before descending to 28 miles, where the density of the air would have permitted its wing controls to guide it on its final glide–path to the target.

3. The A–10 booster would burn to propellant depletion, separate, and the A–9's engine would reach suborbit at 390 km altitude and Mach 10.3 (3,400 m/s) under its own power. After cut-off of its engine the A–9 would re-enter and begin a long glide to extend the range. Upon descending to 45 km (28 mi) the density of the air would have enabled the wing controls and permitted the pilot to be guided by radio beacons on surfaced German submarines in the Atlantic Ocean. After reaching the target area at up to 5,000 km range the pilot would lock in the target with an optical sight, then eject. Death or internment for the unfortunate pilot as a prisoner of war would follow.

As we've seen, work on the A–10 and other futuristic concepts was ordered stopped by Hitler in mid–1943. In late 1944 work on the A–10 resumed under the code name *Projekt Amerika*, but no significant hardware development was possible after the second test of the A–4b which crashed on January 24, 1945.

A–11

Designs beyond the A9/A10 were sketched out as well. Adding an A11 stage would have resulted in a satellite launcher.

A–12

The four–stage A–12 as sketched out at Peenemünde was a true orbital launch vehicle, with the A–9 being a manned orbital space shuttle. Calculations suggest it could have placed 10 tonnes into low Earth orbit. Unless development of a new engine was planned, no fewer than 50 V–2 rocket engines would have been required in the first stage, presaging the Soviet N–1 moon rocket that is thoroughly described in *Avoiding Armageddon*.

It is simply amazing that von Braun's enthusiastic band of rocket engineers sketched out such designs beyond the A-9/A-10. Dr. Roth's Advanced Projects Group also planned to develop, after the war, an unweaponized stratospheric rocket that could travel in 40 minutes from Europe to America. Orbital spaceships were designed that could reach 8 km/sec and 500 km orbital altitude. They planned an observatory in space, the construction of space stations in earth orbit, a space mirror, and the burial in space of the embalmed bodies of the rocket developers and men of the rocket service! Manned expeditions to the moon were also a popular theme for research. Finally, the Advanced Projects Group studied the use of nuclear energy for interstellar flights! These men were a century ahead of their time.

They brought such legacies with them to the United States after the war, at a time when the U.S. Army Air Force considered ballistic missiles to be technically impractical!

Some Germans optimistically believed that had the war lasted another six months, they would have been able to produce the A–10 and strike targets in the U.S. such as New York City. Some also believed that if the war had lasted another two years, they could have developed a 15,000–mile ICBM. However, even at 5,000 km, guidance systems of the time were hopelessly inaccurate. Besides, the Germans had not perfected nuclear weapons. Fortunately for everyone, the war was soon over.

Captured German rocket technology, reports, documents, drawings, spare hardware, test equipment, complete A–4/V–2 missiles, and best of all unmatched rocket science brainpower proved a bonanza for the victorious Allies no sooner had the war ended in May 1945. America and the Soviet Union followed similar paths in refining the rocket technologies pioneered by the multistage Aggregate vehicles.

Ramjet Technology Development Path

Both the United States and the Soviet Union benefited from the ramjet missile technology pioneered by the A–6. Postwar researchers found that it would have been much more efficient for the second stage A-9 to use a ramjet to extend the A–10's range to 10,000 km, presaging the capability of ICBMs that the Cold War adversaries would use to threaten each other with in ensuing decades. After the war, both nations embarked on development programs for supersonic cruise missiles using ramjet propulsion. This path led to the American Navaho and Soviet Burya ramjet missile projects. During the 1950s the U.S. conducted 45 flight tests of the Mach 3 Navaho with a 78% success rate. The Soviets developed and tested the Mach 3.2 Burya as an alternative to

Navaho G–38
29.4 tonnes, 10,200 km range

Burya "Object 350"
97.2 tonnes, 8,500 km range

Comparison between U.S. Navajo and Soviet Burya. The Burya was the first Soviet intercontinental cruise missile. In 1961 it demonstrated a 6,500 km range with a 2,350 kg payload. It was over three times heavier than Navaho because it required two large 27–tonne nitric acid/amine boosters to launch a heavier nuclear warhead.

Korolyov's R–7 ICBM. Burya was launched 18 times between 1957 and 1960, with 7 failures.

However, both countries soon found out that mastering the guidance and materials technology needed for a Mach 3+ cruise vehicle with extended flight through Earth's atmosphere was actually more difficult than for a Mach 22 ballistic missile, so the Navajo and Burya were cancelled before going into production.

Ballistic Missile Technology Development Path

Improvements in rocket structures and engine efficiencies made it possible to design pure ballistic vehicles with cut–off velocities over twice as high as the A–10 and 10,000 km ranges. The Redstone, Thor, Jupiter, and Atlas rockets were flying before their equivalent-range Navaho counterparts. The Soviets developed comparable ICBMs: the UR–100, UR–500, and R–36 rockets. In the end, these faster, uninterceptable designs won out.

The Navaho program provided the engine technology that allowed the U.S. to develop ballistic missiles rapidly and catch up with the Russians in the Cold War space race. The Navaho required a large liquid propellant rocket engine to get its Mach 3 ramjet up to ignition speed. It was *this* engine, derived with assistance from expatriate German rocket engineers—and not the ramjet—that provided the basis for the rockets that would later take Americans into space. The Navaho Program also pioneered chem-milling fuel tank fabrication techniques, inertial and stellar navigation, and a host of other technologies used in space vehicles for the next 40 years. Such technology put North American Aviation and its Rocketdyne Division in a leading position that allowed them to dominate American manned spaceflight by capturing the prime contracts for the X-15, Apollo, and Space Shuttle projects.

The only comparable scientific mobilization since Germany's Peenemünde engineers and scientists was the Apollo Program in the United States to land a man on the Moon and return him safely to Earth. *Not surprisingly, that effort was in large part organized by the former Peenemünders.*

OPERATION PAPERCLIP

By January 1945 everyone knew the end was near. The Battle of the Bulge had failed to stem the inexorable advance of Allied armies from East and West. Germany was caught in a pincer movement from which the Nazis would not survive. The Russian Army had advanced into Pomerania which threatened Peenemünde. The number of surviving Peenemünde staff, a precious asset to Germany, was tremendous: A-4 development, fabrication and modification, 1940; A-4b development, 27; Wasserfall and Taifun development, 1455; support and administration, 760. On January 31 SS General Kammler ordered the entire Peenemünde team to relocate to the Thuringia region in central Germany. *Today's Freistaat Thüringen (Free State of Thuringia) is one of 16 federal states in Germany.* A week later the Peenemünde staff received desperate conflicting orders from the Wehrmacht to join the army and fight. In the panic of the war's inevitable end, Major General Dornberger received more conflicting orders. Not being inclined to suicide, they would obey the original orders.

On February 14, 1945 the last V-2 was launched from Peenemünde (others continued being launched from other locations). Dornberger, Wernher von Braun, and some 500 key scientists and engineers left the site on February 17. They packed all of the equipment, including parts for improved V-2 prototypes, and 14 tons of documentation into cases discreetly marked "EW," for *Elektrotechniches Werk* (electrical factory). Von Braun decided that the V-2 central records would be saved and hidden. All other V–2 files and secondary records were intentionally burned in order to give the V–2 central records value as a bargaining chip with the Allies after the war. Workers destroyed records that could not be evacuated (including Wasserfall and Taifun blueprints) and detonated remaining facilities to keep them out of Russian hands. When Russian infantry finally stormed Peenemünde on May 5, they would find it 75% destroyed. By mid-March, most of the Peenemünde staff had evacuated to the areas of Nordhausen, Bleicherode, Sonderhausen, Lehesten, Witzenhausen, Worbis, and Bad Sachsa.

With Hitler and most of the Nazi leadership still alive, many German missiles of all types were still in the pipeline to the front. Often they encountered damage in shipping or they had been rejected by the troops because of functional problems. The troops had no capability to do repair work in the field, so all faulty missiles were returned to a field repair facility. Towards the end, Germany was running out of manpower faster than airplanes and missiles. There were not enough pilots to fly Luftwaffe aircraft. In spite of the wartime conditions and many ups and downs, more V-1s and V-2s were being produced than could be launched by the troops in the field as their areas of operation kept shrinking.

March 15, 1945

"To whom should we surrender?" von Braun asked his key staff. "Moreover, how do we surrender when the SS is putting the screws to us and watching our every move?" Everyone's eyes were on their leader, who was blessed with a magnetic personality. We must safeguard all our designs and drawings-everything we have struggled with is documented in them." "I heard a radio report yesterday that the Soviet Army is 160km from Berlin," said Gröttrup, von Braun's deputy for guidance systems and controls. "With the vicious Stalin in charge, we dare not surrender to the Russians. Look how they have treated their prisoners of war. After we have been tortured to reveal everything we know, they will summarily execute us by hanging." Little did Gröttrup suspect that months later he would be in Soviet hands.

After some more discussion von Braun, eyeing his friend, summarized the group's decision. "We have created a new means of warfare, and the question as to what nation, to what victorious nation we are willing to entrust this brainchild of ours is a moral decision more than anything else. I know everyone here wants to see the world spared another conflict such as Germany has just been through. Only by surrendering such a weapon to people who are guided by the Bible can such an assurance to the world be best secured. Our future will be brightest with the Americans, who have suffered the least from the war and who might be able to afford to support rocket research. We will surrender to the Americans when the time is right."

The following week von Braun suffered a compound fracture of his left arm when his driver fell asleep at the wheel on a trip to regional army headquarters in Erfurt. His injuries were serious, but he insisted that his arm be set in a cast so he could leave the hospital and insure a safe surrender to the American side.

In April, as the allied forces advanced deeper into Germany, General Kammler ordered 400 members of the Peenemünde science and engineering team to be moved by train into the town of Oberammergau in the Bavarian Alps. But the German security pass printers had made a mistake. The movement order papers had the acronym "PFBV" instead of "PFEB" for Projekt Für Elektrische Betriebsmittel (Project for Electrical Equipment). "This critical error could cause our whole team to be shot by firing squad if the SS guards think we are deserting on forged papers," said a distraught Eberhard Rees, von Braun's closest lieutenant. "That's right, Wernher, the SS has set up roadblocks all over areas still under Wehrmacht control, with orders to shoot suspected deserters," added Dieter Huzel. "Then we come up with a new name that fits PFBV, because that is the acronym that the SS guards will have on their copies of our movement orders," concluded von Braun. So the Peenemünders quickly improvised with Projekt Für Besondere Verwendung (PFBV, for "Project for Special Use"). They assembled a convoy, including improved V-2 prototypes, test equipment and many boxes of precious documents. They managed to make it to Oberammergau. Today the city is famous for its Passion Play of Christ held every 10 years. In Oberammergau they were closely guarded by the SS with orders to execute them if they were about to fall into enemy hands. However, von Braun managed to convince SS Major Kummer to order the dispersion of the group into nearby villages so that they would not be an easy target for U.S. bombers. Anticipating the advance of Allied troops, von Braun directed his men to hide 14 tons of numbered crates containing most of the important Peenemünde documents and blueprints in an abandoned mine shaft in the Harz mountain range. They sealed the opening to the mine with a dynamite explosion. About this time Von Braun had to be hospitalized again because he had not allowed his broken arm to recuperate properly a month earlier. This time his bones had to be painfully re-broken and re-aligned, and the left arm recasted.

April 30 - May 2, 1945

On May 1, von Braun and his associates learned over the radio that Adolph Hitler was dead. They had to act fast to avoid the clutches of the SS and certain death. Everyone gathered their belongings and fled towards the American lines. Germany's unconditional surrender on May 8 was just days away. By now advancing American forces had reached the vicinity of Oberammergau. On the morning of May 2, Wernher's brother Magnus, who could speak broken English, tied a white handkerchief to the handlebars of his bicycle and went out, actually seeking contact with invading American troops. Private First Class Joseph Minto of the U.S. Army 44th Infantry Division was on patrol, carefully watching every bush and behind every tree for German snipers. Magnus saw him first and skidded his bike to a halt so as not to provoke PFC Minto into shooting him. As Minto approached he called out, "my name ist Magnus von Braun. My brudther eenvented the V-2. A group of us vant to surrender." He was taken to the 44th Division's Command Post at Reutte (26km away). While waiting to talk to Colonel Rick Schraeder, Magnus hurriedly wrote down his resume and curriculum vitae (he could write better English than he could talk). The Colonel must have been impressed, because Magnus was allowed to return to his compatriots. He met them at 1400 (2:00 pm) with safe conduct passes and news that von Braun and his entourage were wanted for further interrogation. They all gathered and began their descent down a mountain

pass when they suddenly met an American soldier who ordered them to stop. All were escorted to Reutte where they were processed and met with 40 other Peenemünde personnel. Wernher von Braun, his military supervisor Maj. Gen. Dr. Walter Dornberger, and 126 principal engineers were now under American control. The cadre of outstanding rocket scientists and engineers who would form the nucleus of America's efforts to land men on the moon was already starting to take shape.

To America!

Before the war's end, the United States had already been thinking about what to do with captured German scientists and engineers, lest they fall into Soviet hands, the new enemy of the Cold War. Colonel Gervais William Trichel, the chief of the Rocket Branch of U.S. Army Ordnance, was one of the few Americans who had pondered the disposition of German rocket experts prior to their surrender. He sent Army Major Robert B. Staver to London to work with the British Security Service, aka Military Intelligence, Section 5 (MI-5) in developing

Werhner von Braun (left arm in cast) and his rocket science team surrenders to the U.S. Army 324th Regiment, 44th Infantry Division, May 2, 1945

a list of German rocket engineers, technicians, and scientists; ranking them in order of significance. Wernher von Braun's name headed the list, code-named the "Black List." As source material, Staver was aided by the "Osenberg List," which was found by a Polish laboratory technician in an improperly flushed toilet in March 1945.

In November 1944 Col. Trichel and the Army negotiated a contract with General Electric for *Project Hermes*, an agreement for the development of long-range guided missiles and scientific research using captured V-2s. In March 1945 Trichel directed Colonel Holger N. Toftoy, chief of Ordnance Technical Intelligence, to locate 100 operational V–2s and ship them to the Army's White Sands Missile Range in New Mexico.

As soon as Col. Toftoy learned about the Allied discovery of the V–2 plant at *Mittelwerk* he sent Major Staver to Nordhausen to investigate. *Mittelwerk and Nordhausen were discussed in the* Сергей *chapter.* After verifying the astounding discovery of rows of partially assembled V–2s in the underground facilities, Staver met with members of von Braun's staff and learned of the hidden cache of Peenemünde documents. The peace agreement stipulated that the Soviet Union would occupy Nordhausen, and Britain would control Dornten before the end of May, so Toftoy and Staver had to improvise quickly. Toftoy sent Major James P. Hamill to Nordhausen, where in nine days he supervised shipment of 341 rail cars, containing enough V-2 materials and spare parts to assemble 100 V-2s, to Antwerp in preparation for shipment to the United States. With the Peenemünde documents, these materials would fill the hulls of 14 Liberty Ships. Col. Trichel's original requirement for 100 operational V-2s at White Sands was realized.

Staver convinced the Germans to help him find the hidden documents. He shipped 14 tons of the Peenemünde cache out of Dornten even as the British were erecting roadblocks prior to assuming control. On June 19, 1945, two days before the scheduled handover of the area to the Soviets, Maj. Staver and Lt. Col. R. L. Williams took von Braun and his department chiefs by jeep from Garmisch to Munich. The group was flown to Nordhausen. The next day they were evacuated 64 km (40 mi) southwest to Witzenhausen, a small town in the American Zone.

The question of what to do with German technicians in American custody was laden with political, military, and moral overtones. Some feared that allowing them to continue their research might allow for a rebirth of German militarism. Secretary of the Treasury Henry Morgenthau sought a punitive policy toward Germany, with no room for coddling weapons developers. However, it is unfair to punish everyone who "worked on weapons of war." Of course they did. So did the Americans, and those who worked on the atom bomb project which killed over 100,000 people. Secretary Morgenthau, God bless his departed soul, could assert that America had the moral high ground over Germany and the Axis Powers as aggressors, but he should have made allowances for the "fog of war" and how German citizens felt during the war. Not to condone the Nazis, but the saying, "during the war, everyone was a Nazi; after the war no one was a Nazi" has a kernel of truth in it. Von Braun said it best when he said of Adolph Hitler, "I began to see the shape of the man – his brilliance, the tremendous force of personality. It gripped you somehow. But also you could see his flaw — he was wholly without scruples, a godless man who thought himself the only god, the only authority he needed."

The most compelling moral argument hinged on the involvement of the Germans with either the Nazi Party or slave labor at Mittelwerk. Many German academics, scientists, and technicians had been members of the Nazi Party, often because party membership brought benefits such as research grants and promotions. The Party often bestowed honorary rank as a reward. So Nazi Party membership alone seemed an inadequate criteria, and advocates of using German scientists suggested distinguishing "ardent" Nazis from those who joined the Party out of expediency, like Wernher von Braun. In fact, von Braun unwittingly helped the Americans win the war. The V-2 was ineffective as a war weapon, and helped suck up Germany's precious resources. World-renowned physicist Freeman Dyson summed it well when he said,

"... those of us who were seriously engaged in the war were very grateful to Wernher von Braun. We knew that each V-2 cost as much to produce as a high-performance fighter airplane. We knew that German forces on the fighting fronts were in desperate need of airplanes, and that the V-2 rockets were doing us no military damage, each impact killed only six people on average. From our point of view, the V-2 program was almost as good as if Hitler had adopted a policy of unilateral disarmament!"

Similar ambiguities clouded the issue of responsibility for the slave labor at Nordhausen. To preserve them from Allied bombing attacks, the Germans had dispersed V-2 and V-1 manufacturing facilities far from Peenemünde, under the supervision of Himmler's SS. Himmler and SS-General Kammler dictated production schedules and allocated V–2s for deployment and for testing. Neither Dornberger nor von Braun had direct authority over Mittelwerk, but both men visited the plant several times and observed conditions. Dornberger—and von Braun—could influence V–2 production only indirectly, by lobbying for greater resources. For his part, von Braun later admitted that he had indeed visited Mittelwerk on several occasions, summoned there in response to attempts by Mittelwerk management to hasten the V-2 into production. He insisted that his visits lasted only hours, or at most one or two days, and that he never saw a prisoner beaten, hanged, or otherwise killed. He conceded that in 1944 he learned that many prisoners had been killed, and that others had died from mistreatment, malnutrition, and other causes, that the environment at the production facility was –in his words-- "repulsive."

The only others whose Nazi past came to haunt them publicly included Dr. Hubertus Strughold and Arthur Rudolph. As the head of Nazi Germany's Air Force Institute for Aviation Medicine, Strughold participated in a 1942 conference that discussed "experiments" on human beings carried out by the institute. The experiments included subjecting inmates of the Dachau concentration camp to torture and death by being immersed in water, placed in air pressure chambers, forced to drink sea water and exposed to freezing temperatures. But Dr. Strughold has many merits in his favor. He was a world leader in space medicine for the U.S. Air Force, playing an important role in developing the pressure suit worn by early American astronauts. He co-founded the Space Medicine Branch of the Aerospace Medical Association in 1950, and was named chief scientist of the Aerospace Medical Division in 1961, among other achievements.

Arthur Rudolph's case was probably a miscarriage of justice. Like von Braun, he dreamed of rockets for travel into space, not as weapons of war. There is no evidence that Rudolph was complicit in the mistreatment of the forced labor prisoners at Mittelwerk, he was just sent there by the Nazis to produce V-2s. The issue isn't "did he work on weapons of war," rather did he "participate in, condone, or encourage unwarranted human suffering."

The fact that in December 1944 he only produced four missiles-- that were later returned from Peenemünde as defective-- instead of the 50 that the Nazis wanted, speaks well of him. Rudolph's merits are overwhelming. Technical director for the Redstone missile project. Project manager for the Army's successful Pershing missile program. Project director of the Saturn V rocket program (which performed flawlessly to land men on the Moon), developing system requirements and the mission plan for NASA's Apollo Program. Awarded the NASA Exceptional Service Medal and the NASA Distinguished Service Medal. Does this sound like the work of a Nazi traitor? Not. Yet the Office of Special Investigations (OSI) basically badgered him to leave the United States and renounce his United States citizenship. It was more like coercion and duress. The hapless Rudolph, in poor health and a man without a country, was granted citizenship by West Germany after they questioned a number of witnesses and determined no basis for prosecution. To add insult to injury, he was denied a U.S. visa to return in 1989 for the 20th anniversary celebration of the first Moon landing; and a measure was introduced in the U.S. House of Representatives to strip him of the NASA DSM, not once, but twice.

American strategists argued that the Germans might help bring the war in the Pacific to an end, and pressured the Truman administration to support a program of exploitation of German scientific expertise. Russian and British interest in German scientists raised concern that the United States might miss a historic opportunity. Truman had no reservations about using German expertise as long as the program could be kept secret. The pressing need to secure the cream of enemy assets was obvious, and the Joint Chiefs of Staff codified many different intelligence efforts into a top-secret *Project Overcast* on July 20. However, the name seemed compromised when the housing in Landshut, Germany (just east of Munich) where the Germans were temporarily kept was commonly nicknamed "Camp Overcast." This forced a name change. In March 1946, the effort to gather top-secret Nazi technology became known as ***Project Paperclip***. The term "Paperclip" stemmed from the fact that dossiers of the most highly valued German scientists were flagged with paperclips.

The German intellectual capital was formidable and priceless. German achievements extended beyond mere advances in weapons. They included developments in wind tunnels, materials, and other disciplines necessary to build an advanced scientific infrastructure. Initially, the United States planned to permit only about 100 individuals to enter the country. Ultimately, however, despite the admonitions of Gen. Dwight D. Eisenhower that there were to be no dealings with any Nazi, Washington approved the entry of about 700, with family members.

In 1945, interrogation of von Braun's inner circle, now ensconced in Witzenhausen in the American zone, gave way to negotiations over terms for consultation services. Colonel Toftoy requested authority to bring 300 rocket experts to the United States, but received permission to transfer only 100. Von Braun had insisted that the smallest group that could be transferred was 520, but he helped pare the list to 127, ensuring that they represented a cross-section of his organization. Negotiations did not always proceed smoothly. Questions arose over whether transfers would be permanent, if they could be renewed, whether wives could accompany their husbands, what salary they would be paid—none of which had clear-cut answers, given the ad hoc nature of the program. Persistent French, British, and Russian interest in exploitation gave the Germans some leverage. *The Russians just missed their chance of grabbing von Braun in July, as we learned in the* Серге́й *chapter with the escapades of Lt. Vasiliy Kharchev and Col. Rick Schraeder.* In the end, the von Braun group remained together and stayed with the Americans as the least undesirable alternative. "We despised the French, we were mortally afraid of the Soviets, we did not believe the British could afford us, so that left the Americans," one member of the group explained.

On August 15 Emperor Hirohito announced Japan's acceptance of the terms of the Potsdam Declaration over the radio. World War II officially ended on September 2 when the formal surrender ceremony took place in Tokyo Bay aboard the battleship U.S.S. Missouri. America's use of former Nazi scientists provoked strong political protests on the part of the citizenry. Many thought that, with hostilities over, there was no justification for using them. But better–informed heads prevailed. The copious knowledge already flowing from the captured scientists ruled out any major change in course.

INTERLUDE AT FORT BLISS

Operation Backfire

Like everyone else the British scrounged for German missile hardware, rocket scientists, and engineers for their own missile program. In Northern Germany they captured a surprising amount of specialized V-2 fuel and LOX transporters, handling and launching equipment--basically all the items necessary for firing the missiles. They also salvaged enough parts to build eight V-2s, but some were missing parts. The Brits picked a usable launch pad near Cuxhaven in Germany for Operation Backfire, whose aim was to demonstrate the weapon to Allied personnel. As word spread of the V-2's awesome (for its time) capabilities, all the major partners became interested in learning more about the design, the operations, and the tactical use of the V-2. Led mostly by the British, the allies used captured German troops who had served in V-2 firing units and a small group of Peenemünde and Mittelwerk workers to launch four V-2s in October 1945. The engine of one failed shortly after launch, but it still flew 24 km. The British even initiated an active French missile development program soon after the war ended. A number of personnel on the program eventually migrated to Egypt and spent a few years developing military systems under Colonel Nasser.

On Sept. 18, 1945 von Braun and the top six members of his team departed Orly Field in Paris in a C-54 Skymaster, arriving two days later at New Castle Army Air Field, Delaware. Others arrived by ship on November 16, December 6, and February 3, 1946. The Army soon sent 117 of them, including von Braun, on long-term assignment to Fort Bliss, Texas, a large Army installation just north of El Paso. The German scientists and engineers were given the job of training military, industrial, and university personnel in the intricacies of rockets and guided missiles and helping refurbish, assemble, and launch the V-2s that had been shipped from Germany to the White Sands Proving Ground in New Mexico. Out of expediency all were cleared to work in the U.S. after having their backgrounds "bleached" by the military. Much of the information surrounding Operation Paperclip is still classified. The repatriated German scientists were generally well-treated, but were unable to leave the station without military escort. For the remainder of the decade, the Germans served as consultants to the Army, Navy, and private contractors. As with the Soviets, the mass transfer of personnel and equipment after the war accelerated the U.S. development of ballistic missiles.

The years at Fort Bliss were not blissful for the rocket men from Germany. Unlike the Peenemünde years before or the Saturn V and Apollo Program years later, no clear goal unified them. They were consultants to American military and industrial researchers, advisers to the dreams of other men. Dornberger and von Braun had fostered cooperative enterprise, of course; but no corresponding sense of collective identity emerged from the military-industrial-university complex supporting Peenemünde. But the period was crucial, for at Fort Bliss the members of von Braun's group began to view themselves as members of a team.

The peculiar circumstances of life at Fort Bliss reinforced the sense of a team. New to a foreign country in which many had at best a cursory understanding of the language, separated from their families, sharing professional interests, viewed with suspicion by the people of El Paso, the Germans drew together. They hiked in the nearby Organ Mountains, played chess and read, and played ball games on a makeshift field between the barracks. Pranks reflected a boarding-school atmosphere, as when Major Hamill reprimanded von Braun, "The wall of Mr. Weisemann's room has been broken through. This matter was not reported to this office. The pieces of the wall have evidently been distributed to various occupants of Barracks Number 1." The elite nature of the group that led to charges of arrogance created another common front; one American described them as "a president and 124 vice presidents."

The president, of course, was von Braun. Not only did the other Germans unequivocally accept him as their leader, but von Braun insisted on his prerogatives. Relations with Hamill were often prickly. Von Braun resented it when Hamill questioned his subordinates, issued orders, or transferred personnel without working through him, and threatened to resign several times. Hamill ignored the threats, but acceded to von Braun's control of the team.

Relations between von Braun and Colonel Toftoy remained on a higher plane. Toftoy exerted a calming influence on the group, and worked to meet their needs. Within a year, he had won the right for the Germans to begin bringing their families. During his stay at Fort Bliss, von Braun proposed marriage to Maria Luise von Quistorp, his maternal first cousin. This time his father approved the marriage, which took place in March 1947 in Landshut, Germany. The von Braun's first daughter, Iris Careen, was born at Fort Bliss (1948). The von Brauns would have two more children, Margrit Cécile in 1952 and Peter Constantine in 1960.

In the spring of 1948, Toftoy and Hamill devised a scheme to overcome a legal technicality that troubled the group. Since they had entered the United States without passports or visas, their immigration status was in doubt. They crossed into Mexican territory and returned the same day with papers listing Ciudad Juarez as their port of debarkation, El Paso their port of arrival.

Colonel Toftoy's principal purpose in bringing the Germans to Fort Bliss was Project Hermes, which consisted of many different smaller projects with the goal of developing Americanized experimental versions of the German weapons. The Hermes C–1 subproject involved test firing of the Mittelwerk V–2s intended to give Americans experience in rocket research, testing, and development . However, the V–2 parts were in disarray, having been packed by soldiers, shipped to New Orleans, reloaded on freight cars, repacked once

Von Braun (7th from right) and his team of rocket scientists at Ft. Bliss, El Paso Texas

again on trucks, and finally left in the open on the desert at White Sands. Working with General Electric as the prime contractor, the Germans had to reassemble the parts into complete vehicles, and subjected some of them to static firings on the test stand. The first V-2 launch at White Sands was a flop. It went only 3.5 miles up and exploded. The first successful firing took place on May 10, 1946 and reached an altitude of 71 miles. About 68 V-2s were finally launched, 45 of them successfully, all carrying a wide variety of instruments and all intended as scientific experiments yielding aerodynamic data, information on the composition of the upper atmosphere, and boot-strapping American rocketry research. The Navy launched a V-2 from the deck of the USS Midway. Another single-stage V-2 set an altitude record of 130 miles on August 22, 1951. These successes convinced the Army to fund Hermes C–1 as a full scale missile research and development program, which led to the Redstone launch vehicle. You'll find out more about Project Hermes and how von Braun's team and the Army developed the Redstone in *Avoiding Armageddon.*

The V-2 itself became the baseline from which many later rockets were derived. Eight of them were modified into the world's first two-stage vehicles by stacking a JPL–developed WAC (Without Attitude Control) Corporal missile to create the "Bumper-WACs" that were described in the Theodore chapter. They demonstrated the advantages of staging, which had been so well-defined by Oberth in his 1923 book *Die Rakete zu den Planetenräumen"* (The Rocket into Interplanetary Space) as we saw in the Robert chapter. Two Bumper-WACs were taken to a remote Naval Air Station in Florida for launch. The first launch from Cape Canaveral (then a deserted area) took place on July 24, 1950.

The Air Force opened up the Cape Canaveral Air Force Base (CCAFB) and it grew into the Eastern Test Range (ETR). The Air Force later set up a Western Test Range (WTR) at Vandenberg AFB for launching satellites over the poles (if you try to launch a polar-orbiting satellite from the ETR, you end up flying over the eastern U.S. which is a no-no). Years later NASA established the Cape Canaveral Space Center, later renamed Kennedy Space Center (KSC) after President John Kennedy's assassination in 1963. KSC is often called "The Cape" to this day.

The success of the American V-2 experiments also provided confidence to other firms developing rocket designs that did not use V-2 technology per se. This influence extended for decades. The V-2 benefited Convair's Atlas ICBM and the Douglas Thor IRBM. In 1955, the Air Force launched the Titan program as a backup to the Atlas. The connections made with German engineers at Ft. Bliss continued on into the Titan era. Heavy–lift Titan IV ICBMs carried the highest-megaton warhead of all USAF missiles and served NASA's Project Gemini and later Air Force and NASA satellite launch vehicles.

A FRESH START

What Is Redstone Arsenal?

Huntsville, now a mid-size city in north central Alabama, was founded in 1805 by John Hunt, an early settler from Tennessee. He was attracted to the area by its fertile farmland, plentiful game, and a large spring in what is now downtown. However, Hunt did not properly register his claim, and the area was purchased by Leroy Pope, who imposed the name Twickenham on the area to honor the home village in England of his distant kinsman Alexander Pope. Due to anti-English sentiment during the War of 1812, the name was changed to Huntsville to honor John Hunt, who had been forced to move to other land south of the new city. Cotton, soybeans, and cattle became the driving forces of the local economy. Huntsville had become dominant in the textile industry by the 1920s, but the Depression brought the town's ability to turn raw cotton into fabric to its knees. Fortunately, on June 8, 1941 Army officials visited the cotton town in search of a site for a weapons plant. After visiting other potential sites, the War Department announced on July 3, 1941 that they had settled on a 30,000 acre site west of Huntsville. The new chemical munitions manufacturing and storage plant was given the name Huntsville Arsenal. The Army announced that Huntsville Arsenal would build bombs, grenades and chemical weapons. Workers were given the task of producing toxic chemicals as weapons, then packing them in shells, grenades, and bombs. Huntsvillians saw an Army plant as a welcome boon to the local economy which had been suffering through a Depression-era economic blight. The cost of the vast new complex of buildings was advertised as over $40 million (it exceeded $47 million), which further garnered public support.

On April 15, 1955, von Braun and 40 members of his team and their families took the oath of citizenship in the United States.

Three months later the Army decided it needed another arsenal in Huntsville dedicated to manufacturing conventional shells and bombs. As Major Carroll Hudson was flying around looking for potential sites, it struck him that the limestone hills and soil in the environs of Huntsville were a reddish color, so he proposed the name Redstone for the new arsenal. Army officials selected a 4000 acre tract next to the Huntsville Arsenal for the Redstone Arsenal. Again, the local community was very supportive of the Army's investment. Local citizens by the thousands became employed, including women. Under wartime pressures, construction on both arsenals was completed in early 1942 at a combined cost of $81.5 million. For comparison, the War Department built the Pentagon in approximately 16 months (Sept. 1941 to January 1943) at a comparable cost of $83 million. The main munitions and bomb-making facility—the Redstone Ordnance Plant—made so many weapons that the Army had to build innumerable storage bunkers to store them onsite. The bunkers can still be found today.

In the late 1940s the commander of Fort Bliss rejected Toftoy's plans for expansion, and insufficient funds forced cancellation of research projects. Toftoy believed rocket research had become too decentralized. In 1949 he began to search for a new location at which to conduct Army rocket research, thus initiating the chain of events that would lead to the establishment of the NASA Marshall Space Flight Center. In August he visited Redstone Arsenal and neighboring Huntsville Arsenal, then listed for sale by the Army Chemical Corps because their wartime mission had long since been completed. Toftoy liked the site. Senator John Sparkman, a Huntsville resident and chair of the city's Industrial Expansion Committee, lent support after the city lost a bid for the Arnold Engineering Development Center (AEDC, in Tennessee) which was explained in the Theodore

chapter. After a personal appeal to General Matthew B. Ridgway, now-General Toftoy won approval in October 1949 to incorporate Huntsville Arsenal into Redstone Arsenal and transfer the entire von Braun group to Alabama.

To implement this decision, Dr. Wernher von Braun and almost all members of his team, as well as many American military and industrial members of the group left Fort Bliss in the summer of 1950 and moved to Huntsville. There they formed the nucleus of the U.S. space program. At that time several original members of the team departed and joined private industry, where they eventually took leading positions at such aerospace contractors as General Dynamics/ Convair, North American Aviation, Lockheed, and Aerojet. Major General Walter Dornberger rose to the post of Vice-President at Bell Aircraft Corporation. He played a major role in the creation of the X-15 aircraft and was a key consultant for the X-20 Dyna-Soar project. He also had a role on the creation of ideas and projects which, in the end, led to the creation of the Space Shuttle. Another small group who had worked on the LOKI project moved north in order to continue their development work there for an Army contractor. Loki, named after the Norse god of trouble and mischief, was an unguided anti-aircraft rocket based on the German Taifun. Loki eventually found widespread use as a sounding rocket for scientific research. In 1955 Hermann Oberth himself joined the staff at the Redstone Arsenal.

Huntsville Becomes the Rocket City

Toftoy's shift to Redstone Arsenal began the economic, cultural, and political transformation of Huntsville and surrounding Madison County, Alabama. The first small contingent of Germans arrived in March 1950, and others soon followed. The move to Huntsville involved not only the German rocket experts, but 800 others, including General Electric and Civil Service employees, and 500 military personnel. In mid-1951, the Army transferred the Redstone missile program to the Arsenal. It was the von Braun's team's first major project. By June the Arsenal's working population swelled to more than 5,000 people. Huntsville's population would triple by the end of the decade, and much of the growth was due to the infusion of federal money for the Arsenal.

The Redstone missile's military accomplishments were not as important as the many scientific research and development missions to which it was finally adapted. Due to its high reliability, its dependability and its availability, the Redstone would play a most important role in early space exploration efforts during the U.S. Mercury manned space program. The path to this non-military role lay in the Von Braun team's development of the Jupiter C, which likewise was not destined to become an Army IRBM. As these and subsequent events unfolded, Huntsville earned the moniker, "The Rocket City."

AMERICA MAKES A COMEBACK

Much like Helmut Gröttrup and his team in the Soviet Union (Серréй chapter), Dr. Wernher von Braun and his German rocketeering expatriates provided a tremendous boost to post-war rocket technology advances in the United States. One of the first beneficiaries was the Navy's Viking, of which a series of 12 were fired, each of slightly different design. The Viking stood about as tall as the V-2 but was slimmer. Its diameter was 83 cm (32.5 in) and it weighed 4376 kg (9650 lb) at liftoff. Instead of having heat-resistant graphite vanes in the exhaust for guidance, the Viking's motor was mounted in a gimbal ring, so that steering and balancing were accomplished by tilting the whole motor slightly. Otherwise the arrangement of the components was the same as in the V-2, and it used the same propellants. The first three were marred by premature cutoffs of the engines. Of the remaining nine, seven were complete successes. The Vikings were used to gather weather and scientific data. The highest altitude and speed reached by a Viking rocket (Number 11) was 158 mi (250 km) and 4,000 mph (1.8 km/sec), attained on May 24, 1954.

In its final configuration, the U.S. Navy's Viking rocket was 45 ft long and 45 inches in diameter. Like the V–2, hydrogen peroxide was decomposed to provide steam for its turbopump. It could carry over 1,000 lbs. of instruments to an altitude of 150 miles.

The related U.S. Navy Aerobee rockets and the militarized Redstone and Jupiter–C missiles from a warfighting point of view are described in Avoiding Armageddon. These and other Army missiles were converted to civilian use by replacing their nuclear weapon payloads with civilian payloads, and performing other minor modifications.

By the mid-1950s, the Army's Redstone-derived Jupiter C and the Navy's Vanguard were running practically neck and neck in the race to match the Soviet's Sputnik successes. The principal impetus behind these developments was their potential as weapons in the Cold War against the Soviets. Military men in both the United States and the Soviet Union saw satellites as predecessors for orbiting nuclear launch platforms. However, von Braun and his cohorts were more interested in *civilian* space exploration. They had already witnessed the horrors of war first hand. The thought of Ike's mutually assured destruction (MAD) doctrine drew chills down their spines. Von Braun once said that he wanted to "use rockets to blaze trails to other planets, not to destroy our own." Nevertheless, out of necessity they followed the same path as in Germany, developing the combat missiles for the Army as a means to reach their goal of space travel for peaceful purposes.

The U.S. Army's early missile programs are more thoroughly described in *Avoiding Armageddon* which details missile development, principally in the United States and Soviet Union. It is easy to get all the missiles' names and their roles confused. Let's see if a family allegory can help.

> *Roughly speaking, the German V–2 MRBM was the grand–daddy of all rockets and gave the world's rocket programs on both sides of the Iron Curtain a jump start after World War II. The Army used V–2 technology to develop the "son" of the V–2, the Redstone MRBM. In turn, the Redstone "fathered" three missiles, in order: the twins Jupiter–C and Juno I, and the Mercury–Redstone. They all had different names but bore a family resemblance to their father, and were peace-loving civilians.*
>
> *The Redstone had a couple of distant relatives: the military Jupiter IRBM and its civilian twin brother, Juno II. The point of this story, besides getting you so thoroughly confused that you'll have to read Avoiding Armageddon to figure it all out, is that the Jupiter–C and Jupiter IRBM pair and the Juno I and Juno II pair are unrelated despite sharing the same name. They look different, use different engines and propellants. In any case, military missiles have always come first, civilian versions follow and are derived from them—a theme that will be played over and over again in The Saga of Rocket Science series.*

In mid-1954 the Army began "Project Orbiter" at Redstone Arsenal. This short-lived program was to launch an instrumented satellite into orbit using a three-stage Jupiter C with a fourth stage added. To distinguish the four-stage Jupiter C space launch vehicle (SLV) from the three-stage Jupiter C, it was officially named Juno (later amended to Juno I when Juno II was created as the Jupiter IRBM's twin). The name derived from von Braun's wishes to make it appear as peaceable as the Navy's Vanguard rocket. The Vanguard name was derived from the Viking weather study rocket, which of course was not a weapon. Juno was the queen of the gods and the wife of Jupiter in Roman mythology. However, the name Juno never really stuck.

> *You'll learn lots more about the Redstone, Thor, Jupiter, and Jupiter–C missiles in **Avoiding Armageddon**.*

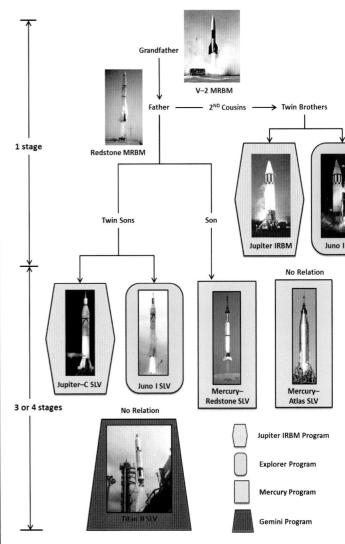

Family analogy among Redstone, Jupiter C/Juno I, Mercury–Redstone, Mercury–Atlas, Jupiter IRBM/Juno II, and Titan II.

ABMA

On February 1, 1956, the Army Ballistic Missile Agency (ABMA) was established at Redstone Arsenal under the command of Major Gen. John "Bruce" Medaris. He was charged with development and production of the Jupiter IRBM and the weaponization of the Redstone ballistic missile. Medaris' energetic leadership and forceful direction as head of the ABMA would play a key role in the adaptation of both the Redstone and Jupiter C to civilian use for scientific research and manned space exploration, although he may not have realized this at the time.

MG Medaris, von Braun, and T. (Thomas) Keith Glennan, the first administrator of NASA. Glennan took immediate steps to give NASA independence from the military, to Medaris's dismay.

done. *In hindsight, the United States could have beaten the Soviets into space by a whole year!*

Mid-Summer, 1957

President Eisenhower's administration wanted to avoid launching the first artificial satellite with a rocket that was quite obviously a military missile. "I don't want the Army launching an unauthorized satellite, and apologize later. Don't underestimate von Braun," Eisenhower was heard to say. The Army was explicitly forbidden to go to orbit in their Jupiter–C flight tests Von Braun's team could only "think" about a satellite, or make paper drawings and paper studies only. Still nurturing his long-held dream of spaceflight, von Braun was forced to work his project in secrecy, against official prohibition by the government to not build a satellite. Unofficially, von Braun kept a Redstone rocket stored, ready to put up a satellite. Officially it was a Redstone out of the production line, with the militarily justifiable objective of testing how a rocket could withstand a long time in storage. Satellite parts, instruments, even a full-size model of a live Jupiter C fourth stage sat in other warehouses at Redstone Arsenal.

"I have put together parts of a satellite in my garage," Dr. Ernst Stuhlinger said. He had been an integral part of von Braun's team from the beginning. "Great," von Braun responded, smiling. "Bring them here so we see how they fit into the satellite shell." With the assistance of Arthur Rudolph, fit them they did. Von Braun disguised the roughly completed satellite under some clothes and drove it to the Cape. They were officially at the Cape to launch the Jupiter C only for reentry nosecone tests in support of the Army's Jupiter IRBM.

"Have you finished the modifications?" von Braun asked Ernst Steinhoff, his GNC (guidance/navigation/control) specialist since the Peenemünde days. "It is done," Steinhoff replied. "Where is it?" queried von Braun. "We took the ST-90 off and installed a modified LEV-3 in the instrument compartment on the bird, like you asked." Telling no one except trusted insiders, Steinhoff had clandestinely switched out the Jupiter C's ST-90 guidance system and substituted a LEV-3 autopilot control system like those used for Jupiter IRBM tests. But he jury-rigged the LEV-3 to shoot the third stage up in a try to make orbit, not down to the horizontal to perform yet another reentry nosecone test.

On September 20, 1956 the ABMA launched a Jupiter C for the first time from Cape Canaveral. The ABMA wanted to test the staging and sequential firing of three of its four stages. Taking the place of a satellite in the fourth stage were 20 pounds of instrumentation and 10 pounds of sand, both carefully weighed to match the anticipated weight of the Explorer satellite (30 pounds). The test went beautifully. Telemetry showed that the third stage reached well into space at an altitude of 1,097 km (681 mi, over twice as high as the Space Shuttle) and a speed of 13,000 mph. One night von Braun wondered, "What if we are allowed to install one of the midget Sergeant solid rocket motors on the fourth stage? Could we accelerate the remaining 5,000 miles per hour and put a satellite into orbit?" "Let's check it and see. I think we can," said Arthur Rudolph, part of von Braun's inner circle. They put pencil to paper (or rather, slide-rule to paper as there were no calculators or personal computers in those days). They figured out that a powered fourth stage could easily boost a 30-pound satellite into orbit at 18,000 mph (over 8 km/sec). All they needed was permission from the Army. They would have to design the satellite, of course, but the hard part—the propulsion, missile, guidance, and stage separation systems were

Unbeknownst to Steinhoff, Bill Touchon, one of the GNC technicians, had noticed the LEV-3 jury-rigged schematic on Steinhoff's drawing board when he went to get coffee. Touchon could tell it was no ST-90 and knew why. He was stridently anti-communist, and had recently attended Senator Joe McCarthy's funeral on May 2. Yes, *that* Joe McCarthy. A few years earlier, Senator Joseph Raymond McCarthy (R-WI) had even leveled charges against the U.S. Army, claiming that Army Secretary Robert Stevens had concealed foreign espionage activities. After 36 days of televised hearings, known as the "Army-McCarthy hearings," McCarthy was unable to substantiate any of his allegations. But Touchon was still adamant that the nation's Cold War RV testing program not be compromised by— as he called them— "spacenuts," especially if they had a Nazi past. He called the Office of the IG (Inspector General) in Washington.

The IG responsible for keeping the Department of the Army in check got furious. "How dare von Braun disobey orders!?" he fumed at his assistants. Within days he dispatched a representative down to the Cape with an associate, hopefully to catch von Braun in the act. But von Braun also had lots of eyes and ears. Practically everyone on his team greatly admired their charismatic leader. In their eyes, he could do no wrong. When the word got to von Braun, he "asked" (depending on his tone and demeanor, asking was the same as ordering) Steinhoff to quickly dismantle the illicit autopilot control apparatus and put the ST-90 guidance system back in. Randy Clinton took it upon himself to hide the real instrumentation in the trunk of his car. The ruse worked. The IG rep looked over all the missile systems and configuration. Everything checked out. "See, there is nothing to this false accusation!" an indignant von Braun exclaimed. Satisfied, the IG representative returned to Washington.

Aware that there was a snake lying in his grass, von Braun had Clinton drive the modified LEV-3 autopilot control unit back to Huntsville and away from prying eyes at the Cape. He was elated when on August 8 a Jupiter C lifted a 1/3-scale Jupiter nose cone to an altitude of 460 km (285 mi) and a range of 1,344 mi (2,163 km). The test successfully demonstrated that Allen and Eggers' reentry nosecone geometry would work, and that the ABMA's composite plastics would create an insulating blanket through ablation to protect the warhead from reentry heating. The Army IG was pleased.

VANGUARD

The Army's competition in rocket launchers was a group of engineers and scientists at the Naval Research Laboratory (NRL) in Washington, D.C. headed by engineer John Hagan. Located off the eastern bank of the Potomac river just south of the old Bolling AFB, NRL is a civilian arm of the U.S. Navy, a think tank run by the Navy, paid for by the Navy, but basically staffed by civilian scientists and engineers who conduct its day-to-day operations. The Navy claimed space as an extension of the sea because it would be traversed by ships— spaceships (for Star Trek fans, who could argue?). The NRL's proposal—called Vanguard because it would lead the United States into space (hah, countered von Braun and the Army folks) — would be a three–stage finless rocket with a liftoff weight of around 22,200 pounds (10.1 tonnes). 88% of this weight was propellant. The first two stages were liquid-propellant rockets, guided by a "strapped-down" (rigidly mounted in place) gyro reference system, and controlled by engine gimbaling and reaction jets. The third stage was a solid-propellant rocket motor, unguided but spin-stabilized. A jettisonable nose cone protected the payload. By Soviet standards the payload was miniscule—varying between 1.3 kg to 23.6 kg (2.9 lb – 52 lb). The Vanguard rocket had a first stage thrust of 27,000 pounds. That was about three percent of the thrust of Korolyov's R-7 rocket that had put the first two Sputniks into orbit!

In early August, 1955 a White House selection team met in secret behind closed doors and picked Vanguard as opposed to Redstone for the nation's first space launch vehicle (SLV). The Navy design had a better tracking system and more scientific growth potential. And it would not interfere with high-priority military missile programs, since it would use a new booster based on the Viking research rocket. The Army's Project Orbiter was cancelled. The Eisenhower administration was laying the groundwork for Ike's network of ISR (Intelligence, Surveillance, Reconnaissance) satellites for national security.

The United States and Soviet Union were locked in a propaganda battle, and President Eisenhower had won the opening salvo by preempting by four days the Soviet IGY announcement of sending a satellite to space (Ike announced it on July 29, 1955). The Soviets won the next two moves with Sputnik 1 (Oct. 5, 1957) and Sputnik 2 (Nov. 3, 1957). In making the Vanguard Program open to the public, Eisenhower could both gain public acceptance of the development cost (over $40M, a small fortune in those days) and not be hindered by public release constraints put on classified missile development. In contrast, the Soviet Union did not have to justify to its citizens the funds spent on going to space, and felt no compelling need to make public announcements of its progress (while conveniently hiding failures).

Vanguard SLV was originally planned to have enough thrust to loft a 30-pound scientific package housed in a 20-pound, 20-inch diameter ball into a much higher orbit than the Redstone missile–derived SLV, giving it a correspondingly longer useful life before it decayed from orbit. Navy planners also touted their use of transistorized electronic circuit boards powered by solar cells located on the skin of the satellite. So it seemed the Army entrant

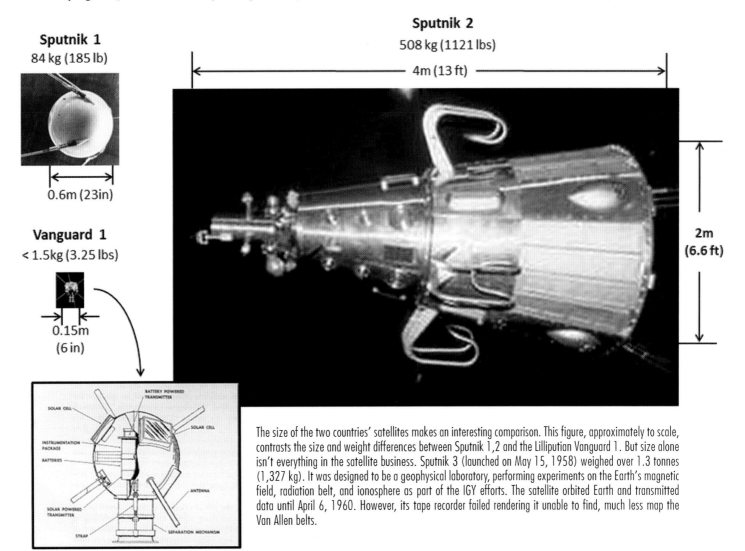

Sputnik 1
84 kg (185 lb)

0.6m (23in)

Vanguard 1
< 1.5kg (3.25 lbs)

0.15m (6 in)

Sputnik 2
508 kg (1121 lbs)

4m (13 ft)

2m (6.6 ft)

The size of the two countries' satellites makes an interesting comparison. This figure, approximately to scale, contrasts the size and weight differences between Sputnik 1,2 and the Lilliputian Vanguard 1. But size alone isn't everything in the satellite business. Sputnik 3 (launched on May 15, 1958) weighed over 1.3 tonnes (1,327 kg). It was designed to be a geophysical laboratory, performing experiments on the Earth's magnetic field, radiation belt, and ionosphere as part of the IGY efforts. The satellite orbited Earth and transmitted data until April 6, 1960. However, its tape recorder failed rendering it unable to find, much less map the Van Allen belts.

One would think that President Eisenhower would have picked the Army's rocket, after all he graduated from the United States Military Academy (USMA) at West Point, New York, and had served a long career in the Army. The decision to pick Vanguard was influenced by von Braun and his team's Nazi past. There was resentment against a German team. Eisenhower was the commanding invading general of D-Day and knew first hand of the Nazi atrocities. In addition the

lacked the pizzazz being promised for Vanguard. But appearances can be deceiving.

The Vanguard team initially adopted the Redstone team's approach because it gave successful results. Von Braun had carefully developed the Redstone SLV over a 4–year period using an ultraconservative engineering process that tested individual rocket components, then tested individual assembled systems, and finally tested the combined systems as a complete rocket. Using a secret new calculating machine called a computer, data from each component and subsystem test was used to build data models predicting how the unit would react under different operating conditions. Because the predictions were only as good as the data, the process had to be slow to build accurate models.

The first two flights of the Vanguard program, designated Test Vehicle (TV)-0 and -1, were actually the Navy's last two remaining Viking rockets. TV-0, launched on December 8, 1956, primarily tested new telemetry systems, while TV-1 on May 1, 1957 was a two-stage vehicle testing separation and ignition of the solid-fueled upper stage of Vanguard. Despite these initial successes, Vanguard engineers faced some really daunting obstacles:

1. NRL was a relative newcomer to the space launch business. NRL engineers had neither the Army's long years of development experience nor the cover afforded by a classified program. Redstone had the best booster and a ready-to-go design. The Redstone had already flown many times, which would minimize risk.
2. In 1955 NRL engineers found out that the Vanguard vehicle's original design lacked the thrust necessary even to reach orbit. Unlike the Army/JPL team which would use mostly existing components, all three Vanguard stages (based on Martin Company missiles) had to be completely redesigned within a year to stay in the competition.
3. The NRL team would have to develop a completely new payload carrier and satellite. Ironically, the fact that Vanguard would use an unproven third stage and payload, costing more and being much more difficult to develop, may have worked in its favor. The cover of a civilian satellite program [under the International Geophysical Year (IGY) of 1957–1958] provided President Eisenhower with the pretext he needed to get ahead of the Soviets in a deadly Cold War arms race and launch an ISR network of satellites for national security. If the Soviets could launch a dog into space, they could launch an H-bomb. At stake was the best delivery system for H-bombs aimed at an implacable enemy. In the end, the Vanguard payload itself was far superior to the Army's Explorer.
4. Vanguard was underfunded and not given priority status, The program didn't have the Air Force's massive funding on the competing Thor IRBM, and Titan and Atlas ICBM programs, nor the Army's well–funded Jupiter IRBM. Even worse, a slow, underfunded approach would have to be abandoned in favor of skipping test steps if the satellite launch deadline was to be met.
5. Vanguard became plagued with manufacturing

errors largely due to poor supervision in the Martin main plant, already working overtime to get the Titan ICBM missile successfully airborne. Inspections at Cape Canaveral conducted by the Navy revealed fine filings, metal chips, and dirt in the tankage and engine. This caused return of parts to the Martin plant for rework and ultimately delayed the launch of Test Vehicle 2. The Navy tried three times to run static tests on the Vanguard TV–2 without success. Project Vanguard boss John Hagan was furious at the Martin Company and put them on notice that contact funds would be withheld until it achieved satisfactory performance.

The U.S. Air Force was essentially experiencing the same kind of problems as Vanguard. However, because their work was classified, these problems were not subjected to public scrutiny. Although the Navy took pains to publicly share the kinds of development problems it was having and the proposed solutions with the entire rocket community, there was little reciprocity.

After several abortive attempts, the Navy launched TV-2 on October 23, 1957 (next page) three weeks after Sputnik. It was the first real Vanguard rocket with three stages. Note the lack of fins on the 1st stage. The flight successfully tested first/second stage separation and spin–up of the third stage at 121 miles (195 km). Early Army rockets like the Redstone and Jupiter also used gyroscopic spin–stabilized guidance in those early days. For this demonstration the 3rd stage was an inert test article with no motor or payload, and wasn't intended to enter Earth orbit.

As an interesting side note, while we take successful rocket launches for granted today, pioneering rocket engineers in the 1950s had to learn all kinds of things that were simply unknown at the time. During static testing of the TV–2, the vent valve in the liquid oxygen (LOX) tank got blocked, building up pressures in the tank past safe levels. After halting the test, an inspection revealed that the LOX had been contaminated with water condensed out of the air in the empty system. Turned to ice at cryogenic temperatures, this ice blocked the LOX tank pressure relief valve. Engineers devised a procedure of purging the empty LOX system (piping and tanks) with a constant flow of inert nitrogen until filling it. This solution became the accepted practice for all LOX–propelled rockets, and solved problems plaguing the Thor and Atlas up to this time.

Vanguard 2 successful launch

in the Theodore chapter, the von Kármán effect refers to the tendency of air flowing past an object to curl or produce eddies on the leeward side. The eddies set up an oscillation that may be reinforced by the natural frequency of the structure. The telephone pole–shaped Vanguard, standing alone after its service tower was removed some 60 minutes prior to flight, would be vulnerable to turbulent eddy forces and strong vibrations. In the worst case, the nation's pride and joy could topple over like a domino. In an effort to minimize the von Kármán effect, Vanguard technicians equipped their 72–foot vehicle (22 m) with black rubber spoilers. These fin-like strips extended down the sides of the structure for about two thirds of its length. At the top of each was a protruding shoe, designed to catch downrushing air and strip off the spoilers at about 1000 feet altitude.

The initial launch attempt on Wednesday, December 4 was scrubbed. Vanguard project officials announced that another countdown would start late Thursday afternoon with liftoff scheduled for Friday morning at 0800. Shortly after 5:00 p.m. Thursday the second countdown was put on hold because of delays in verifying

By December 1957 the Navy was ready to declare victory over von Braun's U.S. Army team. Depressed at having lost the space race, the Vanguard crew worked feverishly to prepare the next Test Vehicle (TV–3) for launch at Cape Canaveral. It would carry Vanguard I's grapefruit–sized (6 inch, 3.25 lb) satellite into orbit. Despite the pint–size payload, Americans felt was it was high time the U.S. got back at the Russians and restore America's shattered pride. The proud NRL team welcomed national TV media to witness the historic event, and correspondents came from as far away as Europe. All the hotel rooms in the vicinity were booked as the public flocked to watch the launch.

December 6, 1957

High winds (20 to 30 mph) had nixed launch attempts earlier in the first week of December. Calculations by NRL and Martin Company aerodynamic specialists showed it was foolhardy to launch in winds over 17 mph because of the von Kármán effect. As discussed

the proper operation of the vehicle controls system. After a few hours the countdown continued with no apparent problems, besides a few short holds during the night. The Vanguard stood clear against a starry sky, two bright white lights glaring at its base and a red beacon shining at its top. Giant searchlights bathed the rocket with a surreal blue–white light.

Vanguard 1 satellite on exhibit at the National Air and Space Museum in Washington, DC.

By 10:30 Friday morning the countdown reached T–60 minutes, signaling the beginning of the final and critical phase of the launch. At T–50 the winds were level at 16 mph on the pad with gusts up to 22 mph, a touch–and–go situation. Gusty winds could cause the rocket engine nozzle to crash against the surrounding piping as it rose from the clearance hole in the launch platform. But the field crew pressed on. Stations of the radio tracking network downrange sent "all clear" signals to the electronics telemetering crew in the backroom of the blockhouse. Like its name implies, a "blockhouse" is a reinforced concrete structure built to shelter launch and operations personnel from the heat, noxious fumes, noise, and flames of a rocket launch. At T–30 minutes fierce blasts from a bullfiddle warning horn sent people scurrying from the area and either into the blockhouse or to a safe distance in their cars. The heavy blockhouse doors clanged shut at T–25 minutes and counting. A report that surface winds were now 15 knots brought a shrug from Dan Mazur, the field manager. The figure was high, but the trend was downward. Countdown minutes changed to seconds. At T–45 seconds the umbilical cords that supply consumables and electrical power to the rocket began dropping away. Bill Escher's voice quavered as he counted down to T–1 second over the public address system. The tension in the blockhouse edged upwards from high to almost unbearable.

Test conductor John Gray commanded, "Fire!" and Paul Karpiscak, a young Martin engineer, flipped the toggle switch on his oblique instrument panel. In the crowded blockhouse control room all eyes were on the big windows overlooking the pad. Sparks at the base of the rocket signaled that the pyrotechnic igniter inside the first stage had kindled the beginning of the oxygen and kerosene fumes. With a howl the engine started, brilliant white flames swiftly filling the nozzle and building up below it as the vehicle lifted off at 11:45 am.

But things were quickly going to hell inside the rocket, in millisecond timeframes. The kerosene fuel tank never built up enough pressure due to a fault in the engine launch sequence. In turn this caused low fuel injector pressure prior to the fuel turbopump spinning up to speed. The low injector pressure allowed some of the burning contents of the thrust chamber to enter the fuel system through the injector head. While this was going on a loose connection in a fuel line above the engine caused a little fuel to leak out on top of a helium vent valve and blow down on the engine. The result? Catastrophe.

Vanguard Catastrophe, December 6, 1957.

On board telemetry later showed that despite these problems the engine developed full thrust. But not for long. With a series of rumbles audible for miles around, the vehicle, having risen about four feet into the air, suddenly sank within two seconds.

In the words of Arthur C. Clarke and the editors of Life magazine,

"It seemed as if all the gates of Hell had opened up. Brilliant stiletto flames shot out from the side of the rocket near the engine. The vehicle rose about three feet, agonizingly hesitated a moment, quivered again, and in front of our unbelieving shocked eyes, began to topple. It also sank like a great flaming sword into its scabbard down into the blast tube. It toppled slowly, breaking apart, hitting part of the test stand and ground with a tremendous roar that could be felt and heard even behind the 2-foot concrete walls of the blockhouse."

Back in the blockhouse, someone shouted: "Look out! Oh God, no!" In the control room, Perry Douglas screamed "Duck!" Nearly everybody did. Then the fire control technician pulled the water deluge lever, loosing thousands of gallons of water onto the steaming wreckage outside, and everybody straightened up. The next voice to be heard in the room was that of Mazur, courageously issuing orders: "O.K., clean up; let's get the next rocket ready." Spoken like a true rocket scientist who knows the hazards of launch. Already the stunned crew had taken in a startling fact. As TV-3 crashed into its bed of flame, the payload in its nosecone had leaped clear, landing apart from the rocket. The satellite's transmitters were still beeping! But the little sphere itself was too damaged for reuse.

The two companies involved pointed fingers at each other. The Glenn L. Martin Company (GLC) blamed its first stage engine subcontractor, General Electric (GE), for an "improper engine start." According to the Martin people a fire started in the fuel injector before liftoff, resulting in its destruction and complete loss of thrust immediately after liftoff. Impossible, said the GE folks. Not only was their no "improper start," but according to telemetered and photographic data the X–405 LOX/RP–1 engine had come to full thrust, only to lose thrust when a little leaked fuel atop a helium vent valve blew down on the engine. The cause: the Martin launch crew had ill–advisedly used the Vanguard's fuel lines as "ladders" while working on the vehicle, causing one of the lines to break loose during launch. Milton Rosen, Project Vanguard technical director, and his managers tentatively sided with GLC. In the end, GE took the decision in good stride, on behalf of U.S. national prestige.

The Army team hated to see this, but were comforted because they knew that theirs was the best bird. "This is a lesson," von Braun observed. "If you work people long hours, they'll make mistakes."

Meanwhile, TV newscasters and an anxious public were horrified. The Germans would call Vanguard *Spätnik* (*spät* being German for late). The British press had a field day with the embarrassing Vanguard explosion. The London Daily Herald called it "FLOPNIK!" while the London Daily Press smeared it with "KAPUTNIK!" Soviet Premier Nikita Khrushchev joked that the Americans should have named their rocket "Rearguard." In New York City, members of the Soviet delegation to the United Nations asked American delegates if the United States would be interested in receiving aid under the U.S.S.R.'s program of technical assistance to backward nations. Needless to say, a wave of outrage swept the country.

Vanguard Successes

On February 5, 1958 the Navy's next Vanguard failed to orbit. The Navy attempted six more launches in 1958-1959. Only three (out of a total of 11 attempts) were successful. But they were good successes. On March 17, 1958 Vanguard I reached a 406 x 2,466 mile (664 x 3,970km) elliptical orbit. Originally it was estimated to last 2,000 years, but solar radiation pressure and atmospheric drag during high levels of solar activity produced significant perturbations in the perigee height of the satellite, which decreased its expected lifetime to "only" 240 years. Packing a powerful punch in a small package (it weighed less than 1.5 kg), Vanguard 1 measured atmospheric density and composition at

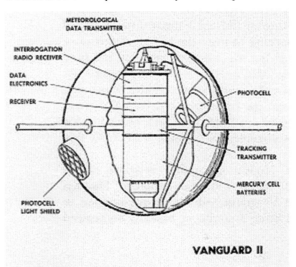

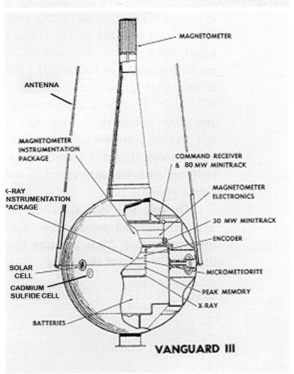

Vanguard 2 and Vanguard 3 satellite schematics

extreme altitudes, and it determined that the Earth is not quite round, but slightly pear–shaped (elevated at the North Pole, flattened at the South Pole, and bulging very slightly in the southern hemisphere). Vanguard 2 (Feb. 17, 1959) continued these measurements and also optically scanned the earth's surface for the first time. Vanguard 3 (Sept. 18, 1959) measured the Earth's magnetic field, the solar X-ray radiation and its effects on the Earth's atmosphere, and the near-earth micrometeoroid environment. The Vanguard program improved U.S. launch vehicle technology, created an important mini-track satellite tracking system, demonstrated the first use of silicon solar cells to provide electric power in a satellite, discovered the Earth's slight pear shape, and obtained a wealth of new scientific data. Vanguards 2 and 3 will last 300 years in orbit. All three satellites are still providing atmospheric density measurements today.

EXPLORER

Dr. Wernher von Braun, Army Generals Bruce Medaris and James M. Gavin, and new Army Secretary Wilbur Marion Brucker had spent the very day that Sputnik 1 launched (October 4, 1957) urging incoming Defense Secretary Neil McElroy to back the Army's Jupiter–C for the satellite program as they showed him around Redstone Arsenal. They adjourned to the Officers' Club to relax over cocktails. "The Soviet approach is to achieve noteworthy development milestones quickly through massive, concentrated effort, then move to the next milestone with minimal continued testing," General Medaris said. "Soviet national policies and the arms race further influence the flow of information necessary for their rocket development." SECARMY Brucker chimed in: "The Soviets present a massive question mark in the minds of the West. With free information flow among the Soviet design bureaus— the scientists and developers working on rockets—there is a unitary Soviet effort that leads to faster problem solving. However, development is being conducted at Top Secret locations—we think there are at least two-- hidden from everyone but the development teams,

and public information about Soviet progress is being understated in their state–controlled press."

"My dear fellows, we also classify missile development," von Braun argued, "but this creates an information gap between the developers who have access to the information and the scientific community—us!— who do not. Furthermore, interservice rivalry and the close holding of corporate secrets impedes the flow of information and isolates our American teams from each other."

Not only was U.S. rocket systems development being classified, but scientific research was also not permitted to be published. Significant scientific experimentation with the shape of nosecones was kept from publication from 1952 until 1958. Open publication was allowed only after it was clear from discussion with Soviet scientists that the principles were already known in the Soviet Union.

Von Braun started spelling out the obstacles facing the Vanguard effort, and vented his frustrations at playing second fiddle to the Navy. Suddenly, Redstone Base Public Relations Officer Gordon Harris rushed in with the news about Sputnik. A red–faced von Braun turned to McElroy. "We could have been in orbit a year ago. We knew they were going to do it! Vanguard will never make it. We have the hardware on the shelf. For God's sake *turn us loose* and let us do something. We can put up a satellite in 60 days, Mr. McElroy, just give us a green light and 60 days!". McElroy just sat there, stone–faced and said nothing. The U.S. Senate had not confirmed him yet, and he felt he had no immediate power to back von Braun's proposal. Medaris took up the challenge. "No Wernher, 90 days. Whatever components you have on that bird, they better be the best you can do"

The men went on to discuss the ramifications of the Sputnik watershed event. Eventually they decided that the Army should make its presence known in Washington as better equipped than the Navy to beat the Russians in the newly inaugurated Space Race. Everyone agreed that the charismatic von Braun would be a key spokesperson.

In Washington before the U.S. Congress, von Braun was asked his opinion about the Russians' ability. He responded: "Based on recent events, I would say that the Soviets have the capability to put a hydrogen warhead on Washington today." Alarmed senators and representatives wanted to know what von Braun, in charge of the Army's missile program, was doing about it. "What can I do?" von Braun answered. "The Vanguard

Program is authorized to launch the first satellite. I can do nothing except read about it in the newspaper."

November 8, 1957. Under public and Congressional pressure to do something, President Dwight Eisenhower had assembled the agencies involved in space launch activities: Air Force, Army, Navy, and NACA. Expecting cooperation, he got none. They had nothing in common except a lot of antagonism. To break the stalemate, Eisenhower decided to meet with the competing agencies separately in the White House's West Wing. Soon it was the Army's turn, represented by ABMA Director Maj. Gen. Bruce Medaris, SECARMY Wilbur Brucker, and Dr. Wernher von Braun. NACA Director Hugh Latimer Dryden was allowed to join them.

"OK, you can go in now," said Shirley Stevenson, Eisenhower's executive assistant. She stopped von Braun. "The President wants to see you separately," she told him.

Eisenhower greeted them with a handshake and a warm smile and asked them to take a seat. He continued a heated discussion with SECDEF Neil McElroy while Senate majority leader Lyndon Baines Johnson (LBJ) looked on. "I refuse to be stampeded into a space race, and I don't want our Germans fighting against the Russians' Germans. We have had enough problems hiding our German rocket men's Nazi past from the general public ..."

"Practically all of them claim to be in compliance with Truman's policy which explicitly forbade recruitment of war criminals or those active in Nazism," McElroy declared in a cynical tone. The President interjected: "Can't you narrow it down better than between 50 to 80 percent of the German rocket team being members of the Nazi Party?" "I would guess it is closer to... maybe 70 percent, but no one is sure. They won't 'fess up, Mr. President," McElroy answered. "Great. Just great. If we have to get rid of 70 percent; no, I'd say 40 percent, of the German scientists and engineers, we **wouldn't have** a civilian space program to provide cover for WS-117L" Eisenhower pronounced. "What about the chief operations director of Mittelwerk, what's his name, Rudolph? And which other of our key rocket scientists was an enthusiastic member of the Nazi Party?"

"We know about Dr. Wernher von Braun," McElroy responded. "He's off the table, Neil!" exclaimed SECARMY Brucker. General Medaris was forced to agree. "That's right. Without him and the rest of the

Germans, *we wouldn't have* a space program," he said. Eisenhower prodded McElroy, "Go on." He did.

"The OSI thinks that Arthur Rudolph not only set the schedules by which the prisoners were worked to death, but that he reported instances of suspected sabotage by prisoners to the SS that led to their subsequent executions. And..." "Hold on, Neil" interrupted Brucker. He was becoming increasingly upset. "Those are merely unproven allegations." "If we let the OSI interrogate him the right way, I think you would get the right answer," McElroy countered.

The two men glared at each other, but Brucker didn't flinch nor give ground. "Surely you wouldn't" he said. "O.K., you too, break it up! Let's let the SECDEF finish!" Eisenhower declared in a commanding tone. The room got quiet. Very quiet.

McElroy went on: "Kurt Debus joined the SS in 1940. Witnesses say that he regularly wore his SS uniform at Peenemünde to display his loyalty. That in 1942 he denounced a colleague to the Gestapo as anti-Nazi. Lt. Col. Herbert Axster, Dornberger's chief of staff, regularly beat foreign slave workers who worked on his estate in Germany. Georg Rickhey, he was the Director General of the Mittelwerk. We don't know much about him. Magnus von Braun was Wernher's little brother. He supervised gyroscope production at Mittelwerk. We don't have anything on him." "Of course you don't," SECARMY Brucker spat back. Because they are *innocent* of all charges!" SECDEF McElroy continued on reading a list of supposed Nazis, and their supposed crimes.

At the end, President Eisenhower, a marvel of efficiency, wrapped up the meeting with specific actions and deadlines. "Gentlemen, we all know that during the war, everyone was a Nazi. After the war, no one was a Nazi. It is very easy for us to sit here and pass judgment. None of us was there. None of us walked in these guys' shoes. NO ONE, and that especially includes you," he said, sternly eyeing NACA Director Hugh Dryden, "is to tell or say ANYTHING about these gentlemen's Nazi past. I want to make that perfectly clear. The Russians have a huge lead over us in booster power." He looked down at the sheet given him by Army Intelligence. "The Vanguard's thrust of 27,500 pounds on the pad is only about 3 percent of the SS-6 Sapwood's 850,000 pounds based on our best intelligence estimate *(actual values were 27,854 to 877,654, showing the inexact nature of intelligence)*. We must have the help of every German on the team to overcome this advantage. This is a matter of utmost national security."

Having garnered everyone's rapt attention, Eisenhower went on. "Wilbur, I know you are on a crusade to lift the Army's ban on satellite launches. But we don't have all the facts yet. Lyndon, by the end of November I want you to convene your committee and review the whole spectrum of American defense and space programs. *Senator Lyndon Johnson was chair of the Preparedness Investigation Subcommittee of the Senate Armed Services Committee.* Give me a report the first week of December. Hugh, work with Stever to establish a steering committee with the mandate to coordinate various branches of the Federal government, private companies, as well as universities within the United States with NACA's objectives. Also harness their expertise in order to develop a space program. I want this done before the end of November. I don't care what you call your committee. Come up with a good name. Report back to me in December, say the second week, with your progress." *Guyford Stever, associate dean of engineering at MIT, chaired the Special Committee on Space Technology, which first met on November 27. NASA's Dryden Flight Research Center in California was named for Hugh Dryden.*

The meeting was about to break up, but Eisenhower was not done yet. "Shirley, please let Dr. von Braun in," he said. "Ah, good afternoon, Mr. President," von Braun said as he shook hands. "It is so nice to see you." Eisenhower got down to business. "Please tell me about the difficulties you think the Vanguard Program is facing," he asked. Von Braun began a five–minute spiel.

"How long before you can get us a launch vehicle ready, to show the Russians we can do it too?" Eisenhower bluntly asked the rocket expert. "You have all those birds stored at the Arsenal. Knowing you, you probably already have one in the wings, ready to go," Eisenhower quipped. "I think I can do it in six weeks," von Braun responded. "No!" General Medaris broke in. "Mr. President, we need nine weeks. At least." "Do it," Eisenhower directed. "At the earliest possible date." "I will get it done by the end of January, Mr. President," von Braun promised.

On the next workday Von Braun told his team, "I want the satellite ready to go. We will call it Explorer." The Army team started taking parts the least bit worn off the Redstones, and putting other parts on the "bird" designated for launching Explorer I. The Jet Propulsion Laboratory (JPL) in Pasadena, California partnered with von Braun's team. As we saw in the Theodore chapter, JPL had pioneered solid fuel motor technology. Under its Director, William H. Pickering, JPL would also build the satellite, as well as a tracking system to pinpoint its location in space. Dr. James Van Allen from the University of Iowa designed Explorer's scientific instruments. The United States was desperate to catch up to the Russians, but they were falling further behind. The Vanguard debacle on December 6 further raised the priority of the Jupiter C launch.

On January 28, 1958 Dr. von Braun left Cape Canaveral for Washington D.C to deal with the media. The Juno I was ready, but for two days bad weather prevented the launch. Stuck in Washington, von Braun could only monitor operations. He only had until midnight January 31. After that, the Navy would get a second chance. On January 31, Kurt Debus called von Braun in Washington. "The weather has moderated. The winds have died down. We can launch in four hours. If things run smoothly, we can launch in two. Do you have any ideas?" "If you can do it in two, do it," von Braun answered. But at 9:45 p.m., von Braun's launch crew ran into a problem. To find out what was wrong, engineer Albert Zadler risked his life by going underneath the fully fueled, live rocket (in today's 'take no risk' environment, NASA would have just scrubbed the launch). "It looks like it was just a spill," Zadler reported.

At 10:48 p.m. everything was ready. The familiar countdown went off without a hitch.

"…Five, four, three, two, one, ignition!" The Juno I and its Explorer I payload thundered off the pad, the North American Aviation/Rocketdyne A-7 rocket engine generating 42,439 kgf (93,562 lbs) of thrust, belching a white-hot flame. It burned LOX and a souped-up 'Hydyne' liquid fuel instead of the alcohol/water previously used to power the Redstone.

> *Hydyne, a mixture of 60% unsymmetrical dimethylhydrazine (UDMH) and 40% diethylenetriamine (DETA), was developed in 1957 at Rocketdyne because it provided 10% higher thrust and higher specific impulse (235 sec) than ethyl alcohol.*

The Jupiter C's GNC systems operated normally. After 155 seconds the Juno's first stage dropped off and the second, third, and fourth stages fired off without a hitch, the 15 midget Sergeant solid motors performed flawlessly. But did the satellite reach orbit?

For expediency--and in the race to beat both the Vanguard and the Russians--von Braun dispensed with guidance on the Juno's upper stages! The Explorer could not be put into a precise orbit. This is like throwing a dart high into the air (or space in this case) and hoping it hits the target. Americans cheered on the orbiting of early Explorer satellites, not realizing it was by a means inferior to that of the Russians—or the Navy's Vanguard, for that matter. No other nation has since used this shortcut method.

Furthermore, in those days telemetry and satellite tracking were primitive. Until it reached the first California tracking station, all everyone could do was wait. The anxious von Braun team gave it a few minutes leeway. Kurt Debus could hear von Braun's heavy breathing over the phone. "O.K. What happened? Do we have a satellite, or don't we?" he asked. The minutes passed. All of a sudden, "We did it! Yeah, yeah!!" they all shouted with glee. The sound was deafening in von Braun's ear, but he was ecstatic. The over-performing rocket had thrown the satellite into a higher orbit than anyone expected. The extra distance (the higher a satellite in orbit, the slower it goes) caused an eight-minute delay. Initially, its perigee was 358 km (222 mi) and apogee was 2,550 km (1,585 mi) giving it a longer than expected period of 114.8 minutes (90 minutes is typical for a satellite in LEO).

Inaugurating the era of space exploration for the United States, Explorer I was a thirty pound satellite that carried instruments to measure temperatures, and micrometeorite impacts, along with an experiment designed by James A. Van Allen to measure the density of electrons and ions in space. The measurements made by Van Allen's experiment led to an unexpected and startling discovery -- an earth-encircling belt of high energy electrons and ions trapped in the magnetosphere now known as the Van Allen Radiation Belt. Explorer I ceased transmitting on February 28, 1958, but remained in orbit until March of 1970. Pioneering space scientist James Van Allen died on August 9th, 2006 at the age of 91.

At 1:30 a.m. on February 1, von Braun held a press conference in the Great Hall of the National Academy of Sciences. "Gentlemen, I am proud to announce the

Explorer I satellite in orbit (artist's concept). It rotated around its long axis at 750 rpm to maintain stability like a gyroscope.

Success! Juno I launches Explorer 1 into low Earth orbit, January 31, 1958. Note the spinning "tub" upper stage.

During the 1940s and 1950s, scientists used the word "computer" to refer to a person rather than a machine. This all-female computer team, many of the members recruited right out of high school, were responsible for doing all the math by hand required to plot satellite trajectories and more. Who said women aren't as capable as men?

successful launch of Explorer 1. America launched its first man-made satellite into Earth's orbit last night. I have long seen my purpose in life as the promotion of spaceflight. Well, today we have answered the Soviet challenge, and in doing so opened the gates to the heavens." At last, von Braun had a satellite. This and subsequent events would transform him from a relatively unknown German rocket scientist into an American folk hero—a man who put the U.S. back in the Space Race.

Americans cheered on the orbiting of these early Explorer satellites, not realizing that von Braun's simple and direct satellite orbit insertion method used no guidance. It got the job done, but by a means inferior to that of the Vanguard or the Russians. No wonder ground control had to guess at where the satellite ended up.

In Russia, when Sergey Korolyov got the news he said, "Apparently the Americans have evened the score. It would appear that we are in a race to space. A race that we, of course, intend to win." The pint-size Explorer I weighed just 14 kg (30.8 lb), 60% of which was instrumentation, only one-sixth as much as Sputnik 1. Sputnik 2 weighed 36.4 times as much as Explorer I. So, obviously Explorer didn't quite "even up the score", right? WRONG. Anyone who owns an IPhone knows that good things come in small packages. Explorer 1 not only evened up the score, but in terms of space science the Americans had soundly beat the Russians in one fell swoop.

Press conference announcing success of Explorer 1, William H. Pickering, James A. Van Allen, and Wernher von Braun. The satellite measured 2m (6.7 ft) in length and weighed 14 kg (31 lbs). It orbited Earth at heights between 360 km and 2520 km (224 mi to 1556 mi) until March 31, 1970.

Taken together, Sputniks 1 and 2 lasted only 9 months in space. Explorer remained in orbit for over 12 years (16 times longer than both Sputniks), enough to orbit the Earth 56,000 times. The U.S. had arrived in space!

The Explorer-1 satellite rotated like a sausage in space at 750 rpm. With nothing to disturb it, aerospace dynamicists and GNC (guidance/navigation/control) engineers expected it to keep rotating about its long axis of <u>least</u> inertia (like a hotdog on a spit). They were surprised when it started precessing due to energy dissipation from flexible structural elements. Scientists and engineers discovered that a body ends up in the spin state that minimizes its *kinetic rotational energy*, this being the <u>maximal</u> inertia axis! After almost 200 years the world's experts had to modify the Eulerian theory of rigid body dynamics to address dissipation.

Another surprise was that once in orbit, the cosmic ray equipment of Explorer-I indicated a much lower cosmic ray count than had been anticipated. Dr. Van Allen theorized that the equipment may have been saturated by very strong radiation caused by the existence of a belt of charged particles trapped in space by the Earth's magnetic field. The existence of these Van Allen Belts, one of the outstanding discoveries of the International Geophysical Year, was confirmed by Explorer-3 which was launched by a Juno I on March 26, 1958. Both satellites also measured the frequency and size of cosmic dust or micrometeorite particles in space by counting impacts. Of the six launch attempts by Juno I in 1958, three were successful (the Explorer 1, 3 and 4 satellites). Its successor the Juno II also dispensed with upper stage guidance. The Juno II used a Jupiter MRBM rather than a Redstone booster as the first stage to launch additional Pioneer and Explorer spacecraft (with upgraded boosters, the Explorer series of spacecraft was never cancelled). Over 100 Explorer spacecraft have conducted a wide variety of scientific experiments.

> *Although we now take successful satellite launches for granted, there were many failures in the beginning of the American and Soviet space programs. For example, the Juno I and II had only a 50% success rate. Three of six launch attempts failed for Juno I and of 10 attempts for Juno II, five failed to reach orbit, and a sixth reached an incorrect orbit. The Navy's poor Vanguard did even worse with a 27% overall success rate.*

First Saturn I on launch pad and liftoff, Oct. 27, 1961

Military Rivalries Persist

Despite the Eisenhower Administration's intent to divorce military from civilian space missions and put the latter under the aegis of a single Agency (NASA), the military Services engaged in bitter rivalries every bit as intense as those taking place in Russian Design Bureaus to assume a predominant role in space.

Army in Space

Thanks to von Braun and his team, the Army led the other Services in civilian space missions. The Army had the Redstone and Jupiter missiles, and had more ambitious plans. In April 1957 von Braun's ABMA group began studies of a large clustered-engine booster to generate 1.5 million pounds of thrust, the Saturn I. In December 1957, right after Sputnik, the ABMA submitted a little-noticed report to the Army brass in Washington entitled "Proposal for a National Integrated Missile and Space Vehicle Development Plan." The proposal had the credentials of von Braun's rocket team behind it. In August 1958, the Advanced Research Projects Agency (ARPA, predecessor to today's DARPA, Defense Advanced Research Projects Agency) approved ABMA's proposal to develop a multi-stage rocket with a clustered-engine (1.5M lbs thrust each) first stage. In logical order the new project was called Juno V. It envisioned a rocket much larger than those used in the Juno/Explorer program, powerful enough to generate 1.5 million pounds thrust—enough to lift payloads weighing tons into orbit. To distinguish its greater scope, the Juno V project was renamed *Saturn I*, after the god of Roman mythology who ruled the universe. It soon became ABMA's most important project. In 1957-1958 ABMA expended about 50,000 man-hours on the Saturn I project alone. Time and money well–spent, as the Saturn I would evolve into the Saturn V which would land 12 men on the moon.

ABMA also proposed using a Redstone as a booster for a manned suborbital flight, calling it Project Adam. Von Braun advocated sealing a man in a cylindrical capsule and blasting him to an altitude of 241km (150 mi), and the same distance in range. His approach had the advantage of being immediately feasible. The Army continued to plan additional Explorer satellite projects and the Pioneer 3 and 4 lunar probe spacecraft using the Juno II.

ABMA also worked on projects sponsored by ARPA, including several satellite missions and deep space probes to be launched by the Juno II series of rockets. Successful missions in this series in 1959 included the launch of Pioneer 4, a deep space probe which went into orbit around the Sun on March 3; and the launch of Explorer VII, which contained x-ray and cosmic ray experiments on October 13. ABMA drew public media attention when a Jupiter booster launched and successfully recovered two monkeys (Able and Baker) on May 28. Finally, ABMA supervised development of the Pershing (a solid fueled tactical missile) by the Martin Company in Orlando, Florida.

Aegis missile launch from the USS Hopper (DDG–70), an Arleigh–Burke–class guided missile destroyer, June 6, 2010.

Navy in Space

The Navy had its Viking and Vanguard programs as described previously. In addition, missiles could be launched from ships into space, the Navy brass argued. Today's Aegis missile defense system with the SM-3 (Standard Missile-3) has proven the point, and given the Navy a preeminent role in ballistic missile defense (BMD) in all theaters.

Air Force in Space

The Air Force had its stable of IRBMs and ICBMs to draw upon for civilian space missions. The Air Force ridiculed the Army's Project Adam as the equivalent of firing a person from a circus cannon. They fought for a more ambitious approach, by adapting an Atlas missile to launch a man into LEO (this was before Yuri Gagarin's historic flight). However, several unknowns would have to be solved, including man's ability to survive in space. The Air Force also planned its own missions to explore space and conduct scientific experiments. Launch vehicles derived from the Air Force's Thor and Atlas missiles were used to launch most of the Pioneer series of lunar probes and interplanetary spacecraft. Launch vehicles derived from the Atlas and Titan ICBMs were used to launch the Mariner, Voyager, and Viking series of interplanetary spacecraft; and the Ranger and Surveyor series of lunar probes. More recently, with the obsolescence of the Titan, the mainstays of Air Force launch capability have become the Atlas family (Lockheed Martin) and the Delta family (Boeing).

Air Force civilian space launch vehicles have included (top row), the Thor and Thor–Agena; (bottom row), Thor–Able, Atlas and, more recently, the Delta in various versions. These will be more thoroughly described in *Avoiding Armageddon*.

NASA TAKES SHAPE

On July 29, 1958 President Eisenhower signed the National Aeronautics and Space Act which formally established NASA into law. Henceforth all civilian space missions of any sort would take place under NASA's aegis, while the Soviets were content to mix military and civilian space exploration missions under the same design bureau. On October 1 Dr. Thomas "Keith" Glennan became NASA's first Administrator and Hugh L. Dryden was named his deputy. NASA immediately absorbed NACA's 8,000 personnel and five laboratories: Wallops Island Station and Langley Research Center in Hampton, Virginia; Lewis Research Center in Cleveland (later renamed Glenn Research Center in honor of John Glenn, the first American to orbit in space); Ames Research Center at Moffett Field and the Flight Research Center at Edwards Air Force Base in California (later renamed the Dryden Flight Research Center). The Space Act also assigned the Navy's Vanguard project and several Air Force projects to NASA, as well as three of ABMA's satellite projects and two of its lunar probes. On December 3 JPL was also transferred to NASA at the Army's expense. None of the Services were willing participants in the dismemberment of their civilian space programs.

Army officials reluctantly transfer the U.S. civilian rocket program to NASA

The biggest controversy roiled about the ABMA. It was the focal point of Army-Air Force wrangling over missile development. As the government's largest R&D enterprise in rocketry, the ABMA prompted the Eisenhower administration to seek a balance between military and civilian space programs. It also focused the debate over whether rocket development should be conducted internally by government specialists, or contracted outside to private business. Von Braun's team became embroiled in this debate, as NASA and the ABMA fought over who should absorb the elite team. Von Braun faced a difficult choice. General Medaris and the Army had sponsored and funded his team, and he feared that NASA might not be as supportive of in-house development. He had seen how the Air Force tended to contract out most of the technical work on major projects, relegating the Air Force's military personnel and in-house government workforce to an oversight role. NASA had drawbacks, too. Eisenhower and his science advisors favored a civilian space program, but one in which space would have to compete with other scientific research programs for federal dollars, so funding could be limited. In contrast, the pressures of the Cold War, which by now included allegations of a 'missile gap' between the U.S. and the Soviet Union, seemed to promise a continued military program. Nonetheless, to Medaris and von Braun, NASA seemed the lesser evil because the rocket team stood to decay under Air Force control.

For its part, NASA leadership viewed the ABMA way of doing business with ambivalence. They considered the Army's hide-bound arsenal system, reliability testing, and engineering conservatism as "hopelessly outmoded." However, they could not ignore the potential of von Braun's elite team of world-class rocket scientists, nor ABMA's boosters and the Saturn program as keys to civilian space exploration.

NASA contracted with ABMA to provide eight Redstones for early Project Mercury (see below) suborbital flights. Reconfigured Mercury-Redstones would be the workhorses of the early manned space program. ABMA continued work on the clustered Saturn booster, which figured prominently in NASA's long-range plans. ***Rocket engine development is usually the longest lead-item in missile design, because it is the most complex and requires the longest and most expensive testing.*** Development of the first stage H–1 engine, which would be clustered to power the first S–1C stage, proceeded as ABMA considered proposed configurations for other stages.

Herbert F. York, the DoD's Director of Research and Engineering, questioned why the Army, whose primary mission was ground warfare, needed a huge booster like the Saturn. SECDEF McElroy and President Eisenhower finally agreed with York. Even SECARMY Brucker agreed with a transfer of Von Braun's team to NASA as the lesser of two evils. By October 6, 1959, negotiators hammered out an agreement to transfer von Braun's team intact to NASA, and assign NASA "responsibility for the development of space booster vehicle systems of any generations beyond those based upon IRBM and ICBM missiles as first stages." After some delays in Congress, President Eisenhower signed an Executive Order on March 15, 1960 making the transfer official. It would take effect on July 1, 1960 (then the start of the fiscal year) at which time NASA would assume full responsibility for the Saturn project. Von Braun was satisfied.

On July 1 the George C. Marshall Space Flight Center (MSFC) became a reality in a quiet ceremony, with von Braun as its first Director (a position he would hold for 10 years). Eisenhower publicly dedicated the new Center on September 8 in front of 20,000 people. The transfer shifted 4,670 people, 1,200 acres, and $100 million worth of facilities from ABMA to MSFC. MSFC would be NASA's lead center for SLV development. Wernher von Braun said with equanimity in a speech in December 1959:

> "…We will no longer be charged with developing long-range missiles for defense. We will be charged with providing the transportation system to carry forward the national space exploration program. For us this is the realization of a dream that dates back to the inception of our rocket development efforts in Europe many years ago."

Construction of the Manned Spacecraft Center (MSC) in Clear Lake City, Texas (on the outskirts of Houston) began in April 1962. MSC officially opened for business in September 1963. It would be the primary center for U.S. space missions involving astronauts. MSC's Mission Control Center (MCC) has been responsible for coordinating and monitoring every crewed NASA mission since Gemini IV. MSC was renamed the Lyndon B. Johnson Space Center (JSC) in 1973, the year President Johnson died. Before passing on he ensured that JSC was located in his home state of Texas. MCC has directed all Space Shuttle missions and to this day manages all activity on board NASA spacecraft including the International Space Station (ISS). MSFC also manages an alternate ISS control center in Huntsville, Alabama.

Where is the 10th NASA center? Although it is internationally owned and operated, the American portion of the International Space Station (ISS) can be considered a 10th NASA "research center" in space.

It bears emphasis again that the former German rocket engineers, taken in the aggregate, exerted an influence far beyond their numbers in the engineering of rockets-- not only in the former Soviet Union; but even more so in the

Von Braun in his office as Director of the NASA Marshall Space Flight Center, May 1964

To recapitulate, today NASA maintains a total of 9 centers including its headquarters:

Headquarters	Washington, D.C.
4 Space Flight Centers (SFC, aka Space Centers (SC))	MSFC, JSC, KSC, and Goddard SFC (GSFC in Greenbelt, Maryland also operates the Wallops Flight Facility (WFF) on the Virginia coast.
4 Research Centers (RC)	Ames (ARC), Langley (LaRC), Glenn (GRC), Dryden Flight RC (DFRC)**
1 Spacecraft Center	JPL (Jet Propulsion Laboratory in Pasadena, California)*

* JPL is operated by NASA but owned by the University of California system **In 2014 DFRC was redesignated the Armstrong Flight Research Center in honor of Neil Armstrong, the first man to set foot on the Moon.

United States. The nucleus of German rocketeering expertise in the U.S. resided in MSFC, where Dr. Wernher von Braun was the undisputed leader and first Director in the crucial decade from 1960-1970. However, other ex-Germans also greatly spread their influence throughout the country--in JSC, KSC, other NASA Centers-- and many held key positions in the engineering of rockets in the aerospace industry. While the influence of the original German rocket pioneers naturally dwindled in the ensuing decades, some stayed active well into the 21st century. They left behind a legacy that will continue in following generations of rocket scientists into the foreseeable future.

On February 14, 1961, James Edwin Webb accepted the position of Administrator of NASA in Washington, DC, succeeding Keith Glennan. Under Webb's tenure the "Agency" (as NASA is often called) undertook one of the most impressive projects in history, the goal of landing an American on the Moon before the end of the decade through the execution of Project Apollo. For seven years after President Kennedy's May 25, 1961 lunar landing announcement, through October 1968, James Webb politicked, coaxed, cajoled, and maneuvered for NASA in Washington. As a longtime Washington insider he was a master at bureaucratic politics. In the end, through a variety of methods Administrator Webb built a seamless web of political liaisons that brought continued support for and resources to accomplish the Apollo Moon landing on the schedule President Kennedy had announced.

The "Mercury Seven" astronauts pose with an Atlas model on July 12, 1962. From left to right: Gus Grissom, Alan Shepard, Scott Carpenter, Wally Schirra, Deke Slayton, John Glenn, and Gordon Cooper.

NASA PUSHES FORWARD WITH PROJECT MERCURY

On October 7, 1958 NASA formally began the U.S. manned spaceflight program with Project Mercury (named for the speedy messenger of the gods in Roman mythology), and put out a nationwide call for experienced jet pilot volunteers. President Eisenhower ordered that they be chosen from a talent pool of college-educated, military test pilots. The education requirement excluded such notables as decorated combat veteran Chuck Yeager, the first person in the world to break the sound barrier in 1947. On April 9, 1959, the U.S. named seven astronauts (the famous "Mercury Seven") to Project Mercury. Following a grueling series of tests, the group was winnowed down to three Air Force pilots-- Leroy Gordon ("Gordo") Cooper Jr., Virgil Ivan ("Gus") Grissom, and Donald Kent ("Deke") Slayton; three Navy fliers—Malcolm Scott Carpenter, Walter Marty ("Wally") Schirra, Jr., and Alan Bartlett Shepard, Jr.; and one Marine pilot, the oldest in the group--John Herschel Glenn, Jr. It says something that four of them were first-born sons.

NASA assigned two broad missions to Project Mercury: (1) to investigate man's ability to survive and perform in the space environment; and (2) to develop the basic space technology and hardware for manned space

flight programs to come. A third objective, of course, was to recover the Mercury capsule and its pilot safely. A small group of engineers from Langley, designated the Space Task Group (STG), assumed a role at the center of NASA planning for manned space flight. Dr. Bob Gilruth headed the STG. Comprised of only 35 members at NASA's founding, STG's numbers swelled to 350 by July 1959.

The STG examined a number of spacecraft designs for early U.S. manned missions. They selected a truncated cone ballistic spacecraft shape because it maximized the habitable volume for a given mass and provided a minimum area blunt surface for ablative cooling during reentry, a shape used to this day. Besides providing adequate volume for crew on short duration missions, the blunt cone shapes lent themselves to the use of escape rockets (on Mercury and Apollo missions) or ejection seats (on Gemini missions) for crew removal in near-pad abort situations. Engineers found that by slightly offsetting the center of mass of the spacecraft from its axis of symmetry, ground control could maneuver the craft--even at hypersonic speeds--so that it wouldn't completely miss the target area for crew recovery after landing. In other words, by intentionally designing early manned spacecraft so they were slightly off-balance—like a spinning top that wobbles—the blunt shapes could generate just enough lift so that they could maneuver sideways or forward and back, as required, rather than just plunging straight down into the atmosphere like a rock.

Capsules designed by Maxime Faget at NASA's Langley Research Center became acknowledged as the best. In fact, it became known that any capsule configuration proposed by the contractors was acceptable as long as it was like one that Faget's team had developed (Faget also developed an early Space Shuttle design). His design looked like a cone with a gently curving shield at the bottom. It would reenter the Earth's atmosphere upside down. The shield provided drag to slow the spacecraft down on reentry through the atmosphere while carrying the intense heat away from the capsule by ablating or charring away. Faget's capsule designs carried the day from the Mercury through the later Gemini and Apollo programs. *Avoiding Armageddon* explains the essentials about reentry heating in plain, simple terms.

After reentry, ballistic spacecraft were decelerated by drogue and main parachutes and were normally recovered at sea. Water landing conditions were more predictable and the fluid medium attenuated a significant fraction of the

impact energy (*Avoiding Armageddon* will describe how Yuri Gagarin had to bail out of his fast-falling spacecraft to avoid getting killed). Moreover, the ocean is a big place with no obstacles, so astronauts could more easily be recovered even if they totally missed the target area. Scott Carpenter would find this out on his Mercury-Atlas 7 flight.

It was rocket men such as these standouts that made the Mercury and Gemini Programs possible. Clockwise from left: Ernst Stuhlinger, MG Holger Toftoy, Robert Lusser, Wernher von Braun, and Hermann Oberth in front.

Mercury-Redstone

The Mercury–Redstone used for the Mercury manned flights was derived from the Jupiter C, a flight-proven design. A fundamental change was to eliminate Jupiter C's staging capability. The Jupiter C had to be modified in a number of ways to accommodate Faget's new capsule design which would contain the pilot-astronaut. In contrast to the Russian approach, American astronauts were involved in the planning and development of their spacecraft, and some even came up with innovations that later saved their lives. For example, the original design lacked a window, making it impossible to steer the craft should the automatic guidance system malfunction or for some other reason. After all, these guys were macho test pilots. Who ever heard of an American astronaut being strapped in with everything controlled from the ground, helpless to steer himself out of trouble?

Other modifications stripped the Jupiter C of its more sophisticated components while permitting it to retain greater performance characteristics than the original single-stage Redstone. To increase its reliability, engineers changed out the sophisticated ST-80 autopilot for the LEV-3, one of the first inertial guidance systems. Since there was no need to provide terminal guidance for the new payload, the aft unit of the payload on the old Redstone, which had contained a pressurized instrument compartment, became the permanent forebody of the main tank assembly. A spacecraft adapter ring likewise had to be designed to simplify interface coordination and to ensure clean separation between capsule and booster. At the other end of the launch vehicle it was necessary to use the most recent engine model, the A-7, to avoid a possible shortage of spare parts. Von Braun's propulsion engineers took the basic responsibility for "manrating" this engine. *Manrating an engine basically involves making it more reliable and less prone to failure because of the precious cargo of humans onboard.*

The Jupiter C's elongated fuel tanks had to be redesigned. To increase safety, engineers decided to revert to alcohol for fuel instead of the more powerful but more toxic hydyne that fueled the Jupiter C. The tanks had to be enlarged to accommodate an extra 20 seconds of engine burn time. Engineers added a high-pressure nitrogen tank to pressurize the larger fuel tank and an auxiliary hydrogen peroxide fuel tank to power the engine turbopump.

The changes kept piling up, about 800 of them. Would the Mercury-Redstone be as flight-worthy as its predecessor? Alan Shepard would soon find out.

Just one day after Soviet cosmonaut Yuri Gagarin became the first person to fly in space on April 12, 1961, the U.S. House Committee on Science and Astronautics met in Washington, DC. Many congressmen pledged their support for a crash program aimed at ensuring that America would catch up. Like President Eisenhower before him, President Kennedy knew that the highly publicized 'Missile Gap' did not exist, and that militarily the Soviets were not as far ahead of the United States as the American public was made to believe. Kennedy was circumspect in his response to the news, refusing to make a commitment on America's response to the Soviets. He knew little about the technical details of

America's civilian space program, and was put off by the massive financial commitment required for a manned moon landing. Nevertheless, on April 20 Kennedy sent a memo to Vice President Lyndon B. Johnson, asking him to look into the status of America's civilian space program, and into programs that could offer NASA the opportunity to catch up. Johnson responded on the following day, concluding that "we are neither making maximum effort nor achieving results necessary if this country is to reach a position of leadership." His memo concluded that a manned moon landing was far enough in the future to make it possible that the United States could achieve it first. Perhaps Johnson was already thinking ahead of building NASA's Manned Spacecraft Center (now Johnson Space Center) in his home state of Texas.

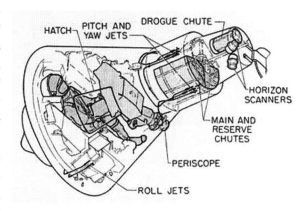

May 5, 1961. Alan Shepard sat enclosed in the cramped Mercury capsule (above), nicknamed the "garbage can," atop the Mercury-Redstone rocket. Al had a reputation as a calm, cool, and collected test pilot. But today he was all excited. He was poised to make history by becoming the first American in space. It was to be a short suborbital flight, just 15 minutes from launch till splashdown. But in those 15 minutes he faced a lot of unknowns: the launch, the g-forces, the reentry. Any one of them could kill him. After two hours of delays, Shepard asked ground control to "...light his candle." He was getting desperate. He wondered, *I really need to empty my bladder. I'm gonna ask if I can open the hatch and go to the bathroom like a normal person would. And I expect a normal answer.* "Hey, guys, I need to take a leak," he called out. "You have to what?" ground control responded. "I gotta piss, dammit!" Shepard replied. He was told that "you're in the nosecone. There will be no door opening." So he told the folks, "I'm going to relieve myself on the spot." And they said, "Oh, you can't do that. You'll short circuit everything." So Al comes back

Alan Shepard sweating it out before launch

Alan Shepard's historic launch, May 5, 1961

with, "well, how about turning the power off?" "O.K., then. Let us check." So they deliberated a few more minutes, and finally turned off the power. A minute ticked by. Ok, Alan, you are cleared to leak." Shepard relieved himself…. Ahhh. If he managed to survive the flight, who cared if his spacesuit was a bit smelly! After a while it started to dry out, so he came back with, "ok, you guys can turn the power back on, and I'll take the risk of a shock."

More waiting. Finally he started to get exasperated. Who wouldn't after interminable delays and five hours in the cockpit. So he said, "Well, I'm getting tired of this, so why don't you tell them to just light the damn candle, 'cause I'm ready to go."

Shortly before the launch, he told himself, "Don't fuck up, Shepard." *Oh God*, he thought. *Every part of this ship was built by the low bidder.* Finally, ground control started the countdown: "…Eight…seven…six…five… four…three… two…one…IGNITION. At 9:34 a.m., Freedom 7 lifted off from Cape Canaveral reaching an altitude of 116.5 miles. Looking through a periscope on the capsule, Shepard exclaimed, "What a beautiful view!

Cloud cover over Florida is three to four tenths near the eastern coast—obscures up to [Cape] Hatteras…I just saw Andros Island, identified the reefs." Three minutes later the Mercury capsule re-entered the Earth's atmosphere. Alan Shepard's historic flight lasted a total of only 15 minutes, 28 seconds. He was weightless for only about 5 minutes. He and the Mercury capsule were recovered 302 mi (490 km) downrange by the aircraft carrier USS Champlain, flawlessly completing the United States' first manned space mission. Al later said how impressed he was that the flight deck of the USS Champlain was just covered with sailors. He was emotional that this was the first sense he got of how much people cared about the Americans' first steps into space.

When John Fitzgerald Kennedy (JFK) was elected as 35th U.S. President in 1960, his attitude toward spaceflight and space exploration in general was unknown. Early on, he had probably not arrived at a firm position in his own mind. President Eisenhower obviously had not considered a national space program and manned space travel to be especially significant in the life and well-being of the country at the time. But there were differences of opinion. The PSAC and the Bureau of the Budget shared Eisenhower's view, while others, particularly members of Congress, NASA, the military Services, many scientists, and industry favored a strong governmental support of space activities.

For the sake of national prestige, President Kennedy decided to back NASA in its far-reaching goals of the exploration of space. On May 25, 1961--just 20 days after Alan Shepard's historic flight--he announced his support for the manned exploration of space as a small part of a speech before a joint session of Congress, "Special Message to the Congress on Urgent National Needs," in the face of recent Communist advances in space. The American effort to send astronauts to the moon had its origins in Kennedy's famous appeal. He said,

> *"If we are to win the battle that is going on around the world between freedom and tyranny, if we are to win the battle for men's minds, the dramatic achievements in space which occurred in recent weeks should have made clear to us all, as did the Sputnik in 1957, the impact of this adventure on the minds of men everywhere who are attempting to make a determination of which road they should take. Now it is time to take longer strides, time for a great new American enterprise, time for this nation to take a clearly leading role in space achievement, which in many ways may hold the key to our future on Earth.* **I believe that this Nation should commit itself to achieving the goal, before this decade is out, of landing a man on the Moon and returning him safely to Earth. No single space project in this period will be more exciting, or more impressive to mankind, or more important for the long-range exploration of space; and none will be so difficult or expensive to accomplish."**

At the time, the United States was still trailing the Soviet Union in space developments. When Kennedy committed the nation to go to the Moon, the United States had a grand total of 15 minutes of human spaceflight experience. Nevertheless, Cold War-era America welcomed Kennedy's bold proposal.

Kennedy's speech ended up galvanizing the country in putting a man on the moon. In national surveys, the public was by no means unanimous in supporting the moon race. The 1960s witnessed priorities and huge expenditures in other areas: the Vietnam War, President Johnson's "Great Society" social programs, and the Cold War. But there was enough public support to spark a burst of technological creativity and harness the largest peacetime commitment

of resources ($24 billion in then-year dollars) of any nation in history. NASA's Langley Research Center estimated that at its peak, the Apollo program employed 400,000 people and required the support of over 20,000 industrial firms and universities.

On July 21, 1961, Gus Grissom became the second American in space aboard the Liberty Bell 7. He beat Shepard's suborbital flight time by only 9 seconds. Minutes after splashdown but before the rescue helicopters could reach him, Grissom heard a dull thud. The entry hatch cover blew away, and the next thing he knew sea water started pouring in because of the ocean swells. Grissom quickly finished unbuckling himself, exited the flooding capsule, and swam away. Four or five interminable minutes later, he was rescued. Another helicopter tried to lift the Liberty Bell out of the water, but it was too heavy and had to be cut loose. It rapidly sank in 2.8 mi (4.5 km) of ocean (it was recovered intact 38 years later from the sea floor). The cause of the mishap: the external release lanyard was held in place by only one screw. It tore loose in the rough seas, triggering the hatch release. The lanyard was better secured on subsequent flights.

Of course on February 20, 1962 John Glenn earned his claim to fame as the first American to orbit the Earth (three times). On the next orbital Mercury flight Scott Carpenter ran into problems, some of his own doing. A balloon experiment failed when the balloon didn't inflate properly and then didn't jettison. Aurora 7 continued to drag the balloon behind it until it burned up on reentry. When he switched from manual control to a fly-by-wire system he forgot to turn off the fuel flow. He mispositioned the capsule by 25° in yaw for reentry, and fired his retros three seconds late. Carpenter barely made it through reentry and splashed 250 miles downrange from his nearest rescuer. Project Mercury's six flights are summarized in the following table and figure. The spacecraft and rockets used to launch them went by the same name.

On Faith 7, the last Mercury mission, Gordo Cooper orbited the Earth for 34 hours, 20 minutes, logging more flight hours than the previous five Mercury astronauts combined. Cooper earned the distinction of being the first U.S. astronaut to sleep in space. The Mercury Program had met its objectives. No serious accidents. The United States was ready to embark on the next phase of manned spaceflight, the Gemini Program.

Project Mercury Flights

Spacecraft	Call Sign	Rocket	Designation	Crew	Launch Date	Duration	Remarks
Mercury-Redstone 3	Freedom 7	Mercury–Redstone	MR-3	Alan Shepard	5 May 1961	15 m 28s	First American in space (suborbital).
Mercury-Redstone 4	Liberty Bell 7	Mercury–Redstone	MR-4	Gus Grissom	21 July 1961	15 m 37s	Second suborbital flight. Spacecraft sank, Grissom rescued.
Mercury-Atlas 6	Friendship 7	Mercury–Atlas	MA-6	John Glenn	20 February 1962	4hr 55 m 23s	First American in orbit (3 orbits). Telemetry falsely indicated heat shield unlatched, so spacecraft's retropack retained during reentry
Mercury-Atlas 7	Aurora 7	Mercury–Atlas	MA-7	Scott Carpenter	24 May 1962	4hr 56 m 15s	3 orbits. Retrofire error caused 250-mile (402 km) splashdown target over-shoot.
Mercury-Atlas 8	Sigma 7	Mercury–Atlas	MA-8	Wally Schirra	3 October 1962	9hr 13 m 11s	6 orbits. Carried out engineering tests.
Mercury-Atlas 9	Faith 7	Mercury–Atlas	MA-9	Gordon Cooper	15 May 1963	1day 10hr 19 m 49s	22 orbits, first American in space for over a day. Last American solo flight. Manual reentry necessitated by systems failure. Closest splashdown to target to date (4 miles).

NOTE:

The Mercury–Redstone and Mercury–Atlas launch vehicles were manrated modifications of the Redstone MRBM and SM–65 Atlas D ICBM, respectively.

| Freedom 7 | Liberty Bell 7 | Friendship 7 | Aurora 7 | Sigma 7 | Faith 7 |

The six Project Mercury launches. All of the Mercury astronauts except Deke Slayton got to go into space. Grounded by a heart murmur, Slayton had to wait until July 15, 1975 when he was assigned as the docking module pilot of the 1975 Apollo–Soyuz Test Project.

PROJECT GEMINI

In January 1962, NASA announced the second U.S. manned space program. Project Gemini, named for the third constellation of the Zodiac and its twin stars, Castor and Pollux, was necessary to bridge the technological gap between Projects Mercury and Apollo. Gemini would be a series of Earth-orbital missions with two-person crews, where astronauts would practice orbital maneuvers, rendezvous, docking, undocking, and reentry while close to Mother Earth. Such maneuvers would be necessary to land men on the moon. but the most critical ones would take place 240,000 miles (386,000 km) away, making rescue impossible. While on the far side of moon and without communications, the CSM (Command and Service Module) engine would have to fire at a precise time for a precise duration to get the three astronauts out of lunar orbit and on a transfer orbit back to Earth.

Project Gemini's major objectives were to 1) subject men and equipment to the rigors of space flight for up to two weeks in duration; 2) rendezvous and dock with an orbiting target vehicle, then maneuver the docked combination by using the target vehicle's propulsion system; and 3) perfect methods of safely reentering the atmosphere and splashing down at a preselected target point.

Of the original Mercury Seven astronauts, three soon dropped out: John Glenn retired from NASA in January 1964. Scott Carpenter had become an aquanaut team leader in the Navy's Man-in-the Sea Project, spending 30 days living and working on the ocean floor as part of the SEALAB II program. NASA grounded him from

NASA·S·65·893

Gemini capsule

flight in July 1964. *In 1967 Carpenter returned to the Navy's Deep Sea Submergence Systems Project (DSSP) as Director of aquanaut operations during the SEALAB III experiment. He earned the unique distinction of being the only human ever to penetrate both inner and outer space, acquiring the dual title "Astronaut/Aquanaut."* The third man, Deke Slayton, was removed from flight status when he developed cardiac arrhythmia during training in the g-loading centrifuge. Keep in mind that the astronauts in those days had demonstrate they could withstand g-forces of up to 8½ g's during Gemini flights without passing out.

NASA again went through a rigorous selection and testing procedure to pick astronauts for the Gemini Project. By September 1962 the second group of nine astronauts— the "New Nine"-- included four Air Force pilots—Frank Frederick Borman II, James Alton McDivitt, Thomas Patten Stafford, and Edward Higgins White II; three Navy pilots—Charles "Pete" Conrad, Jr., James Arthur Lovell, Jr., and John Watts Young; and two civilians—Neil Alden Armstrong and Elliot McKay See, Jr. Again, notice that five of the men were first-born sons.

In October 1963 NASA selected a third group of 14 superbly qualified astronauts. "The Fourteen" included eight Air Force pilots, Edwin Eugene "Buzz" Aldrin, William Alison Anders, Charles Arthur "Art" Bassett II, Donn Fulton Eisele, Theodore Cordy Freeman, Michael Collins, Russell Louis "Rusty" Schweickart and David Randolph Scott; four Navy pilots, Alan LaVern Bean, Eugene Andrew Cernan, Roger Bruce Chaffee and Richard Francis Gordon, Jr.; and two Marine pilots, Ronnie Walter "Walt" Cunningham and Clifton Curtis 'C.C.' Williams. Flying is a risky business. Bassett, Chaffee, Freeman and Williams all died before they could fly in space. The rest of these lucky fellows flew on the Apollo Program.

The two-person Gemini capsule (above) weighed four tons, as compared to the one and a half tons of the Mercury capsule. It was the same shape as Mercury, but accommodated two suited astronauts side-by-side. Again, the astronaut-pilots insisted on having windows. Each had his own window to look out of. Mind you, these were not picture windows like in a home. They were relatively small rectangular affairs, just big enough to look out of and see the Earth or another spacecraft. Unlike Mercury, which could only change its orientation in space, the Gemini spacecraft could alter its orbit. It could also dock with the Agena Target Vehicle, which had its own large rocket engine and was used to perform large orbital changes. Gemini was the first American manned spacecraft to include an onboard computer, the Gemini Guidance Computer, to facilitate management and control of mission maneuvers. It was also unlike other NASA craft in that it used ejection seats, in-flight radar and an artificial horizon - devices borrowed from the aviation industry. Using ejection seats to push astronauts to safety was first employed by the Soviet Union in the Vostok craft manned by cosmonaut Yuri Gagarin.

The Gemini Project consisted of two unmanned test flights, followed by 10 manned flights. The idea behind the Gemini flights was to launch an unmanned Agena spacecraft using the Atlas booster. The Agena would fire its engine shortly after shroud jettison and separation from the Atlas over the Atlantic ocean. Over Ascension Island, a second burn would boost the Agena into a low circular orbit where it would await the Gemini spacecraft for rendezvous. The Atlas ICBM was not powerful enough to send a two-person spacecraft into space, so the largest ICBM in the inventory, the two-stage Titan II, was adapted for civilian missions. A demilitarized version, the Gemini-Titan (Titan II GLV), launched all 12 missions. Both stages burned nitrogen tetroxide (N_2O_4) and Aerozine-50 (A-50) storable propellants. A malfunction detection system was installed to inform the crew of the Titan II GLV's status, and improve response in an emergency. Redundant systems, including a backup flight control system were fitted to reduce the chances of launch failures. The second stage was modified to accommodate the Gemini spacecraft, and unnecessary vernier engines and retro-rockets were removed. An inertial guidance system replaced the radio control system used on the missiles, and modifications were made to the tracking, electrical and hydraulics systems.

Initial flights of the Titan II vehicle took place in early 1962. Flight data showed vibration levels as high as 5 Gs in the 9- to 13-Hz range. Right away, engineers suspected the pesky pogo phenomenon as the culprit. Pogo had been encountered before on the Thor vehicle. Studies indicated that vibration at 2 gs would be very painful for a human and that ½g was the maximum that should be allowed. Hence, the program to launch a two-man Gemini capsule was jeopardized. The Air Force adapted the Thor vibration studies to combat the problem in Titan II.

The first innovation was to add a standpipe and a tee in the oxidizer propellant feed system just upstream of the oxidizer pump. The pipe stood vertically and was partially filled with gaseous nitrogen shortly before launch. The idea was to tune the standpipe with a gaseous nitrogen bubble to absorb flow oscillations in the 11-Hz frequency range. This modification was flown, but it actually increased the violence of the vibration so much that a guidance accelerometer triggered premature engine cutoff for the stage.

At the same time, vibration tests for astronauts were conducted at NASA Ames Research Center. The tests were performed on a centrifuge with superimposed 11-Hz vibration and confirmed the human discomfort limits previously established. Pain was directly associated with motion of the eyeballs and testicles, as well as from internal heating that resulted from sloshing of the brain and viscera. The vibration frequency was also in the range of normal brain waves, adding confusion to decision making, hand and arm movement, and even speech.

Extensive testing was also done to generate test-verified equations describing the dynamics of structure, the propellant feed systems and the engines. Pump tests showed that as inlet pressures were reduced toward cavitation, the pump started acting as an amplifier, causing large oscillations in the thrust chamber pressure.

By late 1963, studies were complete. The fuel pump had been modified slightly, and a piston accumulator had been built for the fuel system and added to the vehicle. The system was flown with suppression devices on both the fuel and oxidizer feed lines with unqualified success.

An interesting side note involves the standpipe. Normally the standpipe was charged with nitrogen right before liftoff. On one flight, however, there was a hold after the gas was inserted. The oxidizer on that vehicle was nitrogen tetroxide which dissolves nitrogen gas. At launch, part of the gas had already been absorbed and the nitrogen bubble was smaller than desired so that the standpipe frequency was too high to do a good job of absorbing the 11-Hz vibrations. For that launch, the oscillations peaked at nearly 0.8 gs (peak-to-peak). The astronauts reported difficulty in reading instruments and their speech was impaired.

NASA chose the grounded Deke Slayton to head up the Astronaut Office in Houston, and made him responsible for selecting the crew members for each Gemini flight. Slayton chose wisely. Keep in mind that Gemini astronauts would have to endure extended stays in space in a very cramped crew cabin. The following table summarizes the Gemini Program's flights. All of them used the Titan II Gemini launch vehicle.

Project Gemini Flights

Spacecraft	Command Pilot Pilot	Mission Dates	Duration	Remarks
Gemini III	Gus Grissom John Young	23 March 1965	4hr 52m 31s	First two-person U.S. crew, first piloted spacecraft to change orbital path, first onboard computer.
Gemini IV	Jim McDivitt Ed White	3-7 June, 1965	4d 1hr 56m 12s	First U.S. EVA (White, 22 minutes), first U.S. four-day mission, computer failure necessitated manual reentry.
Gemini V	Gordo Cooper Chuck Conrad	21-29 August, 1965	7d 22hr 55m 14s	First week-long flight; first use of fuel cells for electrical power; evaluated guidance and navigation system for future rendezvous missions. Completed 120 orbits.
Gemini VII	Frank Borman Jim Lovell	4-18 December 1965	13d 18hr 35m 1s	When the original Gemini VI mission was scrubbed because its Agena target for rendezvous and docking failed, Gemini VII was used for the rendezvous instead. Primary objective was to determine whether humans could live in space for 14 days.
Gemini VIa	Wally Schirra Tom Stafford	15-16 December, 1965	1d 1hr 51m 24s	First space rendezvous accomplished with Gemini VII, station-keeping for over five hours at distances from 0.3 to 90 m (1 to 300 ft).
Gemini VIII	Neil Armstrong Dave Scott	16-17 March, 1966	10hr 41m 26s	Accomplished first docking with another space vehicle, an unmanned Agena B stage. When undocked, a Gemini spacecraft thruster malfunction caused near-fatal tumbling of the craft, which Armstrong was able to overcome; the crew effected the first emergency landing of a manned U.S. space mission.
Gemini IXa	Tom Stafford Gene Cernan	3-6 June, 1966	3d 21m 50s	Rescheduled from May to rendezvous and dock with augmented target docking adapter (ATDA) after original Agena B target vehicle failed to orbit. ATDA shroud did not completely separate, making docking impossible. Three different types of rendezvous, two hours of EVA, and 44 orbits were completed. Splashed down only ½ mile from target.
Gemini X	John Young Mike Collins	18-21 July, 1966	2d 22hr 46m 39s	First use of Agena B target vehicle's propulsion systems. Spacecraft also rendezvoused with Gemini VIII target vehicle. Collins had 49 minutes of EVA standing in the hatch and 39 minutes of EVA to retrieve experiment from Agena B stage. 43 orbits completed.
Gemini XI	Chuck Conrad Dick Gordon	12-15 September, 1966	2d 23h 17m 08s	Gemini record altitude, 1,189.3 km (739.2 mi) reached using Agena propulsion system after rendezvous and docking on first orbit. Gordon made 33-minute EVA and two-hour standup EVA. 44 orbits
Gemini XII	Jim Lovell Ed Aldrin	11-15 November, 1966	3d 22hr 34m 31s	Final Gemini flight. Rendezvoused and docked manually with its target Agena and kept station with it during EVA. Aldrin set an EVA record of 5 hours 30 minutes for one space walk and two stand-up exercises, and demonstrated improvements to previous EVA problems.

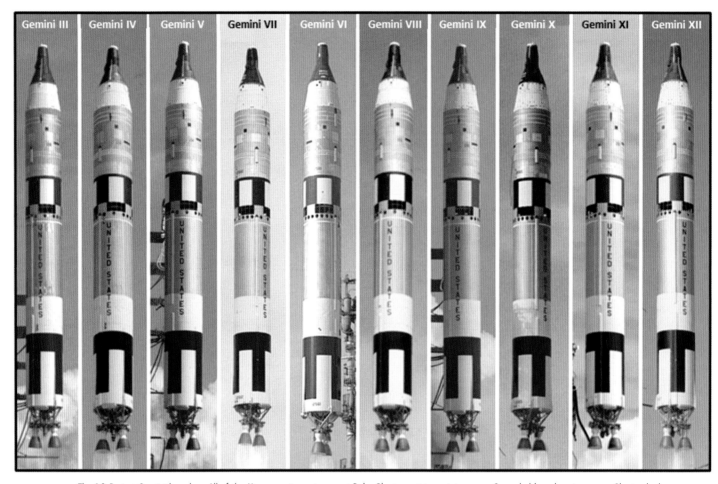

The 10 Project Gemini launches. All of the Mercury astronauts except Deke Slayton got to go into space. Grounded by a heart murmur, Slayton had to wait until July 15, 1975 when he was assigned as the docking module pilot of the 1975 Apollo–Soyuz Test Project.

Not shown are the first two unmanned checkout flights (Gemini 1 and 2) which took place on April 12, 1964 and January 19, 1965. The first flight qualified the demilitarized Titan II for flight, and the second flight tested the reentry heat shield (it passed).

 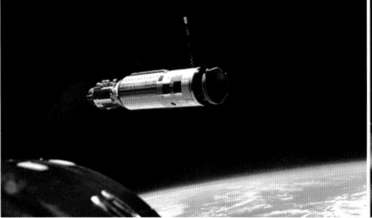 

Left to right: Edward White during spacewalk (Gemini IV); rendezvous of Gemini VI and VII, Dec. 1965; Agena target vehicle as seen from Gemini VIII prior to docking, March 1966.

Maneuvering in Orbit

It takes orders of magnitude less energy to move around where you need to go to on orbit than it does to get off the ground for three reasons: the upper stages or payloads weigh a lot less, in weightlessness the effects of gravity are negligible, and in space aerodynamic drag is absent. Mind you, gravitational force is still present, almost to the same degree as on the ground, but its effects are nullified by centrifugal force (like twirling around on a fast merry-go-round). Once in orbit, the spacecraft must be maneuvered like a car to its final destination and/or simply dock with another system in orbit. Simple, right?

NOT. Rendezvous and docking (RVD) maneuvers are complex and not at all like driving a car. Not only are the distances and speeds much greater, but the laws governing movement are very different. RVD engine reliability is just as important as for space launch. Astronauts marooned in orbit, inserted into the wrong orbit, or reentering at the wrong angle will be just as dead as their counterparts who may have perished in a catastrophic explosion during liftoff.

RVD consists of a series of orbital maneuvers and controlled trajectories which successively bring the *active* vehicle (the chaser) into the vicinity of, and eventually into contact with, the *passive* vehicle (the target). An unmanned Agena spacecraft served as the target for five of the Gemini missions. Accurate timing and precise thrust control are crucial. Even small errors can result in miss distances of tens of kilometers. The last part of the chaser's approach trajectory has to put it inside narrow boundaries of position, speed, and orientation. Because of the rules of orbital mechanics, to rendezvous with a spacecraft in a higher orbit, you have to *slow down*, not speedup. This lofts you to a higher orbit where your orbital velocity is lower, and the trailing spacecraft can gradually catch up. In this case the trailing spacecraft is considered the target, because it remains passive (letting the laws of orbital mechanics do the work). In some cases the target can perform its own maneuvers to speed up the process, but it has to "stand absolutely still" during the final approach phase.

In the early 1960s, the science of maneuvering and orienting things in space was still in its infancy. The Russians had a head start in this department: first dual manned spaceflight and approach with Vostoks 3 and 4 (1962); first unmanned spacecraft rendezvous and docking with Cosmos 186 and 188 (October 30, 1967);

first docking between two manned craft in LEO and exchange of crews with Soyuz 4 and 5 (1969). However, the Americans achieved one important first. On March 16, 1966 Neil Armstrong and Dave Scott manually performed the first *manned* RVD when they docked with an unmanned Agena target vehicle in LEO. It would take decades for America to master the challenges of Automated Rendezvous and Docking (AR&D) where all the functions, even the final docking, are controlled by computer. You'll learn much more about RVD and AR&D in the book *In Space to Stay*.

Orienting in Orbit

It is not enough to get where you want to go by maneuvering in space, one has to also orient the spacecraft correctly. This is called attitude control. It is not just crucial for RVD, but for scientific missions as well. All scientific spacecraft (well, maybe with the exception of the International Space Station which is too big to reorient in space) have to have precise stabilization and attitude control. There are several reasons. High (and/or low) gain antennas must be accurately pointed to Earth for communications and data transmission (aka downlink) and sending commands or software updates to the spacecraft (aka uplink). The Hubble Space Telescope and Chandra X-Ray telescope, for example, would be useless without very precise stabilization, attitude, and aiming control. Attitude control is also used so that the heating and cooling effects of sunlight and shadow may be used intelligently for thermal control. In space, temperatures can range from +250°F in direct sunlight to -250°F in the shade. Solar power arrays must be pointed in the right direction, otherwise you'll end up with a dead spacecraft. For guidance, short propulsive maneuvers must be executed precisely at the right time and in the right direction. An example is spy satellites, which may have to swoop down for a closer look at something.

In guidance and control of spacecraft, there are six what engineers call "degrees of freedom" (DOF) for a spacecraft to move about, and there has to be absolute certainty about them for both the chaser and the target. The first three (3-DOF) are the ones you already know about—in three-dimensional space an object can move up and down, sideways to the right or left, and forward or backwards. Usually it also means that the body is considered a "point mass," as it moves in any direction in space. But a spacecraft of finite extent can also rotate about any of those three directions of movement. Pretend that your spacecraft has three toothpicks stuck through it. They have to be stuck at orthogonal (right) angles corresponding to the first 3-DOF: sideways, forward-back along the longitudinal axis, and vertically (up-down). It doesn't matter if the spacecraft is shaped like an airplane with wings or a sphere. Now imagine you can rotate each toothpick independently about its axis. You have three rotations, giving you degrees of freedom no's 4, 5, and 6. Add them all together, and you have six, or 6-DOF. Engineers usually refer to a 3-DOF body that doesn't rotate at all as a "point mass."

The overall trajectory for a 3-DOF and a 6-DOF body is the same. Where the 6-DOF comes into play is for attitude dynamics and control where engineers have to worry about the details of which way and how fast the spacecraft is rotating. Again, we'll explain more about attitude dynamics and control in the book *In Space to Stay*.

The Gemini spacecraft used a new Orbit Attitude and Maneuvering System (OAMS). The OAMS pushed the spacecraft away from the spent Titan II second stage and performed an orbit insertion maneuver to LEO. In LEO the OAMS let a pilot steer the spacecraft, hold it steady, or in case of trouble perform a deorbit maneuver and reenter the atmosphere.

The OAMS had 16 small engines which also burned storable propellants (N_2O_4/MMH). Eight of the engines allowed the pilot to pitch, roll and yaw the spacecraft about any axis. The other eight engines enabled maneuvering in any direction (forward, back, and sideways).

Close Calls

Although there were no fatalities, not everything went peaches and cream during the 10 manned Gemini flights. On Gemini IV Ed White became the first American to have nothing between him and outer space but his spacesuit. "I'm out," he announced. "This is the greatest experience. Just tremendous" he continued while floating in space on a fully extended 25-foot tether. He tried maneuvering with a handheld "space gun" and found he could do this with a surprising degree of accuracy. For safety reasons, White was supposed to come back in before they entered the night side of the orbit. But he was enjoying himself too much to come in. "Get your butt back inside," CAPCOM Gene Franz had

to tell him after 36 minutes of EVA. However, the reentry hatch would not lock into place as necessary for cabin repressurization and Earth reentry. Commander McDivitt had to take the latch apart, reinstall it, and lock it back into position.

Gemini V accomplished its mission by keeping astronauts in space for two weeks, setting a world record for endurance in space. It would be Gordo Cooper's last flight. NASA saw him as a daredevil due to his love for auto and speedboat racing, and he soon found himself relegated to less glamorous jobs in NASA.

Emboldened by Gemini V's 120 orbits and full week in space, the U.S. decided to try for two weeks. On the next try, however, Gemini VI was called off when the Atlas launching its Agena Target Vehicle (ATV) exploded during a test launch on October 25, 1965. NASA felt pressured for time by the Space Race with the Soviets. They decided to postpone Gemini VI and launch the next spacecraft in the series, Gemini VII, on schedule. Gemini VII would trade places with Gemini VI by becoming the target vehicle instead of the other way around.

Gemini VII launched OK on December 4. The now renamed Gemini VIa was supposed to follow it and launch on December 12. During the Gemini VIa launch, the Titan GLV's first stage engine fired for two seconds, shaking the vehicle on the pad, then suddenly it shut off when an electrical plug fell out. Gemini VIa's onboard computer, sensing no upward motion, aborted the launch. Would the vehicle topple over like happened on Vanguard in December 1957, killing the crew? Astronauts Wally Schirra and Tom Stafford knew that ejection would have cancelled the entire mission, a major setback for the Gemini Program. The target vehicle, Gemini VII, could only wait out there in space so long. So what to do? Commander Schirra coolly waited out the problem until it was apparent the Titan engine had safely shut down. Whew! The launch restraint system just held the rocket on the pad when the booster shut down. Had their launch vehicle risen even a few centimeters, it would have toppled back onto the pad and exploded.

Gemini VIa got back on track with a launch 3 days later. It rendezvoused (but did not dock) with Gemini VII. The crews took many pretty photos. Gemini VII set a new space endurance record of 14 days in space. NASA found out several things about long space flights. Besides about a third of people getting nausea on their first flight into space, bodily fluids tend to drift upward towards the upper torso and even the head. It takes a while for the body to adapt to not having to fight gravity to keep fluids from accumulating in the lower torso and legs. People react differently to weightlessness. Some astronauts got a stuffy feeling in their sinuses and nasal passages, and a ruddy complexion or perhaps even a headache. For most, weightless conditions dull the sense of smell. Astronauts also get taller. Weightless conditions cause fluid to gather in the spinal discs, while the spine itself grows slightly longer because it no longer has to bear the weight of the body. Humans typically gain one to two inches in height. More recently, some astronauts aboard the International Space Station have experienced space-induced ocular impairments during long-duration missions. These changes in eyesight are linked to increased intracranial pressure caused by shifts in bodily fluids from the lower extremities to the upper body in microgravity; and and there are possibly other contributing factors. Another problem can be sleeping until you get used to weightlessness. There is no sensation of weight between your body and the bed. The sensation of one's legs floating up off the mattress can be disconcerting at first. NASA has adapted special restraints and viscoelastic mattresses which mold to one's body shape to resolve these problems. Another effect of extended stays in space is that astronauts lose muscle mass, legs become weaker, and bones lose calcium (becoming weaker as well). That's why exercise has become mandatory for all astronauts during lengthy stays in space.

Gemini VIII's RVD mission went well until the spacecraft's control system malfunctioned while it was docked nose-to-nose with the ATV. One of the Gemini's thrusters locked in the on position, and the Gemini capsule and ATV started to yaw and roll uncontrollably. The tumbling got worse and Commander Neil Armstrong knew he had to do something quick. Armstrong used the OAMS to stop the roll, but the moment he stopped using the thrusters, it started again. Dave Scott noticed that the Gemini's attitude fuel was down 30%, so the problem must be with the Gemini and not the Agena as they originally thought. Thinking quickly, Armstrong shut off the OAMS. He used the Gemini's reentry reaction control system (RCS) to stop the spin. Then the astronauts fired each OAMS thruster in turn and isolated the fault to number 8 which was stuck on. At this point they had to abort the mission short and perform the first emergency landing of a manned U.S. space mission.

Gemini IX got scrubbed for the same reason as Gemini VI (the Agena target failed to orbit), so it launched two weeks

later on June 3, 1966 as Gemini IXa. NASA made do with an off-the-shelf substitute for the Agena, an Augmented Target Docking Adapter (ATDA). When astronauts Tom Stafford and Gene Cernan pulled up alongside the ATDA, they found that its shroud had only come half off. "It looks like an angry alligator out here rotating around," Commander Stafford told mission control. "Could I use the (Gemini) spacecraft to open the jaws?" he gingerly asked. "No," ground control eventually decided. On the third day, Cernan was to attempt the longest spacewalk yet, using the U.S.'s first rocket pack—the Air Force's Manned Maneuvering Unit (MMU) with everything self-contained: oxygen, propulsion system, stabilization system, communications system and telemetry for downlinking biomedical data. But Cernan soon ran into problems. On the EVA he had to go back to the rear of the Gemini spacecraft to fetch the MMU. Lacking any decent hand and foot holds, this took longer than planned. Then his faceplate started fogging up, made worse by tremendous exertions in trying to operate the MMU's valves and perform other movements. Cernan's heart rate shot up to 150, 160, then 180 beats per minute. He might lose consciousness, the flight surgeon at mission control warned. The astronauts mercifully called the spacewalk short and started getting Cernan back in. But as Cernan went through the entry hatch, his spacesuit's cooling system overheated, causing his visor to fog up completely and him to lose all vision. Finally, with effort he and Stafford managed to get the hatch closed and repressurize the Gemini capsule. At least they landed the closest of any of the Gemini's to the target: ½ mile.

Cernan had run into the same problem that Alexei Leonov had on his first EVA aboard Voskhod 2. Cernan's every move had to work against the internal pressurization of his space suit. Although four psi doesn't sound like much, it was enough to make the suit swell up like a balloon and render movement almost impossible. And there was nothing to hold on to. In space, Cernan's body turned away from his work site with every move, and handling the heavy AMU was extremely difficult (Space Shuttle astronauts would use a much-improved Manned Maneuvering Unit (MMU) for shuttle missions in the 1980s-2000s).

Gemini X and XI were successful as shown in the table above. The "Grand Finale," Gemini XII was a huge success, due in part to hard work on the part of Buzz Aldrin. Aldin trained for his EVA by long sessions working underwater in his spacesuit. NASA found that underwater training closely simulated weightless conditions in space, so Aldrin's success made EVA training underwater in huge tanks mandatory for all future spacewalkers. Learning from

Cernan's experience, Aldrin conserved his energy and was able to work outside the spacecraft for 5 ½ hours. Better space tools and installation of a few hand and footholds by the engineers helped tremendously too.

The United States was now ready to embark on the Apollo series of missions to land men on the moon (and return them safely to Earth). A number of contemporaneous factors converged in the tumultuous decade following Explorer I's successful launch (January 1958) that would cement America's lead role in landing 12 men on the Moon and returning them safely to Earth. But the United States had to overcome one huge obstacle among many. The indomitable Sergei Pavlovich Korolyov would not go down easy. Not by a long shot.

Gemini XI launch on a Titan II from LC–19 at
Cape Kennedy, Sept. 12, 1966

Afterword

Thank you so much for reading this, the first book of *The Saga of Rocket Science* series. Please read the next one, *Avoiding Armageddon*, which carries forward the fascinating story of our species' journey into space, and relates personal stories from the many thousands of men and women who made this possible. The book relates how, thanks to these unsung heroes, World War III and a nuclear holocaust were averted, which would have killed untold hundreds of millions, if not more. Enjoy the journey.

ACRONYMS AND GLOSSARY OF KEY TERMS

开始 火箭	Rockets, the Beginning
AAM	Air–to–Air Missile
ABMA	Army Ballistic Missile Agency
ACS	Attitude Control System. Spacecraft and satellites require very precise stabilization, attitude, and aiming control. Attitude control is also necessary for accurate pointing of communications and data link antennas, solar array and thermal control, etc.
Adiabatic flame temperature	The equilibrium temperature of the products of complete combustion when the propellants are fed at the optimum stoichiometric mixture ratio in a rocket engine.
AEDC	Arnold Engineering Development Center
AFB	Air Force Base
AFRC	Armstrong Flight Research Center (Edwards AFB, California). See DFRC
AGM	Air–to–Ground Missile
aka	also known as
ALCM	Air Launched Cruise Missile
AMCOM	Army Missile Command (Redstone Arsenal, Alabama)
Angle of attack	Angle between a wing's chord line and the direction of the undisturbed free air stream in front of the wing
AR&D	Automated Rendezvous and Docking
ARC	Ames Research Center (Moffett Field, California)
ARC	Atlantic Research Corporation
ARPA	Advanced Research Projects Agency
ARPANET	Advanced Research Projects Agency Network
ASM	Air–to–Surface Missile
ATDA	Augmented Target Docking Adapter (used on Agena)
ATV	Agena Target Vehicle
BMDS	Ballistic Missile Defense System
BMW	Bavarian Motor Works
C2	Command and Control
CalTech	California Institute of Technology (Pasadena, California)
Camber	Distance between a wing's upper and lower surfaces, measured in a chordwise direction from the wing's leading to its trailing edge (see mean chord line in text).
CCAFB	Cape Canaveral Air Force Base

CEP	Circular Error Probable, the radius within which 50% of a missile's warheads are predicted to impact. In other words, half the warheads will impact outside a circle of diameter 2xCEP.
Сергéй	Sergei
CERN	Conseil Européen pour la Recherche Nucléaire
CHEOPS	CHaracterizing ExOPlanets Satellite
Chord	Length of a straight line connecting the leading and trailing edges of a wing (see mean chord)
CSAF	Chief of Staff, Air Force
DARPA	Defense Advanced Research Projects Agency
DEW	Directed Energy Weapon
DFRC	Dryden Flight Research Center (Edwards AFB, California), now AFRC
DoD	Department of Defense
DOF	Degree of Freedom. 3–DOF represents a point mass that can move in three directions: up–down, sideways, and forward–back. 6–DOF represents a body that can also rotate or spin like an airplane or rocket in three directions: about the pitch axis, roll axis, and yaw axis. Engineers often start with a simpler 3-DOF model of a rocket or missile in flight, then perform higher fidelity (more accurate) flight dynamics analyses with a 6-DOF model.
ETF	Engine Test Facility
ETR	Eastern Test Range
EVA	Extravehicular Activity
GALCIT	Guggenheim Aeronautical Laboratory, California Institute of Technology
GDF	Gas Dynamics Facility (at AEDC)
Gestapo	Geheime Staatspolizei (Secret State Police)
GIRD	Gruppa issledovaniya reaktivnogo dvizheniya (Group for the Study of Reactive Motion)
GRC	Glenn Research Center (Cleveland, Ohio)
GSFC	Goddard Space Flight Center (Greenbelt, Maryland)
Gulag	Soviet–era prison camps
Heer	Army in Nazi Germany
HVP	Heeresversuchsanstalt Peenemünde (Army Research Center Peenemünde)
IAA	International Academy of Astronautics
ICAS	International Council of the Aeronautical Sciences

ICBM	Intercontinental Ballistic Missile, with maximum range over 5500 km
IGY	International Geophysical Year (1957–1958)
IRBM	Intermediate Range Ballistic Missile, with range from 1000 to 5500 km
IRFNA	Inhibited Red Fuming Nitric Acid
Isp	Specific impulse, a measure of a rocket's efficiency. Specific impulse is the ratio of a rocket's thrust to the weight flow rate of the propellants (through its combustion chamber and exit nozzle) that it takes to produce that thrust. The units are in seconds, because specific impulse represents the kilograms–thrust an engine or motor produces per kilogram of rocket fuel (or solid propellant for motors) per second of operation.
ISS	International Space Station
JATO	Jet–Assisted Takeoff
JFK	John Fitzgerald Kennedy
JPL	Jet Propulsion Lab (Pasadena, California)
JSC	Johnson Space Center (Houston, Texas)
KEW	Kinetic Energy Weapon
kg	kilogram
kgf	kilogram-force, the force exerted by a mass of 1kg on Earth's surface
KGB	Komitet gosudarstvennoy bezopasnosti (Committee for State Security) in former Soviet Union
kN	kilonewton (about 225 pounds of force)
Kriegsmarine	Navy in Nazi Germany
KSC	Kennedy Space Center (Cape Canaveral, Florida)
LaRC	Langley Research Center (Hampton, Virginia)
Launch table	a rudimentary launch pad without a launch tower, commonly used in the early days of rocketry
lb.	pound
lbf	pound-force, the force exerted by a mass of 1lb on Earth's surface
LEO	Low Earth Orbit
LeRC	Lewis Research Center (now GRC)
LH2	Liquid Hydrogen (LH2)
LOX or LO2	Liquid Oxygen (LO2)
Luftwaffe	Air Force in Nazi Germany
MAD	Mutually Assured Destruction, the strategic detterent doctrine of the Cold War

Magnus Effect	A cylinder spinning in a fluid tends to speed up the flow near its surface, in the direction of its spin, and slow it down where its spin is contrary to the flow direction. This creates lift perpendicular to its axis on the side where the flow is flowing fastest. In aircraft design, the Magnus Effect became the basis for the theory of lift. This is because the wing is so shaped as to cause the same type of circulation as obtained by spinning. When the airplane wing is propelled in straight flight the circulation of air around the wing creates the pressures that result in lift.
Manrating	Manrating (aka Human rating) involves making rocket engines, launch vehicles and spacecraft more reliable and safe enough for humans to fly in. Manrated systems must undergo a more rigorous–and expensive–design, fabrication, production, and testing process.
Mass ratio	Ratio of total vehicle weight at burnout (including payload) to its total weight on the pad before liftoff
MCC	Mission Control Center
Mean camber line	Locus of points midway between a wing's upper and the lower surfaces, measured in a chord–wise direction from its leading to its trailing edge (see camber).
Mean chord	The total planform area of a wing divided by its span (see planform)
MGB	Ministerstvo Gosudarstvennoi Bezopasnosti (Ministry for State Security)
MGM	Mobile (Surface Attack) Guided Missile (missile designator)
MMH	Monomethyl Hydrazine
MMU	Manned Maneuvering Unit
MOL	Manned Orbital Laboratory
MON	Mixed Oxides of Nitrogen
Moore's Law	An observation made by Intel co-founder Gordon E. Moore in 1965 that, over the history of computing hardware, the number of transistors on integrated circuits doubles approximately every two years. His prediction has proven to be accurate. Moore's Law is used in the semiconductor industry to guide long-term planning and to set targets for research and development.
MRBM	Medium–Range Ballistic Missile, with range between 500 and 1000 km
MSC	Manned Spacecraft Center (now JSC)
MSFC	Marshall Space Flight Center (Huntsville, Alabama)

MT	Megaton, in reference to the destructive power of nuclear weapons. One megaton equals the destructive power of one million tons of TNT (trinitrotoluene) explosive.
MVD	Ministerstvo Vnutrennikh Del (Ministry of Internal Affairs)
NACA	National Advisory Committee for Aeronautics
NASA	National Aeronautics and Space Administrationi
NATO	North Atlantic Treaty Organization
Navier–Stokes	A set of complex differential equations which form the theoretical basis for the science of fluid mechanics
NK	Narodnyy komissariat (People's Commissariat)
NKVD	Narodnyy komissariat vnutrennykh del (People's Commissariat for Internal Affairs). The NKVD was a precursor to the KGB in the former Soviet Union.
NMD	National Missile Defense
NRL	Naval Research Lab (Washington, DC)
NSGWP	National Socialist German Workers Party, the Nazi party
NTO	Nitrogen Tetroxide (N_2O_4)
OAMS	Orbit Attitude and Maneuvering System
ORDCIT	Ordnance/California Institute of Technology
Payload mass fraction	Ratio of payload weight to the total vehicle weight on pad before liftoff
Planform	The silhouette of a wing when viewed directly from above or below, perpendicular to the direction of flight. Since the leading and trailing edges of most wings are curvy, calculus has to be used to calculate planform area.
Pogo	Pogo oscillation, a reference to the bouncing of a pogo stick, is a potentially dangerous type of self-excited combustion oscillation in liquid fuel rocket engines. This longitudinal oscillation results in variations of thrust from the engines, causing variations of acceleration on the rocket's structure, giving variations in fuel pressure and flow rate. Pogo places stress on the frame of the vehicle which can be severe.
Propellant mass fraction	A rocket's propellant mass fraction is the ratio of propellant weight consumed during flight to the vehicle's total weight on the pad before liftoff.
PWT	Propulsion Wind Tunnel (at AEDC)
R&D	Research and Development
RATO	Rocket–Assisted Takeoff
RC	Research Center

Reicheswehr	National Defense in Nazi Germany
Reichsführer–SS	Highest rank in Nazi Germany's Security Service, the Schutzstaffel (SS)
Revvoyensovet	Revolutionary Military Council in the former Soviet Union
RFNA	Red Fuming Nitric Acid
RLM	Reichsluftfahrtministerium (Nazi Ministry of Aviation)
RNII	Reaktivnyy nauchno–issledovatelskiy institut (Reactive Scientific Research Institute)
RP–1	Rocket Propellant Grade 1 (specially refined kerosene)
rpm	revolutions per minute
RVD	Rendezvous and Docking
SA	Sturmabteilung, paramilitary brownshirts in Nazi Germany
SAB	Scientific Advisory Board
SAM	Surface–to–Air Missile
SATKA	Surveillance, Acquisition, Tracking, and Kill Assessment
Schutzstaffel	Protective Squadron (aka Security Service, SS) in Nazi Germany
SECAF	Secretary of the Air Force
SECARMY	Secretary of the Army
SECDEF	Secretary of Defense
SFC	Space Flight Center (NASA)
Sharashka	Secret research laboratories where scientists and engineers caught up in Stalinist purges were put to work on projects assigned by the Communist party leadership
SLBM	Submarine–Launched Ballistic Missile
SLCM	Submarine–Launched Cruise Missile
SLCM	Surface–Launched Cruise Missile
Slenderness ratio	Ratio of length to width (diameter) of a rocket
SLTK	Survivability, Lethality, and Key Technologies
SLV	Satellite Launch Vehicle
SM	Strategic Missile
SMERSH	Smert Shpionam (Death to Spies), former Soviet armed forces counter–intelligence agency
SOB	Sistema oporozhneniya bakov (Tank Emptying System)
SOBIS	Sistema oporozhneniya bakov i sinkhronizatsii (Tank Emptying and Synchronization System)
Specific impulse	See Isp

SS	Security Service (aka Protective Squadron) in Nazi Germany
SS-Allgemeine	General in Nazi Germany's Security Service, the Schutzstaffel (SS)
SSM	Surface–to–Surface Missile
SSME	Space Shuttle Main Engine
SS-Obergruppenführer	Chief in Nazi Germany's Security Service, the Schutzstaffel (SS)
SS-Standartenführer	Colonel in Nazi Germany's Security Service, the Schutzstaffel (SS)
STG	Space Task Group
Stoichiometric mixture ratio	Ratio of oxidizer to fuel by weight in combustion whereby the chemical reactions fully occur, allowing the propellants to burn completely. A stoichiometric mixture ratio results in the highest performance, however other factors come into play in actual rocket engine designs. For example, operation at lower than optimum stoichiometric mixture ratio may be required to limit exposure of the combustion chamber and nozzle to high temperature extremes.
t/w	Thrust to weight ratio, the amount of thrust a rocket produces at ignition divided by its weight
TASS	Информационное агентство России (Telegraph Agency of the Soviet Union)
TESS	Transiting Exoplanet Survey Satellite
TNT	Trinitrotoluene
Tonne	Metric ton, 1,000 kilograms or 2,205 pounds
TV	Test Vehicle (Vanguard Program)
UDMH	Unsymmetrical Dimethyl Hydrazine
USAF	United States Air Force
USMA	United States Military Academy
USSR	Union of Soviet Socialist Republics
VNIIEF	All–Russian Scientific and Research Institute of Experimental Physics
Waffen SS	Weapons Protective Squadron (aka Security Service, SS) in Nazi Germany
Wehrmacht	Nazi war machine
Werk	Factory or Plant
WFF	Wallops Flight Facility (Wallops Island, Virginia)
WSMF	White Sands Missile Test Range
WTR	Western Test Range

ACKNOWLEDGEMENTS AND CREDITS

The author is indebted to many websites and their caretakers, individuals, and organizations without which this work would not have been possible. Picture and illustration authors deserve particular acknowledgement and a sincere thank you for permission to incorporate their works herein. In many cases I have customized, modified, and/or grouped their invaluable contributions as I felt necessary to communicate to my readers the panorama underlying the birth, development, and maturation of rocket science. Websites referenced below were accessed in September 2016 and are subject to change.

In particular, the following worthy persons and websites generously provided a plethora of source materials and excellent photos, graphics, and illustrations.

- Wikipedia, founded by Jimmy Donal Wales, English and Russian websites at http://en.wikipedia.org/wiki/Main_Page and http://ru.wikipedia.org/wiki/Заглавная_страница; respectively.
- Encyclopedia Astronautica at http://astronautix.com, Mark Wade (curator) and Gary Webster (illustrator)
- Google Company at www.google.com, founded by Larry Page and Sergey Brin[1]
- The National Aeronautics and Space Administration
- National Air and Space Museum, Smithsonian Institution, Washington D.C. at https://airandspace.si.edu/
- The Wright Brothers Aeroplane Company, West Milton, Ohio
- California Institute of Technology archives, Pasadena, California
- Clark University Archives, Worcester, Massachusetts
- United States Air Force Academy Library, Colorado Springs, Colorado.
- Milton Lehman, *This High Man: The Life of Robert H. Goddard*, Farrar, Straus and Company, New York, copyright 1963 by Milton Lehman
- Theodore von Kármán and Lee Edson, *The Wind and Beyond: Theodore von Kármán, Pioneer in Aviation and Pathfinder in Space*, copyright 1967 by Little, Brown and Company, Boston, Massachusetts

- Michael H. Gorn, *The Universal Man: Theodore von Kármán's Life in Astronautics*, copyright 1992 by the Smithsonian Institution, Washington DC.
- Boris Yevseyevich Chertok, *Rockets and People*, Volume I (no subtitle); Volume II: *Creating a Rocket Industry*[2]
- S.P. Korolyov Rocket and Space Corporation Energia at http://www.energia.ru/english/
- Anatoly Zak at www.russianspaceweb.com
- Departments of the U.S. Army and U.S. Air Force

Visual Materials Credits[3]

Figures and graphics obtained from Wikipedia are either in the public domain of the United States or licensed under Wikimedia Creative Commons 2.0, 2.5, or Wikimedia Creative Commons Attribution-Share Alike 3.0 Unported license (http://commons.wikimedia.org/ wiki/Main_Page). Commons is a freely licensed media file repository

1 Headquarters at 1600 Ampitheatre Parkway, Mountain View, CA 94043

2 Hard copies available as NASA SP–2005–4110, SP–2006–4110, SP 2009–4110, SP 2011–4110; respectively from the Superintendent of Documents, U.S. Government Printing Office at bookstore.gpo.gov. Soft copies available from the NASA History Division at http://history.nasa.gov/SP-4110/vol1.pdf; http://history.nasa.gov/SP-4110/vol2.pdf

3 Includes photos, drawings, illustrations, excluding those generated by author or which qualify as fair use under U.S. copyright law. Wikimedia materials are subject to Wikimedia Foundations terms of use at http://wikimediafoundation.org/wiki/Terms_of_Use. Visuals are subject to Wikimedia Creative Commons licenses: Attribution-ShareAlike 2.0 Generic, Attribution–ShareAlike 3.0 Unported, or Attribution–ShareAlike 4.0 International, as applicable.

Picture Credits

Aerospace Legacy Foundation

136 (top right).

Aerospaceweb.org

120 (left - 2).

Anatoly Zak

154 (second row: last 3 photos); 155 (right); 156 (third column, top two drawings); 214(2); 216.

Associated Press

125 (right); 152 (left); 224 (lower left); 247 (lower right).

Barney Elliott

98, 99.

The Boeing Company

136 (lower right).

British Broadcasting Corporation

255.

California Institute of Technology Archives

44 (top row -2); 102; 103; 104; 108; 110 (bottom left, top right); 142 (bottom); 146 (right); 149; 152 (right).

Carlo Kopp, Ph.D.

258 (bottom - 4).

Celestial Pictures, Ltd.

15, 18.

Central German Broadcasting

174 (right).

Clark University Archives, Worcester MA

Front cover, third row (left - 2); 60 (2); 61; 62 (3); 63; 66; 68; 70; 72 (bottom left); 74 (2); 80; 81 (3); 82 (3).

Cranfield University, U.K.

118 (top -4).

Deutsches Bundesarchiv (German Federal Archive)

Front cover, fifth row (leftmost); 245; 246 (2); 247 (lower left - 6 video stills).

Dhammakaya Foundation

26

Don P. Mitchell

154 (top row: first, fourth and fifth photos).

Encylopedia Astronautica

195 (2); 231 (20 thumbnails); 232 (7 thumbnails) ; 235 (4); 241 (3); 256 (lower left); 257; 258 (top left); 261 (9); 262 (2).

Encyclopedia Britannica

2; 14.

Encyclopedia of Science

131 (top left).

Eugen Sänger and Irene Bredt [1]

179

European Space Agency

209

Flume Creek Company, LLC

266.

Goedart Palm, Ph.D.

26

Holocaust Memorial Museum, Washington DC

249 (left).

Jet Propulsion Laboratory, Pasadena CA

143 (top); 286 (top).

Karina Longworth

249 (right).

Library of Congress--Prints and Photographs Division

Front cover, second row (right - 2); 30 (2); 39 (2); 48; 56 (right); 58.

Lilienthal Museum, Anklam Germany

41

Lockheed-Martin Corporation

138 (left).

Milton Caniff

47

Moscow Aviation Institute

156 (bottom right).

NASA

Front cover, fifth row (second, fourth and fifth photos); 37 (8); 38; 40 (bottom row -2); 42; 43 (2); 44 (bottom right); 55; 56 (left); 69; 88 (bottom); 91; 112 (2); 114 (3); 115 (4); 117 (bottom left); 121; 151; 157; 163 (right); 183[2]; 220-221; 222 (2); 224 (top left); 225; 228; 234; 235 (4); 238 (2); 263 (2)[3]; 280; 285 (bottom right); 286 (lower right); 287 (2); 290; 291; 292; 293; 294; 295 (2); 298 (6); 299; 302 (13); 306.

National Air and Space Museum, Smithsonian Institution

Front cover, second row (leftmost); 28; 31; 83 (bottom); 87; 90; 162; 260 (upper left); 279 (lower right).

National Geographic Society

8, 9, 12; 83 (middle).

Northrop Grumman

138 (right); 139 (bottom left).

NOVIK Samara Military-Historical Club Regional Museum

Front cover, fourth row (third photo).

Purdue University, West Lafayette IN

131 (middle left).

Roscosmos State Space Corporation

154 (second row: first photo, third row - 3, fourth row -5); 173; 210; 217 (2); 277 (right).

Ruptly GmBH

Front cover, fourth row (rightmost); 159.

Science Photo Library

Front cover: fourth row (left - 2), fifth row (middle); 156 (bottom left); 165; 285 (top right).

Smithsonian Institution

72 (upper left).

SonicBomb.com

201 (3).

S.P. Korolyov Rocket and Space Corporation Energia

154 (first row: second and third photos, second row: first photo); 155 (left); 156 (second column, top two photos); 158 (left); 174 (left); 196; 207 (4); 208; 215 (bottom right); 277 (upper left).

U.S. Air Force

Front cover, third row (right - 2); 109 (3); 110 (first column -2); 125 (left); 126; 132; 133 (2); 134; 135; 139 (top -2); 140; 144 (left); 239 (top -2); 240 (bottom right); 260 (upper right - 2 video stills); 289 (5).

U.S. Army

92; 143 (bottom right); 144 (right -2); 145 (2); 252; 259 (8 video stills); 270; 271; 274 (9 thumbnails); 275.

U.S. Navy

106; 146 (left); 272; 277 (lower left -2); 279 (top left); 281 (2); 288.

U.S. Space & Rocket Center, Huntsville AL

Back cover.

University of Applied Sciences, Munich Germany

142 (upper left).

University of Wellington, New Zealand

254.

University of Wroklaw, Poland

120 (bottom right).

Wikipedia

Front cover, top row (2); 5; 35; 40 (top); 46 (2); 84; 85 (3); 93; 107; 116 (3); 117 (top left, middle right); 118 (lower left); 119 (upper right, middle -2); 122; 136 (left); 137 (2); 153; 158 (right); 163 (left); 164; 166; 168; 169; 181; 185; 189; 198; 199; 215 (top right); 223; 229; 239 (bottom -2); 240 (top); 253; 256 (lower right -2); 259 (lower right); 260 (lower right).

Willy Ley

83 (top right); 88 (upper left).

Worcester Polytechnic Institute, Worcester MA Academics Library Archives

64, 65 (3).

YouTube video (The History of Space)

247 (top row - 4 video stills).

NOTES

1. Über einen Raketenantrieb für Fernbomber (book published in August 1944, English trans. by M. Hamermesh)

2. Boris Chertok (trans. by Asif Siddiqi), NASA-SP-2005-4110, Rockets and People, Volume I, p.350

3. Soviet Burya photo in Boris Chertok (trans. by Asif Siddiqi) , NASA-SP-2006-4110Rockets and People, Volume II: Creating a Rocket Industry, p.233

Index